Biodiversity in HORTICULTURAL CROPS

Biodiversity in HORTICULTURAL CROPS

— Volume 1 —

Editors
Professor K.V. Peter
Dr. Z. Abraham

2022
Daya Publishing House®
A Division of
Astral International Pvt. Ltd.
New Delhi - 110 002

First Published, 2007

Reprinted, 2022

ISBN: 9788170354901

Disclaimer:

Every possible effort has been made to ensure that the information contained in this book is accurate at the time of going to press, and the publisher and author cannot accept responsibility for any errors or omissions, however caused. No responsibility for loss or damage occasioned to any person acting, or refraining from action, as a result of the material in this publication can be accepted by the editor, the publisher or the author. The Publisher is not associated with any product or vendor mentioned in the book. The contents of this work are intended to further general scientific research, understanding and discussion only. Readers should consult with a specialist where appropriate.

Every effort has been made to trace the owners of copyright material used in this book, if any. The author and the publisher will be grateful for any omission brought to their notice for acknowledgement in the future editions of the book.

Published by : **Daya Publishing House®**
A Division of
Astral International Pvt. Ltd.
– ISO 9001:2015 Certified Company –
4736/23, Ansari Road, Darya Ganj
New Delhi-110 002
Ph. 011-43549197, 23278134
E-mail: info@astralint.com
Website: www.astralint.com

Digitally Printed at : **Replika Press Pvt. Ltd.**

Devotion

The book is devoted to Late Dr. B.P. Pal, Director General, ICAR New Delhi and Late Dr. Harbhajan Singh, Head, Section of Plant Introduction in the Division of Botany. The section was elevated to National Bureau of Plant Genetic Resources (NBPGR). Late Dr. Harbhajan Singh was the first Director of NBPGR, New Delhi who laid a strong foundation to the study of crops biodiversity in India. Dr. K.V. Peter also devotes this book to his Professor and Doctoral guide Dr. B. Rai Former Professor, G.B. Pant University of Agriculture and Technology, Pantnagar who introduced him to the world of Plant Biodiversity.

Acknowledgement

The Editors are grateful to Director, National Bureau of Plant Genetic Resources, New Delhi for the permission given to Dr. Z. Abraham to be associated with the present compilation. The assistance rendered by Dr. R. Keshvachandran, Co-ordinator, Bioinformatics Centre, K.A.U., Miss Alishya Joy, J.R.F. and Miss Snitha, S., J.R.F. are gratefully acknowledged.

डा. गौतम कल्लू
उप महानिदेशक
(बागवानी एवं फसल विज्ञान)
Dr. G. Kalloo
Former Deputy Director General
(Horticulture & Crop Science)

भारतीय कृषि अनुसंधान परिषद
कृषि अनुसंधान भवन-2, पूसा,
नई दिल्ली - 110 012
INDIAN COUNCIL OF AGRICULTURAL RESEARCH
KRISHI ANUSANDHAN BHAVAN-II, PUSA,
NEW DELHI – 110 012
Ph: (Off.) 25842068, 23382545 (Resi.) 25840191
Fax: 91-11-25841976, 91-11-23097003
E-mail: kalloog@icar.org.in and gkalloo.icar@nic.in

Foreword

Horticultural crops–fruits, vegetables, ornamental plants, tuber crops, plantation crops, spices, medicinal and aromatic plants and mushrooms–constitute the foundation of nutritional and livelihood security of India. They occupy 9 per cent of gross area under cropping, contributing 24.5 per cent of gross value of agricultural output and 54.6 per cent of export earnings in agriculture. India ranks first in production of mango, banana, cauliflower, coconut, cashew, tea and spices. If India has to achieve 8 per cent rate of growth in GDP, growth rate in horticulture has to be 6 per cent. National Horticultural Mission launched by Government of India envisages higher productivity, clean and environment friendly production, value addition and above all sustainability of natural resources especially biodiversity and water. Horticultural genetic resources include improved and obsolete varieties, populations, landraces, genetic stocks and breeding materials of crop plants and their wild and weedy relatives. Many fruit crops like mango, citrus and banana have their centres of diversity in tropical India/South East Asia. Minor fruits like *bael*, Indian gooseberry, papaya, jackfruit, custard apple, *karonda*, cordia, *phalsa*, *kokum*, mangosteen, *ber*, strawberry, blackberry and raspberry, currant and kiwi fruit have potential for cultivation. Vegetable crops having rich diversity in India are cucurbits, leaf vegetables, brinjal, okra and chilli. Tubers like yams, aroids and yam beans have importance in starch and bioethanol industry. In India, significant diversity occurs in medicinal and aromatic plants. *Aswagandha*, rauvolfia, *safed musli*, *babchi*, belladonna, atees, Indian barberry, Indian gentian, *kutki* and Indian henbane are a few such crops. Important ornamentals native to India are orchids, rhododendrons, muskrose, lotus and begonias. Ornamental trees are bauhinia, cassia, flame of forest, Indian coral tree, pride of India and tulip trees.

The collection, characterisation, evaluation, conservation and utilization of Horticulture Genetic Resources (HGR) are important for India which has a vast heritage of these bioresources. The present national network is co-ordinated by National Bureau of Plant Genetic Resources (NBPGR), New Delhi and its Regional Stations. There are 13 All India Co-ordinated Research Projects under the

Division of Horticulture, ICAR, 29 National Active Germplasm Sites (NAGS), several Departments of Horticulture/Vegetable Crops/Fruit Crops/Ornamentals/Plantation Crops and Medicinal Plants under 34 State Agricultural Universities (SAUs) and many State Departments of Agriculture/ Horticulture/Floriculture, involved in this network.

The present volume 'Biodiversity in Horticultural Crops' has 18 chapters contributed by working scientists in respective fields. Dr. Z. Abraham, Principal Scientist, NBPGR, Regional Station, Vellanikkara and Professor K.V. Peter, Kerala Agricultural University have edited this much informative book. I compliment the editors/authors and publishers for bringing out this very informative publication.

G. Kalloo

Contents

List of Contributors

Chapter 1

Bhag Mal, V. Ramanatha Rao, R.K. Arora and Percy E, Sajise: International Plant Genetic Resources Institute (IPGRI) Office for South Asia, NASC Complex, Pusa Campus, and New Delhi – 110 012 Email: b.mal@cgiar.org

Chapter 2

Sofi, A., M.K. Verma, R.K. Verma and W. Choudhary: Central Institute of Temperate Horticulture, Srinagar, J and K – 190 007, E-mail: mahenicar@yahoo.com.

Chapter 3

Prakash, G.S. and M.R. Dinesh: Division of Fruit Crops, Indian Institute of Horticultural Research (IIHR), P.O. Hessarghatta Lake, Bangalore, E-mail: director@iihr.ernet.in.

Chapter 4

Uma, S. and S. Sathiamoorthy: National Research Center for Banana, Tiruchirapalli – 620 102, Tamil Nadu, E-mail: try_nrcbaris@sancharnet.in.

Chapter 5

Ram, D., Mathura Rai and Major Singh: Indian Institute of Vegetable Research, Varanasi – 221 005, E-mail: pdveg@up.nic.in

Chapter 6

Swamy, K.R.M. and A.T. Sadashiva: Division of Vegetable Crops, Indian Institute of Horticultural Research (IIHR), P.O. Hessarghatta Lake, Bangalore – 560 089, E-mail: krmswamy@yahoo.co.in.

Chapter 7

Palaniswami, M.S. and Shirly Raichal Anil: All India Co-ordinated Research Project on Tuber Crops, Central Tuber Crops Research Institute, Trivandrum – 690 017, E-mail: palaniswamims@gmail.com.

Chapter 8

Sathish Kumar and S. Ganeshan: Tropical Botanic Garden and Research Institute, Pacha Palode, Trivandrum – 695 562, E-mail: director_tbgri@rediffmail.com.

Chapter 9

1. ***K. Nirmal Babu, S.P. Geetha, D. Minoo, Yamuna, G. and Praveen, K.***: Indian Institute of Spices Research, P.O. Marykunnu, Calicut – 673 012, E-mail: nirmalbabu@iisr.org.
2. ***P.N. Ravindran***: Centre for Medicinal Plants Research, Arya Vaidya Sala, Kottakal, Kerala, E-mail: pnravi2003@yahoo.co.in.
3. ***K.V. Peter***: Kerala Agricultural University, P.O. KAU, Trichur – 680 656, E-mail: kvptr@yahoo.com.

Chapter 10

V. A. Parthasarathy, K. V. Saji and K. Johnson George: Indian Institute of Spices Research, P.O. Marykunnu, Calicut – 673 012, E-mail: parthasarathy@iisr.org.

Chapter 11

Sasikumar, B.: Indian Institute of Spices Research, P.O. Marikunnu, Calicut – 673 012, E-mail: bhaskaransasikumar@yahoo.com.

Chapter 12

B. Krishnamoorthy, J. Rema and P.A. Mathew: Indian Institute of Spices Research, P.O. Marikunnu, Calicut – 673 012, E-mail: bkmoorthy@iisr.org.

Chapter 13

Thomas, J., K.J. Madhusoodanan and V.V. Radhakrishnan: Indian Cardamom Research Institute, Spices Board of India, Myladumpara, Idukki, Kerala, E-mail: jotkotmala@hotmail.com.

Chapter 14

Sudharshan, M.R. and Gupta, U.: Indian Cardamom Research Institute Spices Board, Regional Station, Sakleshpur – 573 134, E-mail: mrs44545@hotmail.com.

Chapter 15

Z. Abraham and R. Senthilkumar: National Bureau of Plant Genetic Resources, Regional Station, Vellanikkara – 680 656, E-mail: nbpgrtsr@gmail.com.

Chapter 16

S.K. Malhotra and B.B. Vashishtha: National Research Centre on Seed Spices, ICAR, Tabiji, Ajmer – 305 206, E-mail: malhotraskraj@yahoo.com; bbveiah@yahoo.com.

Chapter 17

Nayak, M.G. and M.G. Bhatt: National Research Centre for Cashew, Puttur – 574 202, D.K., Karnataka E-mail: nrccaju@sancharnet.in.

Chapter 18

Annamma Varghese, Y. and Saji T. Abraham: Rubber Research Institute of India, Kottayam – 686 009, E-mail: annamma@rubberboard.org.in.

Introduction

Plant biodiversity is the single greatest resource that man garnered from nature during the long cultural development phase. More than 2.5 lakh species of mosses, ferns, conifers and flowering plants and more than 50000 plant species are yet to be documented, primarily in the remote and less studied reaches of tropical forests. Among horticultural crops, tropical fruits received high priority in agricultural development over decades. Asia recorded 66 per cent increase in fruit production, the highest in world. Region of South East and East Asia is important in respect of India, Indonesia, Thailand and China and accounts for 50 per cent of total global fruit production. Tropical fruits account for 30-54 per cent of total farm income especially of small holders and marginal farmers, contributing to better livelihood of rural people, largely the farming community. Over 70 cultivated species of major and minor fruits are currently grown representing native diversity and a few exotic tropical fruits. Better known 20 species include bananas, citrus, mango, pineapple, papaya, durian, rambutan, jackfruit, litchi, longan, tamarind, chanpedak, carambola, longsat, guava, soursop, custard apple, passion fruit, and jujube (Chapter 1). Many of the tropical fruit trees have multipurpose uses for food, fodder, feed, shelter, timber, fuel and medicine. They are also rich sources of vitamins, minerals, and high dietary energy. The International Plant Genetic Resource Institute (IPGRI), renamed as 'Biodiversity International' initiated research on tropical fruit species in 1993 with a project "Promoting Conservation and uses of Tropical Fruit Species in Asia" funded by the Asian Development Bank for 2000-2003. Five detailed status reports on litchi, kunai, pulasan, mangosteen and Indian gooseberry were prepared. Data on 4967 accessions were computerized which include 2368 accessions of mango, 1453 of citrus, 316 of rambutan, 522 of jack, 185 of litchi and 123 of mangosteen. In Bangladesh, four accessions of mango, possessing year- round bearing habit; nine accessions with off-season flowering and one accession possessing sweetness at green stage were identified. In China, Early Novel Fukumoto, Late Navel lanes, Early and Late Ponkan and Seedless Ponkan were identified as promoting for fresh market and processing. The IPGRI also collected indigenous knowledge to improve productivity of fruits and encouraged farmers and communities to participate in conservation of tropical fruits germplasm. The entire North-Western Himalayan region comprising J and K; Himachal Pradesh and Uttaranchal is rich in diversity of temperate fruits like apple, pear, peach, plum, apricot, cherry, almonds, walnut, peach, hazelnut, chestnut, and persimmons. In apple, 2500 cultivars exist all over the world, but only 25 cultivars are grown commercially which include Delicious cultivars. The entire North Western Himalayan region possesses 31 fruits genera (chapter 2). Deciduous fruits are grown in North Western Himalayas. Major biodiversity exists in walnut, almond, chestnut, hazelnut and pistachio nut. Other temperate fruits are quines, wildfig, wild pomegranate, chinese ber, sea buckthorn, wild strawberry and olives. North East region of "Indian centre" houses diversity of native fruits including mango, jackfruit and citrus (chapter 3). In India, about 1000 mango varieties are grown, most of them arose as chance seedlings. The three main centres of cultivation are Lucknow-Saharanpur belt of Uttar Pradesh, Murshidabad of West Bengal and Hyderabad of Uttar Pradesh. The North East hilly region is rich in mandarins and a few other citrus fruits. Sixteen species of citrus, 52 varieties and seven natural hybrids of Assam are described. Bananas are mentioned in Valmiky's Ramayana, Kautily's Arthshastra and in ancient Tamil classic Silappadikaram. These evidences suggest early existence of banana in India. European grapes originated primarily between Caspian and black seas. Persian invaders introduced cultivated groups to North India in 1300 AD, from where they were introduced into Daulatabad of Maharashtra by King Mohamamedbin-Tughlak. Guava of tropical region originated in Tropical America. The genus Psidium of family Myrtaceae comprises of 150 species of small trees and shrubs, out of which 20 have edible fruits. Papaya is a native of tropical America. It came to India in 16th century. Sapota similarly originated in Tropical America, taken to Philippines by Spanish and from there to other Countries. India is bestowed with a large number of native fruits like aonla, bael, jack, jamun, karonda, kokum and phalsa. Banana and plantains are the oldest fruits cultivated by man for his existance. South and South East Asia and recently Papua New Guinea are centres of origin of bananas (chapter 4). Cavendish clones (AAA) form the main stay of banana industry. Grand Naine, Thellachakkarakeli and red banana are AAA clones. AB clones are Kannan, Ney Poovan and unique subgroups. AAB clones are Mysore, Pomes, and Plantain subgroups. The AAB clones are Blugges, Montain and Pisang Awak.

Vegetable crops in which rich diversity occurs in India are Cowpea, Common bean, Cole crops, Okra, Brinjal, Sweet potato, Taros, Yams, Sword bean, Velvet bean and Elephant foot yam. Wide diversity was observed in chives, leeks and other *Allium spp* in Kumaoun and Garhwal Himalayas; cluster bean in Western arid zone; lab lab bean in Deccan planteau; cucurbits in Rajasthan and MP and leaf vegetables like Amaranth and Fagopyrum in Western Himalayan region (chapter 5). Tropical vegetables consist of solanaceus fruits (tomato, hot pepper, brinjal), cucurbits (water melon, muskmelon, cucumber, pumpkins and squashes, bottle gourd, chow-chow, ivy gourd, bitter gourd, ash gourd, round gourd, snake gourd, pointed gourd, sponge gourd, ridge gourd), malvaceous vegetables (Okra), legumes (common bean, lime bean, dolichos bean, cowpea, sword bean, winged bean), allium (onion, shallot, garlic, leek) and leaf vegetables (amaranth, spinach beet, Ceylon spinach). The indigenous vegetables, both cultivated and wild have important role in household nutrition, economy and food security (chapter 6). There are 25 hot spots of biodiversity of which Western Ghats and Eastern Himalayan region are the most important ones. Indo- Burma region is one of the largest hot spots of the world and is the centre of origin of tuber crops like Yams, taro, Amorphophallus and Alocasia. Tropical tuber crops form an important group of subsistence crops especially to those living in tropics and sub tropical zones. Tuber crops are the third important food crops after cereals and grain legumes. Major tuber crops are cassava, sweet potato, Asiatic yam, African yam, lesser yam, elephant foot yam, taro, tannia and country potato. Minor tuber crops are arrowroot, winged bean, yam bean, canna, safed musli and jerusalem artichoke (chapter 7). Sweet potato is one of the world's highest yielding crops/

unit of area/unit of time. The species diversity of Alocasia is spread over China, VietNam, Indonesia and Australia. There are ornamental and medicinal uses of *Alocasia metallica* and *A.fornicata* respectively.

The 1600 km long mountain chain called Western Ghats is known not only for its spices, condiments and timbers, but also for its orchids- unique species of Aerides, Paphiopedilum, Vandas and leafless Vanillas. Western ghats sustains 302 species, 3 sub- species and 4 varieties of orchids in 80 genera. Abraham and Vatsala in their classical 'Introduction to Orchids' describe 150 native orchids of South India. Peninsular India is a rich repository of spices and herbs. The other major spices growing countries are Brazil, China, Guatemala, Indonesia, Madagascar, Nigeria, West Indies, Malaysia, Srilanka, Spain, Turkey, Mediterranean region and the Central America. Black pepper, Cardamom, Ginger, Turmeric, Vanilla, Capsicum, Cinnamon, Cloves, Nutmeg, Tamarind, Coriander, Cumin, Fennel, Fenugreek, dill, Caraway, anise, saffron, lavender, thyme, origano, celery, anise, sage and basil are important as spices (chapter 9). Among spices, black pepper is called "King of Spices". The name Pepper was derived probably from the Greek name for Black pepper, *Peperi*. The Piper species including *P. betle*, *P. longum*, *P.chaba* and *P. cubeba* originated from Southern Asia. In black pepper, more than 100 cultivars are known today. Indian Institute of Spices Research, Calicut possesses worlds largest germplasm collection of black pepper (chapter 10). The current status of species biodiversity assessment and conservation indicates that vast areas of the North East are yet to be explored to get an understanding of the variability available and the need to collect and conserve them. Biodiversity of ginger and turmeric is concentrated in South Asia and South East Asia. Cultivated diversity is maximum in South and South East Asian Countries. India is a world leader in production and export of ginger and turmeric.

Alleppey Turmeric, Rajapuri Turmeric, Cochin Ginger and Wynadan Ginger are synonymous with superior quality in these spices (chapter 11). Genus Curcuma is mainly of South Asian Origin and center of diversity of Curcuma species differs widely in their habitat. Tree species consist of cinnamon, cassia, nutmeg, cloves, allspices, and garcinia. Many species of Cinnamon have medicinal and spices value. Srilanka is the primary center of origin of *Cinnamomum* and several species occur in the country. So far 7 varieties of cinnamon were released. Garcinia (kokum) is a spice of medicinal value against obesity. It includes 1300 species. Nutmeg (*Myristica fragrans*) is a tree spice, yielding two spices namely, the nutmeg (dried seed) and the mace (dried aril surrounding the seed). The family Myristicaceae has about 18 genera and 300 species. Myristica with 120 species is the largest of two genera and New Guinea for the largest number of species (chapter 12). Myristica swamps are endangered habitats, demanding attention. Clove of commerce is the dried, fully grown but unopened flower buds of Syzygium aromaticum The Syzygium has about 500 species. Allspice (Pimenta dioica) is a polygamodioecious evergreen tree, the dried unripe fruits of which provide the culinary spice, pimento of commerce. Leaves are aromatic and are used to distill out oil. Cardamom is grown in the moist evergreen forest of Western Ghats of South India. Based on nature of panicles, three ecotypes are recognized Malabar, Mysore and Vazhuka (chapter 13). Large cardamom (*Amonum subulatum*) is grown mainly in Sikkim and Darjeeling district of West Bengal in India. There are mainly three cultivars of large cardamom, Ramsey, Sawney and Golsey. In addition, cultivars like Ramla, Vaelanga/Barlangly, Bebo, and Seremna are grown (chapter 14). Kokum, Malabar tamarind and Mysore Gambogo are tree spices under exploited and whose medicinal properties less understood. Botanically, kokum is *Garcinia indica*, belonging to Clusiaceae. Trees are either dioecious or polygamous. Food applications of Kokum colour are in processed fruit products, alcoholic and nonalcoholic beverages, preservatives and instant foods. Malabar tamarind (*Garcinia cambogia*) is an indigenous tropical, under-exploited, semi domesticated spice and oil seed crop. Dioecious in nature, productive bisexual or female trees outnumber males, which are not retained once they exhibit sexuality. Mysore Cambodge *(Garcinia xanthochymus)* is an underutilized crop distributed in lower hills of Eastern Himalaya, Western Ghats and Andaman Islands (chapter 15). Out of a total of 20 seed spices grown in India, 10 are important. Crops under major seed spices are coriander, cumin, fennel, and fenugreek where as ajowan, Indian dill; dill, nijella, celery, sage, rosemary, black cumin, horseradish, poppy and anise constitute the minor group. India is bestowed with immensely rich land races diversity in seed spices (chapter 16). Cashew (*Anacardium occidentale L*) a promising crop of neglected and barren lands, is assuming importance as a foreign exchange earner. Brazil is the original home of cashew. Presence of cashew in other countries including India is to be attributed to man's intervention. One of the immediate priorities of the Indian cashew germplasm programmes is to enhance genetic variability in the collection, as all the plantations were raised from the limited initial introduction and genetic base available in India is only moderate (chapter 17). Among 5 plantations crops, Para rubber tree of commerce (*Hevea brasiliensis*) is the most important source of natural rubber, meeting around 98 per cent of global requirement. Amazon rain forest in Brazil is the original home of rubber tree. History of cultivation of rubber in India dates back to 1878 when rubber was cultivated first as a forest crop in Nilambur in Kerala.

'Biodiversity in Horticultural Crops' deals in detail origin, distribution, botany, variability, descriptions, survey, collection, and maintenance of crop germplasm. Eighteen chapters authored by 49 eminent botanists deal with 17 major crops and their related more than 100 species. Eighteen research organizations are involved in the present compilation.

Biodiversity in Horticultural Crops Vol. 1 *Pages* **1–30**
Editors: **Professor K.V. Peter & Dr. Z. Abraham**
Published by: **DAYA PUBLISHING HOUSE**

Chapter 1

Conservation and Use of Tropical Fruit Species Diversity in Asia: IPGRI's Contributions

Bhag Mal, V. Ramanatha Rao, R.K. Arora and Percy E. Sajise
International Plant Genetic Resources Institute (IPGRI), Office for South Asia, Pusa Campus New Delhi – 110 012, India

Among horticultural crops, tropical fruits received high priority in agricultural development and over the last decade, Asia recorded a 66 per cent increase in fruit production, the highest in the world. The region of South, Southeast and East Asia is important in this respect and four countries–India, Indonesia, Thailand and China account for 50 per cent of the total global fruit production. Also in this region, tropical fruits account for 30-59 per cent of total farm income especially for small holders and marginal farmers (Tropical Fruit Net 2004), thus contributing to better livelihood of the rural people largely the farming community. It is also well established that the tropical fruits contribute significantly towards improved human nutrition and health.

The region is characterized by richness of fruit diversity and about 400 species are distributed in diverse ecosystems. Thus, a wide range of natural diversity occurs, well adapted to sub-humid, humid tropical and semi-arid vegetation types exhibiting wide variation in ecology and climate. In addition to the native fruit species that have been domesticated and diversified in this region, a large number of species of tropical American origin introduced in the distant past have developed agro-ecological niches and are well acclimatized (Verheij and Coronel 1991; Arora and Ramanatha Rao 1995). Over 70 cultivated species of major and minor fruits are presently grown largely representing native diversity, and also including some of the promising exotic tropical fruits (Arora and Ramanatha Rao 1995). About 20 species are better known under cultivation and these include banana, citrus, mango, pineapple, papaya, durian, rambutan, jackfruit, litchi, longan, tamarind, chempedak, carambola, langsat, guava, sour sop, custard apple, salak, passion fruit and jujube (Verheij and Coronel 1991; Singh 1993a; Arora and Ramanatha Rao 1995), the predominant fruits being banana, pineapple, citrus, mango and papaya.

Tropical fruits, particularly the native species, are important for economic welfare of small farmers in Asia. In humid tropics, which hold very rich species diversity, tropical fruit trees are a major component of multi-crop farming systems including home gardens. Most of these species have multi-purpose uses for food, fodder, feed, shelter, timber, fuel, medicine and other uses (Bhag Mal and Ramanatha Rao 2003; Hodel and Gessler 1999). Some of these species are well adapted to marginal lands, and in agroforestry and farm-forestry (Verheij and Coronel 1991). In addition, these are also rich sources of nutrition especially vitamins and minerals and possess high dietary energy (Pareek *et al.*, 1998; Hoe and Siong 1999). Overall, in view of their diverse uses, tropical fruits contribute considerably to food and nutritional security, income generation and poverty reduction and also towards ecosystem and environmental sustainability. The rapid agricultural development and industrialization, changing land use, large scale deforestation, accompanied by other social and cultural pressures, resulted in large scale degradation of natural habitats *vis-à-vis* loss of native diversity. Concerted efforts are thus required to conserve these valuable genepools, adopting diverse conservation and management techniques and approaches so as to effectively utilize this diversity for posterity.

Recognizing importance of conservation and utilization of tropical fruit diversity, the International Plant Genetic Resources Institute (IPGRI), which will become the Bioversity International from 1 December 2006, initiated research on tropical fruit species in 1993 in collaboration with several of its national partners with a small project entitled 'Promoting Conservation and Use of Tropical Fruit Species in Asia' and this was followed by a large elaborate regional project on 'Conservation and Use of Native Tropical Fruit Biodiversity in Asia' funded by the Asian Development Bank (ADB) for the period January 2000–September 2003. Yet another project 'Development of Advanced Technologies for Germplasm Conservation of Tropical Fruit Species' was undertaken during this period with support from the Australian Centre for International Agricultural Research (ACIAR) that strengthened IPGRI's work on tropical fruit genetic resources. Thus, IPGRI's role towards tropical fruit genetic resources conservation and use had been quite significant (Ramantha Rao and Arora 1999; Ramanatha Rao and Bhag Mal 2002; Bhag Mal and Ramanatha Rao 2003; IPGRI 2003a, b; Bhag Mal *et al.*, 2004; Ramanatha Rao *et al.*, 2005).

This chapter presents an overall synthesis of work carried out under the auspices of IPGRI including the three projects funded by different donor agencies and reflects the research thrust provided by IPGRI on strengthening conservation and use of tropical fruit species in Asia involving national programmes. The topic is elaborately covered, providing details on: information gathering, documentation of diversity collected and that available in genebanks, characterization and evaluation of this diversity, identification of promising/elite germplasm for use and application of diverse conservation approaches – both *in situ* and *ex situ* including *in vitro* conservation and cryopreservation with emphasis on complementary conservation strategies.

Need for Intensified Work on Priority Fruit Species

Work on information synthesis in tropical fruits was initiated in 1993 and priority was assigned to identify a few major and minor fruit species for detailed research. Based on the collaborative efforts between IPGRI-APO and International Centre for Underutilized Crops (ICUC), a questionnaire survey was conducted in 15 countries in South Asia (Bangladesh, Bhutan, India, Pakistan, Nepal and Sri Lanka), Southeast Asia (Brunei, Indonesia, Malaysia, Myanmar, Papua New Guinea (PNG), the Philippines, Thailand and Vietnam) and East Asia (China) and a synthesis report containing priority species was prepared.

Subsequently in the 'Expert Consultation on Tropical Fruits' held at Malaysian Agricultural Research and Development Institute (MARDI), Kuala Lumpur, Malaysia in 1994, six priority species of major (mango, citrus and rambutan) and minor fruits (durian, jackfruit and litchi) and a few others, namely, carambola, mangosteen and longan were identified (Arora and Ramanatha Rao 1995). With these aforementioned species genepools, research and development activities with emphasis on genetic resources conservation and use were initiated by IPGRI in Asia. There are several considerations which warrant intensified work on the identified priority fruit species. As already mentioned, these include nutritional and other diversified uses, economic/higher monetary return and agro-ecological considerations, and providing better employment and income generation. Fruit species have higher adaptation to variable agro-ecological conditions and are suitable for diverse agri-horticultural systems and fruit based cropping systems (Singh 2003a). Conservation efforts for fruit crops in national genebanks as well as international research and development efforts (for example, the international agricultural research centres and the Consultative Group on International Agricultural Research (CGIAR) Centres had been very limited (Bhag Mal *et al.*, 2004) (Table 1.1). Recently, countries in the region have also identified tropical fruits as having a viable export potential. Hence, intensification of research and development activities on tropical fruit species assumes high priority.

Table 1.1: *Ex situ* Collections Held in Genebanks

Crops	*National Genebanks*	*CG Centres*	*Total*
Fruit crops	2 79 000	2 500	2 81 500
Vegetables	4 81 000	–	4 81 000
Roots and tubers	77 000	24 000	1 01 000
Condiments	17 700	–	17 700
Total	**8 54 700**	**26 500**	**8 81 200**
All crops	54 00 000	6 00 000	60 00 000
% Hort. crops	15.83	4.42	14.69
% Fruit crops	5.16	0.42	4.7

Major Activities Undertaken

Major activities undertaken by IPGRI include information synthesis on extent and distribution of genetic diversity, genetic resources available, their conservation and utilization, development of complementary conservation strategies including development of *in vitro* conservation and cry preservation protocols, promoting *in situ* conservation, and genetic diversity studies, etc. Major emphasis has been on promoting the conservation and utilization of native tropical fruit species diversity through national, regional and international collaboration and coordination, focusing on the following plant genetic resources (PGR) activities (Arora 1995; Arora and Ramanatha Rao, 1995; Ramanatha Rao and Bhag Mal 2002; Bhag Mal and Ramanatha Rao 2003; IPGRI 2003a, b; Bhag Mal *et al.*, 2004):

1. Identification of priority genepools of major and minor fruit tree species of national and regional importance.
2. Gathering information on distribution and diversity, status of collections, genetic resources activities including characterization, evaluation, documentation, maintenance, conservation and utilization of priority species.

3. Supporting research on: (a) conservation techniques and their adoption in national programmes and (b) genetic diversity studies to assess the status of diversity, genetic threat and need to promote *in situ* conservation.
4. Strengthening human resource development, organizing training courses on conservation and use of tropical fruit species, consultation meetings and workshops.
5. Promoting national, regional and international collaboration to facilitate networking and dissemination of information.

Salient Achievements

IPGRI took several initiatives aimed at conservation and use of native tropical fruit species diversity in Asia, the Pacific and Oceania (APO) region. In 1990s, the activities were mostly supported from IPGRI's core budget. The work gained momentum from January 2000 onwards with initiation of Asian Development Bank (ADB) funded project on 'Conservation and Use of Native Tropical Fruit Species Biodiversity in Asia' in which ten Asian countries, namely, Bangladesh, China, India, Indonesia, Malaysia, Nepal, the Philippines, Sri Lanka, Thailand and Vietnam participated and worked on six priority species, *viz.*, mango, citrus, rambutan, jackfruit, litchi and mangosteen (Budathoki 2003; Dela Cruz 2003; Hue 2003; Idris 2003; Kalloo 2003; Shantha Peiris 2003; Purnomo, 2003; Somsri 2003a; Uddin 2003; Zushang 2003). The country-wise and crop-wise details of accessions collected, characterized, documented, conserved and the elite lines identified are given in Table 1.2 and Table 1.3, respectively. Subsequently, the work on cryopreservation of citrus was undertaken in collaboration with Universiti Kebangsaan Malaysia (UKM) and Universiti Putra Malaysia (UPM) in Malaysia. To promote the work on *in vitro* conservation and cryopreservation, a 3 year project on 'Development of Advanced Technologies for Germplasm Conservation of Tropical Fruit Species' funded by the Australian Centre for International Agricultural Research (ACIAR) was implemented in five countries, namely, Australia, Malaysia, the Philippines, Thailand and Vietnam (IPGRI 2003b). Major accomplishments are given below:

Table 1.2: Country-wise Details of Accessions Collected, Characterized, Documented, Conserved and Elite Lines Identified

Country	*Existing Accessions Documented*	*Accessions Database in CD-ROM*	*New Accessions Collected*	*Accessions Characterized*	*Accessions Added to Field Genebanks*	*Elite Lines Identified*
Bangladesh	857	159	206	196	199	26
China	833	833	129	326	129	28
India	404	552*	316	513*	316	30
Indonesia	110	47	127	107	52	17
Malaysia	165	108	100	140	45	6
Nepal	265	134	96	265	72	44
Philippines	419	519*	419	299	232	9
Sri Lanka	271	185	271	185	120	18
Thailand	982	982	279	799	279	3
Vietnam	661	548	241	529	241	9
Total	**4967**	**4067**	**2184**	**3359**	**1685**	**190**

* includes additional accessions.

Table 1.3: Crop-wise Details of Accessions Collected, Characterized, Documented, Conserved and Elite Lines Identified

Fruit Crops	Existing Accessions Documented	Accessions Database in CD-ROM	New Accessions Collected	Accessions Characterized	Accessions Added to Field Genebanks	Elite Lines Identified
Mango	2368	1665	1003	1804	734	93
Citrus	1453	1775*	555	983	511	51
Rambutan	316	265	288	285	195	17
Jackfruit	522	131	180	152	126	20
Litchi	185	143	60	40	60	1
Mangosteen	123	88	98	95	59	8
Total	**4967**	**4067**	**2184**	**3359**	**1685**	**190**

Information Gathering

Country Status Reports on Priority Tropical Fruit Species

Work on information gathering was initiated in 1995, based on the recommendations of Expert Consultation on Tropical Fruit Species in Asia and identification of six priority species for intensified work in national and regional context (Haq 1994; Arora and Ramanatha 1995). During 1995-1998, IPGRI-APO brought out 23 country status reports on genetic resources of six priority fruit species of three major fruits namely, mango, citrus and rambutan, and three minor fruits, litchi, jackfruit and durian in collaboration with the key national programmes that hold rich diversity in these fruits (Arora *et al.*, 1996a, b; IPGRI 2003a; Bhag Mal *et al.*, 2004). These reports contain valuable information on distribution; extent of genetic diversity of cultivated and wild genepools; the status of germplasm collection, characterization, evaluation, documentation, conservation, improvement and utilization; diseases and pests; and production, processing and marketing aspects. The national programme experts from respective countries contributed to synthesis of information on:

1. Mango (Bangladesh, India, Sri Lanka, China, Thailand, Indonesia and the Philippines)
2. Citrus (India, Nepal, Sri Lanka, China and Japan)
3. Rambutan (Thailand, Malaysia and Indonesia)
4. Durian (Thailand, Malaysia and Indonesia). A monograph on durian was also published (Brown 1997)
5. Jackfruit (Bangladesh, Sri Lanka and India)
6. Litchi (China, India).

Subsequently, during 2000-2002, five more status reports were added providing similar detailed information on selected minor fruits of Southeast and South Asian region, namely:

1. Litchi (*Litchi chinensis*) in Thailand and Myanmar
2. Kuini (*Mangifera odorata*) in Southeast Asia (Idris and Mat Lin 2002a)
3. Pulasan (*Nephelium ramboutan-ake*) in Southeast Asia (Idris and Mat Lin 2002b)

4. Mangosteen (*Garcinia mangostana*) in Southeast Asia (Dela Cruz 2001)
5. Indian gooseberry–aonla (*Emblica officinalis*) in Southeast and South Asia (Pathak 2003).

Based on the country status reports, regional synthesis reports were also brought out for the two most important crops in the region, namely, mango (Sant Ram and Rajan 2003) and citrus (Nito 2003). These reports contain well synthesized consolidated information on origin and distribution, uses, genetic resources, genetic improvement, agronomy, production, processing and marketing, human resource development, limitations/constraints, research and development needs and future prospects. These reports are widely distributed by IPGRI to national programmes and other partners in the region so that this synthesized information can be used in promoting further research and development, facilitating exchange of materials for better use of genetic diversity and also in developing complementary conservation strategies providing an overall regional perspective for Southeast Asia.

Also, to provide a wider awareness, information on distribution of species of underutilized fruits and nuts in their regions of diversity was synthesized. Inventory of Underutilized Edible Fruits and Nuts was published (Pareek *et al.*, 1998) and distributed widely to national programmes/partners.

Locating and Collecting Diversity

Documentation of Information on Existing Collections

Documentation of information on existing collections was mostly carried out under the ADB funded tropical fruit project. The 10 participating countries gathered useful data and information on extant collections and such information was synthesized and shared among partners. Amount of information available for different countries varied greatly, for example, India and China had detailed passport and characterization information, while Malaysia had very limited data and other countries were in between. Data on existing and newly collected accessions were compiled using MS Excel spreadsheet software in all partner countries. National partners were trained in using Excel and the data in the Excel files were easily converted into other formats such as the data interchange protocol (DIP) developed by IPGRI, hardcopy catalogues and analysis on genetic diversity using software such as DIVA-GIS and FloraMap. The data were standardized according to descriptors published by IPGRI for mango, citrus, jackfruit, litchi, rambutan and mangosteen as well as those developed by relevant national programmes (IPGRI 2003a).

Data on existing 4967 accessions were computerized which included 2368 accessions of mango, 1453 of citrus, 316 of rambutan, 522 of jackfruit, 185 of litchi, and 123 of mangosteen (Table 1.4). IPGRI staff assisted national partners in setting up databases, providing guidelines and descriptors, providing training to documentation staff, providing tools and software and worked closely with national staff in developing appropriate documentation systems in different countries.

CD-ROM on TFT Species Developed

IPGRI compiled all available databases of tropical fruit tree genetic resources from 10 countries and produced a CD-ROM, which was made available to all partners. This database contains the passport and characterization data showing significant variation amongst accessions for different fruit tree species and contains information on 4067 accessions of six tropical fruit crops, *viz.*, mango (1665), citrus (1775), rambutan (265), jackfruit (131), litchi (143), and mangosteen (88) and details are given in Table 1.5.

Table 1.4: Existing Accessions Documented for Passport and Characterization Data

Countries	*Fruit Crops*						*Total*
	Mango	*Citrus*	*Rambutan*	*Jackfruit*	*Litchi*	*Mangosteen*	
Bangladesh	321	96	–	440	–	–	857
China	80	653	–	–	100	–	833
India	404	–	–	–	–	–	404
Indonesia	–	–	48	–	–	62	110
Malaysia	99	–	66	–	–	–	165
Nepal	130	135	–	–	–	–	265
Philippines	265	93	–	–	–	61	419
Sri Lanka	189	–	–	82	–	–	271
Thailand	780	–	202	–	–	–	982
Vietnam	100	476	–	–	85	–	661
Total	**2368**	**1453**	**316**	**522**	**185**	**123**	**4967**

Table 1.5: Accessions in Databases Compiled in the CD-ROM for Information Sharing

Countries	*Fruit Crops*						*Total*
	Mango	*Citrus*	*Rambutan*	*Jackfruit*	*Litchi*	*Mangosteen*	
Bangladesh	–	89	–	70	–	–	159
China	82	651	–	–	100	–	833
India	151	384	–	–	17	–	552
Indonesia	–	–	21	–	–	26	47
Malaysia	66	–	42	–	–	–	108
Nepal	87	47	–	–	–	–	134
Philippines	356	101	–	–	–	62	519
Sri Lanka	124	–	–	61	–	–	185
Thailand	780	–	202	–	–	–	982
Vietnam	19	503	–	–	26	–	548
Total	**1665**	**1775**	**265**	**131**	**143**	**88**	**4067**

Ecogeographic Studies and Distribution of Genetic Diversity

Choreographic studies were carried out with focus on surveys based on ecological and climatic parameters of major distribution areas of tropical fruit tree (TFT) species in different countries. These studies provided better understanding of the distribution patterns, and agroecosystems of TFT. Status maps for altitude, the minimum and maximum temperature and rainfall in the region were produced. The maps highlighted the variation in climatic parameters such as altitude, temperature and rainfall in which different TFT species were distributed. By mapping collected accessions of different tropical fruit trees, it was possible to map the areas with high diversity, predict the possible distribution of tropical fruit tree diversity and identify gaps in existing collections. The countries were able to produce the distribution maps for their tropical fruit collections with the geographic information system (GIS) tools such as DIVA-GIS and FloraMap. One such good example is for *Citrus* in India (Singh and Singh, 2003). Regional distribution maps were produced for six crops, namely, citrus, jackfruit, litchi, mango, mangosteen and rambutan showing overlapping distribution of several species. These distribution maps were very useful in identifying diversity-rich areas for further collecting and *in situ* conservation and also predicting new areas with similar climatic conditions for cultivation of these tropical fruit trees (IPGRI 2003a).

Exploration and Germplasm Collection

The areas for further exploration were identified and germplasm collection was done based on ecogeographic studies, diversity distribution maps and information on existing collections. Collecting strategies were suggested for different species and made available to national partners. A total of 2184 accessions of six tropical fruit tree species were collected. This included 1003 accessions of mango, 555 of citrus, 288 of rambutan, 180 of jackfruit, 60 of litchi, and 98 accessions of mangosteen (IPGRI 2003a). Number of accessions of different fruit crops collected in different countries is given in Table 1.6.

Table 1.6: New Accessions of Target Fruit Crops Collected in Different Countries

Countries	*Fruit Crops*						*Total*
	Mango	*Citrus*	*Rambutan*	*Jackfruit*	*Litchi*	*Mangosteen*	
Bangladesh	49	59	–	98	–	–	206
China	14	115	–	–	–	–	129
India	213	68	–	–	35	–	316
Indonesia	–	–	90	–	–	37	127
Malaysia	63	–	37	–	–	–	100
Nepal	64	32	–	–	–	–	96
Philippines	265	93	–	–	–	61	419
Sri Lanka	189		–	82	–	–	271
Thailand	118		161	–	–	–	279
Vietnam	28	188	–	–	25	–	241
Total	**1003**	**555**	**288**	**180**	**60**	**98**	**2184**

Evaluation and Utilization of Diversity

Germplasm Characterization

A total of 3359 accessions of target fruit species, namely, mango, citrus, rambutan, jackfruit, litchi and mangosteen held in the field genebanks in different countries were characterized. These included 1804 accessions of mango in 9 countries, 983 of citrus in 6 countries, 285 of rambutan in 3 countries, 152 of jackfruit in 2 countries, 40 of litchi in 2 countries and 95 of mangosteen in 2 countries as given in Table 1.7 (IPGRI 2003a).

Table 1.7: Accessions Characterized in Target fruit crops in different countries

Countries	*Fruit Crops*						*Total*
	Mango	*Citrus*	*Rambutan*	*Jackfruit*	*Litchi*	*Mangosteen*	
Bangladesh	55	50	–	91	–	–	196
China	80	246	–	–	–	–	326
India	421	75	–	–	17	–	513
Indonesia	–	–	56	–	–	51	107
Malaysia	74	–	66	–	–	–	140
Nepal	130	135	–	–	–	–	265
Philippines	214	41	–	–	–	44	299
Sri Lanka	124	–	–	61	–	–	185
Thailand	636	–	163	–	–	–	799
Vietnam	70	436	–	–	23	–	529
Total	**1804**	**983**	**285**	**152**	**40**	**95**	**3359**

New Descriptor Lists Published

To facilitate extensive characterization and evaluation effort, new descriptor lists were developed, published and widely distributed for crops on which no standardized descriptors were available, such as jackfruit, litchi, rambutan and mangosteen, and descriptors for citrus and mango were revised (IPGRI 1999, 2000, 2002, 2003c, 2003d, 2006).

Germplasm Catalogues Developed

All the participating countries identified and catalogued important varieties of relevant fruit species and these catalogues contained information on passport data and characterization. For example, India produced two catalogues on mango with information on 404 accessions and the Philippines prepared a catalogue of 265 accessions of mango, 93 accessions of citrus and 61 accessions of mangosteen. The catalogues were produced in hardcopy and/or electronic versions and were made available to the partners (IPGRI 2003a; IPB 2003a, b; Rajan 2003; Rajan *et al.*, 2002; Dinesh and Vasugi 2002; Somsri 2003b, c).

Identification of Elite Lines with Specific Desirable Traits

Based on specific desirable attributes, 190 elite lines/promising accessions were identified for further evaluation and use by the breeders in the breeding programmes or directly by the farmers for commercial cultivation. These included 93 promising lines in mango, 51 in citrus, 17 in rambutan, 20 in jackfruit, 8 in mangosteen and 1 in litchi (IPGRI 2003a). Details are given in Table 1.8.

Some of the elite lines or promising accessions identified based on the evaluation studies carried out in different participating countries (IPGRI 2003a) are listed below:

1. In Bangladesh, four accessions of mango possessing year-round bearing, nine accessions with off-season bearing and one accession possessing sweetness at green stage were identified as promising.

Table 1.8: Elite Lines/Promising Accessions Identified in Different Fruit Crops in Different Countries

Countries	Fruit Crops						Total
	Mango	Citrus	Rambutan	Jackfruit	Litchi	Mangosteen	
Bangladesh	3	7	–	16	–	–	26
China	10	18	–	–	–	–	28
India	30	–	–	–	–	–	30
Indonesia	–	–	13	–	–	4	17
Malaysia	3	–	3	–	–	–	6
Nepal	23	21	–	–	–	–	44
Philippines	2	3	–	–	–	4	9
Sri Lanka	14	–	–	4	–	–	18
Thailand	2	–	1	–	–	–	3
Vietnam	6	2	–	–	1	–	9
Total	**93**	**51**	**17**	**20**	**1**	**8**	**190**

2. In China, Early Navel Fukumoto, Late Navel Lane, Early and Late Ponkan, Seedless Ponkan, Late Tangor Wanmi No.3 and Kiyomi were identified as promising and suitable for fresh fruit market and processing. These have been included in the recommendation packages for farmers. Seedless Xuegan, a mutant of common Xuegan, possessing high productivity, high juice and soluble solid content and good flavour was found promising for the processing industry. Gonton Chen and Swingle citrumelo (*P. trifoliata* x *C. grandis*) were observed to be the superior rootstocks for pummelo owing to their vigorous growth, high productivity, large fruit size and better fruit quality.
3. In India, the elite lines identified in mango were Samparpatti, Totapuri, Neelam and Amrapali for high yield; Rumani, Ladavio and Neelam for short height; Bombay 1, Saheb Pasand and Dashehari, Mallika, Amrapali for high total soluble solids (TSS); Borgha, Hilario, Rosa, Dori, Sardar, Pulihara, Vanraj, Janardhan for red peel with good consumer appeal; Bombay, Alphonso, Kharji, Langra, Dashehari, Mallika, Prairi, Vanraj for high pulp recovery; Kensington, Fazli, Langra, Malagoa, Mallika for large fruits; Lazzat Baksh, Manibhatta Appe, Willard, Kuddus for regular bearing; Lazzat Baksh, Guruvam, Gaurjit, Charakaran for early maturity; Kesar, Ratna, Neelkiran for high tolerance to shoot borer and the hybrids Arka Anmol and Arka Puneet possessing medium size fruits with high TSS, high acidity and high pulp recovery were found suitable for export purposes. In Rangpur lime, the elite lines identified were Accession No. 7247 for large fruit size, Tirupati for thinnest peel, cv. Australia for less seedy fruits and strain USA for very high TSS. In rough lemon, the promising lines identified were Meethi Tulia for large fruit size, Sohmyndong Jallandhari Khatti, Limoneria and Sohsarkar for low seed number. Elite lines identified in trifoliate orange were Chethali for their peel, Swingle citrumelo for high TSS. Citrus accession IPS 231 was found free from major graft transmissible pathogens and was identified good for field planting. In litchi, the promising accessions identified were Bombay 1 for high fruit weight and D'Rose for low peel percentage. Some cultivars exhibiting high values of quality traits included D'Rose for high sugar percentage and cv. Dehradan for titrable acidity and ascorbic acid content.

4. In Indonesia, two varieties of rambutan, Garuda and Antalagi were released as superior varieties based on fruit size, flesh quality and ease of rind opening. Two elite lines of mangosteen, Wanayasa and Kalingesing were identified as promising and the latter was released as a superior variety based on its dark purple rind.
5. In Malaysia, three elite lines No.7, 22, 62 of pulasan and three elite lines No.2, 33, 66 of kuini were identified as promising.
6. In Nepal, seedless varieties, Okitsu Wase (KIR004) and Miyagwa Wase of mandarin, Washington Navel (DHK 043) of sweet orange and Biju 4, an early variety of mango with very high TSS were identified as promising.
7. In the Philippines, two elite lines of mango, *viz.*, 2003-024 and 2003-021 having large fruit with small stone, three elite lines of citrus, namely, P03 of *C. maxima*, LAES cultivar of *C. reticulata* and NPGRL cultivar of *C. hystrix* and three elite lines of *Garcinia mangostana, viz.*, 2002-12, 2002-176, 2002-181 were identified as promising based on large fruit size, high TSS and regular bearing.
8. In Sri Lanka, two elite lines of mango, No.28 (edible at green stage) and No.89 (for pickle) and two lines of jackfruit, Rasokas and Father Long with very good fruit quality and attractiveness were identified as promising.
9. In Thailand, two lines of mango, CB0007 and CB0010, possessing several desirable traits such as high fruit weight, firm and thick pulp texture, sweet, yellow peel, small stone and longer shelf life were found outstanding and can be used by farmers for direct commercial cultivation. In rambutan, one outstanding line CB0023 possessed firm aril texture, thick aril, normal skin thickness, orange red skin, high TSS, high fruit weight and long shelf life. Striking feature of this cultivar was that the rind and hair colour did not fade even after 30 days after maturity on the tree.
10. In Vietnam, two of the best rootstocks, namely, Buoi dang and Cam 31a of *Poncirus trifoliata* and one rootstock, Tu hu of litchi were very promising.

Promoting Diversity in Orchards and Home Gardens

To promote use of diversity in orchards and home gardens, assessment of diversity of different crops was considered important. In Nepal, assessment of diversity of citrus and mango was carried out with the active participation of grower groups and individual orchardists and home gardeners. These groups provided detailed information on the fruit quality, maturity, tree morphology, tree canopy, tolerance to diseases, pests and abiotic stress conditions with respect to genotypes, they had in their villages and neighbouring areas. This fruit growers' evaluation was based on their long experience of cultivation, fruit consumption and fruit selling, which proved to be very useful in identifying diversity to be collected (Budathoki 2003). Similar assessment of diversity with communities and home gardens was done in Sri Lanka (Shantha Peiris 2003), Bangladesh (Uddin 2003) and Vietnam (Hue 2003).

Inclusion of a large number of available cultivars and species in the semi-commercial and commercial orchards is a recent practice and was further encouraged and enhanced through IPGRI's project. This was done to reduce the risks from natural biotic and abiotic stress conditions, consumer preferences and market choices. It was discussed that cultivars used by the growers on commercial and semi-commercial scales were mostly improved varieties and very a few local landraces were grown. In contrast, indigenous and unique types of fruit germplasm were found in the home gardens.

Elite lines identified in different fruits will be distributed to semi-commercial orchards and home gardens to promote diversity deployment and enhance production.

Women growers predominantly contributed to growers' evaluation. Through their help, it was possible to identity the high diversity areas of citrus and mango, high diversity orchards, home gardens and also the location of orchards, home gardens/owners with very popular elite lines in the villages.

Using Indigenous Knowledge to Improve Productivity

Documenting indigenous knowledge is extremely important as it gives very useful information on growers' preferences and uses, indigenous production methods and on role of growers in conservation and use, which could be effectively used in developing appropriate research and development programmes on target fruit tree species. In Nepal, indigenous knowledge was collected on citrus, which included varied types of information (Budathoki 2003) such as: (*i*) *in situ* manuring by keeping animals in the orchards, (*ii*) use of temporary 'mobile' toilets near the trees for manuring, (*iii*) planting of whole fruits in the field after removing the peel, (*iv*) smoking treatment for trees in the evenings and nights to repel the insect pests and (*v*) spraying small seedlings with slurry of fresh cow dung and urine to repel animals and protect the plants. Such information could be very useful in increasing fruit production and thereby enhancing farmers' income and can be validated by formal and more controlled research.

In Sri Lanka, indigenous knowledge on mango and jackfruit was collected from different communities (Maharouf 2003a, b). This indicated that there existed varieties that were used for a long time in specific food items and treatment of various diseases. The indigenous knowledge on mango was gathered by interviewing priests, village leaders, Ayurvedic doctors and ethnic communities, Singhalese and Tamils. Information pertained to the use of mango and its plant parts in food, medicinal, festivities, wedding and rituals was gathered. Jackfruit in Sri Lanka is named as *'Bath Gasa'* reflecting its prime value as a food which is similar to that of the staple food rice known as *Bath* in Sinhala. In Malaysia, decoction of roots of pulasan has been reported to be used as vermifuge for fever; and roots are boiled and used for bathing patients inflicted with fever (Idris 2003). Dried seed kernels with faintly sweet smell oil are suitable for use in food. In the Philippines, decoction of dried mangosteen rind is used to cure stomach-ache (Dela Cruz 2003).

In Vietnam, litchi is reported to have medicinal value. It is traditionally used to treat stomach-ache and small intestine pain. Traditional uses included the use of litchi flesh to prevent tiredness and to treat bronchocele or growth on the neck, fruit skin to treat diarrhoea and leaves to treat animal bites. Litchi tree productivity was enhanced with the use of poultry manure. The other traditional practices included the cutting of trunk vertically in middle to induce branching and slicing of trunk for inducing fruiting (Hue 2003).

Conservation of Fruit Crop Genetic Resources

In situ Conservation

Under ADB funded project on tropical fruits, IPGRI promoted the concept of on-farm conservation of fruit genetic resources (Nares *et al.*, 2001). *In situ* conservation activities were initiated in a few countries, namely Bangladesh, India, Nepal, the Philippines, Sri Lanka and Vietnam and suitable *in situ* conservation sites were identified (IPGRI 2003a). These included:

1. Ipilan-Alitao for mangosteen in the Philippines

2. Ramnagar Mirchaiya for mango, and Karki Gaon and Banskharka village for *Citrus* spp. in Nepal
3. Gazipur for jackfruit, Chapai Nawabganj for mango and Maulibazar for citrus in Bangladesh
4. Garo Hills for citrus, Paiyur in Tamil Nadu and Lucknow in Uttar Pradesh for mango in India
5. Bac son district, Lang son province for mandarin; Phuc Trach commune, Huong Khe district, Ha Tinh province for pummelo; Thanh son commune, Thanh Ha district, Hai duong province for litchi; and Chiengpan commune, Yenchau district, Son la province for mango in Vietnam
6. Several home gardens for jackfruit and mango in Sri Lanka and semi- domesticated, natural diversity sites for litchi in Southern China.

Ex situ conservation

National programmes in the collaborating countries are maintaining germplasm of target species in the field genebanks at different locations in respective countries. Fifty two field genebanks were identified/established for different crops. Twenty one field genebanks were identified/established for mango, 13 for citrus, 8 for rambutan, 4 for jackfruit, 4 for litchi and 2 for mangosteen. These field genebanks were enriched with more accessions added to the existing collections, thus enriching the accessible genetic diversity. This included mango (734 accessions), citrus (511 accessions), rambutan (195 accessions), jackfruit (126 accessions), litchi (60 accessions) and mangosteen (59 accessions). Out of a total of 2184 accessions collected, 1685 accessions of six target crops were added to field genebanks in different countries. Remaining collections were in nursery for subsequent transfer to field genebanks at appropriate growth stage (Table 1.9).

Table 1.9: Number of Accessions of Different Fruit Crops Added to the Field Genebanks in Different Countries

Countries	*Fruit Crops*						*Total*
	Mango	*Citrus*	*Rambutan*	*Jackfruit*	*Litchi*	*Mangosteen*	
Bangladesh	49	59	–	91	–	–	199
China	14	115	–	–	–	–	129
India	213	68	–	–	35	–	316
Indonesia	–	–	24	–	–	28	52
Malaysia	35	–	10	–	–	–	45
Nopal	40	32	–	–	–	–	72
Philippines	152	49	–	–	–	31	232
Sri Lanka	85	–	–	35	–	–	120
Thailand	118	–	161	–	–	–	279
Vietnam	28	188	–	–	25	–	241
Total	**734**	**511**	**195**	**126**	**60**	**59**	**1685**

Development of Advanced Technologies for Conservation

IPGRI assisted national programmes in developing advanced techniques for conserving tropical fruit tree germplasm in different countries. Studies on *in vitro* conservation of citrus conducted during

1993-96 at Citrus Research Institute, Chongqing, China resulted in *in vitro* storage of 60 accessions of citrus and further development of protocols for citrus under slow growth conditions. Conservation of shoot tips of trifoliate orange (*Poncirus trifoliata*) at 5°C showed 90 per cent survival, while in red tangerine, there was 60 per cent survival at 10°C. Cryopreservation of shoot tips with 6-day preculture followed by 30 minutes loading and 1-hour dehydration proved good with recovery of growth in 90 per cent shoot tips of trifoliate orange and in 10-30 per cent shoot tips of sweet orange. Cryopreservation studies in mango conducted in 1995-96 in collaboration with the School of Life Sciences, Zhongshan University, Guangzhou, China confirmed recalcitrant nature of mango seeds and encapsulation-dehydration method proved to be more effective (Bhag Mal and Ramanatha Rao 2003; Bhag Mal *et al.*, 2004). Cryopreservation studies in different fruit species (Chaudhury *et al.*, 2000a; Sudarmonowati 2000) revealed varying degrees of success with different techniques and types of material used (Table 1.10).

Table 1.10: Success of Cryopreservation Using Various Techniques in Different Fruit Species

Species	*Technique*	*Type of Material*	*Survival (%)*
Litchi chinensis	Vitrification	Embryonic axes	50
	Air desiccation	Embryonic axes	22–35
Euphoria longan	Vitrification	Embryonic callus	30
		Anthers	15
Nephelium lappaceum	Two step freezing	Shoot tips	10
Citrus sinensis	Encapsulation, dehydration	Shoot tips	25
	Vitrification	Embryonic axes	62.5

Work on development of cryopreservation techniques for long-term conservation of recalcitrant seeds of jackfruit and litchi was done at National Bureau of Plant Genetic Resources (NBPGR), New Delhi, India, in collaboration with IPGRI under the project funded by Department for International Development (DFID), UK. For the first time, in litchi (*Litchi chinensis* Sonn.), embryonic axes from seeds harvested 60-80 days after anthesis survived liquid nitrogen exposure (22-35 per cent recovery) at low moisture content of 12.1-18.3 per cent (Chaudhury *et al.*, 2000b, 2001; Bhag Mal and Ramanatha Rao 2003). Successful cryopreservation was reported in jackfruit (*Artocarpus heterophyllus* Lamk.) with 25-30 per cent survival of cryopreserved axes from partially mature and fully mature seeds (Chandel *et al.*, 1995). With the use of vitrification method of cryopreservation, there was an improvement in recovery percentage of cryopreserved axes in jackfruit and litchi compared to air desiccation method.

Studies were carried out in collaboration with Universiti Kebangsaan Malaysia (UKM) and Universiti Putra Malaysia (UPM), Malaysia to determine seed desiccation and freezing tolerance, and to develop storage protocols for *Citrus* species with support from Rural Development Administration (RDA), Republic of Korea. These studies revealed that in *Citrus aurantifolia* (Cho *et al.*, 2002), survival of seeds decreased with moisture content below 16 per cent. For seeds cryopreserved with testa, the highest survival (41 per cent) was achieved after desiccation to around 7.3 per cent moisture, while for seeds cryopreserved without testa, the highest survival (85 per cent) was achieved after desiccation to around 7 per cent moisture. Zygotic embryonic axes of *Citrus madurensis* could be cryopreserved using vitrification. The optimal conditions were 3-day treatment with 0.1 M sucrose and 24 h preculture on medium with 0.3 M sucrose + 0.5 M glycerol (Cho *et al.*, 2001). Using these pregrowth and preculture treatments followed by treatment with PVS2 vitrification solution and direct immersion in liquid nitrogen, a survival of 83 per cent was obtained with shoot apices (Cho *et al.*, 2001).

IPGRI's work on *in vitro* conservation and cryopreservation work was further strengthened through funding from ACIAR. Under this project, which included Australian native fruit species, conservation techniques were developed for target tropical fruit species (IPGRI 2003b; Drew *et al.*, 2006). In papaya, protocols for vitrification-based shoot tip cryopreservation were refined and applied successfully to a range of papaya genotypes and *Vasconcellea pubescens* (a wild relative). Studies indicated that somatic embryos in papaya were recovered after cryopreservation and that the seed at any moisture content can be germinated with gibberelic acid (GA3) treatment or heat shock; former was more effective. This also showed that papaya seeds may have dormancy right from the beginning. Papaya seeds were stored up to 12 months at a range of moisture content and a range of temperature including cryostorage. Protocols for slow-growth of papaya *in vitro* were developed using a modified medium for papaya microcutting system. Plants were held under normal incubation conditions for 8 to 12 months before transfer.

In *Citrus australasica,* seeds demonstrated tolerance to desiccation and ultra low temperatures and had normal morphology post cryostorage. Seed storage in *C. inodora* and *C. garrawayi* showed tolerance of seeds to desiccation and cryostorage and reduced seedling vigour. Seeds of both species survived cryopreservation with growth and acclimatization of plants post cryostorage. A micropropagation protocol was established for three Australian native *Citrus* species (*C. australasica, C. inodora* and *C. garrawayi*) suitable for mass multiplication and medium-term storage. Encapsulation-dehydration protocols were applied to shoot tips of *C. australasica.* Survival of encapsulated shoot tips was minimal after desiccation to moisture contents suitable for cryopreservation. However, shoot tips of *C. australasica* survived and grew well with post cryopreservation using a standard vitrification protocol (IPGRI 2003b).

In citrus, protocols for adventitious root formation and regeneration of shoot for *C. hystrix* were developed (IPGRI 2000b). Regeneration *via* somatic (nucellar) embryogenesis was developed for calamansi (*Citrofortunella macrocarpa*) and mandarin (*C. reticulata*) using immature and mature seeds. For pummelo (*C. grandis*), callus was induced from juice vesicles and albedo tissues but somatic embryogenesis and shoot regeneration were observed only in callus from albedo. Regeneration system *via* somatic embryogenesis was developed for lime using undeveloped ovules (immature seeds). Immature seeds of citron (*C. medica*), limon (*C. limon*) and native lime 'Dalayap' (*C. aurantifolia*) showed varying degrees of callus and root formation. An effective slow growth medium was identified for mandarin (*C. reticulata*).

Cryopreservation of desiccated seed studied for calamansi, mandarin, pummelo, native lime, limon and kubot (Philippine native *Citrus* sp.) indicated feasibility of low temperature seed storage for short-term conservation. For cryopreservation of embryogenic callus using encapsulation dehydration technique, a suitable pretreatment and desiccation period was identified for *C. reticulata* and *C. sinensis.* For *C. hystrix,* vitrification method was modified to obtain acceptable level of survival. Effects of desiccation and cryopreservation were assessed on lime, pummelo, calamansi, kubot, Tai Cat (mandarin type) and calamandarin (*C. reticulata*) using enzyme systems and no variants were observed. Random amplified polymorphic DNA (RAPD) primers were identified to study genetic stability on *Citrus* regenerants (IPGRI 2000b).

In mango, somatic embryogenesis was obtained and substantial progress was made with induction and maintenance of somatic embryos of mango suspension cultures and protocols for secondary embryogenesis. Successful cryopreservation of somatic embryos was obtained by preculturing embryo masses (EMs) in sucrose and PVS2. Even though 70 per cent recovery was obtained, a major problem is repeatability of these results thus more research is required to optimize the protocol.

A micropropagation system was developed in *Davidsonia* spp. an underutilized native fruit of Australia, through the production of microcuttings *in vitro*. Protocols were developed for shoot and plantlet regeneration *via* organogenesis from a range of explants of *D. pruriens* and *D. jerseyana*. Work on organogenesis is on-going for *D. johnsonii*. Genetic diversity studies indicated that these three species are distinct. Media for litchi and longan micropropagation were developed. Seed desiccation studies of litchi and longan identified optimum desiccation periods. Longan and litchi germplasm can be conserved using the technique of encapsulation-dehydration of shoot tips. In persimmon, a suitable medium was identified for embryo culture and nodal cutting. Cryopreservation of embryonic axes (EA) through vitrification was successful. Results concluded that vitrification was not suitable for persimmon shoot tips (IPGRI 2003b).

Protocols for adventitious root formation and regeneration of shoot in *Nephelium* spp. were developed. Slow growth technique showed the potential for short to medium-term storage of germplasm. Suitable primers for RAPD were identified to be used for genetic stability studies on regenerants.

It must be realized that a lot of research still needs to be done to refine and standardize the existing cryopreservation protocols so that these could be effectively used for conservation for tropical fruit genetic resources.

Promoting Community Participation for Effective Germplasm Conservation

Efforts were made to encourage farmers and communities to participate in conservation of tropical fruit species germplasm. In Nepal, farmer group meetings were organized with help of Agriculture Development Department and awareness was created about the value of their fruit germplasm materials in terms of income generation, socioeconomic aspects and conservation for future use. Fruit diversity fairs, fruit shows and festivals were organized in Nepal, Bangladesh, Vietnam, Thailand and the Philippines during socioeconomic surveys and germplasm collection missions. Close interactions with farmers and communities also motivated them to participate in the germplasm conservation efforts. Farmer organization leaders, village leaders, monks in village temples, women and youth organizations helped in organizing the socioeconomic activities in the villages and provided valuable information during the socioeconomic surveys. In Bangladesh, fruit tree plantation fortnight was organized in which mango growers were trained in propagation techniques and production practices. In India, a biodiversity workshop was organized in Karnataka State to create awareness amongst the people to participate in conservation of mango diversity and Snehakunj, an NGO, helped in collecting germplasm through the help of communities. A national exhibition was organized in Indonesia to create awareness among the communities for conservation and use of tropical fruit tree species. In Vietnam, under the project in collaboration with Centro Internazionale Crocevia (CIC), two community nurseries for propagation of tropical fruit species were established in Phuson, Ninh binh and Chiengpan, Sonla. Grafted seedlings of elite varieties of litchi, mango and pummelo were produced with the participation of local farmers. Two training programmes on grafting and pruning for mango and citrus were organized for these communities. As a result, a community nursery run by farmers produced materials and distributed these to other communities (IPGRI 2003a). It is obvious, from the experiences in the project, that interest of farmers to conserve fruit genetic diversity will effectively take place as they find greater use of these materials.

Need for Complementary Conservation Strategy

Above conservation approaches lay emphasis on need for a complementary conservation strategy (CCS) with different conservation options required for different species – *in situ* and *ex situ* conservation, *in vitro* conservation, cryopreservation, etc. To assess this need and its implementation in wider context, the status of germplasm conservation of target fruit species in the region was reviewed (Drew and Ashmore 2003). Various conservation options such as *in situ* conservation in nature reserves and on-farm/home gardens and *ex situ* methods such as field genebanks, seed genebanks, *in vitro* genebanks, cryobanks and pollen genebanks and their advantages and disadvantages were considered to crystallize ideas to develop appropriate complementary conservation strategies. A number of issues that should be considered while developing complementary conservation strategies for tropical fruit genetic resources include: (*i*) biological factors, (*ii*) methodologies available, (*iii*) conservation objectives, (*iv*) socioeconomic issues and (*v*) organizational issues (Ramanatha Rao and Arora 1999; Ramantha Rao and Bhag Mal 2002; Drew and Ashmore 2003).

Framework for developing a complementary conservation strategy for tropical fruits requires concrete and appropriate actions to be taken at different stages, namely, (*i*) networking of stakeholders at national, regional or international level, (*ii*) defining of objectives and sub-objectives in accordance with 'Global Strategy for Plant Conservation' and determination of species of major importance for each of the participating countries, (*iii*) analysis of feasibility of each option for each objective or sub-objective in terms of infrastructure needs, costs and risks involved, (*iv*) decision on conservation options for each objective or sub-objective relevant to each species of major importance, (*v*) setting up enabling environment – policy and legal issues and funding source, (*vi*) elaboration of strategic action plan by stakeholders and vii) implementation process (Drew and Ashmore 2003; IPGRI 2003a).

Training and Capacity Building

Skill Enhancement/Human Resource Development

IPGRI accorded the highest importance to human resource development in the region to improve effective conservation and sustainable utilization of tropical fruit genetic resources in Asia. Human resource development at national level was brought about through short-term regional and in-country training courses for researchers, short trainings with more practical bias for farmers and some other training for scientific–technical interactions among senior staff for managing and coordinating project programme. In all, 382 personnel were trained on various aspects of tropical fruit genetic resources (Arora and Ramanatha Rao 1998; Stapleton 2002; IPGRI 2003a). Details are given below:

1. Ten regional training courses involving hands-on technical training for the research staff were organized and 113 researchers from 14 countries were trained. Topics covered included characterization, evaluation, conservation and utilization of tropical fruit genetic resources and several related topics such as molecular marker studies, *in vitro* conservation and cryopreservation techniques, ecogeographic surveys and use of GIS tools, socioeconomic studies and data and information management (Table 1.11). Skills acquired and knowledge gained was directly used in effective implementation of project activities.
2. Four in-country training courses were organized in Vietnam, Peoples Republic of China and Indonesia involving 75 researchers. The trainees were exposed to germplasm collection techniques, and methods applied in documentation and data management (Table 1.12).

Table 1.11: Regional Training Courses Organized

Training Courses	*Date*	*Institution*	*No. of Trainees*
Conservation and Use of Germplasm of Tropical Fruits in Asia	18–31 May 1997	IIHR, Bangalore, India	17
Strengthening National Capacity to Manage Information on Tropical Fruit Genetic Resources	9–21 October 2000	IPB, UPLB, Philippines	10
Characterization, Evaluation and Conservation of Tropical Fruit Genetic Resources	15–26 May 2001	IIHR, Bangalore, India	13
Use of Molecular Markers for Characterization of Tropical Fruit Genetic Resources	11–23 October 2001	HAU, Wuhan, China	9
In Vitro Conservation and Cryopreservation of Tropical Fruit Genetic Resources	4–16 February 2002	NBPGR, New Delhi, India	15
Scientific Writing and Proposal Development	4–8 March 2002	VASI, Hanoi, Vietnam	11
Socioeconomics of Conservation and Use of Native Tropical Fruit Species Biodiversity in Asia	12–21 December 2002	HORDI, Peradeniya, Sri Lanka	10
Ecogeographic Survey on Genetic Diversity of Tropical Fruit Species and GIS Tools	24 February–1 March 2003	IPGRI-APO Serdang, Malaysia	15
Molecular Data Analysis of Tropical Fruit Genetic Resources	31 March–8 April 2003	HAU, Wuhan, China	9
Use of Micropropagation, Cryopreservation and Molecular Markers Techniques in Tropical Fruit Genetic Resources	1–26 September 2003	Griffith University, Brisbane, Australia	4
Total (Participants trained)			**113**

IIHR: Indian Institute of Horticultural Research; IPB: Institute of Plant Breeding, UPLB- University of Philippines, Los Baños; HAU: Huazhong Agricultural University; NBPGR: National Bureau of Plant Genetic Resources; VASI: Vietnam Agricultural Science Institute; HORDI: Horticultural Crop Research and Development Institute; IPGRI-APO: International Plant Genetic Resources Institute, Asia: Pacific and Oceania.

Table 1.12: In Country Training Courses Organized

Training Courses	*Date*	*Institution*	*No. of Trainees*
Exploration and Collection Methods for Tropical Fruits	15–18 September 2000	VASI, Hanoi, Vietnam	12
Documentation in Tropical Fruits	4–12 December 2000	FRI, Guangdong and CRI, Chongqing, China	6
Data Management of Plant Genetic Resources	31 October–2 November 2001	VASI, Hanoi, Vietnam	25
Plant Genetic Resources Documentation	8-13 July 2002	IFRURI, Solok, Indonesia	13
Plant Genetic Resources Documentation and Data Cleaning	5-11 August 2003	IFRURI, Solok, Indonesia	19
Total (Participants trained)			**75**

VASI*:* Vietnam Agricultural Science Institute; FRI*:* Fruit Research Institute; CRI*:* Citrus Research Institute; IFRURI*:* Indonesian Fruit Research Institute.

3. Two training programmes on propagation and nursery management of identified elite materials were organized in Vietnam involving 47 farmers. Also, farmers' mango festival was conducted in the Philippines, which involved 100 farmers, who were apprised of the variability and characteristics of the varieties grown. Interaction and group discussions among farmers and scientists proved very useful. These were specially designed to promote farmers practical know-how on techniques in propagation, grafting, pruning and fruit nursery management (Table 1.13).

Table 1.13: Farmers Training and Festivals Organized

Training/Festivals	*Date*	*Place*	*No. of Trainees*
Grafting Methods of Fruit Tree Species	March 2002	Phusom, Ninh Binh and Chiengpan Soula, Vietnam	22
Pruning and Fruit Management	August 2002	Phuson, Ninh Binh and Chiengpan Soula, Vietnam	25
Dinamulag Mango Festival	April 2003	Iba Ambales, Philippines	100
Total (Farmers/participants trained)			**147**

4. Other types of trainings were meant for scientists coordinating the programmes, such as visits, annual technical sessions conducted back to back with Annual Project Meetings and Writeshop Session involving 47 participants who were primarily the national project coordinators (Table 1.14). It is significant to note that in almost all of these trainings, counterpart funding was always provided by local host institutions and an NGO in case of farmer's training and by the local government in the case of the farmers' forum and fruit festival.

Table 1.14: Other Trainings Conducted During 2000–2003

Training	*Date*	*Institution/Place*	*No. of Trainees*
Study Visits of Country Coordinators	10–24 August 2003	NBPGR, New Delhi/ IIHR, Bangalore/ CISH, Lucknow/ NRCC, Nagpur	7
Annual Technical Meetings	6–9 February 2001	Pattaya, Thailand 28 February–2 March 2002	10
	25–28 March 2003	VASI, Hanoi, Vietnam	10
		Chongqing, China	10
Writeshop	16–22 September 2003	IPGRI-APO, Kuala Lumpur, Malaysia	10
Total (Participants trained)			**47**

NBPGR: National Bureau of Plant Genetic Resources; IIHR: Indian Institute of Horticultural Research; CISH: Central Institute for Sub-tropical Horticulture; NRCC: National Research Centre on Citrus; IPGRI-APO: International Plant Genetic Resources Institute, Asia-Pacific and Oceania

Strong national programmes were identified as Centres of Excellence (CoE) for regional training on various aspects of plant genetic resources. These include National Bureau of Plant Genetic Resources

(NBPGR), New Delhi, India for training on *in vitro* conservation and cryopreservation; the Huazhong Agricultural University (HAU), Wuhan, China for training on molecular characterization and Institute of Plant Breeding, University of Philippines, Los Baños (UPLB), the Philippines for training on data documentation. These CoEs will play a significant role in improving the capacity for plant genetic resources education and training in the region.

Strengthening Laboratories and Other Infrastructure Support

Fifteen laboratories in 10 participating countries were strengthened by providing them with critical equipments such as global positioning system (GPS), laboratory equipment, software for geographic information system (GIS), STAT packages, important publications and electronic descriptor software to facilitate their work in ecogeographic studies, mapping, diversity analysis and database development. This assistance significantly improved the capacity of the national programmes to conduct systematic collection and evaluation of the collected germplasm. Twenty-one field genebanks in ten countries were assisted to enable the national programmes to secure their collected materials. Assistance was also provided in terms of critical funds needed for field genebank establishment and maintenance. Technical consultancy services were provided from the developed laboratories to developing national programmes in the region, such as using expertise of NBPGR for developing tissue culture facilities in Nepal Agricutural Research Council (NARC), Kathamandu, Nepal.

Socioeconomic Studies

Socioeconomic studies were conducted in 10 countries and across 6 priority fruits so as to understand grower and consumer preferences *vis-à-vis* production and marketing constraints. For this, participatory research appraisals (PRA) and rapid market appraisals (RMA) were carried out in selected locations. A purposive multistage stratified sampling framework was followed in selecting appropriate location and farmer respondents for these surveys. Analytical methods used in the study included simple tabular analysis and percentages, costs and returns analysis and conventional methods of economic appraisal of investments made such as net present value (NPV), benefit to cost ratio (BCR) and internal rate of return (IRR). Salient findings of socioeconomic studies and the policy perspectives (Chengappa 2003; IPGRI 2003a) are as follows:

1. Low investment and high return make the target tropical fruit trees ideal intercrops especially in home gardens which are important production systems prevalent at the sites studied, including the marginal lands.
2. Commercial orchards growing native tropical fruits have relatively less maintenance cost. For example, the investment on cultivation of litchi in India was found to be profitable and economically viable with IRR of over 48 per cent.
3. Linking farmers to market constraints was analyzed that included location of main markets far from the production centres to keep highly perishable marketable produce of fruits needed for cold storages, transportation and distribution/supply for marketing practices. This process is cumbersome and forces the growers into over dependence on market intermediaries such as pre-harvest contractors and commission agents. Organizing farmers' association/ cooperatives and establishment of common facilities such as pre-cooling unit, preferably cold storage, processing units and refrigerated vans for transportation will help in improving marketing facilities. Thus, encouraging farmers' organizations to undertake marketing activities will provide a viable alternative to the farmers to realize greater share of the sale

price due. Appropriate institutional support will help the growers to realize the benefit of high prices in the markets despite short fruiting season of crops.

4. Several processed products can be made from the tropical fruits under study and particularly such value addition was considered more important for the minor fruits. Appropriate linkages of processing units with the growers will result in improving supply of such products to consumers. The use of diversity to meet the demand for different products is a feasible option, generating more income for the rural sector.
5. Establishment of processing centres in high production areas is vital in addressing the economic viability of this sector through improved utilization, product diversification of fruit genetic resources.
6. The documentation and dissemination of knowledge on nutritive and medicinal values of native tropical fruits will help to improve the level of consumption and thereby create the much needed demand pull for these fruits.
7. Training and awareness programmes to disseminate technology and knowledge on consumption patterns, post-harvest management and processing would be essential to enhance utilization and conservation of diversity of these fruit crops.

Socioeconomic surveys and analyses carried out indicated a vast scope for expansion of area under native tropical fruits with good varieties and exploiting the untapped potential diversity to meet the domestic and international demand both for fresh and processed fruits (Chengappa 2003).

These studies clearly established that the narrow genetic base, limited number of varieties in commercial cultivation, lack of classification and indexing of local varieties, and the need to conserve rare types possessing desirable characters are the important issues of concern and need immediate attention (IPGRI 2003a).

Information Dissemination

Information dissemination is extremely important for their use by diverse stakeholders. The outputs and achievements of IPGRI supported programmes and activities are disseminated through research papers, in scientific journals, status reports, descriptors, germplasm catalogues, inventories, monographs, books and proceedings of the meetings relating to tropical fruit species which have been widely distributed to partners and have also been placed on AFGRN (Asia Fruit Genetic Resource Network) website http://www.afgrn.net. A CD-ROM was developed by IPGRI which contained available databases of fruit tree genetic resources from 10 countries and was made available to all the partners.

The website of AFGRN is established at Central Institute of Subtropical Horticulture (CISH), Lucknow that was given responsibility of maintaining and regular updating this website. The following information can be accessed from the website: (i) projects and priorities in different countries; (ii) activities of the project in different countries; (iii) priority tropical crops, participating institutions, country coordinators and project team members in different countries; (iv) expected outcome of the project with respect to different aspects of tropical fruit tree genetic resources (TFTGR) management; (v) human resource development activities; (vi) list of the TFTGR workers involved in projects; (vii) useful links with TFTGR; (viii) mailing lists; (ix) information on target crops, namely, mango, citrus, rambutan, jackfruit, litchi, and mangosteen.

Collaboration and Linkages

Diverse activities carried out on tropical fruits as described above could be possible through IPGRI's established links with a number of national, regional and international organizations. Some of these collaborative activities are as follows:

Exchange of Information and Germplasm

Bangladesh shared the passport and characterization data on mango, citrus and jackfruit with the partners in other countries. Information on elite accessions, namely, BARI Mango 1, BARI Mango 2, BARI Mango 3 and BARI Mango 4 were also shared with participating countries. Database on citrus, mango and litchi developed in China are available for information to researchers interested in exchange of materials. Malaysia exchanged information on characterization of the underutilized fruit species, pulasan and kuini, with the partner countries.

Lists of material in mango and citrus that can be exchanged from Nepal for research purpose under material transfer agreement (MTA) were developed for placing it on the AFGRN website. The Philippines exchanged passport and characterization information of priority fruit species with several countries in the region. Accession information and databases developed in Sri Lanka were shared with other countries. Germplasm of mango (3 cultivars) and lemon (one cultivar) were received by Sri Lanka from India under MTA between the Indian Council of Agricultural Research (ICAR), India and the Council for Agricultural Research Policy (CARP), Sri Lanka. Thailand shared accession information, catalogue and databases with partners. The database and catalogues of relevant fruit species are placed on AFGRN website. The information on exchangeable varieties of litchi, mango and citrus was made available by Vietnam to other countries in the region. Also, the seminars and workshops organized in Indonesia in collaboration with the International Tropical Fruits Network (TFNet) programme facilitated exchange of information. Accession information, catalogue and databases were shared with partners.

Collaboration with Regional and International Organizations

IPGRI's programme on tropical fruit species over the past two decades was very successful in establishing close collaboration among the participating countries. It also facilitated its linkages and interaction with national partners and donors and international organizations such as Food and Agriculture Organization of the United Nations (FAO), Centre de Cooperation International en Recherché Agronomique pour le Developpement (CIRAD), Asian Development Bank (ADB), Australian Centre for International Agricultural Research (ACIAR), International Centre for Underutilized Crops (ICUC), Underutilized Tropical Fruits in Asia Network (UTFANET), and the International Tropical Fruits Network (TFNet) for promoting research on plant genetic resources aspects in fruit crops. Collaboration with sub-regional networks, such as the South Asia Network on Plant Genetic Resources (SANPGR), Regional Cooperation in Southeast Asia for Plant Genetic Resources (RECSEA-PGR) and the East Asia Network on Plant Genetic Resources (EA-PGR) was established for exchange of technology and information. Collaborative links were also established between different research institutions within each of the ten countries.

Such regional and/or international collaboration proved particularly useful in capacity building of the participating organizations. IPGRI-APO and ACIAR jointly organized Citrus Germplasm Conservation Workshop at Griffith University, Brisbane, Australia on 6-7 October 1997. Participants from Australia, China, India, Japan, Malaysia, Nepal, New Zealand, the Philippines, Thailand, Vietnam and IPGRI attended the workshop. A working group to promote conservation and use of citrus genetic resources was recommended for promoting research and development needs of the region. The proceedings were published and widely distributed and IPGRI laid stress particularly on complementary conservation strategies (Ramanatha Rao and Arora 1999; Ramanatha Rao and Bhag Mal 2002; and Drew and Ashmore 2003).

IPGRI's collaboration with ICUC in identifying priority fruit species for the Asia-Pacific region had been very significant in order to give research and development thrust on these underutilized fruit species (Arora and Ramanatha Rao, 1995). IPGRI also collaborated with DFID-funded project on 'Promoting Tropical Fruits through Dissemination of Information'. Under this project, several monographs, namely, tamarind (*Tamarindus indica*) ber (*Ziziphus mauritiana*), baobab (*Adansonia digitata*), ndjanssang (*Ricinodendron heudelotii*); safou (*Dacrylodes edulis*) and *Annona* species were published (Gunasena and Hughes 2000; Pareek 2001; Sidibe and Williams 2002; Kengue 2002; Pinto *et al.*, 2005; Tchoundjeu and Atangana 2006). Dr R.K. Arora, IPGRI Office for South Asia, New Delhi was in the Editorial Committee and participated in its meetings.

IPGRI and Underutilized Tropical Fruits in Asia Network (UTFANET) supported each other in the training and meetings organized from time to time. In view of the common interest in conservation and use of tropical fruit species, IPGRI and UTFANET agreed to collaborate and signed the Memorandum of Understanding (MOU) between the two organizations. IPGRI-APO participated in the meetings of UTFANET Steering Committee held during the past several years. The agreed activities included (i) development of specific projects by national programmes on three priority species, jackfruit (Bangladesh), pummelo (India) and mangosteen (the Philippines); (ii) documentation of information and its dissemination in electronic format; (iii) development of descriptors for jackfruit, pummelo and mangosteen; and (iv) training in areas relating to collection, conservation and dissemination of genetic resources.

Integrating Biodiversity Conservation and Livelihood Improvement

A pilot study 'Integrating biodiversity conservation and livelihood improvement' was undertaken by IPGRI to study the role of markets of mango and mangosteen in collaboration with Indian Institute of Horticultural Research (IIHR), Bangalore, India. Much of this study's focus was on linking farmers to markets *vis-à-vis* improving their livelihoods through better income generation. Goal of the project was to increase knowledge about roles of market and livelihood strategies in conserving agricultural biodiversity and improving livelihoods, which lead to more effective management and use of agrobiodiversity. The immediate purpose is to gain understanding of the socio-economic characteristics of farm households utilizing tropical fruits and market forces that play a role in their decision making for on- farm conservation and the sustainable livelihoods. The pilot study has made it very clear that the socioeconomic aspects will need a lot of attention and the stakeholders will need in-depth training to be able to effectively implement the projects on tropical fruit tree species. Outcomes of study can be used for more effective implementation of the marketing and value adding activities of the proposed UNEP/GEF project on 'Conservation and Sustainable Use of Cultivated and Wild Tropical Fruit Diversity: Promoting Sustainable Livelihoods, Food security and Ecosystem Service'.

Need for Establishing a Regional Genebank of Tropical Fruit Species

IPGRI has been supporting the idea of its partners, in either geographic sense or on crop-basis, to come together, and develop a concept of regional or subregional genebanks and share the responsibilities and benefits from such an effort. For example, IPGRI, through its International Coconut Genetic Resources Network (COGENT), has promoted establishment of a multi-site International

Coconut Genebank (ICG) consisting of a regional genebank in each of the five COGENT regions (Batugal and Oliver 2004).

This collaborative activity involves conservation and evaluation of valuable germplasm in a designated host country, and the germplasm collected is conserved, evaluated and exchanged on mutually agreed terms. Once the countries agree to contribute germplasm and share the germplasm conserved in the regional collections, requests for germplasm need be made only to the host country of the regional collection. It requires documentation of available national information on the performance of specific germplasm and generation of additional data by all member countries; development and use of standard methods of characterization and evaluation by all members of the network; and development and sharing of database, which will serve as basis of selecting germplasm of interest for conservation and exchange. It is a system that needs to be established through networking (Sajise *et al.*, 2002). This concept has been proposed to Asia Fruit Genetic Resources Network (AFGRN) members, and assumes importance in view of the activities undertaken on tropical fruits and their potentials and focus on improved livelihood of the rural poor; as well as the sustainability of regional efforts put forth towards conservation and use of tropical fruits.

Impact and Sustainability

IPGRI's efforts on tropical fruits had significant impacts and successfully promoted awareness about the conservation and use of tropical fruit tree genetic resources at various levels including researchers, farmers, administrators and policy makers. As a result, there is now a greater commitment by the national governments to support fruit crops research and development programmes. For example, the project on tropical fruit genetic resources conservation and use was funded by the Governments in Malaysia and the Philippines. In Vietnam, research on tropical fruit crops has been identified as one of the ten priority areas to be given due attention by the Government of Vietnam. In India, fruit crop research programme was financially supported under the National Agriculture Technology Project (NATP) and was included in the National Agricultural Innovation Project (NAIP) funded by World Bank to carry forward the ongoing activities. Thus, IPGRI's tropical fruit tree programme and activities triggered placement of fruit crops on higher development agenda of national governments in collaborating countries.

The countries have been able to collect the threatened and rare diversity of identified priority TFT crops including their wild relatives and related species and the use of new tools such as DIVA-GIS and FloraMap will be extremely helpful in identifying the hot spots of diversity for future collecting. The diversity collected has been saved from loss and is safely conserved in genebanks for current and future use, which will have long-term impact on the fruit crop research and development programme in the collaborating countries. The advanced technologies developed for germplasm conservation in some fruit species will be available for use in other fruit crops in different countries.

Useful diversity comprising several elite lines and germplasm with specific desirable attributes is now available for use by the researchers in the breeding programmes. The promising elite lines of the priority fruit crops were multiplied and given to farmers for cultivation in Vietnam, Nepal, Sri Lanka and China, thus demonstrating direct application of the findings in the field.

IPGRI projects on tropical fruits facilitated the establishment of the Asian Fruit Genetic Resources Network (AFGRN) of 10 Asian countries. The Network helped to promote regional cooperation to access and share information through its website http://www.afgrn.net. The website serves as a platform for exchange of information on the progress of work on tropical fruit tree species in the collaborating countries and thus sustains collaboration between partners in Asia region.

Conclusion

IPGRI through its programmes and activities in priority tropical fruit tree species in the region has strengthened the national capacity, promoted collaboration at the regional and international level, and improved sharing of information, technologies and germplasm among partners in the region. IPGRI and its partners are focussing on enhancing national capacity to exploit the native tropical and subtropical fruit diversity in the region. Networking is identified as an essential component for success to harness maximum benefits and future thrusts for exploiting and using species diversity for sustainable livelihood of the poor people. In order to address the issues of *in situ* conservation, management and use of tropical fruits, IPGRI in collaboration with partners from India, Indonesia, Malaysia and Thailand is continuing its efforts towards conservation and sustainable use of cultivated and wild tropical fruit species diversity with funding supoort from different sources. The programme aims at achieving important outputs such as enhancement of farmers and users knowledge and practices on *in situ* conservation and on-farm management; identification, testing and implementation of methodologies and good practices for management of diversity and strengthening the capacity of different stakeholders to effectively use the methodologies and practices.

Involvement of national governments together with NGOs and farmers' organizations is important for conservation and use of genetic diversity if native tropical fruits have to play an important role for sustainable livelihood of rural poor. An integrated approach is needed to improve conservation and utilization of fruit genetic resources in the Asian region. This would improve productivity and quality and enhance income from cultivation of native tropical fruits, and contribute to food and nutritional security, poverty alleviation and the protection of environment in the rural sector.

Acknowledgement

The authors are highly grateful for help and support provided by the national partners in the collaborating countries, namely, Bangladesh, China, India, Indonesia, Malaysia, Nepal, Philippines, Sri Lanka, Thailand, Vietnam and Australia in implementing the tropical fruit tree projects. The authors also express sincere thanks to the Asian Development Bank (ADB), Australian Centre for International Agricultural Research (ACIAR), Department for International Development (DFID) and United Nations Environment Programme (UNEP)/Global Environment Facility (GEF) for funding support for IPGRI programmes and to the IPGRI management to implement the tropical fruit tree projects in the Asia-Pacific and Oceania region.

References

Arora, R.K., 1995. Role of IPGRI in the collection, conservation and utilization of tropical and subtropical fruit tree genetic resources. Pp7-15 (*in*) Fruit Production in the Tropics and Subtropics, JIRCAS International Symposium Series No.3. Ist International Symposium of JIRCAS. Symposium of the XXIVth International Horticultural Congress, Kyoto International Conference Hall, Japan, 22-23 August 1994.

Arora, R.K. and V. Ramanatha Rao (eds.), 1995. Proceedings of Expert Consultation on Tropical Fruit Species of Asia, 17-19 May 1994, Malaysian Agricultural Research and Development Institute, Serdang, Kuala Lumpur, Malaysia. IPGRI Office for South Asia, New Delhi, India. 116 p.

Arora, R.K., V. Ramanatha Rao and A.N. Rao (eds.), 1996(a). Status Reports on Genetic Resources of Mango in South, Southeast and East Asia. IPGRI Regional Office for Asia, the Pacific and Oceania, Singapore.

Arora, R.K., V. Ramanatha Rao and A.N. Rao (eds.), 1996(b). Status Reports on Genetic Resources of Citrus in East (China, Japan) and South Asia (India). IPGRI Regional Office for Asia, the Pacific and Oceania, Singapore.

Arora, R.K. and V. Ramanatha Rao (eds.), 1998. Tropical Fruits in Asia, Diversity, Maintenance, Conservation and Use, Proceeding of the IPGRI/ICAR/UTFANET Regional Training Course on the Conservation and Use of Germplasm of Tropical Fruit in Asia, 19-31 May 1997, IIHR, Hessaraghatta, Bangalore, India. IPGRI for South Asia, New Delhi, India.

Batugal, P.A. and J. Oliver, 2004. The International Coconut Genetic Resources Network. Pp 106-111 *in* South Asia Network on Plant Genetic Resources (SANPGR), Proceedings of the sixth meeting (Bhag Mal, P.N. Mathur, V. Ramanatha Rao and A.H.M. Jayasuriya (eds.). Peradeniya, Sri Lanka, 9-11 December 2002. IPGRI South Asia Office, New Delhi.

Bhag Mal and V. Ramanatha Rao, 2003. Tropical Fruit Genetic Resources Activities in Asia, the Pacific and Oceania Region: Role of IPGRI. Pp.27-35 *in In Vitro* Conservation and Cryopreservation of Tropical Fruit Species. Bhag Mal and V. Ramanatha Rao 2002 (Chaudhary, R, R. Pandey, S.K. Malik and Bhag Mal (eds.). IPGRI Office for South Asia, New Delhi/NBPGR, New Delhi, India. 293 p.

Bhag Mal, V. Ramantha Rao and Y.S. Ramamani, 2004. Status of Tropical Fruit Genetic Resources Activities in Asia, the Pacific and Oceania: IPGRI's efforts. Pp.92-100 *in* Proceeding of Sixth Meeting of South Asia Plant Genetic Resources (SANPGR) held at PGRC, Peradeniya, Sri Lanka, 9-11 December 2002. IPGRI Office for South Asia, New Delhi, India 166 p.

Brown, M.J., 1997. Durion: A Bibliographic Review (R.K.Arora, V.Ramanatha Rao and A.N.Rao, eds.). IPGRI Office for South Asia, New Delhi, India. 188 p.

Budathoki, K., 2003. Conservation and Use of Mango and Citrus Species Biodiversity in Nepal. Final Country Report for Nepal submitted to IPGRI under the ADB-funded project on Conservation and Use of Native Tropical Fruit Species Biodiversity in Asia. Nepal Agricultural Research Council, Kathmandu, Nepal. 90 p.

Chandel, K.P.S., R. Chaudhury, J. Radhamani and S.K. Malik, 1995. Desiccation and freezing sensitivity in recalcitrant seeds of tea, cocoa and jackfruit. Annals of Botany. 76: 443-450.

Chaudhury, R., B.B. Mandal and S.K. Malik, 2000a. Development of cryopreservation techniques for the long-term conservation of recalcitrant seed and vegetatively propagated plant species in India. IPGRI-NBPGR project report (1996-2000), IPGRI, Rome, Italy.

Chaudhary, R.S., K. Malik and F. Engelmann, 2000b. Application of cryopreservation technology for storage of recalcitrant and intermediate seed species. P.3 (*in*) Abstract of International Conference on Science and Technology for Managing Plant Genetic Diversity in the 21st century, Kuala Lumpur, Malaysia, 12-16 June 2000.

Chaudhury, R., R. Pandey, B.B. Mandal, N. Sharma, R.K. Tyagi, A. Agarwal, S. Gupta, S.K. Malik and Z. Hussain, 2001. Germplasm conservation: *In vitro* repository and cryobank. Pp. 209-223 *in* A Compendium of Achievements (B.S. Dhillon, K.S. Varaprasad, K. Srinivasan, M. Singh, S. Archak, U. Srivastava, and G.D. Sharma, eds.). National Bureau of Plant Genetic Resources, New Delhi 329 p.

Chengappa, P.G., 2003. Socioeconomic Studies on Conservation and Use of Native Tropical Fruit Genetic Resources in Asia. Consultancy Report. IPGRI-ADB-TFT Project 'Conservation and Use of Native Tropical Fruit Species Biodiversity in Asia'. IPGRI Office for South Asia, New Delhi, India. 53 p.

Cho, E.G., Y.L. Hor, H.H. Kim, V. Ramanatha Rao and F. Engelmann, 2001. Cryopreservation of *Citrus madurensis* zygotic embryonic axes by vitrification: Importance of pregrowth and preculture conditions. CryoLetters. 22: 391-396.

Cho, E.G., M.N. Normah, H.H. Kim, V. Ramanatha Rao and F. Engelmann, 2002. Cryopreservation of *Citrus aurantifolia* seeds and embryonic axes using a desiccation protocol. CryoLetters. 23: 309-316.

Dela Cruz, F.S., 2001. Status Report on Genetic Resources of Mangosteen (*Garcinia mangostana* L.) in Southeast Asia. IPGRI Office for South Asia, New Delhi, India. 24 p.

Dela Cruz, F.S., 2003. Conservation and Utilization of Germplasm of Mango, *Citrus* spp. and *Garcinia* spp. in the Philippines. Final Country Report for Philippines submitted to IPGRI under the ADB-funded project on 'Conservation and Use of Native Tropical Fruit Species Biodiversity in Asia'. National Plant Genetic Resources Laboratory, Institute of Plant Breeding, College of Agriculture, University of the Philippines, Los Banos (UPLB), Laguna, Philippines.

Dinesh, M.R. and C.S. Vasugi, 2002. Catalogue of Mango Germplasm. Indian Institute of Horticulture Research, Bangalore, India. 160 p.

Drew, R. and S. Ashmore, 2003. Complementary Conservation Strategy for Tropical Fruit Species. Consultancy Report under ADB funded project on 'Conservation and Use of Native Tropical Fruit Species Biodiversity in Asia'. IPGRI South Asia Office, New Delhi, India. 68 p.

Drew, R., A.Ashmore, S.Somsri, N.Noor, T.T.Hoa, O.Damasco, and V.Ramanatha Rao, 2006. Development of Advanced Technologies for Germplasm Conservation of Tropical Fruit Species. Paper presented at the International Horticultural Congress. October 2006, Suwon, R. Korea.

Gunasena, H.P.M. and A. Hughes, 2000. Tamarind. International Centre for Underutilized Crops, Southampton, UK. 171 p.

Haq, N., 1994. Identification of priority species of tropical fruits for research and development. Final Report, IPGRI Project No.93/11 (AOG-APO) ICUC in collaboration with IPGRI-APO. International Centre for Underutilized Crops, Department of Biology, Southampton University, UK.

Hodel, Urs and M. Gessler, 1999. *In situ* conservation of plant genetic resources in home gardens of Southern Vietnam. A report of home gardens surveys in Southern Vietnam. December 1996-May 1997. 106 p.

Hoe, V.B. and K.H. Siong, 1999. The nutritional value of indigenous fruits and vegetables in Sarawak. Asia Pacific Journal of Clinical Nutrition 8(1): 24-31.

Hue, N.T.N., 2003. Conservation and Use of Diversity in Citrus, Mango and Litchi in Vietnam. Final Country Report for Vietnam submitted to IPGRI under the ADB-funded project on "Conservation and Use of Native Tropical Fruit Species Biodiversity in Asia". Vietnam Agricultural Science Institute, Thanh Tri, Hanoi, VietnamIdris, S. and R. Mat Lin 2002a Status Report on Genetic Resources of Kuini (*Mangifera odorata*) in Southeast Asia. Bhag Mal, V. Ramanatha Rao and R.K. Arora (eds.). IPGRI Office for South Asia, New Delhi, India. 23 p.

Idris, S and R. Mat Lin, 2002b. Status Report on Genetic Resources of Pulasan (*Nephelium ramboutan-ake* (Labill.) Leech in Southeast Asia. IPGRI Office for South Asia, New Delhi, India. 28 p

Idris, S., 2003. Genetic Conservation and Utilization of *Nephelium ramboutan-ake* and *Mangifera odorata* in Malaysia. Final Country Report for Malaysia submitted to IPGRI under ADB-funded Project on "Conservation and Use of Native Tropical Fruit Species Biodiversity in Asia'. Malaysian Agricultural Research and Development Institute, Kuala Lumpur, Malaysia.

IPB, 2003(a). Germplasm Catalogue on mango. Institute of Plant Breeding, College of Agriculture, University of the Philippines, Los Banos (UPLB), Laguna, Philippines.

IPB, 2003(b). Germplasm Catalogue on *Garcina* spp. Institute of Plant Breeding, College of Agriculture, University of the Philippines, Los Banos (UPLB), Laguna, Philippines.

IPGRI, 1999. Descriptors for Citrus. International Plant Genetic Resources Institute, Rome, Italy. 66 p.

IPGRI, 2000. Descriptors for Jackfruit (*Artocarpus heterophyllus*). International Plant Genetic Resources Institute, Rome, Italy. 58 p.

IPGRI, 2002. Descriptors for Litchi (*Litchi chinensis*). International Plant Genetic Resources Institute, Rome, Italy. 64 p.

IPGRI, 2003(a). Conservation and Use of Native Tropical Fruit Species Biodiversity in Asia. Final Project Report, Asian Development Bank, RETA 5866, September 2003. IPGRI-APO, Serdang, Malaysia. 89 p.

IPGRI, 2003(b). Development of Advanced Technologies for Germplasm Conservation of Tropical Fruit Species. Final Project Report. Australian Centre for International Agricultural Research. CP/2000/002. IPGRI-APO, Serdang, Malaysia. 12 p.

IPGRI, 2003(c). Descriptors for Rambutan (*Nephelium lappaceum*). International Plant Genetic Resources Institute, Rome, Italy. 54 p.

IPGRI, 2003 (d). Descriptors for Mangosteen (*Garcinia mangostana*). International Plant Genetic Resources Institute, Rome, Italy. 56 p.

IPGRI, 2006. Descriptors for Mango (*Mangifera indica*). International Plant Genetic Resources Institute, Rome, Italy. 57p.

Kalloo, G., 2003. Conservation and use of diversity in mango, citrus and litchi in India. Final Country Report for India submitted to IPGRI under ADB-funded Project on "Conservation and Use of Native Tropical Fruit Species Biodiversity in Asia'. Indian Council of Agricultural Research, Krishi Anusandhan Bhavan, Pusa Campus, New Delhi.

Kengue, J., 2003. Safou (*Dacrylodes edulis*). International Centre for Underutilized Crops, Southampton, UK. 147 p.

Mahrouf, A.R.M., 2003a. Socioeconomic Dimensions in Conservation and Use of Mango Species Biodiversity in Sri Lanka. Final Report on Socioeconomic Aspects studied under IPGRI-ADB-TFT Project "Conservation and Use of Native Tropical Fruit Species Biodiversity in Asia'. Socioeconomics and Planning Centre, Department of Agriculture, Peradeniya, Sri Lanka. 75 p.

Mahrouf, A.R.M., 2003b. Socioeconomic Dimensions in Conservation and Use of Jackfruit Species Biodiversity in Sri Lanka. Final Report on Socioeconomic Aspects studied under IPGRI-ADB-TFT Project "Conservation and Use of Native Tropical Fruit Species Biodiversity in Asia'. Socioeconomics and Planning Centre, Department of Agriculture, Peradeniya, Sri Lanka. 81 p.

Nares, D., V. Ramanatha Rao, and D. Jarvis, 2001. *In situ* conservation of tropical fruit trees in Southeastern Asia. Pp. 51-60 *in* Proceedings of the First Annual Meeting of the Project, 'Conservation and Use of Native Tropical Fruit Species Biodiversity in Asia (Bhag Mal, Y. S. Ramamani and V. Ramanatha Rao, eds.). 6-9 February 2001. IPGRI South Asia Office, New Delhi. 51-60 p.

Nito, N., 2003. Status Report on Genetic Resources of Citrus in Asia-Pacific Region. Bhag Mal, V. Ramanatha Rao and R.K. Arora (eds.). IPGRI South Asia Office, New Delhi, India. 104 p.

Pathak, R.K., 2003. Status Report on Genetic Resources of Indian Gooseberry–Aonla (*Emblica officinalis* Gaertn.) in South and Southeast Asia. IPGRI Office for South Asia, New Delhi, India. 99 p.

Pareek, O.P., Suneel Sharma and R.K. Arora, 1998. Underutilized Edible Fruits and Nuts: An Inventory of Genetic Resources in Their Regions of Diversity. IPGRI Office for South Asia, New Delhi, India. 235 p.

Pareek, O.P., 2001. Ber. International Centre for Underutilized Crops, Southampton, UK. 292 p.

Pinto, A.C. de Q., M.C.R. Cordeiro, S.R.M. de Andrade, F.R. Ferreira, H.A. de C. Filgueiras, R.E. Alves and D.I. Kimpara, 2005. *Annona* species. International Centre for Underutilized Crops, University of Southampton, Southampton, UK. 263 p

Pumomo, S., 2003. Conservation and Use of Diversity on Rambutan and Mangosteen in Indonesia. Final Country Report for Indonesia submitted to IPGRI under ADB-funded Project on ''Conservation and Use of Native Tropical Fruit Species Biodiversity in Asia'. Indonesian Fruit Research Institute, Solok, West Sumatra, Indonesia.

Rajan, S., S.S. Negi and Ram Kumar, 2002. Catalogue on Mango (*Mangifera indica* L.) Germplasm. Central Institute for Subtropical Horticulture, Lucknow, India. 66 p.

Rajan, S., 2003. Electronic Catalogue of Mango (*Mangifera indica*). Central Institute for Subtropical Horticulture, Lucknow, India.

Ramanatha Rao,V. and R.K. Arora, 1999. Conservation Strategy. Pp 7-21 *in* Proceedings Citrus Germplasm Conservation Workshop (P.Broadbent, S.R.Sykes, K.B.Kenington and D. Hailstones, eds.) Brisbane, Australia. 6-7 October 1997. 158 p.

Ramanatha Rao, V. and Bhag Mal, 2002. Tropical fruit species in Asia: diversity and conservation strategies. Pp. 179-190 *in* Proceedings of the International Symposium on Tropical and Subtropical Fruits (Rod Drew ed.). Acta Horticulture 575. International Society for Horticulture Science. 405 p.

Ramanatha Rao, V., Bhag Mal and Z. Zongwen, 2005. Exploration of Native Tropical and Subtropical Fruit Species Diversity: Asian Perspective. Pp. 36-56 *in* Crop Improvement and Production Technology of Horticultural Crops. K.L. Chadha, B.S. Ahloowalia, K.V. Prasad and S.K. Sing (eds.). Horticulture Society of India, New Delhi. 764 p.

Sajise, P.E., Ramanatha Rao, V. and Batugal, Pons, 2001. Tropical fruit trees: Enhancing germplasm exchange and use. Pp. 35-41. In Proceedings of the First Annual Meeting of the Project, 'Conservation and Use of Native Tropical Fruit Species Biodiversity in Asia, Pattaya, Thailand 6-9 February 2001 (Bhag Mal, Y.S. Ramamani and V. Ramanatha Rao, eds.). IPGRI South Asia Office, New Delhi.

Sant Ram and Shailendra Rajan, 2003. Status Report on Genetic Resources of Mango in Asia-Pacific region. Bhag Mal, V. Ramanatha Rao and R.K. Arora (eds.). IPGRI Office for South Asia, New Delhi, India. 196 p.

Shantha Peiris, K.H., 2003. Exploration, Collection, and Conservation of Genetic Diversity of Mango and Jackfruit in Sri Lanka. Final Country Report for Sri Lanka submitted to IPGRI under the ADB-funded project on 'Conservation and Use of Native Tropical Fruit Species Biodiversity in Asia'. Horticultural Crop Research and Development Institute, Gannoruwa, Peradeniya, Sri Lanka.

Sidibe, M. and Williams, J.T., 2002. Baobab (*Adansonia digitata*). International Centre for Underutilized Crops, Southampton, UK. 96 p.

Singh, R.B., 1993(a). Fruit production in Asia-Pacific region. Pp. 1-26 *in* Research and Development of fruits in Asia-Pacific Region. R.B. Singh (ed.) RAPA/FAO, Bangkok. 207 p.

Singh, R.B., 1993b. Conservation and utilization of fruit germplasm in the Asia-Pacific region. Pp. 57-68 *in* Research and Development of Fruits in the Asia-Pacific Region (R.B. Singh, ed.). RAPA/FAO, Bangkok. 207 p.

Singh, I.P. and Shyam Singh, 2003. Exploration, Collection and Mapping of Citrus Genetic Diversity in India. Technical Bulletin No.7. National Research Centre for Citrus, Nagpur, Maharashtra.

Somsri. S., 2003(a). Conservation and Use of Mango and Rambutan Diversity in Thailand. Final Country Report for Thailand submitted to IPGRI under ADB-funded Project on Conservation and Use of Native Tropical Fruit Species Biodiversity in Asia'. Horticulture Research Institute, Department of Agriculture, Chatuchak, Bangkok, Thailand.

Somsri, S., 2003(b). Electronic Catalogue of mango germplasm. Horticulture Research Institute, Bangkok, Thailand.

Somsri, S.2003(c). Electronic Catalogue of rambutan germplasm. Horticulture Research Institute, Bangkok, Thailand.

Sudarmonowati, E. 2000. Cryopreservation of tropiocal plants: current reserach status in Indonesia. Pp. 291-296 *in* Cryopreservation of tropical plant germpalsm. Current research progress and application (F. Engelmann and H. Takagi, eds.). JIRCAS, Tsukuba, Japan/IPGRI, Rome Italy.

Stapleton, P., 2002. Project Proposal Development and Scientific/Technical Report Writing. Consultancy Report prepared under IPGRI-ADB-TFT Project 'Final Report on Socioeconomic Aspects studied under IPGRI-ADB-TFT Project 'Conservation and Use of Native Tropical Fruit Species Biodiversity in Asia'. Bogor, Indonesia. 65 p.

Tchoundjeu, Z. and A.R. Atangana, 2006. Ndjanssang. (*Ricinodendron heudelotii*). Southampton Centre for Underutilized Crops, University of Southampton, Southampton, UK. 74 p.

Tropical Fruit Net., 2004. Global production of tropical and sub-tropical fruits. Issue No. 3.

Uddin, Md. Nazim, 2003. Survey, Collection, Evaluation, Conservation and Utilization of Jackfruit, Mango and Citrus of Bangladesh. Final Country Report for Bangladesh submitted to IPGRI under the ADB-funded project on 'Conservation and Use of Native Tropical Fruit Species Biodiversity in Asia'. Bangladesh Agricultural Research Institute, Dhaka, Bangladesh.

Verheij, E.W.M. and Coronel, R.E., 1991. Plant Resources of South-East Asia No. 2. Edible Fruits and Nuts. Pudoc. Wageningen, the Netherlands.

Zushang, C., 2003. Conservation and Utilization of citrus, mango and litchi in China. Final Country Report for China submitted to IPGRI under the ADB-funded project on 'Conservation and Use of Native Tropical Fruit Species Biodiversity in Asia'. Citrus Research Institute, Chinese Academy of Agricultural Sciences, Beibei, Chongquing, China.

Biodiversity in Horticultural Crops Vol. 1 *Pages* ***31–38***
Editors: **Professor K.V. Peter & Dr. Z. Abraham**
Published by: **DAYA PUBLISHING HOUSE**

Chapter 2

Temperate Fruit Crops

A. Sofi, M.K. Verma, R.K. Verma and H. Choudhary
Central Institute of Temperate Horticulture, Srinagar – 190 007, J&K

The entire North-Western Himalayan region comprising of J&K, Himachal Pradesh and Uttarakhand is very rich in temperate fruit plant diversity. Biodiversity is the key for sustainable fruit production. It has ecological, economical and cultural significance. It helps to ensure sustainability. There are three independent levels of biodiversity, namely ecosystem level, species level and genomic level. Genetic diversity represents variation within a species which represents varietal diversity in a crop plant. It enables species to adapt to new agro-ecosystem and environments through natural and/or human selection.

Among temperate fruits, commercially grown are apple, pear, peach, plum, apricot, cherry, almond, walnut, pecan, hazelnut, chestnut, persimmon and chinese gooseberry. These become cash crops for hill people. Their biodiversity ensures stable productivity, high remuneration and regulated market supply. This biodiversity could be of different fruit species or of different varieties of the same species or of various rootstocks. In case of apple, approximately 2500 cultivars are existing all over the world but only 25 cultivars are commercially grown which mainly include Delicious cultivars. It represents extremely narrow genetic base of cultivated apple. Similarly the entire North Western Himalayan region possesses around 31 fruits genera. For enrichment of germplasm, 482 cultivars/varieties/clones of temperate fruits from 12 advanced countries were introduced through NBPGR to India.

Hot Spots of Biodiversity in Different Fruit Crops

Depending upon concentration of species and richness of plant species at several locations, several sites have been identified as natural habitats of many crop species and these can be protected as *in situ* conservation sites in J&K, H.P and Uttarakhand.

Table 2.1: Occurrence of Species in Different Genera

Fruits (Genera)	*Fruit Genera/Species*
Pome fruits (*Malus* and *Pyrus*)	*Malus pumila, M. baccata, M. sikkimensis, M. dirangensis, Pyrus jaqumontiana, P. polycarpa, P. pyrifolia, P. pashia, P. communis*
Stone fruits (*Prunus*)	*Prunus domestica, P. salicina, P. cerasifera, P. armenicana, Prunus nepaulensis, P. dulcis, P. persica, P.prostrata, P. wallichii, P. cornuta, P. cerasoides, P. mira*
Nut fruits	*Juglans regia, J. nigra, Carya illinoesis, Corylus avellena, C. colourna, C. ferox*
Soft fruits	*Actinidia deliciosa, A. purpurea, A.arguta, Vitis himalayana, Rubus elipticus, R. nivus, R. biflorus, R. fruitcosus, R.lasiocarpus, Ribes nigrus, R. rubrum, R. furmosanium, R.roezli, R. wolfi, R. aureum, R. triste,. R. burejens, R. petraeum*
Minor fruits	*Sorbus lanata, Crategus crenutata, C. oxycantha, Feijoa sellowiana, Cydonia oblonga, Diospyrus kaki, Docyinia hookeriana, D. indica, Punica granatum, Vibernum contifolium, V. lanata, Cornus capitata, Elaegnus umbellate, Castenia sativa, Olea europaea, Zizyphus jujube, Aesandra butracea, Pistachio Pinus girardiana, Hippophae rhamanoides, Ficus species*
Wild fruits	*Aegle marmelos, Cordioblique, Emblica officinalis, Ficus palmata, Ficus roburghii, Myrica nagi, Opuntia dillenii, Ficus indica, Grewia subinaequalis, Passiflora edulis, Sorbus lonata, Rubus niveus, Fragaria nubicola, Rubus ellipticus, Vitis lanata, Sorbushiu spidata, Zizyphus jujube, Spondias, Mangifera, Buchanania latifolia, Corylus colourna, Ribes rubrum, Hippophae salicifolia, H. rhamnoides and Cotoneaster baccillaris.*

Table 2.2: Crop Target Diversity Available at Different Micro Locations in North Western Himalayas

Crop Target Diversity Crops	*J&K*	*H.P*	*Uttaranchal (Uttarakhand)*
Pome fruits (Apple, pear)	Srinagar (*Harwan, Newtheed, Shalimar, Ganderbal*), Anantnag (*Kulgam, Qazigund Devsar, Guddar, Garran Malwan*), Baramulla (*Sopore, Pattan, Bandipora, Rafaiabad*), Pulwama (*Shopian, Rajpora, Drabgam, Kelar*), Budgam (*Arizal, Khag, Chadoora Khansahib, Chararesharif*) Kupwara (*Handwara, Lolab, Rajwar, Ramhal*) Leh (*Kargil and Leh*)	Kullu, Kotgarh, Kinnaur, Bajoura, Kandaghat, Thanedhar	Nainital, Uttarkashi, Paudi, Dehradun, Almora, Pithoragarh
Stone fruits (Cherry, plum, peach, apricot)	Srinagar (*Newtheed, Shalimar, Laar*), Anantnag (*Qazigund, Malwan, Guddar, Wattu*), Baramulla (Uri, *Chandosa, Chandal, Wanigam*), Pulwama (*Tral, Noorpora, Shopian, Rajpora*), Budgam (*Charareshasharief Chadoora*), Kupwara (Rafiabad, Meelyal), Laddakh (Leh, Kargil)	Narkanda, Pangi, Powari of Kinnaur	Nainital Uttarkashi Pithoragarh and Dehradun
Nut crops (walnut, almond, chestnut, hazelnut, pistachio nut)	Srinagar (*Badamwari, Shalimar, Ganderbal, Kangan, Wusan*); Anantnag (*Lehnwan, Kukarnag, Achabal, Qazigund, Kulgam, Dersar*), Baramulla (*Tangmarg, Pattan, Bandipora, Ajas, Safapora, Uri, Karna*), Pulwama (*Shopian, Rajpora, Drabgam Tahab, Chookra*), Budgam (*Chadoora, Yarikha, Surasyar, Talpora, Charangam, Arizal*), Kupwara (*Lolab, Varnaw, Meerlyal Afun*)	Pangi valley, Ropa valley, Moorang of Kinnaur	Uttarkashi, Paudi, Almora and Tehri
Others (Pomegranate, *Sorbus, Cydonia, Docynia, Hippophae, Vitis*)	Doda (Ramban, Batote), Udhampur	Solan (Darlaghat, Narajj, Mandi (Bhaomla, Rewalsob, Lahaul, Spiti, Reeba, Skiba,	Dehradun, Nainital, Almora and Uttarkashi

Malus

Two species of *Malus* namely *M. baccata* var. Himalaica and *M. sikkimensis* are reported from the Himalayas (Hooker, 1879 and Anon., 1962). Distinct biotypes of *M. baccata* growing wild in different agro climatic regions of the Himalayas have been identified (Kishore, 2001). Details are described as follows:

M. baccata var. Himalaica (Maximum) Schneid

It is available in wild at an altitude of 3000m amsl in Himachal Pradesh and 1303m amsl in Meghalaya. It also grows wild in Uttaranchal (Srivastava *et. al.*, 1977), where it is used as dwarfing rootstock for apple. Tree is small producing red coloured fruits, but not worthy for fruit production. This species is partially resistant to collar rot (*Phytophthora cactorum*), root rot (*Dematophera necatrix*) and woolly aphid (*Eriosoma lanigerum*).

M. baccata (Linn.) Borkh (Shillong)

It is seen wild in Shillong (Meghalya) at an altitude of 1300m amsl (Randhawa, 1987). It is locally called 'Soh Sheur' in Himachal Pradesh. It forms a small low branching tree, producing small scarlet red fruits. It is also used locally as dwarfing rootstock. It is propagated by mound layering. It has high degree of resistance to powdery mildew (*Podosphaera leucotrica*), woolly aphid and field resistance to apple scab (*Venturia inaequalis*).

M. baccata (Linn.) Borkh (Khrot)

It is available wild in Himachal Pradesh at an altitude of 2000m amsl and locally called 'Banphal'. Trees are low and spreading in nature, produce small yellow fruits with red stripes. It is dwarfing as M-9. It is easy to propagate, having good graft compatibility with commercial cultivars of apple.

Malus baccata (Linn.) Borkh (Giabung)

This species was collected from Giabung (Himachal Pradesh) at an altitude of 2060m amsl. It is locally known as 'Lead' producing small scarlet red fruits. It can be easily propagated by mound layering.

Malus baccata (Linn.) Borkh (Dhak)

It is collected from Dhak at an altitude of 2060m amsl in Himachal Pradesh, locally known as 'Bambtu'. Trees are tall and spreading. It produces semi-vigorous apple rootstocks. It can be propagated by mound layering

Malus baccata (Linn.) Borkh (Rohru)

This is collected from Rohru (Himachal Pradesh) at an altitude of 2100m amsl. Trees are small and bushy. Based on stomatal density it is revealed that this rootstocks may be more dwarfing than M-9. Under field conditions, it is moderately resistant to powdery mildew.

Malus baccata (Linn.) Borkh (Srinagar)

This species is available wild in Srinagar (J&K) at an altitude of 1390m amsl. Trees are semi dwarfing in vigour and produce small yellow coloured fruits.

Malus sikkimensis (Hook F.) Kochneex Schneider

It is available wild in Sikkim at an altitude of 1970m amsl and locally known as 'Tipsi'(Randhawa 1987). Owing to occurrence of apomixis (Anon. 1962), its true to type propagation is possible through mould layering. It has field resistance to collar rot and powdery mildew. Brown (1975) observed that it is highly sensitive to latent viruses present in scion wood.

Pyrus

At present there is little information available about the *Pyrus* species growing wild in India. A few noticed from Himalayas are briefly described below:

Pyrus pashia Buch-Ham.ex.D.Don

It is locally known as 'Kainth' and found abundantly wild all over Himalayas from 700 to 2000m amsl. It originated from Himalayas (Bailey, 1953). The trees are vigorous and bear small round gritty fruits of very poor quality. It gives compatible graft union with pear cultivars but susceptible to fire blight (*Erwinia amylovora*), collar rot (*Phytophthora cactorum*) and powdery mildew (*Podosphaera leucotricha*).

Pyrus pashia Ham. Cx. D. Don. Var. Kumaonii (Scne Stapt.) syn. *Pyrus kumaonii* Deve.

It grows wild in Uttaranchal and Himachal Pradesh and locally called 'Shiara' or 'Jarainth'. Trees are more vigorous, yielding larger fruits than that of *Pyrus pashia*. Fruit quality is poor but superior to *P. pashia*. It is easy to propagate by stooling but poor through air layering. It is resistant to powdery mildew but susceptible to fire blight.

Pyrus pyrifolia (Burm. F.) Nakai. Var. Culta (Makins) Nakai Syn. *Pyrus serotina* Rend. Syn. *Pyrus sinensis* Auct. Non Poir.

It grows wild in Khasi hills of Meghalaya and other parts of North Western Himalayas between 788 to 1980m amsl. Trees are vigorous, bear larger, crispy, juicy, sweet and aromatic fruits. It is locally known as 'Nakh' in Himachal Pradesh, J&K, Uttranchal and 'Nashpati' in Meghalaya. It can be easily propagated by mound layering. It showed resistance to collar rot and powdery mildew in field.

Pyrus jacquimontiana Dcne. Jard

It grows wild in Kochli (1538m amsl) and Powari (1538m amsl) areas of Himachal Pradesh. It is locally known as 'Besar'. The trees are open headed, bears small roundish to oblong and brownish colour fruits with numerous white raised lenticels.

Pyrus polycarpa Hook.F.

It grows wild in groups in Shillong (1300m amsl) and Somarin (1350m amsl) areas of Meghalaya. Locally it is known as 'Deingloporam'. It bears small globose fruits in clusters. It is a good rootstock for pear. Under field conditions, it is free from powdery mildew and fire blight.

Prunus

Prunus includes the wild apricot locally known as chulli (*Prunus armeniaca*), bitter almond (*Prunus amygdalus*), Behmi (*Prunus mira*-natural hybrid between almond and peach), wild peach (*Prunus persica Batsch*) and sour cherry (*Prunus cerasus*).

Prunus armeniaca (Chulli/Wild Apricot)

It is locally known as 'chulli' found abundantly wild in North Western Himalayas and is the traditional crop of dry temperate regions like Laddakh (J&K) and Kinnour (HP). Laddakh and Kinnour are home of wild apricot. Large degree of variability existed for future exploitation in the form of trees and fruit parameters. A few of the seedlings bear good quality fruits with sweet kernels and can be released as selection. Fruits are consumed fresh and sun dried. The seedlings are also used as rootstock for cultivated apricot and plum (Sofi and Verma, 2003).

Prunus amygdalus (Bitter Almond)

Wild forms of almond generally bear bitter kernels. Bitterness in almond results from hydrolysis of glucoside 'amygdalin' which produces benzylaldehyde (a bitter substance) and cyanide (poisonous). Sweet kernels, originated in these plants by mutation, led to domestication of almond. Oil is generally extracted from bitter kernels. These are generally found in all parts of North Western Himalayas (Sofi and Verma, 2004).

Prunus mira (Behmi)

Generally known as Behmi, a natural hybrid between almond and peach, it grows in the Himalayas. Fruits are sweet when ripen and used by locals for making alcoholic drink. Seeds are used for raising seedlings on which almond is budded/grafted.

Prunus persica Batsch (Wild Peach)

Many wild forms of peach are found in North Western Himalayas. Fruits are very late in ripening, small in size and taste is very sweet. They may be free from stone type. These are resistant to peach leaf curl, a serious problem in commercial peach cultivars.

Prunus cerasus (Sour Cherry)

This is known as wild cherry used as rootstock for sweet cherry. *Prunus padus* (bird cherry) is also used in Himachal Pradesh as root stock. Trees are usually 6-9m tall, open headed spreading or round toped and generally without a central leader. Branches are more or less dropping or slender. Fruit is roundish oblate or cordate with slightly compressed sides with an average diameter of 2 cm. Leaves are much smaller and dark green in colour than sweet cherry. It is hardier than sweet cherry. It is used for making juice, fruit salads and for cooking. It can also be preserved by canning, freezing and sun drying. Cherry syrup is made by concentrating the juice. It is rich in protein and sugars and has more calories than apple. It has higher carotene, folic acid and a rich source of calcium, potassium, magnesium, iron and zinc.

Nut Crops

Deciduous nuts are only grown in North Western Himalayas. Major biodiversity among nut crops is available in walnut, almond, chestnut, hazelnut, pistachionut and chilgoza.

Juglans regia (Walnut)

Walnuts are found abundantly in J&K, HP and Uttarakhand in natural habitat and cultivated by farmers. It is the second important fruit crop of J&K state according to the area under occupancy. Wild types have very hard shell and less commercial importance. Most of the existing population is of seedling origin and there exists vast genetic variability which is being exploited for commercial cultivation. It is a rich source of fats, proteins, minerals and vitamins. Under NATP projects all the three walnut growing states have been explored and a total of 198 diverse genotypes identified based on nut and kernel characters. These are of seedling origin and can be released as cultivars/selections after further evaluation. Fifty four elite genotypes were planted at CITH, Srinagar for evaluation.

Desirable characters to look for in the selection of good quality nuts are big size, light colour, medium to good shelled with good shell seal, white coloured kernel that separates easily from shell and having good taste and flavour (Sofi and Verma 2004).

Castanea species (Chestnut)

It is a tall deciduous tree found in Himalayan forests. Chestnuts are monoecious with flowers borne on long spikes (Catkins) on current seasons growth.

Corylus avellana (Hazelnut)

This is a new crop for temperate zone of India. Its wild type (*Corylus colurna*) grows in Pangi area of Chamba and in high hills of Kullu and Shimla districts of Himachal Pradesh. It is also known as Filberts, Cobnuts, Barcelona nut, European hazel and belongs to family Betulaceae.

Pinus gerardiana Wall (Chilgoza)

It is also known as Neoza. It grows in North Western Himalayas ranging from Bank of Sutlej and Tapri upto Pooh in Kinnaur, Pangi sub division of Chamba and forests of Doda. In its natural habitat, the tree is usually found in dry regions outside the influence of monsoon, where rainfall is scanty and heavy snowfall during winter.

Pistachio species (Pistachio nut)

It is dioecious. It produces brown and green flowers without petals and borne on auxiliary racemes or panicles in early summer. Reddish wrinkled fruits are borne in clusters like grapes. Fruit is botanically a drupe and edible portion is seed. Colour of kernel varies from yellowish through shades of green which extents through kernel.

Carya illinoensis (Pecan nut)

Pecans are new crops for temperate zone, having good scope for its cultivation. It can be grown at an elevation of 500-1500m amsl. Having long tap root system, it has tolerance to drought conditions. Therefore it can be grown under drought prone areas. Promising cultivars of pecan are Mahan, Nellis, Burkett and Western Schley.

Other Temperate Fruits

Cydonia oblong (Quince)

Quince is sub-acidic and is a very popular fruit tree in Kashmir valley. Its fruits are generally cooked. Tree is grown for its seed. The small trees are planted as ornamental trees for attractive flowers and leaves. It survives under wide range of soils.

Ficus palmata (Wild Fig)

It is one of the tastiest fruit trees growing wild in the mid Himalayan region at an elevation of 1550m AMSL. Fruits are juicy containing 45.2 per cent extractable juice and 80.5 per cent moisture. Raw fruits alongwith new growth are used as vegetable by villagers.

Punica granatum (Wild Pomegranate)

It grows in waste tract of hill slopes of J&K, Himachal Pradesh and Uttarakhand. It is available abundantly wild in Doda and Udhampur districts of J&K; Solan, Shimla, Mandi, Shirmour and Bilaspur of HP, and Nainital, Almora and Dehradun of Uttarakhand. Every year Anardana worth tens of thousands of rupees is collected from hills. A large extent of biodiversity is reported from the region for future exploitation. Wild pomegranates are locally known as 'Dadu', 'Daram', 'Dolima', 'Daen' and 'Dadim'. Fruits are harvested from its natural habitat and kernels are dehulled and dried to make 'Anardana'. It is a shrub or small tree often armed with spines, rarely grows over 6m high. It is found abundant at elevations between 600m to 1800m in H.P., J&K and Uttarakhand. It bears scarlet red colored showy flowers with lush green foliage. The Anardana is rich in vitamin C (36.6 mg/100g) and contains 2.57 per cent protein, 0.01 per cent phosphorus, 0.47 per cent potassium, 0.03 per cent calcium and 0.13 per cent magnesium.

Zizyphus jujube (Chinese Ber)

This fruit tree/shrub is found commonly in wild forest and wastelands throughout mid hills upto elevation of 1400m amsl. Trees are spiny and bushy, bear small globular sweet to sub-acid fruits. Spines are comparatively larger. Wood of this plant is very hard and durable, used in making agricultural implements. It has resistance against powdery mildew and fruit borer.

Hippophae rhamnoides (Sea Buckthorn)

It is a dioecious usually spinescent small shrub, leaves small, lanceolate, bears small flowers, greenish or yellowish fruits enclosed by succulent receptacle ovoid, orange yellow or scarlet. Locally it is known as 'Treskas', 'Chharma', 'Sorma', 'Fasru' and 'Dhurchule. It occurs in river beds of drier ranges of Northwest Himalaya at an altitude of 2300 to 4000 m amsl. Fruits are acidic and rich in vitamin-C (1000mg/100g). Another species *H. salcifolia* D. Don also occurs in the region and has almost similar potential due to its nutritional value and other multifarions uses in the trans Himalayan region.

Rubus ellipticus (Rubus)

This is commonly known as "Hisara"/"Hisalu". It bears one of the tastiest wild fruits in hills. The plant grows in abundance throughout mid hills near water source. Fruits are highly nutritive. Juice has attractive colour and rich flavour, can be preserved as such and can also be used for squash making. A very good jam can also be prepared from this fruit. This fruit offers excellent opportunity for cultivation as a hedge or fence plant.

Trees are bushy type, 3-10m in height and are wild available throughout temperate zones of country ranging from North East to North Western Himalayas. A large degree of variability is available in the country for its exploitation.

Fragaria nubicola (Wild Strawberry)

It is found wild in J&K, HP and Uttarakhand. The wild type is smaller in size but is free from diseases. It is mild in taste compared to cultivated type. Fruits are rich in vitamins, minerals and pectin. It has sweet flavour. Major acids are citric and malleic acid, but it also contains quince, succinic, glycolic acid and oxaloacetate. Pigment responsible for red colour is anthocyanin, pelargonidin-3-monoglucoside and traces of cyandin. Major tannins are glycostlated leucoanthocyanidins.

Olea europeaea L. (Olive)

It is an evergreen plant requiring chilling for fruiting as in other deciduous fruit plants. Olive is used mainly for oil extraction. It is also used for pickling and for medicinal purposes. The olive has a wide range of adaptability. Its cultivation can be undertaken in areas ranging from 650-2300m amsl. It is found wild in J&K, (Doda, Udhampur, Kupwara and Baramulla), HP (Chamba, Kullu, Mandi,

Sirmour and Solan) and Uttarakhand (Dehradun). In olive, fruit flesh is the main source of oil. Olive oil is not only tasty and nutritious but it is free from cholesterol having polyunsaturated fatty acids. It is useful for regulating high blood pressure and is safe for patients suffering from heart or other related ailments. It is good for human skin. Oil massaging is useful in muscular pains and provides warmth during winter.

Enhancement of Productivity through Use of Biodiversity

In India, wild germplasm of temperate fruits and nuts appear to be introductions done over the long past and have undergone eco-geographical modifications/changes in plant and fruit characters. In genus *Malus*, there are two indigenous species *viz.*, *Malus baccata* var. Himalaica and *M. sikkimensis*. However, *Malus baccata* (Shillong), *M. baccata* (Khrot), *M. baccata* (Giabung), *M. baccata* (Dhak), *M. baccata* (Rohru), *M. baccata* (Srinagar) and *M. deerangenesis* from Arunachal Pradesh, Similarly in genus *Pyrus*, several ecotypes are identified like *Pyrus pashia*, *P. pashia* var. Kumaonii, *P. pyrifolia* (Burm. F) Nakai var. Culta Nakai (*P. serotina* Rend), *P. vulgaris*, *P. jacquimontiana* and *P. polycarpa*.

The wild sources are important sources for resistant/elite traits and are used for crop improvement programmes.

References

Anonymous, 1962. *The Wealth of India: Raw material*. CSIR, New Delhi, 6: 234-249.

Bailey, L.H., 1953. *Standard Encyclopedia of Horticulture*. MacMillon Co. New York, Purdue University Press, USA. Pp-3-38.

Brown, A.G., 1975. Apple. In: *Advances in Fruit Breeding*, (Eds.) Moore and Janick. Purdue University Press, USA, Pp3-38.

Hooker, J.D., 1879. *The Flora of British India* II, L. Reeve and Co. LTD, NR Ashford Kent, England.

Jindal, K.K. and Gautam, D.R. (Eds.), 2001. Compendium of winter school on "Productivity of Temperate Fruits-Issues and strategies" published by Department of Pomology, Dr. Y.S. Parmar University of Horticulture and Forestry, Nauni, Solan (H.P). 498 Pp.

Kishore, D., 2001. Utility of *Malus* and *Pyrus* species for improvement of productivity of temperate fruits. In: *Productivity of Temperate Fruits* (Eds.) Jindal and Gautam. Department of Pomology, Dr. Y.S. Parmar University of Horticulture and Forestry, Nauni, Solan, H.P., pp. 145-150.

Randhawa, S.S., 1987. *Wild Germplasm of Pome and Stone Fruits*, IARI, New Delhi-110012.

Sharma, B.D., Rana, J.C. and Yadav, S.K. (Eds), 2001. A Profile, NBPGR Regional Station, Shimla published by NBPGR, New Delhi-12. 45 Pp.

Sofi, A.A. and M.K. Verma, 2003 Technical Project Progress report from 1999-2003 Project on "sustainable management of plant biodiversity (NATP)" CITH, Srinagar, J&K 35. Pp.

Sofi, A.A. and M.K. Verma, 2004. Final Project Progress report on "Clonal selection and quality planting material production of almonds and walnut (NATP)". CITH, Srinagar, J&K 38 Pp.

Srivastava, R.P., Pathak, R.K., Bana, D. S. and Pandey, V.S., 1977. Utilization of some important fruit trees growing wild in the Himalayan region. In: *Fruit Breeding in India*, (Ed.) Nijjar, G.S. Oxford and IBH Publishing Co., New Delhi, pp. 1-8.

Biodiversity in Horticultural Crops Vol. 1 *Pages* **39–56**
Editors: **Professor K.V. Peter & Dr. Z. Abraham**
Published by: **DAYA PUBLISHING HOUSE**

Chapter 3

Tropical Fruits

G.S. Prakash and M.R. Dinesh
Division of Fruit Crops, Indian Institute of Horticultural Research,
P O Hissarghatta Lake – 560 089

Concept of origin of cultivated plants was first developed by A de Candolle and geographic centres of variability were developed by Vavilov. He identified Asia as a major centre with "Indian centre" of North East region as primary or secondary centre of origin for many crop plants. This region is centre of diversity for several important native fruits including Mango, Jackfruit and Citrus. Apart from major fruit crops, India is home to several underutilized fruit crops. Due to increased pressure on land use, several of wild types, which are a great source of beneficial alleles, are disappearing. There is urgent need to conserve them in both *in situ* and *ex situ* conditions. Plant genetic resources are of great importance as they form the basic raw materials to meet current and future needs of crop improvement programmes. A wider genetic base assumes priority in plant breeding research aimed at developing new varieties for increased crop production (Paroda, 1991). This diversity comprises of native landraces, local selections, elite cultivars and wild relatives of crop plants. Collection and conservation of this diversity in a systematic manner are the primary responsibility of all plant genetic resources institutes/centres. Mention on use and cultivation of fruits is seen in epics like 'Ramayana'. Plant genetic resources are thus our heritage, which need conservation for posterity.

During long period of domestication, utilization and cultivation, a wide array of fruit crop variability got generated by natural means and through both conscious and unconscious selection. Huge wealth of variability also got generated/adapted and diversified by crop introductions in exotic environment or through migration of human population.

Although, mankind used only about 5,000 plant species worldwide to meet food and other needs, this number is just a fraction of total world flora. With population growth, we are increasingly dependent on most productive plants. Today, only about 150 plant species are important in meeting the food (calories) needs of humans worldwide. There is a greater dependence on fewer plant species; 20 to 30 in global context (Harlan, 1975). This gradually, resulted in the loss of native genetic resources which are otherwise essential as building blocks of genetic diversity. It is estimated that there are about 500 species of tropical fruit trees under 30 families and 59 genera in Asia Pacific Oceania region (Arora, 1998). In Southeast Asia alone, there are 120 major fruit species and 275 minor fruit species (Verheij and Coronel 1992). In Asia, 50-60 species belong to the most important indigenous fruits (Arora and Rao 1996). Citrus, mango, banana, rambutan, jackfruit, litchi and durian occupy 80 per cent of total fruit production in the region.

Wild Species and Diversity

Role of wild species in fruit improvement programme is becoming important as donor source for many of the disease and pest resistance. It is to be mentioned here that in most of the perennial trees, evaluation of wild species or the indigenous germplasm has not been done extensively either by morphological or by molecular means. Some areas may be richer in biodiversity than other areas, and some species may also have more variation than others in a particular area.

Conservation is very important, because many species are becoming extinct and many others are threatened and endangered. Diversity of some fruits is well collected, while for other fruits, relatively little has been done (Arora, 1994). Gaps in collections are found both in and between species and between regions. This is especially true for both underutilized species and wild crop relatives, where big gaps are noted. Kostermans and Bompard (1993) indicated that *Mangifera blommesteinii, M. leschenaultii, M. superba* and *M. paludosa* are in real danger of extinction. High genetic erosion is noticed for jackfruit, *Citrus spp.* and *Litchi chinensis* in a survey carried out by the International Centre for Underutilized Crops (ICUC) and IPGRI (Haq, 1994). Collection and utilization of wild species are not easy tasks, as they require specific climate and do not so easily get acclimatized to *ex situ* conditions on introduction.

Biodiversity of Fruits

Vavilov identified Asia as a major centre with "Indian centre" of North East region as primary or secondary centre of origin for many crop plants. This region is centre of diversity for several important native fruits including mango, jackfruit and citrus. Plant genetic resources represent sum total of the diversity that come from wild species and primitive forms, accumulated through evolution and natural selection, plant introduction, migration and domestication and also materials developed by artificial selection and breeding. The region remained isolated for a long time, even today accessibility is poor to many parts of this region. This undisturbed condition with suitable environmental conditions and variable altitudes is one of the major reasons for genetic diversity of the region through natural selection and evolution although primary causes being the agro-climate, which are wet tropics with rain forests.

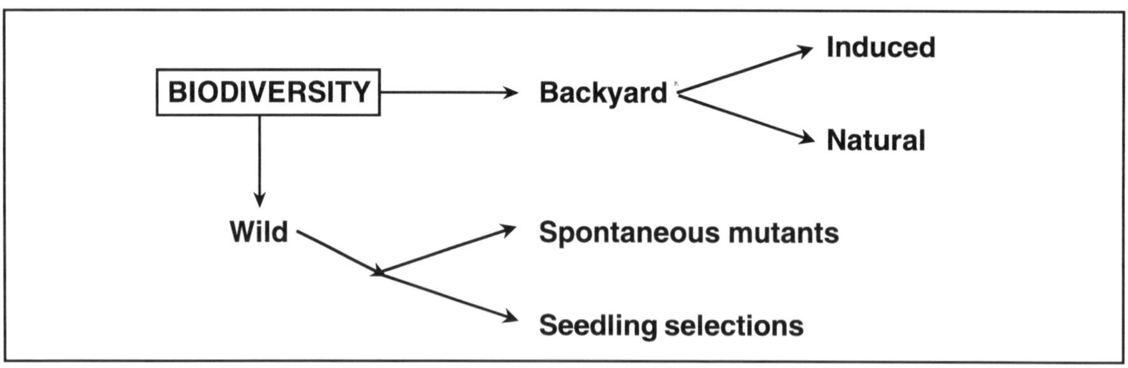

Figure 3.1: Biodiversity of Tropical Fruit Species

Biodiversity can be located either in wild or in backyard. With regard to many of tropical fruit species, variability can be traced in wild, wherein many species grow naturally even today *viz.*, occurrence of *Mangifera sylvatica* in North-Eastern parts of India or *M. andamanica* in Andaman group of islands. In wild, this diversity has come about mainly because of spontaneous mutants over a period and due to dispersal of seeds and seedling population. In backyard, mainly seedling populations are cause of diversity as noticed in fruits like mango and jackfruit. Diversity due to natural means has come about because of seed dispersal as in pickling types of mango *viz.*, Appemidi types in Uttara Kannada of Karnataka or be it the varietal wealth in Western Ghat regions.

Characteristic Features of Tropical Fruit Tree Diversity

Main causes for tropical fruit diversity in India be it mango or a underutilized fruit like jamun, whether in the wild or in the cultivated types are:

1. Presence of high heterozygosity
2. Cross pollination
3. Seed propagation
4. Absence of vegetative propagation in earlier days
5. Indiscriminate multiplication.

Unlike other crops, wherein there is need to create variability, in tropical fruit species, it is management of diversity, which is the more challenging task. In mango, varietal diversity itself is considered a hindrance to improvement (Naik, *et al.*, 1958).

Main centres of diversity for fruits in India are shown in Table 3.1.

Table 3.1

Region	*Species*
Western Himalayas	*Elaeagnus hortensis*, *Ficus palmata*, *Fragaria indica*, *Moms spp.*, *Prunus acuminata*, *P. cerasiodes*, *P. cornuta*, *P. napaulensis*, *P. prostrata*, *P. tomentosa*, *Pyrus baccata*, *P. communis*, *P. kumaoni*, *P. pashia*, *Ribes graciale*, *R. nigrum*, *Rubus ellipticus*, *R. moluccanus*, *R. fruticosus*, *R. lasiocarpus*, *R. lanatus*, *R. niveus*, *R. reticulatus*, *Zizyphus vulgaris.*
Eastern Himalayas	*Fragaria indica*, *Morus* spp., *Myrica esculenta*, *Prunus acuminata*, *P. cerasiodes*, *P. cornuta*, *P. jenkinsii*, *P. napaulensis*, *Pyrus pashia*, *Ribes graciale*, *Rubus lineatus*, *R. ellipticus*, *R. lasiocarpus*, *R. moluccanus*, *R. reticulatus*
North-eastern region	*Citrus assamensis*, *C. ichangensis*, *C. indica*, *C. jambiri*, *C. latipes*, *C. macroptera*, *C. media*, *C. aurantium*, *Docynia indica*, *D. hookeriana*, *Eriobotrya angustifolia*, *Mangifera sylvatica*, *Musa accuminata/M. balbisiana* complex, *M. manii*, *M. nagensium*, *M. sikkimensis*, *M. superba*, *M. velutina*, *Pyrus pyrifolia*, *P. pashia*, *Prunus cerasiodes*, *P. cornuta*, *P. jenkinsii*, *Ribes graciale*, *Rubus ellipticus*, *R. moluccanus*, *R. reticulatus*, *R. lasiocarpus*, *Myrica esculenta.*
Gangetic plains	*Aegle marmelos*, *Cordia myxa*, *C. rothii*, *Emblica officinalis*, *Grewia asiatica*, *Morus* spp.; *Phoenix* spp.; *Syzygium* spp.; *Zizyphus nummularia and other* spp.; and *Manilkara hexandra* (more in North-Western plains).
Indus plains	Meagre occurrence of *Syzygium*, rich variaton in *Carissa congesta.*
Western peninsular tract	*Artocarpus heterophyllus*, *A. lakoocha*, *Garcinia indica*, *Diospyros* spp., *Ensete superba*, *Mangifera indica*, *Mimosops elengii*, *Spondias pinnata*, *Vitis* spp., *Zizyphus oenoplia*, *Z. rugosa*, *Rubus ellipticus*, *R. lasiocarpus*, *R. moluccanus.*

Source: Arora and Nayar, 1984.

Mango

Mango is native to India. Mukherjee (1949, 1985) opined that this genus might have originated in the region covering Burma, Siam, Indo-China and Malayan Peninsula. Genetic diversity of mango available in India is very rich and at present more than one thousand vegetatively propagated varieties exist in the country. Clonal selection, selections from chance seedlings and breeding efforts resulted in identification of many elite and improved varieties of mango for commercial cultivation in the country.

All these varieties belong to *Mangifera indica*. Apart from *M. indica*, India is home of four other species *viz., M. andamanica, M. khasiana, M. sylvatica* and *M. camptosperma* (Mukherjee *et al.*, 1985). The species of *Mangifera* occur mainly as complex biotic community in tropical humid forests, sub-tropical rain forests and tropical dry forests/woodlands of Indo-Malayan biogeographic realm (Mukherjee, 1985).

The *Mangifera* germplasm falls under two categories;

1. *Seedling races*: This group includes both wild and cultivated types. Under this category, the cultivated ones come under monoembryonic types. Polyembryonic types are seen generally in Western Ghats of Peninsular India.
2. *Horticultural races*: They include varieties, which when grown under different agro-climatic conditions and propagated vegetatively from parent materials showed clonal differences. Varieties like Alphonso, Dashehari and Langra have clones resembling them in a few of the morphological characters. Yadav and Singh (1985) opined that mango varieties of Northern and Southern regions belong to two different eco-geographic regions.

Centres of Mango Diversity and Distribution in India

In India, seven centres of mango diversity recognized, one being a sub-centre (Yadav and Rajan, 1993). These are the places where maximum diversity is noticed for species as well as varietal diversity. They are:

1. Humid Tropical region-Manipur, Tripura, Mizoram and S. Assam
2. Chota Nagpur Plateau-Trijunction of Orissa, Bihar and Madhya Pradesh
3. Santal Paraganas in Bihar
4. South Madhya Pradesh adjoining Orissa and Andhra Pradesh
5. Dhar Plateau of Madhya Pradesh adjoining Gujarat and Maharashtra
6. Humid Tropical South Peninsular India
7. Andaman and Nicobar Islands

Varietal Diversity

In India, about thousand varieties of mango are grown. Most of these varieties arose as chance seedlings. Each mango-growing region in India grows a different variety. Although, thousand varieties are grown, one characteristic feature is growing of hardly twenty-five varieties on a commercial scale in different states. Most of the commercial varieties arose as a result of selection of seedling types for fruit characteristics like colour, taste, flavour, size and bearing habit. In India, there are three main centres of development of mango *i.e.* Lucknow–Saharanpur belt of Uttar Pradesh, Murshidabad area of West Bengal and Hyderabad area of Andhra Pradesh. Most of these varieties have specific characteristics, require specific climate for optimum performance and have strong regional consumer preference. Although, growth in mango is genetically controlled, environmental interaction has brought about change in growth pattern under different agro climatic conditions, which also contributed for its biodiversity.

Citrus

The North East hilly region is rich in fruits, vegetables and flowers, especially orchids. It is a centre of origin of Mandarins and a few other citrus fruits. Sixteen species of Citrus, 52 varieties, and seven natural hybrids of Assam were described by Bhattacharya and Dutta as early as 1956. They also reported two species of sub genus *Eueitrus viz., C. indica* and *C. assamensis* and three species of sub genus of Papeda *viz., C. ichangensis, C. latipes* and *C. microptera* which grow at high altitudes. *C. indica* is the most primitive species of citrus and probable progenitor of cultivated species. It is found in various fruit forms in NEH region. Diverse forms of Pummelo, Sour Orange, Rough Lemon, Sour Pummelo and Adajamir Sweet Lime are found in this region.

Varietal Diversity

Mandarin orange is concentrated in Maharashtra (Nagpur, Amaravathi, Wardha and Yavatmal) in North East region of India (Assam, Arunachal Pradesh and Meghalaya), a limited area of Karnataka (Kodagu), Tamil Nadu (Nilgris, Palani and Shevroy hills) and Kerala (Wynad). Sathpura hills of Madhya Pradesh adjoining Vidharbha region of Maharashtra also grow good quality mandarins. Kinnow Orange, a hybrid of King X Willow Leaf Mandarin has recently spread in North West India, especially in Punjab, parts of Himachal Pradesh, Uttar Pradesh and Rajasthan. The main advantage of cultivation of introduced varieties/hybrids is that they also add to varietal diversity by throwing spontaneous mutants over a period of time.

Sweet oranges are adapted well to arid tropics and sub tropics. They are commercially grown in Andhra Pradesh, Maharashtra, Punjab and parts of Tamil Nadu, Rajasthan and Uttar Pradesh. Sweet orange cultivar in Andhra Pradesh is Sathgudi, whereas in Western and Central India sweet orange cultivar Mosambi is popular. In North Western India the cultivars Malta, Jaffa and Valencia are popular.

Acid lime is grown on a commercial scale in Tamil Nadu, Andhra Pradesh, Maharashtra and Karnataka. Lemons are grown commercially only on a limited scale. Eureka lemon in North-Western regions and Assam lemon in North Eastern India are popular.

Banana

Bananas are one of the ancient fruits cultivated by man. It could be assumed that the fruit has evolved with the civilization (Krishnamurthi and Seshadri 1958) and found in Indus valley in 327 B.C. Apart from its mention in Valmiki's Ramayana, it also finds a mention in Kautilya's Arthshastra and ancient Tamil classic Silappadikaram. These evidences suggested early existence of banana in India. The wild *Musa acuminata* occurs in Assam, Burma, Siam, Indo-China, the Malayan Peninsula and Archipelago and the Philippines.

North-Eastern region of India – including the states of Assam, Arunchal Pradesh, Tripura and Mizoram, lie at a point where *Musa balbisiana* from Indian subcontinent meets *Musa acuminata* from Southeast Asia. These two species and other wild relatives have mingled to form a distinctive concentration of genetic diversity, occurring in semi-evergreen, sub-tropical forests of hill slopes. Further sources of diversity occur in valleys and plains where bananas are common as a backyard crop. Habitat of wild bananas is shared with tribal groups who practice a form of shifting cultivation.

Unfortunately, a growing number of sites where wild bananas once grew, are now denuded. Wild species found in this part of India include *Musa acuminata, Musa balbisiana,* several species from the section *Rhodochlamys* as well as *Ensete glaucum.*

Diploid and triploid *acuminata* cultivars were taken by man to native areas of *balbisiana,* which resulted in natural hybridization and formation of hybrid progeny with genomes: AA, AAB, and ABB. It is thought that subsequent dispersal of edible bananas out of Asia was brought about by man. Secondary diversification within groups of cultivated bananas are results of somatic mutations.

Eumusa and Rhodochlamys are found in Assam region of India and Thailand whereas Callimusa and Rhodochlamys in Borneo, its surrounding islands, and Indonesia. Australimusa is largely found in Malayan islands, and Indonesia. It is also found in Assam, Indo-China, Malayan and Papua New Guinea, which is a primary centre of cultivated AA types. *M. balbisiana* occurred in Sri Lanka, India, Myanmar, Indonesia and Malaysia where A × B hybrids have evolved.

Table 3.2

Region/State	*Cultivars*
Andhra Pradesh	Amrit Pani (Rasthali, AAB), Thella Chekkarakeli (AAA), Karpura Chekkarakeli (Poovan, ABB), Monthan (ABB), Robusta (AAA)
Assam	Bhimkol, Manohar (ABB), Chini Champa (AB), Malbhog (AAB), Jahaji (AAA), BorJahaji (AAA), Kanch Kol (ABB)
Bihar	Alpan (AAB), Chini Champa (AB), Basrai (AAA), Kothia/Muthia (ABB), Bateesa (ABB), Malbhog (Rasthali, ABB)
Gujarat	Dwarf Cavendish (AAA), Lacatan (AAA), Harichal (AAA)
Karnataka	Dwarf Cavendish (AAA), Robusta (AAA), Poovan (AAB), Rasbale (AAB, Rasthali), Marabale (Pome, AAB), Monthan (ABB), Elekki Bale (AB, Ney Poovan)
Kerala	Nendran (ABB, plantain), Palayankondan (AAB, Poovan), Kunnan (AB), Rasthali (AAB), Monthan (ABB) and Red Banana (AAA)
Maharashtra	Basrai (Dwarf Cavendish, AAA), Robusta (AAA), Safed Velchi (AB), Monthan (ABB), Rajeli (Plantain, AAB)
Tamil Nadu	Virupakshi (Pome, AAB), Poovan (AAB), Rasthali (AAB), Nendran (AAB), Monthan (ABB), Dwarf Cavendish (AAA), Robusta (AAA), Peyan (ABB)
Bengal and Orissa	Champa (AAB), Morthaban (AAB, Rasthali), Amrit Sagar (AAA), Giant Grover (AAA), Lacatan (AAA), Monthan (ABB)

Source: Singh, 1996.

Grape (*Vitis* spp)

European grape *V. vinifera* originated primarily between Caspian and Black sea. It is a hybrid between two American spp *V. vulpine* and *V. labrusca,* which resemble *V. parviflora* and *V. lanata* found in Himalayan region. This region is a secondary centre of origin. Famous Indian medicine scholars, Sushruta and Charaka in their medical treatises entitled 'Sushruta Samhita' and 'Charaka Samhita', respectively, written during 1356-1220 BC, mentioned the medicinal properties of grapes. Kautilya in his 'Arthashastra' written in fourth century BC mentioned the type of land suitable for grape cultivation. Native spp. resembling *Vitis lanata* and *Vitis palmata* grow wild in North-Western Himalayan foothills. Indigenous varieties known as 'Rangspay', 'Shonltu White' and 'Shonltu Red' are grown in Himachal Pradesh even today.

Cultivated grapes are introduced into North of India by Persian invaders in 1300 AD, from where they were introduced into South (Daulatabad in Aurangabad district of Maharashtra) during historic event of changing the capital from Delhi to Daulatabad by King Mohammed-bin-Tughlak. Ibn Batuta, a Moorish traveller who visited Daulatabad in 1430 AD, saw flourishing vineyards in South India.

Table 3.3

Species	*Region*	*Salient Characters*
V. riparia	North Western Himalayan region	Small berries, purple black in colour, cold hardy, early flowering
V. lanata	Himalayan region	Purple black berries known for crack resistance and plant resistant to diseases
V. barbata	Parts of Assam, Khasi hills and Bengal	
V. parviflora	North West Himalayas from Kashmir to Nepal	Small berries, delicately flavoured
V. tomentosa	Greater part of Deccan peninsula	

The spp *V. vinifera*, has not originated in India. Thompson Seedless, which belongs to spp *V. vinifera* grown in several parts of the world and in India contributed for diversity in varieties; several mutants came about in grapes due to its extensive cultivation over years.

Guava

Guava is an important fruit crop of India. It originated in tropical America and grows from Mexico to Peru. It is distributed widely all over equatorial regions of tropical and sub-tropical climate. Guava was introduced during seventeenth century into India.

It gained considerable prominence on account of its high nutritive value, availability at moderate prices, pleasant aroma and good flavour. It is one of the commonest fruits liked by rich and poor alike and is popularly known as 'apple of tropics'. At present, it is grown throughout length and breadth of the country from sea level to 1300 m altitude and is so much acclimatized that it appears to be native to India. The most important guava growing states are Uttar Pradesh, Bihar, Madhya Pradesh and Maharashtra.

The genus *Psidium* of Myrtacease family comprises of 150 species of small trees and shrubs. About 20 species have edible fruits, of which the most commonly cultivated is the common guava *i.e., Psidium guajava* L. It is reported that the value of wild *Psidium* species lies mainly in their utility as rootstocks for regulation of vigour, fruit quality and resistance to pests and diseases.

Varietal Diversity

Guava is mainly a self-pollinated crop but cross-pollination is also common. This resulted in large variability in seedling population from which promising genotypes are selected in different agro-climatic regions of country. In India, different workers in different regions described guava varieties. Main centre of variability in guava is the Allahabad area in Uttar Pradesh. Promising cultivars of different states are shown in Table 3.4.

In India, due to seed propagation, varietal diversity is seen for guava, but species diversity is not seen.

Table 3.4

State	Cultivars
Andhra Pradesh	Allahabad Safeda, Anakapalli Banarasi, Chittidar, Hafsi (Red Fleshed), Lucknow-46, Sardar, Seedless, Smooth Green, Smooth White
Assam	Amsophri, Madhuriam, Safrior payele
Bihar	Allahabad Safeda, Chittidar, Hafsi (Red Fleshed), Harijha, Seedless
Maharashtra	Dharwar, Dholka, Kothrud, Lucknow-24, Sardar
Gujarat	Nasik, Seedless, Sindh
Tamil Nadu	Anakapalli, Banarasi, Bangalore, Chittidar, Hafsi, Nagpur Seedless, Smooth Green
Uttar Pradesh	Allahabad Safeda, Apple Colour, Chittidar, Red Fleshed, Banarasi Surkha, Sardar, Mirzapuri Seedless
West Bengal	Behrampur and cvs. of Uttar Pradesh

Papaya

Papaya (*Carica papaya* L.) is one of the most important fruit crops valued for its rich nutrient content. It is a rich source of Vitamin A (2020 I.U), Vitamin B1 (40 mg), Vitamin C (46mg), protein (0.5 per cent) and mineral matters (0.4 per cent). Papaya is native to tropical America, its place of origin is Southern Mexico and Costa Rica. It was taken to Manila by Spanish in mid 16th century, reached Malacca thereafter. It was introduced to India during 16th century. It is grown both in tropical and sub-tropical parts of world. Wild diversity is not reported for papayas in India.

Varietal Diversity

In India, varietal diversity is seen for papaya. Variability seen is more because of open pollination and indiscriminate multiplication using these seeds. In papaya, there are two basic types of varieties. Those varieties, which are dioecious, produce only female and male plants, and 'gynodioecious' which produce both female and hermaphrodite plants. A few of the varieties that are grown in different states are hereunder.

Sapota

Sapota (*Achras sapota* L.) is a popular dessert fruit belonging to family Sapotaceae. It originated in tropical America, taken to Philippines by Spanish and from there spread to other countries (Purseglove, 1968).

Table 3.5

State	Cultivars
Andhra Pradesh	CO 2, CO 5, Sunrise Solo, Taiwanese lines
Bihar	Pusa Dwarf, Pusa Majesty, Pusa Nanha, Pusa Giant, Pusa Delicious and Ranchi
Karnataka	Coorg Honey Dew, Washington, Sunrise Solo, CO2, Surya and Taiwanese lines
Maharashtra	Washington, CO2, Pusa Delicious, Pusa Majesty, Ranchi and Taiwanese lines
Orissa	CO2, Coorg Honey Dew, Washington, Ranchi, Pusa Dwarf and Pusa Delicious
Tamil Nadu	CO2, CO3, CO4, CO5, CO6, CO7 and Coorg Honey Dew
Uttar Pradesh	Coorg Honey Dew, Pusa Dwarf, Pusa Delicious, CO3 and Barwani Red

In India, it is grown in the states of Andhra Pradesh, Gujarat, Karnataka and Orissa. About 30 varieties are reported in India at various places. A number of locally grown genotypes identified include Bhuri patti, Morabba, Kali patti, Turi patti, Gole patti, Singapuri, Khabari and Chhumukia type. Among these, a genotype identified in Navsari in Gujarat locally known as 'Morabba' bears fruits of bigger size and superior quality in comparison to Kalipatti, a local genotype grown in about 80 per cent area of Gujarat. It is a selection from grafted plants collected from nursery located in Golwal. It may have originated as bud mutant and is now being propagated vegetatively. Another somatic mutant having desirable characteristics of fruit was identified in Paria (Rai, 1995). Wild diversity is not observed for sapota, as it is grown over the years by using grafts.

Biodiversity of Underutilized Fruits

In India various native fruits, like aonla (*Emblica officinalis*), bael fruit (*Aegle marmelos*), jackfruit (*Artocarpus heterophyllus*), jamun (*Syzygium cuminii*), karonda (*Carissa congesta*), Kokum (*Garcinia indica*) and phalsa (*Grewia subinaequalis*) with a lot of diversity are grown in a wide range of agro-ecological situations, under tropics, subtropics and temperates. They are grouped as underutilized. A few of these fruits yield juice with excellent flavour, which can be converted into blended beverages and these could play an important role in meeting demand for nutritious, pleasantly flavoured and attractive natural food of high therapeutic value. Encouraging local people to produce these fruits helps in significantly checking uncontrolled harvesting from wild and assisting in retention of various species in their native habitats where they perform the best.

Jackfruit

Artocarpus is a genus of small to large evergreen trees, distributed from Sri Lanka and India to South China and through Malaysia to Solomon Islands. Nine species are recorded in India. The spp, *A. heterophyllus* Lam. is grown for its edible fruits, and *A. chaplasha* Roxb., *A. hirsutus* Lam. and *A. lakoocha* Roxb., are important timber trees.

A. chaplasha Roxb, is distributed in the moist deciduous and evergreen forests of sub-Himalayan tracts from Nepal eastwards to West Bengal, Assam and Tripura. In West Bengal and Assam, it occurs in moist types of mixed deciduous and evergreen forests. In Andaman and Nicobar islands, it is an important constituent of evergreen and deciduous forests.

A. cummunis J.R. and G.Frost, commonly known as bread fruit is found mainly in West Coast and Western Ghats, Wynad, in the Nilgris, Lower Plains, the Courtallam hills and the Annaimalais. There are two distinct varieties in this species. One is a seeded type and the other entirely seedless. Seeded types are found in wild state in its native land and are of little economic value. It is not useful in culinary preparations but the seeds, which resemble chestnut, are relished when roasted or boiled.

A. heterophyllus Lam. commonly called as jackfruit is one of the most popular fruits of South India. The tree is indigenous to evergreen forests of Western Ghats at altitudes of 450–1200m, but seen growing throughout other hotter parts of India too. Because of seed propagation, existing population of jack comprises innumerable trees differing from each other in fruit characters of shape, size and quality.

A. hirsutum Lam is commonly found in the evergreen forests of Western Ghats from Konkan Southwards, is fairly common in North Kanara and Kodagu in Karnataka to Kerala where it is an important tree. It requires heavy rain fall, not less than 174 cm annually and thrives well on lateritic soils at foot of Ghats. Tree can stand shade, but thrives the best with a fair amount of light. It does equally well in open and withstands exposure to sun after first a few years.

A. lakoocha Roxb, is commonly known as monkey jack. In its wild state, it is chiefly found in moist or deciduous forests along banks of streams and along site of moist ravines. It thrives the best in deep laterite soils and generally comes to bear after about eight years. It is commonly cultivated throughout greater part of India as a shade or ornamental tree. It is perhaps one of the foremost among neglected but useful trees. It is found distributed in evergreen, semi-evergreen and moist deciduous forests upto an altitude of 1800 m in Eastern and Northern India. On the West Coast, it is found from Konkan southwards to Kerala, and in Tamil Nadu. It is also found growing in many localities in Andaman Islands.

Aonla

Aonla or Indian gooseberry (*Emblica officinalis* Gaertn.) is a wonder fruit for health conscious population. It is grown in India for more than 3500 years. Sushruta, the father of ancient medicine (during 1500 BC-1300 BC), mentioned about its usefulness in 'Ayurveda' in great detail. It belongs to family Euphorbiaceae and is one of the important indigenous fruits of Indian subcontinent. In different parts of India, it is known by different vernacular names such as Amla or Aonla in Hindi, Dhatri, Dhatriphala, or Amalaki in Sanskrit, Amla or Amalaki in Bengali and Oriya, Nelli in Malayalam and Tamil, Amlakamu in Telugu, Amolphal in Punjabi (Gurmukhi) and Aonla, Myrobalan or Indian gooseberry in English (Pathak, 2003). Plant and fruit of aonla are regarded sacred by 'Hindus' and have great mythological significance.

Aonla tree is native to tropical Southeast Asia, particularly Central or Southern India, Pakistan, Bangladesh, Sri Lanka, Malaysia, Southern China and to Mascarene Islands. Seedling trees are of common occurrence in mixed deciduous dry forests of India, ascending from sea level (Western and Eastern Ghats, Aravali and Vindhya hills) to 1300 m above sea level, from North-West Himalayas (Jammu and Kashmir, Himachal Pradesh, Uttarakhand) to Eastern Himalayas in Assam, Meghalaya, Mizoram, Manipur and Tripura (Pathak, 2003). Natural distribution of wild aonla is found on the Himalayas, Chota Nagpur, Bihar, Orissa, West Bengal, North Circars, Deccan, Karnataka and in Western Ghats (Rawat and Uniyal 2003).

In India, the homeland of aonla, domestication began first in Varanasi (earlier known as Benaras) district of Uttar Pradesh with the initiative of Maharaja of Kashi. Banarasi, a superior genotype was selected from wild aonla trees available in large number in nearby Vindhyan hills. Authentic information regarding its cultivation dates back to 1881-82 in Pratapgarh district of Uttar Pradesh.

The wild aonla germplasm is confined mostly in mixed forests with sloppy topography and sometimes even difficult to approach. A rich genetic diversity of aonla exists in North-Eastern region of India, particularly lower Assam, Maghalaya, Mizoram and Tripura (Yadav *et al.*, 2001). Aonla grows abundantly in forest of Khasi and Garo hills of Meghalaya and locally known as "Sohmylleng" (Pandey *et al.*, 1993). Natural population of aonla in West Khasi hills (Nongkhyllum, Rajaju, Khonjoy area) of Meghalaya warrants *in situ* conservation, which may even be declared as gene sanctuary for this species (Hore 1998). Mizoram is homeland of wild aonla and star gooseberry (*Phyllanthus acidus*), which has potential as dwarfing rootstock for aonla. Wild Star gooseberry trees are found in forests of Kolasib, Thingdawl and Champhai in Mizoram. Madhya Pradesh forests have rich diversity of aonla. Jharkhand and adjoining areas of Chhatisgarh have rich diversity of aonla in the native forest. Important sites in Jharkhand are Lali forest near Ranchi, Dalma range of Jamshedpur, Theo Ghat forest of West Singhbhoom, Tiamara valley area between Ranchi and Jamshedpur, Ramgarh area of Hazaribagh, Parasnath hills of Girideeh, Kodemera and Jaomi areas of Bihar border, Simdega and

Netarhat forest areas of Gumla, Belta forest of Daltonganj, Palamu and Garhwa of Jharkhand and adjoining areas of Sarguja and Ambikapur districts of Chhatisgarh and Sahdol district of Madhya Pradesh. Belta forest (Daltonganj), Netarhat range in Gumla and Dalma range of Jamshedpur have comparatively high plant population of aonla in natural habitat. In Western and Eastern ghats, three species of *Phyllanthus, P. emblica, P. indofisheri* and *P. acidus* are of common occurrence. A wild strain of aonla grows in the Himalayas up to an altitude of 1600 m amsl. Fruits of wild aonla are relatively smaller. In mid Himalayas, almost wild aonla is distributed right from Western to Eastern Himalayas including Nepal (Pathak, 2003).

Bael

Bael (*Aegle marmelos*) is native to India and cultivated throughout South East Asia and East Indian Archipelago. The genus consists of 2 to 3 species. It is found in UP, Bihar and West Bengal. Some important types selected in different regions are UP: NB1, NB5, NB6: Bihar: Etawah Kagzi, Sewan Large, Mirzapuri and Deoria.

Ber

Out of 50 reported species under Zizyphus nearly 18 to 20 are native to India. A resume of species availability in different locations in India are given in Table 3.6:

Table 3.5

Species	*Location*
Z.apatala, Z.funiculosa, Z.incurva	North-Eastern hills
Z.mauritiana, Z.nummularia	All over the drier tracts, particularly in North-West India and UP
Z.oenoplia, Z.rugosa	Throughout India except in drier tracts, particularly in Central and Eastern India
Z.vulgaris	North-Western Himalayas
Z.rupicola	Central and Eastern India
Z.xylocarpus	MP and peninsular region

There are more than 100 named varieties in ber and areas rich in variability are identified in several places in UP, Rajasthan, Haryana, Gujarat, MP, Maharashtra, AP, Karnataka and Tamil Nadu. A few popular cultivars are Umran, Gola Reshmi and Illaichi.

Custard Apple

Genus *Annona* contains 120 species originating from warm countries but a few important species became integral part of Indian culture bearing names of great heroes of the epic Ramayana. A few important species and their natural distribution are given in Table 3.7.

Fig

Original home of origin of fig (*Ficus carica*) is South Arabia. There are four horticultural types in this crop *viz.*, Smyrna, Capri, Sanpedro and Adriatic. This crop has very narrow range of diversity in India and there is scope for introducing exotic germplasm. There are wild relatives found in India and a few of them are given in Table 3.8.

Table 3.7

Species	*Common Name*	*Varieties*	*Location*
A. squamosa	Sweet sop, Sharifa, Sitaphal	(a) Green types: Balanagar, Mammoth, British Guinea, Washington-95, Barbadose seedling, Arka Sahan, (An F1 hybrid) (b) Red types: Red Sitaphal	Low and medium elevations throughout tropics
A. cherimola	Lakshmanphal	Cherimoyar	Cooler places in India
A. atemoya		Pinks mammoth, Bradley, Keller, Page, African Pride, Island Gem	Adopted to colder climate and alkali soils
A. reticulata	Ramphal or Bullock's Heart	Used mostly as a root stock	
A. glabra	Pond apple	Root stock	Flooded areas
A. muricata	Sour sop, Hanumanphal	Root stock	Mountain regions of India
A. mantna	Mountain soursop	Used in breeding programme for quality	
A. purpurea	Soncoya	Used as resistant source against fruit cracking	
A. scleroderma	Eoshto	Used in breeding programmes for thick hard shell	

Table 3.8

Species	*Common Name*	*Distribution*
F. auriculata	Timla	Bihar,Orissa, Khashi hills, Manipur
F. benghalensis	Banyan	All over India
F. benjamina	–	–
F. carica	Fig	Uttar Pradesh, Rajasthan, Andhra Pradesh, Maharashtra, Karnataka
F. elastica	Indian Rubber tree	Assam and Khasi hills
F. glomerata	Cluster fig	–
F. hispida	–	Through out India
F. krishnae	Krishna's fig	–
F. lucescens	–	North India, MP and W peninsula
F. palmata	–	N.W India and Rajasthan
F. religiosa	Peepal tree	All over India
F. rumphii	–	Punjab, MP, Assam
F. samicordata	–	Punjab, Assam, Bengal, Khasi hills and Manipur

Jamun

The genus *Syzygium* comprises about 1000 species of evergreen trees and shrubs; most of them are tropical in origin. Jamun is found in Western Ghats and very extensively in the tropical region. Diversity is due to high heterozygosity and seed multiplication. Some of the species are described in Table 3.9.

Table 3.9

Species	*Common Name*	*Distribution*
S. aqueum	Watery Rose-apple, Fruits edible	A small tree distributed in Assam and Meghalaya
S. arnottianum	Produces edible fruits	W.Ghats, The Nilgris, Palni and Anamalai hills
S. aromaticum	Clove, dried flower buds are of commercial importance	Evergreen trees cultivated in Tamil Nadu and Kerala
S. claviflorum	Fruits are acidic and edible	Andamans
S. cumini	Java plum, Jamun, Jambu	Throughout India
S. fruticosum	Wild Jamun	Avenue tree
S. jambos	Rose-apple	Many parts of India
S. mappaceum	Grown as ornamental plant	Assam, Meghalaya, Arunachal Pradesh and Tamil Nadu
S. samarangense	Wax Jumbu, fruits edible	Andamans and many parts in India
S. zeylanicum	Aromatic fruits are edible	Maharashtra, Karnataka, Orissa, Kerala and Andamans

Pomegranate

Pomegranate (*Punica granatum*) is an ancient fruit, which originated in Persia, Afghanistan and Baluchistan and naturalized in Western India very early. Its wild forms are found in lower hills of Himachal Pradesh. Most of the pomegranate types cultivated in India are of seedling origin and thus providing a wide range of variability with respect to fruit shape, size, and mellowness of seed, aril colour, rind colour, sweetness and acidity of juice. Some popular varieties in different regions are furnished in Table 3.10.

Table 3.10

Region/	*Variety/Type*
Maharashtra	Ganesh, Bhagwa, Mridula, Aaraktha,G-137,P-23, P26, Muskat
Karnataka	Ganesh, Ruby, Bassein Seedless
Gujarat	Dholka
Rajasthan	Jodhpur Red, Jodhpuri White
Tamil Nadu	Yercaud, Vellodu, Kabul Red, CO-1

Apart from above mentioned fruit species, there are several other species of fruits for which considerable diversity exists in wild and conservation of such fruits needs to be carried out both *in situ* and in *ex situ*. There is also need to work out diversity using molecular means, so that the concept of 'core collection' can be practiced effectively.

In situ Conservation of Tropical Fruit Trees

In situ conservation of tropical fruit tree species is one of the most important aspects in the overall conservation of fruit diversity (Dinesh 2001). It is well known that many of the species of mango, when introduced to other areas do not perform well or die. It is observed that *Mangifera andamanica, M. camptosperma* and *M. griffithi* when introduced to mild tropics could not survive (Prakash, 2001).

Inspite of innumerable problems that are faced in the *ex situ* conservation, it is still advantageous to maintain them in field genebank, as it keeps biodiversity of a particular species safe when plants are destroyed in wild. Hence to rationalize, the concept of core collection was introduced, which in a limited set represents genetic spectrum in whole collections (Brown, 1989). It is proposed that landraces should be preserved for future generations because they harbour a diversity of interesting traits for future breeding work and for developing new farming systems and moreover, reflect cultural identity of certain groups of people (Altieri and Merrick, 1987).

Until recently, germplasm conservation of crop landraces and their wild relatives, relied on *ex situ* methods (*i.e.* conservation of biological materials outside its natural habitat, UNCED 1992), mostly in germplasm banks. More recently *in situ* (on-farm) conservation (*i.e.* conservation of biological diversity in its natural habitat) is proposed as a conservation strategy which allows evolutionary processes to continue rather than being halted as occurs to a large extent in *ex situ* conservation (Frankel *et al.*, 1995; Maxted *et al.*, 1997).

Pollen Storage as a Means to Conserve Diversity in Tropical Fruit Trees

Genetic conservation through pollen storage is desirable for a variety of horticultural plant species, since pollen is known to transmit important genetically heritable characters. Pollen is a product of genetic recombination and can provide a reliable source of nuclear genetic diversity at the haploid stage.

Fruit tree pollen is generally required to be stored for controlled crossings, either to achieve a desired breeding objective, or to overcome a constraint involved in commercial fruit production.

Gene pool conservation at the haploid stage can, therefore, be effectively accomplished through pollen, which can provide a rich source of nuclear genetic diversity. A major emphasis on research needs include pollen storage (Arora and Rao 1995) for citrus, mango, rambutan, jackfruit, durian and litchi. Pollen cryopreservation research has been recently recognized by IPGRI (IPGRI 1995). As one of the genepool components in an integrated PGR conservation programme, pollen can serve as an alternative or additional *ex situ* conservation method. Alexander and Ganeshan (1993) have extensively reviewed pollen storage research in fruit crops. Grout and Roberts (1995) elaborately described the methodology involved in pollen cryopreservation. Hoekstra (1995) assessed merits and demerits of pollen storage for genetic resources conservation.

One of the main lacunae so far in the survey and collection of tropical fruit tree species has been the non-adoption of 'Geographic Information System' (GIS) tools, with the result that many of the regions with rich diversity were excluded. Use of Geographic Information System tools of late has helped in avoiding duplication of surveys and carrying out surveys with fair degree of accuracy. These tools can also help in isolating genotypes with exceptionally good characteristics.

GIS Tools and Preparation of Distribution Maps

Geographic Information System (GIS) is widely used in management of natural resources. Presently GIS is being widely used in mapping biodiversity by different organizations. GIS is a database management system with specific functions to handle spatial data, *i.e.* latitude and longitude. Many applications of GIS are developed for commercial purposes or for specific management purposes, for example, Atlas, MapInfo for Windows, Arc/Info, etc. for commercial use, and GRID, FloraMap, DIVA, etc. for specific purpose of mapping biodiversity. For mapping biodiversity and its assessment for tropical fruit tree species, softwares such as FloraMap and DIVA were developed by the International Potato Centre (CIP) and International Centre for Tropical Agriculture (CIAT) for research purposes.

GIS has two kinds of softwares, *viz.* rector-based system and raster-based system. Rector-based system stores geographic data as points, while raster-based system stores data as grid cell. For mapping genetic diversity, rector-based system is popularly used. Using DIVA and FloraMap maps have been generated for fruit species like *Mangifera* and *Citrus.*

Biodiversity Assessment Using Molecular Techniques

With advent of molecular biology techniques, DNA-based markers have replaced enzyme markers in germplasm identification and characterization. Because of its plasticity, ubiquity and stability, DNA is the ideal molecule for such analysis in fruit crops (Lalitha Anand, 2001). Although, molecular markers are used to create extensive linkage maps for many annual plants (Bernatzky and Tanksley 1986; Helentjaris 1987; McCouch *et al.*, 1988), a few attempts are made for their use in fruit crops. Genetic analysis and breeding of woody perennial fruit species can be complicated by many factors including long periods of juvenility, high ploidy levels and lack of described Mendelian markers (Rowland and Levi, 1994). Task of developing molecular marker-based genetic maps is both challenging and important.

Basic premise is that variation in nucleotide sequence of DNA can be exploited to produce characteristic fingerprints. Systems available are:

1. Restriction fragment length polymorphism (RFLP)
2. DNA amplification fingerprinting (DAF which comprises AFLP- amplified fragment length polymorphism/RAPD-random amplified polymorphic DNA)
3. Microsatellites
4. Cleaved amplified polymorphic sequences (CAPS).
5. Future Needs

Given the nature of tropical fruit tree species, diversity assessment, conservation and utilization are a tough task. Although, commendable work has been done in tropical fruit tree species, still a lot remain to be done. Due to very large diversity and ambiguity in nomenclature, the task is really very challenging. Keeping this in mind, some of the areas that need to be looked into are given below:

1. Exploration of different fruit species and preparation of distribution maps is need of the hour. It helps in locating diversity and the unique characters in germplasm and also helps in adaptation of tree fruit species in *in situ* conservation by giving distribution pattern.
2. Characterization of genepool in tropical fruit crops has not been done for many of fruit species. There still remains ambiguity in nomenclature; as a result, conservation also has become a problem. Morphological characterization followed by molecular characterization need to be taken up.
3. Collection and evaluation of wild species and using them in breeding programme help in conserving them *in situ* and *ex situ.* In many of the tropical fruit species, this has not been done.
4. Screening biodiversity available for different diseases and pests will help isolating and conserving desirable genes, which ultimately help in conserving species and varieties.

References

A De Candolle, 1884. *Origin of Cultivated Plants.* Hafner, New York.

Alexander, M.P. and Ganeshan, S., 1993. Pollen storage. In: *Advances in Horticulture, Vol 1: Fruit Crops, Part I* (Eds.) Chadha, K.L. and Pareek, O.P. Malhotra Publishing House, New Delhi, pp. 481–486.

Altieri, M.A. and Merrick, L.C., 1988. Agroecology and *in situ* conservation of native crop diversity in the third world. In: *Biodiversity*, (Ed.) Wilson, E.O. Washington, D.C., National Academy Press, pp. 361–369.

Arora, R.K. and Rao, V., 1995. *Proceedings of the expert consultation on Tropical Fruit Species of Asia*, 17-19 May 1994, Malaysian Agricultural Research and Development Institute, Serdang, Kuala Lumpur IPGRI Office for South Asia, New Delhi, 116 p.

Arora, R.K., 1994. Promoting conservation and use of tropical fruit species in Asia. *Proceedings of Expert Consultation on Tropical Fruit Species of Asia* held at MARDI, Kuala Lumpur, Malaysia 17-19 May.

Arora, R.K. and Nayar, E.R., 1984. *Wild Relatives of Crop Plants in India.* NBPGR Sci. Monogr. No. 9: 90.

Bernatzky, R. and Tanksley, S.D., 1986. Towards a saturated linkage map in tomatobased on isozymes and random cDNA sequences. *Genetics*, 112: 887–898.

Borthakur, D.N., 1993. Plant genetic resources of North East. *Ind.J. Hill Farming*, 6(1): 1–18.

Brown, A.H.D., 1989. Core Collections: A practical approach to genetic resources management. *Genome*, 31: 818–824.

Brush, S.B., 1995. *In situ* conservation of landraces in centers of crop diversity. Crop Sci., 32: 346–354.

Dodds, K.S. and Simmonds, N.W., 1948. Genetical and cytological studies of Musa. The origin of an edible diploid and the significance of inter-specific hybridizatioin in the banana complex. An addendum on the nomenclature of edible banana by W.W. Cheesman. *J. Gen.*, 48: 285–296.

Dinesh, M.R., 2001. Field genebank management in tropical fruit species. Regional Training Course on "Characterization, evaluation and conservation of tropical fruits genetic resources" held at Bangalore from May 15–26th, 2001.

Frankel, O.H., Brown, A.D.H. and Burdon, J.J., 1995. *The Conservation of Plant Biodiversity.* Cambridge University Press.

Grout, B.W.W. and Roberts, A.V., 1995. Storage of free pollen, pollen embryos and zygotic embryos of seed by cryopreservation and freeze drying. In: *Genetic Preservation of Plant Cells In vitro*, (Ed.) B. Grout. Springer Verlag, Berlin, pp 63–74.

Haq, N., 1994. Identification of priority species of tropical fruits for research and development [Final Report IPGRI Project No. 93/11 (AOG–APO), ICUC in collaboration with IPGRI–APO]. International Centre for Underutilized Crops, Department of Biology, Southampton University, UK.

Harlan, J.R., 1975. *Crops and Man.* American Society of Agronomy, Madison, Wisconsin, 295 p.

Hayes, W.B., 1953. *Fruit Growing in India*, 2nd Revised edition. Kitabistan, Allahabad, India.

Helentjaris, T., 1987. A genetic linkage map of maize based on RFLP's. *Trends Genet.*, 3: 217–221.

Hoekstra, F.A., 1995. Collecting pollen for genetic resources conservation. In: *Collecting Plant Genetic Diversity: Technical Guidelines*, (Eds.) L. Guarino, V. Ramanatha Rao and R. Reid. CAB International, Wallingford, UK, pp. 527–550.

Hore, D.K., 1998. Fruit germplasm diversity and its development in North-Eastern India. *Proc. Nat. Conf. Sci. and Tech.*, pp. 101–106.

IPGRI, 1995. *Annual Report*, IPGRI, Rome.

Kostermans, A.I.G.H. and Bompard, J.M., 1993. *The Mangoes: Their Botany, Nomenclature/Horticulture and Utilization*. Academic Press, London.

Krishnamurthi, S. and Seshadri, V.S., 1958. Origin and evaluation of cultivated bananas. *Ind. J. Hort.*, 15: 135–145.

Lalitha Anand, 2001. Assessing genetic diversity using molecular techniques in tropical fruit species. In: *Proc. Regional Training Course on Characterization, Evaluation and Conservation of Tropical Fruits Genetic Resources* held at Bangalore 15–26th May.

Maxted, N., Ford-Lloyd, B.V. and Hawkes, J.G., 1997. Complementary conservation strategies. In: *Plant Genetic Conservation*, (Eds.) N. Maxted, B.V. Ford-Lloyd and J.G. Hawkes. Chapman and Hall, London, pp. 15–39.

McCouch, S.R. Kochert G., Yu, Z.H., Khush, G.S., Coffman, W.R. and Tanksley, S.D., 1988. Molecular mapping of rice chromosomes. *Theor. Appl. Genet.*, 76: 815–829.

Mitra, S.K. and Bose, T.K., 1985. Guava. In: *Tropical and Sub-tropical Fruits of India*. (Ed) T.K. Bose. Naya Prokash, Calcutta.

Mukherjee, S.K., 1949. A monograph on the genus *Mangifera* L. *Lloydia*, (12): 73–136.

Mukherjee, S.K., 1985. Systematic and eco-geographic studies on crop gene pools 1. *Mangifera* L. IBPGR, Rome.

Naik, K.C., Bhujanga Rao, C. and Raman, V.S., 1958. Problems in crop improvement in mango. *Indian J. Hort.*, 15: 159–167.

Pandey, G. Sharma, B.D., Hore, D.K. and Rao, N.V., 1993. Indigenous minor fruit genetic resource and their marketing status in North-Eastern India. *J. Hill Res.*, 6(1): 1–4.

Paroda, R.S. and Arora, R.K., 1991. *Plant Genetic Resources Conservation and Management Concepts and Approaches*. Published by International Board for Plant Genetic Resources, Regional Office for South and Southeast Asia, c/o NBPGR, Pusa Campus, New Delhi, India.

Pathak, R.K., 2003. Status Report on Genetic Resources of Indian Gooseberry–Aonla (*Emblica officinalis* Gaertn.) in South and Southeast Asia. APO Publication.

Pathak, R.K., 2003. Twenty five years of research and development of aonla. Lead paper presented in National Seminar on Production and Utilization in Aonla during August 2003 at Salem, Tamilnadu, Organized by Aonla Grower Association of India.

Prakash, G.S., 2001. *In situ* conservation of tropical fruit species. Regional Training Course on "Characterization, evaluation and conservation of tropical fruits genetic resources" held at Bangalore from May 15–26th, 2001.

Rai Mathura, 1995. *Genetic Diversity in Sapota* (*Manilkara achras* Mill) in India. IPGRI Newsletter for Asia, the Pacific and Oceania, No. 17: 12–13.

Rowland, L.J. and Levi, A., 1994. RAPD-based genetic linkage map of blueberry derived from a cross between diploid species (*Vaccinium danowi* and *V. elliotti*). *Theor. Appl. Genet.*, 87: 863–868.

Rawat, R.B.S. and Uniyal, R.C., 2003. Amla: An important species for Indian system of medicine. Paper presented in *National Seminar on Production and Utilization in Aonla* during August 2003 at Salem, Tamilnadu, organized by Aonla Grower Association of India.

Purseglove, J.W., 1968. *Tropical Crops: Dicotyledons*. Essex: Longman Group and Co. Ltd. BurntMill, United Kingdom.

Schroeder, C.A., 1958, The origin, spread and improvement of the avocado, sapodilla and papaya. *Indian J. Horticulture*, 15(3&4): 160–164.

Simmonds, N.W., 1962. *Evolution of Bananas*. Longman Publ. U.K.

Simmonds, N.W. and Weatherup, T.C., 1990. Numerical taxonomy of the wild bananas. *New Phytol.*, 115: 561–571.

Singh, H.P., 1996. Perspective of banana production and utilization in India. In: *Banana: Improvement, Production, and Utilization,* (Eds.) H.P. Singh and K.L. Chadha. Proceedings of the conference on 'Challenges from banana production and utilization' in 21st century held in Trichy, India 24–25 September, 1996.

Singh, I.D. and Sirohi, S.C., 1977. Breeding new varieties of papaya in India. In: *Fruit Breeding in India,* (Ed.) Nijjar. IBH Publ., pp. 187–296.

Singh, R.N., Yadav, I.C. and Pandey, S.N., 1977. Grape improvement work at IARI. In: *Viticulture in Tropics,* (Ed.) K.L. Chadha, G.S. Randhawa and R.N. Pal, pp. 63–70.

Singh, R., 1992. Improvement of grapes by hybridization. In: *Proceedings of "Staff course on Grapes"*, p. 3.

Singh, H.P. and Chadha, K.L., 1993. Genetic resources of banana. In: *Advances in Horticulture, Vol. I: Fruit, Part I,* (Eds.) K.L. Chadha and O.P. Pareek eds. Malhotra Publishing House, New Delhi, pp. 123–142.

Singh, H.P. and Chadha, K.L., 1993. Genetic resources of citrus. In: *Advances in Horticulture, Vol. I: Fruit, Part I,* (Eds.) K.L. Chadha and O.P. Pareek eds. Malhotra Publishing House, New Delhi, pp. 95–122.

UNCED, 1992. *Convention on Biological Diversity*. United Nations Conference on Environment and Development, Geneva. http://www.biodiv.org/convention/default.shtml).

Verheij, E.W.M. and Coronel, R.E., 1991. Plant Resources of South-East Asia No. 2. Edible Fruits and Nuts, Pudoc, Wageningen.

Yadav, I.S. and Singh, H.P., 1985. Evaluation of different ecological groups of mango cultivars for flowering and fruiting under subtropics. *Prog. Hort.*, 17(3): 168–175.

Yadav, I.S. and Rajan, S., 1993. Genetic resources of mango. *Advances in Horticulture, Vol. I, Fruit , Part I,* (Eds.) K.L. Chadha and O.P. Pareek eds. Malhotra Publishing House, New Delhi, pp. 77–93.

Biodiversity in Horticultural Crops Vol. 1 *Pages* **57–70**
Editors: **Professor K.V. Peter & Dr. Z. Abraham**
Published by: **DAYA PUBLISHING HOUSE**

Chapter 4

The Genus *Musa* (Banana and plantains)

S. Uma and S. Sathiamoorthy
National Research Center for Banana, Tiruchirapalli – 620 102, Tamil Nadu

Banana and plantains are the oldest food fruits cultivated by man. Although plenty of historical references are available, Indian reference in Ramayana (2020 BC) is the oldest. They are referred as divine fruit and to attain this divinity it traversed a million of years to transform themselves from seedy, non pulpy, non-edible wild types into pulpy parthenocarpic sweet fruited types. It is well quoted by Steve Connor (2001) as 'ergonomically shaped to fit the human hand, with a non-slip surface. It has outward indicator of ripeness–green, yellow and black. Its disposable wrapper has a tab at one end for removal and perforated edges for easy peeling. The fact that banana has a pointed end and a curved shape for easy entry into the mouth, and who could argue that it was indeed an act of divine inspiration.

Banana belongs to family *Musaceae*, of the 6 families in the order zingeberales, which includes around 1000 species. Musaceae has two major genera, *Musa* and *Ensete*. *Ensete* is unigeneric and has no sections. While Eumusa has four major selections namely, *Australimusa*, *Callimusa*, *Rhodochlamys* and *Eumusa*. Details of the sections, members and uses are represented in Table 4.1.

Origin and Evolution of Banana and Plantain

Banana and Plantain originated from south and south East Asia. Recent explorations discovered involvement of Pacific Islands especially Papua New Guinea (PNG). Extent of diversity witnessed for AA diploids especially for Plantain diploids led to rewriting of *Musa* evolution theory. Thus centre of origin of wild banana stretches from India up to Pacific Islands including Papua New Guinea(PNG), Micronesia, Melanesia and Samoa. Beyond India, though banana especially Plantains are cultivated extensively in African countries, wild forms are seldom found in forests. It is assumed that plantains must have domesticated primarily in PNG as evidenned by cultivation of seedless diploids in gardens, occurrence of wild diploids in adjoining forests and semi-wild variants growing in between the two (De Langhe, 2005).

Table 4.1: Conspectus of the Family *Musaceae*, Classification, Distribution and Uses of *Musa* Species

Genus	*Basic Chromosome No.*	*Section*	*Distribution*	*Selected Species*	*Uses*
Ensete	9	-	W.Africa, P.N.Guinea, India	*Ensete superbum, Ensete glaucum, Ensete ventricosum, Ensete livingstoniana, Ensete proboscidea, Ensete gilletti, Ensete buchanani, Ensete homblei, Ensete perrier*	Fibre vegetable Vegetable ornamental
Musa	10	Australimusa	Queenslaand New Caledonia Philippines Australia	*M. textilis M. maclayi M. lolodensis M.peekeelii M. fehi*	Fibre Fruit & vegetables
		Callimusa	Indo-China Indonesia	*M. coccinea M. violascens M. gracilis*	Ornamental
	11	Eumusa	India	*M. acuminata M. balbisiana M. schizocarpa M. itinerans M. flavicarpa, M. sikkimensis M. cheesmanii, M. nagensium, M. halabanensis, M. ochracea*	Fruit, Vegetable, Fibre and medicinal
	11	Rhodochlamys	India Indo-China Philippines Thailand Malaysia	*M. ornata, M. velutina, M. laterita, M. sanguinea, M. mannii, M. aurantiaca, M. aurantiaca M. rosacea, M. rosea, M. rubra*	Ornamental

Wild *acuminata* diploids, *Musa acuminata* contributing A genome to present day bananas are predominantly spread in Pacific Islands, Indonesia, Malaysia, Philippines, Myanmar and India. The other progenitor of present day bananas, *Musa balbisiana* is less spread than its counterpart, *Musa acuminata*, in India, Myanmar, China and Philippines. In these regions, introgression occurred across national boundaries and diversified based on utility and production systems. Thus resulted a number of interspecific hybrids with various genomic constitutions like AB, AAB and ABB at secondary centers of origin.

Domestication and Diversification

In the *Musa* family, genera *Ensete* and among the four major sections, only *Eumusa* and *Australimusa* members have edibility and thus underwent human domestication over several years resulting several forms. Man mended this crop from its original seedy state to his favoured parthenocarpic state (non-seedy) against nature's laws of selection.

Even during this process, it is assumed that man cultivated the seeded, non-edible wild forms of members of *Musaceae*, especially genera *Ensete*. The corms (underground stem–extension of the pseudostem) are routinely used as fodder for livestock. But under famine conditions it very well substituted for food.

Transcontinental introgression of original *Ensete* and most other primitive types was along with the earliest traders throughe Indian Ocean. *Ensete* acclimatized and diversified to suit local needs and now Africa being considered as one secondary centres of origin.

Similarly *Musa acuminata* originated in South-East Asia especially in Malaysia, Indonesia and Papuva New Guinea, where it diversified profusely into a number of subspecies. Extent of variation was from leaf pigmentation to flower colour to extended female phase.

A parallel evolutionary development was witnessed with respect to *Musa balbisiana* where seedy species developed into soft seeded types like Bhimkol (Uma *et al.*, 2003 of FAO report) of India, Parthenocarpic diploids (BB) like Abuhon and vigorous triploid cvs like Saba (ABB) of Philippines and Pisang Kepok of Indonesia (Nasution 1982). Both seeded and parthenocarpic diploid *balbisianas* freely hybridized with *Musa acuminata* to evolve diverse forms of B rich genomic clones like AB, AAB, ABB and ABBB etc. Thus India has more diversity for B rich genomes (ABB) which predominantly constitutes the hardy culinary types.

In the areas of tertiary diversity, where cultivated clones are commercially grown, mutations are a regular phenomenon leading to selection of better types through human interventions. Most of present day export oriented Cavendish clones (AAA) are result of tertiary evolution.

Agro-ecologies in Areas of *Musa* Diversity and Status of *Musa* Diversity (Wild) in India

Wild *Musa* species are largely distributed in tropical rain forests and wet evergreen forests to moist deciduous forests of low rainfall zones. The hilly tracks of these areas harbour a mosaic of tropical forests conducive for *Musa* species.

Tropical rain forests are found in North Eastern Indian Himalayas including Assam, Arunachal Pradesh, Meghalaya and parts of Nagaland. They are also the features of Andaman and Nicobar Islands. Tropical moist forests stretch in Manipur, Mizoram, Tripura of North-Eastern states, interior areas of Western Ghats of Karnataka and Kerala (Rao, 1999; Arora and Nayar 1983).

These ecological zones are characterized by wide climatic variations ranging from tropical, and subtropical temperature to alpine zones. (The temperature ranges from 8-15°C (mean min) to 30-35°C (mean max.) and an average annual temperature is between 18 and 22°C. March-April are the hottest months while November-January are the cold months. Rainfall is well distributed throughout the year except for February-April. Tropical rain forests receive a rainfall of 3200-2800 mm/year and Chirapunji of Meghalaya state received the world's highest annual rainfall till a few years back.

Tropical moist forests receive an annual rainfall of 1200-1600mm. South western monsoon accounts for maximum rain in North-Eastern states, Western Ghats and Andaman and Nicobar

Islands from April-August. Depressions during November-December also add on to annual rainfall in Eastern Ghats and Andaman and Nicobar Islands. Relative humidity is usually high between 82 and 85 per cent.

Domesticated Banana is basically a tropical crop but successfully acclimatized to subtropical regions. Similarly their progenitors, wild *Musa* species are also found to occur from tropical to subtropical climate with altitudes ranging from mean sea level to 1800m above MSL. Most of wild species are sheltered at an altitude between 500-1000m above MSL in the sub Himalayan mountains and Western Ghats. Survey in India suggests that *Musa* diversified conveniently at medium altitudes of 500-700 m over several years and moved either way towards higher and lower elevations and distributed without much problems. This has been the case especially in North-Eastern India in Khasi, Jaintia, Naga, Patkai and Garo hills. While in Western Ghats, two centres of diversity, namely Agasthiarmalai and Silent Valley are under moist, tropical evergreen forests where a parallel evolution of *Musa* species took place along with North-Eastern zone.

Similarly Andaman and Nicobar Islands, consisting of 350 islands and islets, lie in the Bay of Bengal (latitude 6°14°N and longitude of 92-94° E). Within a land area of 8500 sq.km, a lot of flora have evolved over millions of years due to insular nature of territory, physical isolation between islands and also from neighboring continental landmasses. The major islands are, North Andaman, Middle Andaman, South Andaman, Little Andaman, Car Nicobar, Teressa, Katchal, Kamorta, Noncowry, Little Nicobar and Great Nicobar.

Eastern Ghats are located along the coastal line of Bay of Bengal of India. Originally it extended from Northern Orissa down up to Andhra Pradesh. But due to faster deforestation in this zone, hardly any forests exist except for some pockets of northern Andhra Pradesh in Arakku Valley and Southern Orissa where, wild *Musa* sp. are known to occur. They are low hills with an altitude of 300-400m above MSL, annual rainfall ranging from 900-1300 mm and RH of 78-80 per cent.

Genetic Diversity in Wild Species

Musaceae has more than 30 species known scientifically spread across Asia and Pacific. The details of wild species available in each section are provided in Table 4.1.

Of the 11 species of the section Eumusa, 56 per cent of total diversity is found in India.

M. acuminata

M. acuminata is one of the ancestral species of present day bananas and contributed 'A' genome. It has a number of distinct subspecies identified from Malaysia and Indonesia. In India, *Musa acuminata* ssp. *burmannicoides* (Calcutta-4) is predominant in Southern Peninsula. There are several forms of *Musa acuminata* found in North-East India which have not been characterized and identified systematically. Basically *Musa acuminata* subspecies harbour resistant genes for a number of biotic and abiotic stresses. *Musa acuminat ssp* burmannicoides otherwise refered as Calcutta-4 is the only resistant gene source for Sigatoka leaf spot and being extensively used in global breeding programmes. Unfortunately this subspecies, neither exists in any of the areas explored nor in any of the Indian genebanks.

Andaman and Nicobar Islands harbour *Musa acuminata* subspecies very specific to those islands, in addition to Western Ghats of Kerala, Karnataka, North-Eastern Mizoram, Arunachal Pradesh, Meghalaya, Tripura, Manipur and Nagaland.

Musa acuminata is slender, medium statured, short duration plants which generally prefers mild climate. Fruits are also slender, long and number of hands depends on subspecies. By nature, they are highly female fertile with more than 100 seeds/fruit. Pulp is mucilaginous and usually acidic.

Descriptions of various subspecies of *Musa acuminata* are attempted by Markku and de Langhe (2001).

Musa balbisiana

A very hardy species in the section Eumusa with good acclimatization to the harsh climatic conditions. Tolerant to cold, drought and immune to most of the fungal and bacterial diseases. Plants are characterized by robust stature (3-5 m), longer crop duration (18-26 months), short and stocky fruits. True *Musa balbisiana* is female fertile and has more than 200 seeds/fruit. But one of its evolutionary relatives, with soft seeds, Bhimkol, was domesticated on a large scale in Assam. Comparatively restricted diversity over *Musa acuminata*, this species is distributed widely in most forest zones of India.

Bhimkol

This is the most popular *Musa balbisiana* semi wild, seeded type (BB) which crept into large scale cultivation in N-E India. Over years, Bhimkol is accepted as a household variety and no more considered as wild. From utility point of view, it is referred as a 'Pheral' (a wild variety under cultivation in its area of diversity). Bhimkol has its maximum distribution in Assam and to some extent in its neighbouring states like West Bengal, Meghalaya etc.

Bhimkol surprisingly has no localised cultivation on a stretch of land but each and every Assamese household has a minimum of 4-5 Bhimkol clumps in their backyards and its cultivation is approximately worked out to be more than 8,000 ha. For a wild variety, this popularity is worth mentioning, which is attributed to the several of its good traits mentioned under ethnobotany.

Discussion with people and the available literature have no specific mention of its era of commercial recognition. It is evident that Bhimkol is grown from time immemorial by Assamese tribes. Complete survey and study for diversity using morphological and molecular characterization has led to the understanding that the diversity among Bhimkol is almost nil. This owes to the fact that Bhimkol has strong pollination and fertilization barriers and the seed development is always incomplete. The ovules enlarge to become seeds but they are devoid of any endosperm and well developed embryo. Hence they are termed as empty seeded fruits. This trait has attributed to the soft seeded nature of Bhimkol and in turn, its acceptable edibility.

Elavazhai

It is a common seeded, diploid and wild type grown in all rural house holds of Western Ghats of Karnataka. The main uses of Elavazhai in day to day life are usage of leaves as dining plates as a part of their culture of having food only on banana leaves and as a shade crop in areca (*Areca catechu*) plantations.

Musa nagensium

It is most widely distributed in all seven North Eastern states but the least described and studied. Because of its conspicuously black pseudostem, easily, it gets confused for *Musa balbisiana* by amateurs. Majestically tall, plants are only next to *M.lolodensis*. Pitch Black pseudostem, maroon, shiny male bud (watery, bold seeds) are its characteristics. 8-15 hands, self-pollinated, highly female fertile with 80-100 big, shiny and warty seeds. Its orange flesh is very sweet (32°B) and widely used as animal feed along with other plant parts like rhizome. Male bud fasciation and multiple male buds in a bunch are often noticed in this species.

M. itinerans

One of the equally wide spread species of Eumusa, is popularly referred as walking banana. It has rhizomatious stem, buds crawl away from mother plant and emerge at a distance of 1–1.5m. Because of their itinerant nature of Wandering, it is rightly referred as *M.itinerans*. Highly drought tolerant, plant owing to its dormant, rhizomatous stem, medium stature and conspicuous violet pigmentation in juvenile phase. Male bud is brown black with yellow stripes. Widely distributed in Mizoram, Arunachal Pradesh, Meghalaya and Nagaland, it has only 4-6 hands with stocky short, beaked fruits often dull green and with more than 150 seeds/fruit.

M. sikkimensis

Its distribution is restricted to Arunachal Pradesh and mostly seen sympatrically growing with *M.nagensium*. Robust plants (3-4m) with maroon pseudostem. Bunch is small with 4-6 hands, characteristically angular (3) short, compressed fruits with equally long petioles. Bud is maroon, bold and conspicuously top (twister) shaped.

Unidentified *Musa* Species

India harbours plenty of *Musa* gene wealth but accessibility and difficult terrain make it tough to travel extensively in the interior forest zones and collect. With efforts from NRCB, 3 new species were identified in the genus Eumusa.

M. swarnaphalya

The *M.swarnaphalya* is characterized by robust stature of 8.0–8.5 m in height with maroon coloured pseudostem. Leaves are dark green, 150–170 cm long having round laminar bases. Intensely wax coated petioles (50–60 cm) are tightly clasped on the pseudostem. Petiolar canal is closed and the margins are erect, sometimes overlapped. Peduncle is short, 15–20 cm long, glabrous, green, slightly angular in position. Male axis is pendulous, barren and bract scars are not predominant.

Male bud is lanceolate in shape, greenish yellow in colour and heavily wax coated, deeply imbricate, outer face of the bract is greenish yellow and inner face is pale orange yellow in colour. Bract tip is blunt as in case of *M. balbisiana* and tinted with purplish brown on its outer side. Bracts open 1–2 at a time and do not revolute. Male flowers are yellow, 3.5–4.0 cm long, compound tepal and lobes are yellow.

Free tepal is rectangular in shape, 1.6–1.8 cm long with smooth margins; its apex is also smooth and pointed.

Ovary is straight 1.2–1.3cm long, yellow in colour, style is straight and translucent white and stigma is orange yellow. Fruits are ashy green (before maturity), biseriate, 3.0–3.5 cm long, pedicel is unusually long with 1.2–1.5 cm and 6–8 fruits/hand. Pollen is fertile and germinability is more than 80 per cent.

New Species 1

Was identified in the Seppa area above Tippi of west Arunachal Pradesh. It has restricted distribution in this zone and growing sympatrically with *M.itinerans*, *M.nagensium* and *M.rosaceae* (of the section rhodochlamys). Robust and tall plant, inflorescence hang is parallel with persistent bracts just above the male bud since the bracts only reflex (lift) but not revolute (twist backwards), bud has the lotus flower like appearance. At the time of bunch maturity, plants shed all its leaves and stands bare with only the peduncle bearing the bunch. This is a very unusual criterion not noticed in any of the *Musaceae* members.

Male flowers are orange in colour and so also the stigmatic tip. Bract is pink in colour with profuse ash coating. Male bud is very bold and grows until the ground level. Peduncle is also bold till its rear end and resembles that of an elephant's trunk.

Genetic Diversity for the Cultivated Species

AA Diploid

Though South-East Asia is the centre of origin for *acuminata* diploids, India has a few but distinctive members, some of which have assumed commercial significance. Most of its members are choice cultivars grown in backyard gardens.

e.g. Matti, Kadali, Surya Kadali, Sannachenkadali, Namarai, Anai Komban, Hatidat, Chingan, etc.

Acuminata diploids have distinct characteristics and cannot be classified into sub-groups. Among members, Matti assumes commercial significance in southern tips of Tamil Nadu, where in hot and humid climate,its full potential to yield 30-35 kg bunch with 18-20 hands is exhibited. Fingers are short and curved with sweet pulp known for its medicinal properties. Finger orientation has a typical wind-blown appearance. This lacks any synonyms and is well known in all southern states. Kanai Bansi or Anaikomban is the choice cultivar of southern districts of Tamil Nadu, Kerala and parts of North-Eastern India owing to its tolerance to leaf spot disease and yield potential of 13-15 kg. Straight finger orientation and fruit smoothness without angularity makes it one of the desirable quality parameters.

Kadali is a small fruited, sweet variety restricted to hot and humid regions of southern tips. A mutant, Naivedya Kadali, is grown mainly in temple backyards to offer it to god. Sannachenkadali is another choice variety restricted to backyard orchards. It is distinct from other diploids due to red pigmentation of its pseudostem and fruits. Apart from indigenous diploids, exotic diploids like Pisang Lilin, Tongat, Cultivar Rose are also maintained for use in breeding programmes.

AAA Clones

Cavendish clones form mainstay of global banana industry both at domestic and export fronts. This domination is witnessed from the last century owing to sweet fruits, wider adaptation and suitability to long distance transportation. Originally Gros Michel (AAA) but a non-Cavendish clone was the ruling export variety but devastated by wide spread Fusarium wilt (*Fusarium oxysporum* fsp. *cubense*). Attempts to breed resistant or tolerant Gros Michel resistant to Fusarium wilt was not an instant success. This led to poor taste of Gros Michel. Cavendish clones immune to Fusarium wilt entered the export scenario. Now, Grand Naine is the leading variety.

Grand Naine

Most of Cavendish clones arose from a common base clone through mutations. It is similar to Robusta, Pedda Pacha Arati, Jahaji etc. Grand Naine is the most accepted international variety of commercial cultivar in recent days and has occupied the premier export slot. In India, its popularity increased replacing existing varieties like Shrimanti and Robusta.

It belongs to Giant Cavendish sub group. It is tall statured plant, heavy yielder with a long cylindrical bunch. Duration of the crop is 11-12 months. On an average, it produces a bunch weighing 25 kg and goes upto 32-35 kg. Number of hands varies from 8-10 with 200-220 fingers/bunch. The length of finger is 14-15 cm and girth 12-13 cm. Internationally, it is acceptable both as a dessert type and for processing. It has a better pulp to peel ratio and has more market acceptability. It is highly susceptible to Sigatoka leaf diseases and nematodes with immunity to Fusarium wilt (Race 1 and 2). Tissue cultured plants of 3-4 months are highly susceptible to Erwinia rot.

Thellachakkarakeli

It is the choicest backyard cultivar of Tamil Nadu and North-Eastern region and occupies commercial significance in Andhra Pradesh especially in Godavari delta.

Plant is medium statured and delicate. Grows to a height of 2.25–2.4 m above ground with greenish yellow pseudostem covered with uneven array of brown blotches. Leaves are pale, attached to pseudostem with a yellowish pedicel. Petiolar wings are broad with a conspicuous reddish pink margin. Inflorescence is put forth 10-12 months after planting. Female phase has 4-6 hands each with 12-14 short, stout and stocky fruits. Male bud is bold, covered with unequal pink coloured bracts. Male flower is cream coloured with lot of nectar. Fruits are characteristically greenish yellow in colour which when ripe turn golden yellow. Pulp enclosed in a thick skin is firm, very sweet, juicy with characteristic aroma.

Red Banana

Red banana is a choice variety. It is popular in Tamil Nadu, Kerala, Karnataka, Andhra Pradesh and to some extent in Western and central India. The plant is tall and roubust statured with 2.5-3.5 m height and 70-80 cm in girth. Crop duration of this variety is 16-18 months. The bunch weight is 20-25 kg with 6-7 hands and 80 fruits/bunch. The length of the fruit is 16-18 cm and girth is 15-16 cm. Fruits are sweet and pulp is orange yellow in colour with a pleasant aroma. The fruit skin is thick.

It is popularly grown in perennial plantations and becomes vulnerable for the attack of Fusarium wilt, weevils, borer, nematodes apart from viruses like Banana Bunchy Top Mosaic Virus and Banana Bract Mosaic Virus.

It often produces a green mutant popularly referred as Pacha Kappa, Sandana Vazhai, Venkadali and Vella Vazhai. The red colour of the pseudostem, petiole and fruit skin is replaced by green. Fruit skin turns attractive yellow upon ripening. Rest of the characters is same as red banana.

AB Clones

These are characterized by medium stature, erect leaf orientation and long petioles. The basal colour of male flower is pink with 2-5 irregular flowers/hand. Bunch has a short peduncle and orientation is either horizontal or at an angle but never pendulous. This genomic group is subdivided into Kunnan, Ney Poovan and Unique sub-groups.

Kunnan has members like Adukka Kunnan, Thaen Kunnan, Naattu Poovan, Adukkan and Narmine with slight variability for fruit size, waxiness of fruit and plant stature. Till date Ney Poovan was the sole member of this group. But in an explortion to southern tips of Tamil Nadu, two ecotypes were identified. Commercial Ney Poovan has an angular bunch and bold fruits, whereas its ecotype has horizontal bunch, and small fruits with strong wind – blown appearance. This has tastier fruits than bold fruits. Kunnan sub-group is distinguishable from Ney Poovan sub-group by its erect leaf orientation, black blotches at petiolar junction and angular bunch orientation. Though Kunnan and

Ney Poovan sub-groups have common genomic constitution, many of characters of Ney Poovan sub-group like smoothness of leaves, light pigmentation, sweet pulp, presence of blotches etc. bring it more nearer to *acuminata* genome. Under the Unique sub-group, Kodapanilla Kunnan and Padali Moongil find their place owing to absence of male bud and elongated plantain-like fruits. Pulp quality of Padali Moongil is very sweet, powdery from pulp and has unique taste. Kodapanilla Kunnan, by virtue of its high-cost fruits, finds commercial cultivation in Tirunelveli district of Southern Tamil Nadu.

AAB Clones

Many dessert clones belong to AAB genomic status and they differ morphologically thus distributed in over a number of subgroups like Silk, Mysore, Pome, Plantain etc. Silk encompasses mainly Rasthali or Rasabale or Malbhog of various regions. Diversity at morphological and molecular levels was worked out (Uma *et. al.*, 2003).

Mysore

Mysore group is also a subgroup with one universal clone, Mysore has many synonyms, very widely grown in various continents and in India both in mainland and in Andaman and Nicobar Islands.

Pome

Pome is another subgroup with a number of clones mostly grown for markets. Most famous ones are Hill Banana of Tamil Nadu with intense aroma. They perform very well at higher altitudes. Though most clones are more or less similar in taste with apple flavour, differ morphologically to a greater extent. Varieties like Nendra Padathi, Jwari Bale of Karnataka fall into unique groups. They exhibit unique persistent male flowers on male axis resembling that of an elephant's trunk.

Plantain Sub Group

Comprises plantains of Africa and Asia, Maia Maoli/Popoulou of Asia Pacific. They are starchy and need to be cooked before eating. Plantains of Asia and Africa are also used as dessert fruits. In India, diversity for plantains is restricted to 3 other types like French, Horn, and False Horn types.

ABB Group

ABB group is mainly constituted by cooking or culinary banana, while Asia has even dessert bananas in this group. Pisang Awak is one major subgroup spread world wide and with diversified uses as culinary, dessert and for wine making. Owing to tolerant gene sources for drought and cold, it has become popular in higher altitudes of North-Eastern India.

Peyan

Peyan is another noteworthy group restricted to Tamil Nadu and nurtured for medicinal properties. Acidic and sweet fruits are two different ecotypes.

Bluggoe

Bluggoe and Monthan are major culinary types in the world apart from plantains and Popoulou groups. A lot of diversity in morphological traits is described and diversity at molecular level is being worked out.

Pisang Awak

Pisang Awak is one of the dual purpose varieties of this ABB group, more specifically used as dessert bananas.

Section Rhodochlamys

Section Rhodochlamys is the only other section of *Eumusa* found widespread in India, apart from *Eumusa* members. Its wider distribution in terms of species and spread is prevalent in North-Eastern India, especially in Arunachal Pradesh. This section comprises of 11 species (Hakkinen *et al.*, 2002) of which seven occur in India (Uma *et al.*, 2005).

Table 4.2: Species Occurring in States of India

Sl.No.	*State*	*Occurring Species*
1.	Assam	*Musa laterita*
2.	Arunachal Pradesh	*Musa velutina, Musa rosacea, Musa aurantiaca, Musa laterita, Musa sanguinea*
3.	Megalaya	*Musa laterita*
4.	Manipur	*Musa rosacea, Musa ornata*
5.	Nagaland	*Musa laterita*
6.	Tripura	*Musa ornata* (Chittang forest)
7.	Mizoram	*Musa rosasea, Musa rubra**
8.	Andhra Pradesh	*Musa ornata***
9.	Tamil Nadu	*Musa laterita*
10.	Karnataka	*Musa laterita*
11.	Kerala	*Musa laterita*

* Our explorations could not locate *M.rubra* in any of the N-E states. But Indian Institute of Horticultural Research (IIHR), Bangalore, India has a collection from Indo-Myanmar border.

** As per Jacob (1952), occurrence is noted in moist evergreen forests of Western Ghats of earlier Madras province (inclusive of Tamil Nadu, Kerala and Karnataka). It was mentioned not as *M. laterita* but as Kattu Vazha. Based on above reported observations, and on discussions with local tribes, forest department personnel as well as earlier reports, cartography was drawn for the distribution.

Diagnostic Features of Rhodochlamys Members

In general, Rhodochlamys members are slender, plants highly acclimatized to mild climatic conditions. Unlike *Eumusa* members, they have erect bunching and uniseriate fruits. Flowers are both male and female fertile and when pollinated produce 20-30 seeds/fruit.

Genetic relatedness of Rhodochlamys with *Eumusa* and proximity in distribution resulted in development of intersectional hybrids in nature, some exhibiting basic *Eumusa* traits while others, of Rhodochlamys. They exhibit a number of resistant genes, both for biotic and abiotic traits. *Musa laterita*, is the most drought tolerant species owing to its rhizomatous roots, which remain dormant in summer and sprout under congenial environment conditions. Similarly most of them are immune to Sigatoka leaf spot and Fusarium wilt diseases. This resistant trait can be successfully exploited through classical breeding and serve as a good material in Marker Assisted Selections (MAS) programmes.

Straight and translucent white, stigma is orange yellow. Fruits are ashy green (before maturity), biseriate, 3.0–3.5 cm long, pedicel is unusually long with 1.2–1.5 cm and 6–8 fruits/hand. Pollen is fertile and germinability is more than 80 per cent.

At the time of our study, only a few plants located 500- 800 m apart were in flowering and in early stages of fruit development. Hence the study on pulp and seed characters, germinability etc., could not be observed. Suckers brought and planted at NRCB field gene bank have not yet come to bear. Our discussions with local Nitshi tribes of Arunachal Pradesh also revealed the prolonged vegetative phase and seed fertile nature of this new species.

Distribution Pattern of Rhodochlamys Members in India

Detailed explorations in various states revealed endemic nature of most of Rhodochlamys species in North East India. Jacob (1952) reported *M. laterita* in Western Ghats of Kerala and Karnataka, occurrence and distribution of *M. ornata* in Arakkuvalley of Andhra Pradesh of Eastern Ghats (Sundararaj, 1952) and our explorations suggest occurrence and distribution of section Rhodochlamys even in southern states of India.

Surprisingly, detailed exploration in North and Central Andaman islands revealed absence of section Rhodochlamys in these islands. Though Nicobar islands have not been covered, but discussion with local personnel from Botanical Survey of India and State Agricultural departments strengthened the authors' views that the Rhodochlamys has no occurrence and distribution in Andaman and Nicobar islands. Their geographical isolation from the Indian mainland could be one of the probable reasons.

Genus *Ensete*

Genus *Ensete* was kept as *Musa* in the Physocaulis sub genera by Baker (1893). But recent Banana taxonomists formed a new term *Ensete* as Genus and given the biname Ensete *superbum* (Rox)., *Ensete glaucum* (Rox)., to *Musa superbum* Rox., *Musa glaucum* Rox., respectively. Actually there are 6 main species in *Ensete*, but more than 26 species names were given by different authors. Now it is clearly known that they are not different species, but are synonyms for main 6 species of *Ensete*.

Ensete species (selected) in Africa (Ethiopia) are considered as 'tree of the poor'. Since it is substituted for food and it is also referred as 'tree against hunger' (Lockhart 1984) (ref. Pg. 45 in *Ensete* book). Originally a tropical species, it enjoys heat and moisture tolerance but has diversified its acclimatization to dry, rocky and moist regions. In India, genus *Ensete* has 2 species. In Southern India, especially Western Ghats harbour *Ensete superbum* while North Eastern India has *Ensete glaucum*. The latter species seem to have Thailand, Myanmar, and Bangladesh as its natural habitat.

Diagnostic Features of *Ensete*

Monocarpic, single stemmed plant, non stoloniferous, clasping leaf sheaths, bracts and flowers are integral with one another, first one or two hands bear hermaphrodite flowers. Pollen grains have warty exine wall, comparatively bigger than *Musa* pollens. Seeds are bold, more than 7 mm diameter, globose to subglobose with smooth surface.

Table 4.3: Distribution Pattern of Cultivated Banana Clones in Different Indian States

State	AA/AB	AAA	AAB	ABB
Andhra Pradesh		Dwarf Cavendish Robusta Thellachakkarakeli	Amrithapani Karpurachakkarakeli	Monthan
Assam		Jahaji Borjahaji Manjahaji Robusta Dwarf Cavendish Honda, Agniswar Threk Marang Bharath Moni There Kanchi Tulsi Manohar	Chenichampa Jatikol Digjowa Kulpait Malbhog Amritman Alpon Garomoina Dasaman	Manohar Kanchkol Bharatmoni Ankur, Geda Gera, Bankel, Sabji kela, Nute pong, Kait Khullung, Kait Shejeng, Singhalaji
Bihar		Dwarf Cavendish Jahaji	Chenichampa Alpon, Chinia Malbhog	Muthia, Kothia, Gauria, Kanthali
Gujarat		Dwarf Cavendish Lacatan Harichal Gandevi collection		
Karnataka	Elakki Bale	Robusta	Poovan, Rasa Bale Jwari Bale Mundugod Bale	Karibale Embale Sakkara Bale
Kerala	Njali Poovan Kunnan Poomgalli Poovilla Chundan Adukkan Kadali, Neyvadya Kadali, Matti, Namarai Siguzani	Sevvazhai Morris Robusta	Nendran Palayankodan Poovan Kali	Monthan, Peyan Peykunnan, Vella Palayankodan
Maharashtra	Safed Velchi	Dwarf Cavendish Robusta Grand Naine Sindhurni Hanuman Ardhapuri	Lalvelchi Rajeli	
Manipur		Jahaji Borjahai Torbat Dwarf Cavendish	Champa Sabri Thera Hao Malbhog Somai	Banria Kachkola Haong Ong Hangou
Mizoram		Cavendish Jahaji Dwarf Cavendish	Malbhog Champa Sabri Bai bhal-al-their	Kachkeli Banria Banpawl

Contd..

Table 4.3–Contd..

State	*AA/AB*	*AAA*	*AAB*	*ABB*
Nagaland		Jahaji Borjahaji Dwarf Cavendish	Malbhog Champa Sabri	Banria Sabji Kola Kechulepa Kacha Kola
Orissa	Mohan Bansi Moinsi Singha	Robusta Pot Kapura	Mortman Mysore	Poussia Chokua
Tamil Nadu	Matti Kadai, Namarai Anaikomban Sannachenkadali Ney Poovan Kunnan Poovilla Chundan	Sevvazhai Robusta Dwarf Cavendish Manoranjitham	Poovan Rasthali Virupakshi Pachanadan Nendran	Karpuravalli Sakkai Monthan Peyan
West Bengal		Amrit Sagar Giant Governor Lacatan	Mortman Champa	Monthan Kanthali
Tripura		Jahaji	Chompa Sabri Kulpait Borchampa	Banria Sagarkol Anaji Kola Bangla Kola Shail Kola

Table 4.3: Banana in Different Languages in India

Tamil Nadu	:	Vazhai	Kerala	:	Vazha
Karnataka	:	Bale	Anthra Pradesh	:	Aratti
Maharashtra	:	Kela	Uttar Pradesh	:	Kela/Mouze
Madhya Pradesh	:	Kela/Mouze	Bihar	:	Kela/Mouze
West Bengal	:	Kol	Orissa	:	Kol, Bonthal
Assam	:	Kola	Meghalaya	:	Kait
Manipur	:	Hai Lophu/Changbui Lophu	Mizoram	:	Changel/Bahl-al
Tripura	:	Sema tribe Oucho Angami tribe – Jeatusa			

References

Bhakthavatsalu, C.M. and Sathiamoorthy, S. 1979. Banana clonal situation in India: A resume. *Fruits*, 34(2): 99-105.

Baker, 1893. A synopsis of the genera and species of *Musaceae*. *Ann. Bot.*, 7: 189-222.

Chakravorti, A.K., 1948a. On the occurrence of non-stoloniferous species of *Musa, M.agharkarii* sp. nov. in the Chittagong Hill tracts (Bengal). *J. Indian Bot. Soc.*, 27: 90–95.

Chakaravorti, A.K., 1951. Origin of cultivated banana of S. E. Asia. *Indian J. Plant Breed.*, 11: 34–46.

Dutta, S., 1952. Some bananas of Assam. *Indian J. Hort.*, 9: 26–35.

Daniells, J., Jenny, C., Karamura, D. and Tompkepe, T., 2001. Musalogue: A Catalogue of *Musa* Germplasm. Diversity in the Genus *Musa* (E. Arnaud and S. Sharrock, Comp.) INIBAP, Montpellier, France, pp. 213.

Gandhi, S.R., 1952. Banana culture in western India. *Poona Agric. Coll. Mag.*, 42: 180–209.

Hakkinen, M. and De Lange, E., 2001. *Musa acuminata* in Northern Borneo-Preliminary report. INIBAP, France, pp. 27.

Hakkinen, M. and Sharock, S., 2002. Diversity in the genus *Musa*. Focus on Rhodochlamys. International Network for the Improvement of Banana and Plantain (INIBAP) Annual report 2001. Montpellier. France, pp. 16–23.

Jacob, K.C., 1952. *Madras Bananas: A Monograph*. Government Press, Madras, pp. 228.

Krishnamurthi, S. and Seshadri,V.S., 1958. Origin and evolution of cultivated banana. *Indian J. Hort.*, 15: 135–145.

Nayar, T.G., 1952. On the occurrence of *Musa banksii* (*F. musella*) var. Sangamapatti. *Indian J. Hort.*, 2: 146–50.

Nayar, T.G., Sundararaj, J.S. and Seshadri, V. S., 1957b. Nendran banana, its culture and taxonomical status. *Indian J. Hort.*, 14: 69–76.

Simmonds, N.W., 1956. Botanical results of the banana collecting expedition 1954–1955. *Kew Bull.*, 3: 463–489.

Simmonds, N.W., 1959. *Bananas*. Longmans, Green and Co. Ltd. London. Pp. 170.

Simmonds, N.W., 1962. *The Evolution of the Bananas*. Longmans, Green and Co. Ltd. London, UK, pp. 170.

Simmonds, N.W. and Shepherd, K., 1955. The taxonomy and origins of the cultivated bananas. *J. Linn. Soc. (Botany)*, 55: 302–312.

Singh, H.P., Uma, S. and Sathiamoorthy, S., 2001. *A Tentative Key for Identification and Classification of Indian Bananas*. National Research Centre for Banana (ICAR), Tiruchirapalli, India, pp. 61.

Singh, H.P. and Uma, S., 2000. Genetic diversity of Banana in India. In: *Banana Improvement, Production and Utilization*, (Eds.) Singh, H.P. and Chadha, K.L., pp. 503.

Sundararaj, D.D., 1955. *Musa acuminata*: Wild seeded banana. *Indian Hort.*, 3: 16–21.

Uma, S. and Sathiamoorthy, S., 2002. *Names and Synonyms of Bananas and Plantains in India.* National Research Centre for Banana, Tiruchirappalli, India (Publishers), pp. 60.

Uma, S., Shyam, B., Selvarajan, R. and Sathiamoorthy, S., 2001. Collection, characterization of banana and planting in North-Eastern India: Exploration Programme. In: *Advancing Banana and Plantain R&D in Asia and Pacific*, (Eds.) Molina, A.B., Roa, V., 10:103–109.

Venkataramani, K.S., 1948. A note on the banana variety "Moongil". *Madras Agric. J.*, 35: 402–405.

Venkataramani, K.S., 1949. On the occurrence of *Musa balbisiana*. *Madras Agric. J.*, 36: 552–554.

Biodiversity in Horticultural Crops Vol. 1 *Pages* ***71–108***
Editors: **Professor K.V. Peter & Dr. Z. Abraham**
Published by: **DAYA PUBLISHING HOUSE**

Chapter 5

Temperate and Sub-tropical Vegetables

D. Ram, Mathura Rai and Major Singh
Indian Institute of Vegetable Research, Varanasi – 221 005

Plant genetic resources can be described as the total genetic diversity of cultivated species and their wild relatives (Ford-Lloyd, 2001). 'Biological Diversity' means variability among living organisms from all sources including, *inter alia,* terrestrial, marine and other aquatic ecosystems and the ecological complexes in which they are part. This includes diversity within species, between species and ecosystems (Article 2 of the Convention on Biological Diversity) whereas, biodiversity can be defined as the quality, range or extent of differences between biological entities in a given situation. Around 2,50,000 plant species were described so far and a much larger number still remain to be duly recognized. Yet only about 3,000 of them were picked up from wild by human being for their use. They are grown on varying scales since beginning of agriculture, considered nearly ten thousand years old. Man depends on about 30 major food crops that 'Feed the World' (Harlan, 1975). In this way, biodiversity is of considerable importance in daily diet of human being, who depends on bark, young shoots, buds, flowers, fruits, leaves, roots, tubers and mushrooms, especially during hunger, when other cultivated crops are unavailable. Wild plants have often played an important role in many diets due to higher nutritional value than cultivated species. These are too directly or indirectly consumed/used as vegetables in form of roots, tubers, bulbs, rhizomes, leaves, stems/shoots/culms, fruits and in a very few cases as flowers. Prized genetic materials along with their wild and close relatives are used by plant breeders for development of improved crop varieties/hybrids. They also constitute a priceless reservoir which contains genes conferring better adaptation to stress environments and resistance to diseases and pests.

Centres of Origin and Regions of Diversity

A Russian plant explorer and geneticist, Nikolai I. Vavilov, was one of the first scientists to recognize importance of plant genetic diversity in nature. On the basis of geographical survey, he summarized that cultivated plants originated from eight basic geographical centres of origin and two subcentres (Vavilov, 1951). His studies revealed that out of the wide range of plant diversity in tropical and subtropical regions of the world, the major food crops have come mainly from high mountain valleys, isolated from each other to a large extent and with a great habitat range. The world's diversity in cultivated plants is distributed in twelve regions (Zeven and de Wet, 1982). Over 90 per cent of plant species for food and agriculture are located in economically developing parts of the world, namely, the African, Asian and Latin American continents and the Far East Islands (Madeley, 1996).

For underutilized, neglected and lesser known domesticated/cultivated vegetables, this diversity is mainly confined to seven regions, Chinese-Japanese, Indo-Chinese-Indonesian, Hindustani, Mediterranean, European-Siberian, African and South American regions (Arora, 1985), where approximately 495 species of 540 species occur. The region possessing maximum diversity at global levels is tropical America, tropical Asia and Mediterranean. In tropical Asian region, both India and China hold maximum diversity. Indian subcontinent is well known since long as an important centre of origin and diversity of a large number of vegetable crops.

India is a homeland of 167 cultivated species and 329 wild relatives of crop plants (Arora, 1991).

It was observed that people made useful selection in several vegetable crops, which were eventually domesticated and cultivated. Several weedy species were never or only temporarily domesticated, remaining as weeds but often hybridizing by chance with the cultivated ones and thus, enhancing the diversity in cultivated plants. India is 'Primary centre' for crops like eggplant, smooth gourd, ridge gourd, cucumber etc. and secondary center for cowpea, okra, chillies, pumpkin and several Cole crops (Table 5.1).

Table 5.1: Centres of Diversity of Major Vegetable Crops

Gene Centre	*Primary Centre*	*Secondary Centre*
Chinese-Japanese	Egg plant, wax gourd, chinese cabbage, Kangkong, welsh onion	Watermelon, amaranth
Indo-Chinese	Wax gourd, Sponge gourd, ridge gourd, bitter gourd, sword bean, winged bean, taro, chayote, cucumber, bottle gourd, yams	Chinesecabbage, bottlegourd, cucumber, yam, bean, amaranth, yard longbean, Kangkong
Hindustani centre	Egg plant, wax gourd, cucumber, ridge gourd, bitter gourd, sponge gourd, hyacinth bean, drumstick, okra, Kangkong	Water melon, melon, Roselle, bottle gourd, amaranth
Central Asia	Onion, garlic, carrot, spinach	Eggplant, water melon, melon, cauliflower
Near East	Onion, garlic, leek, beet	Okra
Mediterranean	White cabbage, cauliflower, water melon, broccoli, radish	Sweet pepper, garlic, okra
African	Egg plant, water melon, melon, bottle gourd, cowpea, okra, Roselle	Onion, shallot, lima bean, White cabbage amaranth
European-Siberian	Lettuce	Onion, White cabbage, common bean, cauliflower, spinach, carrot
Central America and Mexican region	Tomato, hot pepper, pumpkin, squash, yam bean, sweet potato, common bean	—
South American region	Tomato, hot pepper, cassava, pumpkin, lima bean, chayote, amaranth sweet potato	Common bean
North American region	—	Tomato, Egg plant, melon, water melon, pepper, squashes, onion, lettuce, lima bean, okra, pumpkin

Agro-climatic Zones of India

India is located between 8° N to 38° N latitude and 68° E to 97.5° E-longitude and exhibits extreme variations in edapho-climatic situations, agro-climatic regions and floristic diversity. Altitudinal variations are observed from below sea level to above vegetational limits in the Himalayas *i.e.* more than 3500 m above average sea level; the climate change from monsoon to temperate/alpine in Northern/Himalayan zone. Sehgal *et al.* (1992) reported occurrence of 20 agro-climatic regions in India based on physiographic, climatic and cultural features. Around 80 species of major and minor vegetables, apart from several wild/gathered kinds occur (Choudhury, 1967 and Seshadri, 1987) in India. Concentration of genetic diversity comprising native species and landraces occurs more in Western Ghats and North-Eastern Himalayas. Richness of plant diversity is largely due to ecological diversity superimposed with tribal and ethnic diversification, plant usages and religious rituals. Crops in which rich diversity still occurs in India include cowpea (*Vigna unguiculata*), common bean (*Lablab purpureus*), cole crops (*Brassica* species), okra (*Abelmoschus esculentus* and related species), brinjal (*Solanum melongena* and related species), sweet potato (*Ipomoea batatas*), taros (*Colocasia* and *Alocasia*), yams (*Dioscorea esculenta, Dioscorea alata, Dioscorea deltoidea*), sword bean (*Canavalia species*), velvet bean (*Mucuna* species), and elephant foot yam (*Amorphophalus* species). *Solanum spp.* are widely distributed in the North Eastern region; yams in Western ghats and North Eastern states, chives, leeks and other wild *Allium* spp. in Kumaon and Garhwal Himalayas; cluster bean in Western arid zone; Lablab bean in Deccan plateau; cucurbits in Rajasthan and M.P; and leaf vegetables like *Amaranths*, and Fagopyrum spp. in Western Himalayan region.

North-Western and Eastern Himalayan Region

Under Western and Eastern Himalayan region, only temperate vegetables are predominantly found. Enormous diversity occurs in *Allium* species- leek, shallot and other introductions of *Allium sativum* and *Allium cepa.* Sporadic diversity also occurs in asparagus, spinach, chenopods, amaranths, and *Beta vulgaris.* Other vegetables like Brassicas, squashes, Cucurbita spp., Cucumis spp., chillies, bell pepper, peas, faba bean, cowpea, horse radish, artichoke, potato, Colocasia, tomato, parsley, coriander, ginger, *Sechium edule* and *Cyclanthera pedata* do express a lot of variability.

North-Eastern Region Including Assam

In this region also, most of the temperate vegetables predominate. Maximum diversity occurs in leaf vegetables like amaranth and *Brassica* species. Other vegetables like chillies, tomato, brinjal, okra, taros, yams and cucurbits are grown in this tract. Several kinds of beans like winged bean, French bean and lima bean are specialized due to edaphic and climatic factors. In the lower tract, rich diversity occurs in cucumber, pointed gourd, chayote, bitter gourd, spine gourd (*Momordica dioica* Roxb) and meetha karela (*Cyclanthera pedata*). Among wild species *Abelmoschus manihot* (pungens form), *Alocasia macrorrhiza, Amorphophallus bulbifera, Colocasia esculenta, Cucumis hystrix, Cucumis trigonus, Dioscorea alata, Luffa graveolens, Moghania vestita, Momordica cochinchinensis, Momordica macrophylla, Momordica subangulata, Trichosanthes cucumerina, Trichosanthes dioica, Trichosanthes dicaelosperma, Trichosanthes khasiana, Trichosanthes ovata, Trichosanthes truncata and Solanum indicum* are prevalent.

The Northern Plains/Gangetic Plains Including Tarai Region

Northern Plains/Gangetic Plains including Tarai Region is one of the richest pockets of diversity in major vegetable crops. Due to year round cultivation, more diversity emerges and it is well fitted into cropping patterns. Rich diversity can be observed in *Cucumis* species, *Luffa* species, *Cucurbita* species,

Benincasa hispida, *Lagenaria siceraria*, *Momordica* species, *Trichosanthes* species, *Solanum* species, *Capsicum* species, *Abelmoschus* species, *Brassicas* and tuber crops.

North Western/Indus Plains

In this region, variability exists in species of *Cucumis*, *Momordica*, *Citrullus*, *Solanum*, *Amaranthus*, *Chenopodium*, *Abelmoschus*, *Capsicum* and *Allium*. There are certain sporadic pockets, rich in indigenous germplasm like cucurbits, okra, eggplant and garlic. Wild species of *Momordica balsamina*, *Citrullus colocynthis* and *Cucumis prophetarum* are prevalent.

The Central Region/Plateau

More diversity occurs in *Cucurbita* species, ash gourd, round gourd, bitter gourd, pointed gourd, ridge gourd, okra, eggplant, chillies, tomato, root and bulbous crops and sporadically in leaf vegetables. As indigenous vegetables, more diversity exists in cucurbits, eggplant, okra and chillies. Wild species like *Cucumis setosus*, *Cucumis trigonus*, *Luffa acutangula var. amar and Momordica cymbalaria* are more common in this region.

The Western and Eastern Peninsular Region

Western and Eastern Peninsular Region is an extremely rich region for cucurbits (cucumber, bitter gourd, bottle gourd and squashes), eggplant, okra and chillies (both annual and perennial types). More landrace diversity occurs in snake gourd in the western and for Luffa species and eggplant in eastern region. Sporadic diversity can be spotted for leaf vegetables like *Amaranth*, *Brassicas*, *Chenopods*, *Spinach*, *Beta vulgaris*, *Basella rubra* and *Basella alba*. Several wild species like-*Abelmoschus angulosus*, *A. moschatus*, *A.manihot*, *A. ficulneous*, *Amorphophallus campanulatus*, *Colocasia antiquorum*, *Cucumis hystrix*, *C. setosus*, *C. trigonus*, *Luffa acutangula var amara*, *Luffa graveolens*, *Luffa umberrata*, *Momordica cymbalaria*, *Momordica denticulate*, *Momordica dioica*, *Momordica cochinchinensis*, *Momordica subangulata*, *Solanum indicum*, *Solanum melongena*, *Trichosanthes anamalaiensis*, *Trichosanthes bracteata*, *Trichosanthes cordata*, *Trichosanthes cuspidata*, *Trichosanthes horsfieldii*, *Trichosanthes piniana*, *Trichosanthes perottitiana*, *Trichosanthes nerifolia*, *Trichosanthes himalensis*, *Trichosanthes multiloba and Trichosanthes villosa* are quite prevalent in this region.

Variability distribution of different vegetable crops to different agro-ecological regions of India are presented in Table 5.2.

Variability, Distribution and Genetic Erosion of Vegetable Crops

Valuable genetic resources of vegetable crops are vanishing rapidly. Three main processes which caused loss of genetic diversity of cultivated crop species are genetic erosion, genetic vulnerability and genetic wipe out. Several other factors like shrinking of natural resources, population pressure, urbanization; deforestation, monoculture and changing cropping pattern including use of hybrids/ improved varieties are responsible for loss of genetic diversity. These are not mutually exclusive but are, in fact inter-locked by demand of increasing population and rising expectations. This genetic erosion is taking place at a time when new tools of biological research enable scientists to focus as much on diversity of genes as on diversity of genotypes. There is considerable success in protecting and preserving agro-biodiversity albeit under *ex situ* conditions during past one quarter century but much remains to be accomplished.

Table 5.2: Distribution of Variability of Major Vegetable Crops to Different Agro-ecological Regions of India

Sl.No.	*Agro Ecological Regions*	*Geographical Ranges*	*Variability in Vegetable Crops*
1.	Humid Western Himalayan region	J&K, H.P. and parts of U.P.	Cucurbits, radish, carrot, turnip, peas, cowpea, chillies, brinjal, okra, spinach, fenugreek, amaranth, *Solanum khasianum, Solanum hirsutum, Sechium edule, Basella rubra*
2.	Humid Bengal/ Assam basin	W.B. and Assam	Cucurbits, radish, cowpea, chillies, brinjal, okra, spinach, beet, *Abelmoschus manihot ssp. manihot,* amaranth, *Solanum indicum, Solanum khasianum, Solanum surattense, Cucumis sativus var. vikkimensis, Edgeria dargelingensis, Melothria assamica, Momordica cochinchinensis, Sechium edule, Tuladiantha coordifolia, Basella rubra*
3.	Humid Eastern Himalayan region and Bay Lands	Arunachal Pradesh, Nagaland, Manipur, Mizoram, Tripura, Meghalaya, Andaman and Nicobar islands	Cucurbits, radish, cowpea, pea, chillies, brinjal, okra, spinach, amaranth, *Abelmoschus manihot ssp. tetraphyllus, Solanum khasianum, Solanum torvum, Solanum sisymbrifolium, Solanum ferox, Solanum verbasifolium, Cucumis hystrix, Luffa echinata, Sechium edule*
4.	Sub-humid Sutlej Ganga Alluvial plains	Punjab, U.P. and Bihar	Cucurbits, radish, peas, brinjal, okra, spinach beet, fenugreek, onion, garlic, *Abelmoschus manihot ssp. tetraphyllus var. pungens, Abelmoschus tuberculatus, Solanum indicum, Solanum khasianum, Solanum torvum, Solanum surattense, Solanum hispidum, Cucumis hardwickii, Cucumis trigonus*
5.	Humid Eastern and South Eastern uplands	East M.P., Orissa and A.P.	Cucurbits, radish, carrot, cowpea, chillies, brinjal, okra spinach, amaranth, garlic, *Abelmoschus manihot ssp. manihot, Solanum surattense, Solanum torvum*
6.	Arid Western plains	Haryana, Rajasthan and Gujarat	Cucurbits, cauliflower, radish, carrot, peas, cowpea, chillies, brinjal, okra, spinach beet, fenugreek, onion, garlic, amaranth, *Abelmoschus tuberculatus, Abelmoschus ficuleneus, Abelmoschus manihot ssp. tetraphyllus, Solanum torvum, Solanum nigrum, Citrullus colocynthes*
7.	Semi Arid Lava Plateau and Central Highlands	Maharashtra and West M.P.	Cucurbits, cauliflower, radish, carrot, cowpea, chillies, brinjal, okra, spinach, fenugreek, amaranth, onion, *Solanum surattense, Solanum torvum, Solanum nigrum, Solanum khasianum, Cucumis setosus, Luffa acutangula var. acutangula*
8.	Humid to Semi Arid Western Ghats and Karnataka Plateau	Karnataka, Tamil Nadu, Kerala and Lakshadweep islands	Cucurbits, chillies, brinjal, okra, *Abelmoschus crinitus, Abelmoschusangulosus, Abelmoschus ficuleneus, Abelmoschus moschatus, Abelmoschus manihot var. tetraphyllus, Solanum trilobatum, Solanum indicum, Solanum incanum, Solanum pubescens, Solanum surattense, Solanum torvum, Luffa acutangula var. acutangula, Melothria angulata, Basella rubra*

Table 5.3: Variability and Distribution Status of Vegetable Crops

Crops	*Genera*	*CS*	*DS*	*GVS*	*GES*	*GCP*
Brinjal	*Solanum*	C	W	W	M	H
Chillies	*Capsicum*	C	W	H	M	H
Tomato	*Lycoperscon*	C	W	H	M	H
Cluster bean	*Cyamopsis*	C	R	M	M	M
Cowpea	*Vigna*	C	W	H	M	H
French bean	*Phaseolus*	C	W	M	M	M
Lablab bean	*Lablab*	C	W	H	M	M
Mucuna bean	*Mucuna*	C	R	H	M	M
Pea	*Pisum*	C	R	M	M	H
Carrot	*Daucus*	C	R	M	M	H
Elephant foot yam	*Amorphophallus*	C	R	H	M	M
Giant taro	*Alocasia*	W	R	H	M	M
Radish	*Raphanus*	C	W	H	M	H
Taro	*Colocasia*	W	W	H	M	M
Yam	*Dioscoria*	W	R	H	M	M
Garlic	*Allium*	C	W	M	M	H
Onion	*Allium*	C	R	M	M	H
Okra	*Abelmoschus*	C	W	H	M	H
Amaranth	*Amaranthus*	C	W	H	M	M
Climbing spinach	*Basella*	C	W	M	M	M
Fenugreek	*Trigonella*	W	R	M	M	M
Palak	*Beta*	C	R	M	M	H
Bitter gourd	*Momordica*	C	W	H	M	H
Bottle gourd	*Lagenaria*	C	W	H	M	N
Cucumber	*Cucumis*	C	W	H	M	M
Pointed gourd	*Trichosanthes*	C	L	H	H	M
Pumpkin	*Cucurbita*	C	W	H	M	H
Luffa	*Luffa*	C	W	H	H	H
Tinda	*Prae-citrullus*	C	L	M	M	M
Water melon	*Citrullus*	C	R	H	M	M
Ash gourd	*Benincasa*	C	W	H	M	M
Snake gourd	*Trichosanthes*	C	W	H	M	H

CS: Crop status (C: Cultivated, W: Wild); DS: Distribution status (W: Wide spread distribution, R: Regional distribution, L: Localized distribution); GVS: Germplasm variability status (H: High, M: Medium, L: Low); GES: Genetic erosion status (H: High, M: Medium, L: Low); GCP: General Crop priorities (H: High, M: Medium).

Genetic Resources of Vegetable Crops in Indian Gene Centre

Wild plants have often played an important role in many diets due to higher nutritional value than cultivated species. Wild species and putative ancestral forms of vegetable crops contain valuable genes that are of immense need in vegetable breeding programme using conventional methods or modern biotechnology. These genetic resources can be utilized in development of new cultivars, strains and hybrids and also in restructuring of existing ones that lack one or other attribute. Distributional pattern of wild plant genetic resources in different botanical/phyto geographical regions and areas of their concentration where rich diversity of wild species still continues to perpetuate, are of special significance for undertaking programmes on collection as well as for *in situ* conservation of biodiversity. Important families possessing wild genetic diversity are Brassicaceae (*Brassica*), Malvaceae (*Abelmoschus*), Leguminoceae (*Canavalia, Lablab, Trigonella, Vigna*), Cucurbitaceae (*Citrullus, Coccinia, Luffa, Momordica, Neoluffa, Trichosanthes*), Solanaceae (*Solanum*), Amaranthaceae (*Amaranthus*), Dioscoreaceae (*Dioscorea*), Amaryllidaceae (*Allium*) and Araceae (*Alocasia, Amorphophallus, Colocasia*). The bulb of greater galangal (*Alpinia galangal* family Zingiberaceae) can also be eaten raw. In the Garwhal hills, *Cornus capitata* and *Cornus controversa* (family Cornaceae) are eaten raw. In Rajasthan, the whole plant of *Gisekia pharnaceoide* (family Mulluginaceae) is widely consumed during food shortage but in the South and West (Deccan Region) the leaves are used as greens, as are leaves of *Glinus trianthemoides*.

Genetic Resources of Solanaceous Vegetables

Tomato, brinjal and chillies are the important Solanaceous vegetables, widely grown all over the country throughout the year for their edible fruits.

Lycopersicon Species

Tomato is a warm season crop, sensitive to frost and usually cultivated in sub-tropical and mild cold climate. It is native of South America and Mexico, however, domestication took place in Mexico (Boswell, 1949) and now grown all over the world. The greatest variability occurs in Mexico, Central America and Coastal Peru. The wild species of *Lycopersicon* have shown to be a valuable source of resistance to various diseases and pests to quality traits in addition, which is vital requirement in any breeding programme. *Lycopersicon pimpinellifolium* is resistant to *Fusarium* wilt, early blight and leaf curl disease and is rich source of ascorbic acid; *Lycopersicon cheesmanii* is resistant to salinity and a source of high total soluble solids (TSS); *Lycopersicon chilense* is tolerant to leaf curl disease; *Lycopersicon chmielewskii* has high content of TSS; *Lycopersicon hirsutum* is an excellent source of resistance to several pests and to leaf curl disease; *Lycopersicon peruvianum* is resistant to root knot, tobacco mosaic virus, leaf curl and has high ascorbic acid content and *Lycopersicon chilense* is a good source of drought resistance.

Solanum Species

Brinjal or eggplant (*Solanum melongena* L.) being the crop of Indian Origin has developed some secondary variability in China (Vavilov, 1926). A large number of land races/traditionally grown cultivars are developed in different agro-ecological zones of India and these landraces possess valuable genes for resistance to biotic and abiotic stress and adaptation to various environments. Maximum diversity and distribution of *Solanum* species was noted in southern India, foothills of Himalaya and northeastern region. Widely distributed species in the region include *Solanum torvum, Solanum indicum, Solanum insanum, Solanum surattense, Solanum pubescens* and *Solanum khasianum*. A sum of 36 accessions of cultivated eggplant and 91 other *Solanum* species from Sri Lanka were also acquired. Some distinct species include *Solanum verbascifolium, Solanum surattense, Solanum indicum, Solanum nigrum* and *Solanum viarum* from Nepal.

Chillies and Bell Pepper

Capsicum Species (Chillies or Pungent Pepper and Bell Pepper or Simla Mirch)

Chillies (*Capsicum annuum* L.) are usually grown in warm to hot and humid climate throughout the country. The large fruited bell pepper of today originated from tiny pungent and pointed fruited wild species *Capsicum annuum*. It is a cool season crop and grown in sub-tropical and temperate region. There are four other species (*viz. Capsicum frutescens, Capsicum chinense, Capsicum baccatum and Capsicum pubescens*), which are commonly grown in southern and Central America except *Capsicum frutescens,* which is also grown in some parts in USA. Due to continuous selection, long history of cultivation and popularity of the crop, sufficient genetic variability has been generated and now India is treated as secondary center of diversity for chillies. In hot chillies (*Capsicum annuum* L.), rich variability in plant and fruit morphological traits occurs throughout the country particularly in south peninsular region, northeastern region, in foothills of Himalayas and Gangetic plains. Variability includes in plant type, fruit size/shape (long, short, pointed, smooth, wrinkled), bearing habit (fruits facing upwards, horizontal, downwards) and pungency in addition to annual and perennial types. However, in bell pepper less variability has been observed. Mostly introduced lines are grown particularly in mid hill of Himalaya and parts of hilly region of Central India including Karnataka and Maharashtra.

Genetic Resources of Cucurbitaceous Vegetables

Cucurbita Species

The genus *Cucurbita* comprised of about 27 species, both wild and cultivated mostly concentrated in tropical and sub tropical regions of Central and South America. In this genus,a few species are of commercial importance *viz.* Field pumpkin (*Cucurbita moschata* Duch. Ex. Poir.), winter squash/ pumpkins (*Cucurbita maxima* Duch.), Summer squash or vegetable Marrow (*Cucurbita pepo* L), winter squash pumpkins (*Cucurbita mixta* Pang. syn. *C. angyrosperma*) and Malabar gourd or Fig leaf gourd (*Cucurbita ficifolia* Boucha). Vegetable Marrow (*Cucurbita pepo* L), is an introduced crop which withstands cooler climate. *Cucurbita texa* Gray grows wild in texas (Chadha and Lal, 1993).

Cucumis Species

The genus *Cucumis* comprises of about 26 species. Major crops of economic importance are cucumber (*Cucumis sativus*), muskmelon (*Cucumis melo*), snap melon (*Cucumis melo* var. *momordica*), longmelon (*Cucumis melo* var. utilissimus). Indian sub-continent is the center of origin for *Cucumis sativus* and center of diversity for *Cucumis melo* (Zeven and de Wet, 1982). The wild species (*Cucumis hardwickii*) is found growing in natural habitats in the foothills of Himalayas. The free hybridization with cultivated *Cucumis sativus,* with no reduction of fertility in F_2 generation suggested that *Cucumis hardwickii* is likely progenator of cultivated cucumber. The cucumber *Cucumis sativus var. sativus* has 3-5 lobed leaves, ovary usually 3 placentiferous with fruits oblong, obscurely trigonus or cylindric. However, *Cucumis sativus* var. *sikkimensis* has 7-9 lobed leaves, ovary often 5 placentiferous, fruits ovoid-oblong, adapted in temperate and humid climate. Muskmelon (*Cucumis melo*) based on the distribution of diversity can be grouped into seven sub sets (Munger and Robinson, 1991).

Cucumis melo var. agrestis (Kachri)–It is wild type with slender vines and small inedible fruits, probably synonym of *Cucumis melo var. callosus* (*Cucumis callosus.*) and *Cucumis melo var. trigonus* (*Cucumis trigonus*).

Cucumis melo var. cantaloupensis Naud. (Musk melon or cantaloupe)–Medium sized fruits with netted, warty or scaly surface, flesh usually orange but sometimes green, aromatic or musky flavour. Fruits dehiscent at maturity, usually andromonoecious.

Cucumis melo var. flexuosus Naud. (*Cucumis melo var. utilissimus*)–Snake melon, snake cucumber, Tar-kakari, fruits long and slender consumed at immature stage, monoecious.

Cucumis melo var. momordica–Snap-melon or phut is a monoecious crop grown in India and other Asian countries. It has white to pale orange, less sweet pulp. The smooth surface of fruit starts cracking at the time of maturity.

Cucumis melo var. conomon (sweet or pickling melon)–It is generally andromonoecious in nature and bears small fruits with white flesh.

Cucumis melo var. inodorous Naud. (Winter melon)–Fruit is smooth or wrinkled surface with white or green flesh, and lacking musky odor. It is also andromonoecious in nature and usually requires more time for maturity.

Cucumis melo var. dudain Naud. (Mango melon, Pomegranate melon)–Fruits small, globular, smooth motted but not netted, flesh with acid flavour, pickling type.

Out of the seven melons *Cucumis melo var. agretis*, *Cucumos melo* var. *cantaloupensis* Naud. (musk melon), *Cucumis melo var. utilissimus*, *Cucumis melo* var. *momordica*, and *Cucumis melo var. flexuosus* Naud are available in Indian Sub-continent.

Luffa Species

The Indian gene center has rich diversity in genetic resources of *Luffa* species. The genus comprised of 9 species in the world and out of this 7 species (*Luffa acutangula, Luffa cylindrica, Luffa echinata, Luffa graveolens, Luffa hermaphrodita, Luffa tuberosa* and *Luffa umbellata*) are native to India. There is ambiguity with regards to *Luffa tuberosa* and *Luffa umbellate* because they are synonyms to species of *Momordica* and *Cucurbita* respectively (Chadha and Lal, 1993). Sponge gourd (*Luffa cylindrica*) and ridge gourd (*Luffa acutangula*) have rich diversity in vine and fruit morphological characteristics throughout India particularly in North-Eastern region including Sikkim, West Bengal, Western, Central and Southern India. *Luffa hermaphrodita* originated from *Luffa graveolans*, is another potential species distributed in parts of North-Central India. *Luffa acutangula* var. *amara* grown in peninsular India, is a wild relative of *Luffa cylindrica* and *Luffa echinata* in natural habitats in Western Himalayas, central India and Gangetic plains.

Lagenaria Species

Under genus *Lagenaria*, 6 species are reported, out of these *Lagenaria abyssinica, Lagenaria siceraria,* and *Lagenaria leucantha* are common. Among them *L. siceraria* is generally cultivated in all tropical parts of the world, especially in India and a few African countries. Remaining species are wild, perennial and diocious in nature. Two wild species *i.e. Lagenaria abyssinica*, and *Lagenaria bravifolia* are perennial in nature. There are suggestions that *Lagenaria* occurs in wild form in South America and in India (de Candolle 1982).

Citrullus Species

The genus, *Citrullus* has two species of economic importance *viz.* water melon (*Citrullus lanatus* (Thunb.) and round gourd (*Citrullus vulgaris* Schrad var.*fistulosus*). Based on evidence (de Candolle 1982) and linguistic data (Filov and Vilenskaya, 1972), *Citrullus lanatus* originated in Africa and reached India in prehistoric times. Cultivation of large watermelon *Citrullus lanatus* var. *citrodes,* is comparatively recent and Soviet varieties grown today still have shape of their African ancestors (*Citrullus lanatus* var. *caffer*). Shimotsuma (1963) reported that *Citrullus vulgaris, Citrullus colocynthis, Citrullus ecirrhosus* and *Citrullus nandinianus* are related and cross compatible with each other. Whitaker (1933) considered *Citrullus colocynthoides* as probable ancestor of watermelon.

Trichosanthes Species

Trichosanthes has 22 species and reported to be of Indian origin/parts of Tropical Asia or Indo-Malayan region. Among these, *Trichosanthes anguina L.* (snake gourd), *Trichosanthes dioica Roxb.* (pointed gourd) are cultivated throughout India. Major zone for distribution of diversity for *Trichosanthes dioica* is North-Central and NorthEastern India including West Bengal whereas *Trichosanthes anguina* is distributed throughout the country. Rich diversity in *Trichosanthes anguina* is observed in north-eastern states, West Bengal, Malabar Coast and in Eastern ghats in low and mid hills.

Momordica Species

The genus *Momordica* has 60 species reported to occur in the world and 7 species in India. The cultivated species, *Momordica charantia* (Bitter gourd) is grown all over the country in tropical and sub-tropical climate. *Momordica dioica* (Kartoli) grows all over West Bengal, Assam, parts of Bihar and adjoining areas; *Momordica cochinchinensis* (kakrol or sweet gourd or golkakora) is the most popular particularly in Tripura, Assam and West Bengal. *Momordica cymbalaria occurs* mainly in Western Ghats, Maharashtra, Southwards and Eastern peninsular tract, *Momordica denudate* in peninsular tract, *Momordica macrophylla* in North-eastern region and *Momordica subangulata* in North-Eastern hills, Eastern Ghats and Deccan peninsula.

Benincasa Species

Benincasa hispida Cogn. known as ash gourd or wax gourd, is a native of Java and Japan. It was domesticated in India during pre-historic time. Widely grown all over the country in tropical and sub-tropical climates, it possesses variability in morphological characteristics and quality of fruit.

Coccinia Species

Coccinia has about 35 species distributed in tropical Africa and Asia. Out of these, only one species *i.e. Coccinia grandis* is under cultivation. Related species are *Coccinia histella and Coccinia sessilifolia,* widely cultivated in Africa, Central America, China, Malaysia, Australia and other tropical Asian countries (PIER, 2001).

Sechium Species

Genus *Sechium* originated in mountainous region of America and Mexico. It includes *Microsechium compositum, Microsechium gintonii, Sechium compositum, Sechium edule, Sechium hintoni* and *Sechium jamaicense.* Principally it was confined to Tropical and Sub-tropical climate and particuarly in mid hill conditions. Maximum variability occurs in Sikkim. In Meghalaya and Mezoram, it is grown on a commercial scale and is popular in Darjeeling Hills, also.

Leaf Vegetables

The Indian sub-continent is the place of origin for a number of leaf vegetables like *Beta vulgaris* var *bengalensis, Basella rubra, Cichorium endivia, Enhyndra fluctuant, Lactuca indica* and *Amaranthus spp.* Most of the leaf vegetables possess good amount of diversity, which is acclimatized under extreme variation in altitude with long history of its cultivation. Wild relatives of different leaf vegetables possess valuable gene particularly for resistance to biotic and abiotic stresses.

Beta Species

Beta species include *Beta adanensis, Beta altissima, Beta atriplicifolia, Beta bengalensis, Beta bourgaei, Beta brasiliensis, Beta campanulata, Beta chilensis, Beta cicla, Beta corolliflora, Beta lomatogona, Beta macrocarpa, Beta macrorhiza, Beta maritime, Beta maritima subsp. danica, Beta maritima var. atriplicifolia, Beta maritima var. erecta, Beta maritima var. prostrata, Beta nana, Beta orientalis, Beta palonga, Beta patellaris, Beta patula, Beta perennis, Beta procumbens, Beta trigyna, Beta trojana, Beta vulgaris, Beta vulgaris cv. conditiva, Beta vulgaris cv. saccharifera, Beta vulgaris f. rhodopleura, Beta hybrida, Beta vulgaris subsp. adanensis, Beta vulgaris subsp. cicla, Beta vulgaris subsp. Flavescens, Beta vulgaris subsp. Lomatogonoides, Beta vulgaris var. glabra, Beta vulgaris var. cicla, Beta vulgaris var. crassa, Beta vulgaris var. atriplicifolia, Beta vulgaris subsp. macrocarpa, Beta vulgaris var. altissima, Beta vulgaris subsp. vulgaris, Beta vulgaris subsp. provulgaris, Beta vulgaris subsp. patula, Beta vulgaris subsp. maritime, Beta vulgaris subsp. orientalis, Beta vulgaris var. erecta, Beta vulgaris var. flavescens, Beta vulgaris var. foliosa, Beta vulgaris var. grisea, Beta vulgaris var. macrocarpa, Beta vulgaris var. maritime, Beta vulgaris var. orientalis, Beta vulgaris var. perenni,s Beta vulgaris var. pilosa, Beta vulgaris var. prostrata, Beta webbiana, Beta x intermedia and Beta vulgaris var. trojana.* The spinach beet also known as Indian spinach beet (*Beta vulgaris* var. *bengalensis* Hort.) is a close relative of beet root. It is grown in the plains and hills of India from Punjab to North Eastern Region including West Bengal, Orissa, M.P., Rajasthan, Gujarat and Maharashtra.

Amaranthus Species

Origin of various species of cultivated amaranth is complicated, because wild ancestors are pantropical cosmopolitan weeds (Mohideen and Irulappan, 1993). *Amaranthus spinosus, Amaranthus hybridus* and *Amaranthus dubius* are tropical types. *Amaranthus retroflexus, Amaranthus viridis, Amaranthus lividus* and *Amaranthus graecizans* are more temperate hot season weeds. In all probably, it might have originated in India and spread to neighbouring countries by traders, budhist monks and others. In India a number of domesticated forms are available in North Eastern States, West Bengal, Orissa, Tamil Nadu, Andhra Pradesh, Karnataka and Kerala. Distribution pattern of *Amaranthus blitum* (weed throughout India), *Amaranthus caudatus* (wild throughout India), *Amaranthus gangeticus* (weed in warmer parts of India, variants as *Amaranthus mangostanus, Amaranthus paniculatus* (warmer parts of India), and *Amaranthus spinosus* (weed throughout India) are reported.Wide ranges of variability in plant morphological characteristics are available in different parts of the country. In Orissa and North Eastern States, a special type is grown where stem is consumed after full growth.

Spinacia Species

Genus *Spinacia* belongs to family Chenopodiaceae. It originated in South West Asia. There is description about four types of plants with reference to sex expression, which are: (i) extreme males (ii) vegetative males (iii) monoecious and (iv) females. The monoecious plants may predominantly be staminate, pistillate or pure pistillate, early but with some staminate flowers late or almost equally staminate or pistillate throughout season. Greater variability can be spotted in plains and in foothills. Variability occurs in plant morphological characteristics, seed type and foliage yield.

Chenopodium Species

Genus *Chenopodium* belongs to family Chenopodiaceae. *Chenopodium album is* found as weed in plains and *Chenopodium murale* in hills upto 3000 m. In bothe species, maximum variability can be observed in upper Gangetic plains, extending to Northern hills; elsewhere only sporadic distribution.

Basella Species

Climbing spinach or Poi is grown all over the country particularly in plains and mid-hills during hot humid season The genus *Basella* includes *Basella alba, Basella cordifolia, Basella rubra, Basella tuberosa* and *Basella vesicaria.* Rich varability exists in plant morphological characteristics including stem fleshiness, leaf shape and size in South India, North Eastern region including Assam, West Bengal and Orissa.

Trigonella Species

Genus *Trigonella* has 22 species, having different chromosomes numbers. The 15 species having 2n=16 chromosomes include *Trigonella angeuna, Trigonella arabica, Trigonella bolansae, Trigonella calliceras, Trigonella coerulea, Trigonella corniculata, Trigonella cretica, Trigonella foenum-graecum, Trigonella gladiata, Trigonella glomerata, Trigonella gordej, Trigonella hamosa, Trigonella lypokyi, Trigonella monspeliach and Trigonella radiate.* Three species have 2n=18 (*Trigonella stellata, Trigonella striata, Trigonella ornithopodioides*); two species have 2n=44, (*Trigonella geminiflora, Trigonella grandiflora*) (Fedorov, 1974) and another species has 2n=44 (*Trigonella polycerata*). Two species of *Trigonella* are of economic importance. These are *Trigonella foenum-graecum* L., the common Methi and the other is *Trigonella corniculata* Kasuri or Champa methi. It grows wild in North Western parts of India (Bailey, 1950).

Portulaca Species

Parslane (*Portulaca oleracea* L.) is indigenous to Himalaya. It occurs in several forms during hot weather, in cultivated and wild state. Cultivated forms have broad leaves whereas in wild; both broad and small leaf types are seen growing. It is a tiny creeper with small, succulent fleshy leaves and tender stems, which are cooked like spinach.

Brassica Species

Chinese cabbage (*Brassica pekinensis* and *Brassica chinensis*) is indigenous to China and Eastern Asia where it is in cultivation since 5th century. There are two types of Chinese cabbage- heading type (*Brassica pekinensis*) and open leaf type (*Brassica chinensis*). The former forms an erect, moderately compact, usually cylindrical head, while later develops clusters of succulent leaves with forming head. Rich diversity in plant type occurs in foot hills of Himalaya and North Eastern states including Sikkim.

Allium Species

The Alliums include onion, garlic and their relatives–leeks, shallots and chives that are prized vegetables due to food, medicine or religious purposes. The genus Allium has about 750 species widely distributed over warm temperate and temperate zones of Northern hemisphere. Regions of high species diversity occur in Turkey, Iran, North Iraq, Afghanistan, Soviet Socialist Republic, middle Asia and West Pakistan. A second less pronounced center of species diversity occurs in Western North America. About 40 species of *Allium* are reported in India of which seven are cultivated, the main being onion and garlic. Much diversity has built up in India in these crops as they have been cultivated from ancient times and selections have been made in different agro-ecological zones. Both these crops are known only in cultivation and while no known wild ancester is known for *Allium cepa. Allium longicuspis,* endemic to central Asia is a progenitor for *Allium sativum.* Within *Allium cepa,* two horticultural varieties are recognized: the dry bulb onion and the multiplier onion (shallots). In *Allium sativum,* despite it being propagated exclusively by the division of ground cloves or aerial bulbils, a wide range of variability is available in several traits. The leeks (*Allium ampeloprasum*) succeed fairly well in India thriving at elevations above 300 m and often remain fit for use until the end of rainy season. The chives (*Allium shoenoprasum*) are distributed in the Northern Himalayas from Kashmir to Kumaon. Cultivated and wild types interbreed and there is plenty of variation in cultivated forms. Attention is being given on these and other wild species, as they possess many disease resistant genes potentially useful in improvement programmes. *Allium ampeloprasum* for white rot resistance, leek yellow stripe virus, leek rust; *Allium roylei* for downey mildew; *Allium fistulosum* for pink rot, smut and onion fly, and *Allium tuberosum* for *Fusarium oxysporum* and leek moth are particularly noteworthy.

Raphanus Species

Genus *Raphanus* includes *Raphanus caudatus, Raphanus landra, Raphanus lyratus, Raphanus pterocarpus, Raphanus raphanistrum, Raphanus raphanistrum subsp. Landra, Raphanus sativus, Raphanus sativus var. caudatus, Raphanus sativus var. longipinnatus, Raphanus sativus var. niger, Raphanus sativus var. oleiferus, Raphanus sativus var. oleiformis, Raphanus sativus var. sativus, Raphanus violaceus* and X *Brassicoraphanus* species.

Daucus Species

Genus includes several species like *Daucus aureus, D. bicolor, D. bocconei, D. brachiatus, D. broteri, D. capillifolius, D. carota, D. carota cv. atrorubens, D. carota subsp. carota, D. carota subsp. commutatus, D. carota subsp. drepanensis, D. carota subsp. Gadecaei, Daucus carota subsp. gummifer, Daucus carota subsp. hispanicus, D. carota subsp. hispidus, D. carota subsp. maritimus, D. carota subsp. maximus, D. carota var. atrorubens, D. carota var. boissieri, D. carota var. commutatus, D. carota var. sativus, D. crinitus, D. durieua, D. gingidium,* etc. In these genera two types *i.e.* Asiatic and European types are more popular.

Abelmoschus Species

Centre of diversity of *Abelmoschus* includes West Africa (Benin, Togo, and Guinea), India, and Southeast Asian countries, *i.e.*, Burma, Indochina, Indonesia, and Thailand (Chevalier 1940; Van Borssum Waalkes, 1966; Siemonsma, 1982a, b, Hamon 1988; Hamon and Hamon 1991; Bisht *et al.*, 1995). *Abelmoschus esculentus* is the commercially cultivated species of okra. In West Africa *A. caillei* is also cultivated for its fresh fruit. Geographical distribution of *A. caillei* is from Northern Guinean extending to Southeast Cameroon. *Abelmoschus esculentus* is cultivated in most humid parts of Guinea, mostly in an intercropped system with *A. caillei* (Hamon and Charier 1983; Hamon 1988). The genus *Abelmoschus* comprises nine species (IBPGR, 1991) and among them *Abelmoschus moschatus* and *A. manihot* are semi wild, and show a greater diversity than cultivated forms (Hamon *et al.*, 1991). *Abelmoschus moschatus* is morphologically and genetically the most different from other species (Hamon and Yapo, 1986) with probably the widest geographical distribution (Hamon and Charrier, 1983), which extends from the South Pacific Islands through Indochina and India and over to Central and Western parts of Africa. Cultivated form (*Abelmoschus moschatus* var. *moschatus*), better known as "*musk mallow,*" "*Jew's mallow,*" or "*ambrette*" in Africa, Asia and America has fragrant seeds used for making perfume (Hamon and Charrier 1983). The subspecies *biakensis* is endemic to Papua New Guinea (Charrier, 1984). Cultivation of *A. manihot* is mainly in Southeast Asia and is known under the vernacular name "*qibika*", which is also found in India, Northern Australia and less frequently in the American continent and tropical West Africa. Its cultivation is exclusively for leaf consumption. The species *A. tetraphyllus* (IBPGR, 1991) *A. ficulneus, A. crinitus, A. angulosus* (Van Borssum Waalkes, 1966), and *A. tuberculatus* (Pal *et al.*, 1952) are truly wild. *Abelmoschus ficulneus* is from the northern parts of Australia to Southeast Asia and from India to most parts of Africa. *Abelmoschus tetraphyllus, A. crinitus, A. angulosus* and *A. tuberculatus* are exclusively of Asian origin. *Abelmoschus tetraphyllus* var. *tetraphyllus* and var.

pungens are endemic to Southeast Asia with extension of the former to Papua New Guinea and New Ireland. *Abelmoschus crinitus* is found at low altitudes in India and Southeast Asia and *A. angulosus* at high altitudes in India, Sri Lanka, Indochina, and Indonesia. *Abelmoschus tuberculatus* is endemic to the northern and western parts of India (Charrier, 1984; Velayudhan and Upadhyay, 1994).

Based on differeznt Floras, Floristic accounts and other published evidence expressed main areas of concentration of wild species are as follows:

1. Semi-arid tracts of North and Northwestern India are for *A. ficulneus* and *A. tuberculatus.*
2. *Tarai* range and the foothills of the Himalayas are for *A. crinitus, A. manihot* and *A. tetraphyllus or A. tetraphyllus* var. pungens.
3. Western and Eastern Ghats and also peninsular tracts are for *A. manihot* (including *tetraphyllus* type), *A. angulosus* mainly confined to hilly tracts up to 2000 m in south India with only meager distribution elsewhere.
4. Northeastern regions are for *A. crinitis* and *A. manihot*, mostly var. *pungens.*

In some of wild taxa, intra-specific variation exists and is taxonomically identified as in *A. manihot* var. tetraphyllus and A. *pungens, A. angulosus* var. *purpureus* and *grandiflorus*, in *A. moschatus* var. *biakensis* and *rugosus*, and in *A. tuberculatus* var. *deltoidefolius.*

Augmentation/Collection of Vegetable Germplasm

There are various ways to augment vegetable germplasm.

Introduction/Collection from Abroad

During the past, introductions of new vegetable varieties in every crop are becoming very popular among farmers or being used in selection. Example can be cited in every crop in vegetable group where more than half the varieties grown in India are either direct introductions or selections thereof. Some of the notable introductions during recent years are bacterial wilt and fusarium wilt resistant lines of tomato from Taiwan, nematode resistant varieties Quinte and Early Rouge from Canada. Carrot variety Beta III has high carotene content. This is used by many human nutrition programmes in India to cope up Vit A deficiencies. In watermelon high yielding varieties tolerant to *Fusarium* wilt were introduced from USA. In muskmelon a multiple disease resistant line AC70-54 has resistance to Gummy stem blight, powdery and downy mildew. In cucumber also gynoecious line Gy-4 is introduced which is being used for breeding high yielding multiple disease resistance including mildews and *Fusarium* wilt.

In Chinese cabbage, heat tolerant and black rot resistant lines are introduced from Taiwan. In cabbage, lines resistant to club rot, black rot and heat tolerances have also come from Taiwan. In capsicum, TMV tolerant, nematode tolerance and high yielding varieties World Beater and Yolo Wonder have all proved useful introductions. In chillies, hot pepper lines from AVRDC, Taiwan are very successful at several locations in India. A number of accessions of cucurbit vegetables and their wild relatives were introduced from different countries. This includes *Citrullus lanatus* (8, USA), *Cucumis melo* (15, France, Japan), *Cucumis sativus* (15, Japan, USA), *Cucumis heptadactylus* (1 USA), *Cucumis metuliferus* (16 USA), *Cucumis anguria* (94 USA), *Cucurbita pepo* (2, USA), *Cucurbita species* (4, Algeria, Japan), *Luffa cylindrica* (1 Japan) and *Lagenaria siceraria* (1 Japan).

A number of vegetable germplasm were introduced from other countries are given in Table 5.4.

Table 5.4: A Few Promising and Useful Introduction

Crop Species	*Accessions*	*Country*	*Promising and Useful Attributes*	*Indentor*
Cucumber sativus	EC-399914-37	UK	Resistant to CMV, downy mildew	PAU, Ludhiana
	EC-398030	China	Early maturing determinate type, bears fruits in cluster	IIVR, Varanasi
	EC-398966-67	USA	Resistant to angular leaf spot	Sandoz India
	EC-398968-70	USA	Resistant to Anthracnose	Sandoz India
	EC-398971 73	USA	Resistant to fruit rot	Sandoz India
	EC-398974-90	USA	Resistant to downy mildew	Sandoz India
	EC-398991-9007	USA	Resistant to leaf spot	Sandoz India
Citrullus lanatus	EC-393240-43	USA	Small medium round fruits, red fleshed for breeding purpose	IIVR, Varanasi
Citrullus lanatus	EC-402549	USA	Multiple disease resistant to race-2, anthracnose, fusarium wilt and gummy stem blight with early maturity	—
Cucumis melo	EC-399866-212	UK	Resistant to downy mildew, powdery mildew and CMV	—
Cucumis melo	EC- 382726-36	France	Germplasm with male sterility genes MS1 to MS5, white fruit (gene), Fusarium wilt resistant (genes form-1, form-3)	IARI, New Delhi
Cucumis sativus	EC-382737-39	USA	Gynoecious lines	IARI, New Delhi PAU, Ludhiana
Cucumis species	EC-382500-69	USA	Different *Cucumis* species with nematode resistance	PAU, Ludhiana
Cucurbita pepo	EC-380995	USA	Unique, small, oblong early maturing, flesh comes out in strings, can be baked or used in breads	IIVR, Varanasi
	EC-380996	USA	Hull-less seeded pumpkin; seeds can be popped and eaten like snack, high in nutrition and good source of Zinc	IIVR, Varanasi
Citrullus lanatus	EC-380989-91	USA	Yellow fleshed water melon, early maturing, sweet crisp flesh	IIVR, Varanasi
	EC-382753	USA	Breeding line SSDL, resistant to Fusarium wilt and anthracnose	IIVR, Varanasi
	EC-378523-23	USA	Fusarium wilt resistance	IIVR, Varanasi
Bitter gourd	EC-399808			
Sponge gourd	EC-305586			
Ridge gourd	EC-284347			
Bottle gourd	EC-305378			
Palak	EC-284349			
Tomato	Sioux, Fireball, Marglobe, Best of All, Roma, Money Maker			

Contd..

Table 5.4–Contd..

Crop Species	Accessions	Country	Promising and Useful Attributes	Indentor
C. pepo	EC-516625-96			
Citrullus lanatus	EC-509468, EC-477972, EC-477973, EC-477974			
C. melo	EC-468986, EC-477977, EC-477978, EC-477979, EC-477980, EC-477981			
C. sativus	EC-497645, EC-497646, EC-497647			
Bell pepper	California Wonder, Yolo Wonder, Oshkosh, Ruby King, King of North, Early Giant, Chinese Giant, World Beater			
Onion	Texas Early Grano, EC-4144	USA	Mild flavoured, large sized, yellow, suitable for fresh use	
	Bermuda Yellow, EC-109123	Philippines	Non-bolting salad onion	
	Giza-6, Giza-20, EC-200936, EC-200937	Sudan	High yielding, large sized, dark red pungent types	
	Nu Mex Br-1, EC-169372	USA	Short day, resistant to bolting and pink root disease	
	Sweet Sandwich, EC-169373	USA	Dehydrator/Salad type	
Garlic	EC-244949, EC-244858	Taiwan	Large bulbs, purple cloves	
	EC-210991	Egypt	Compact white bulbs, bold cloves	
	EC-158250	Taiwan	Large light pink bulbs, bold cloves, Bold white cloves, field tolerant to purple blotch	
Sweet potato	EC-332805	Puerto Rico	Red skin and yellow flesh	
	C-12	Peru	Deep orange, semi erect, resistant to leaf scales, drought tolerant	
Okra	Ghana Red	Ghana (Africa)		
Radish	Japanese White, China Red, Rat Tail Radish			

Contd..

Table 5.4–Contd..

Crop Species	*Accessions*	*Country*	*Promising and Useful Attributes*	*Indentor*
Turnip	Snow Ball, Purple Top, White Globe			
Peas	Early Badger, Bonneville, Sylvia, Asauji, Arkel			
Cauliflower	Snowball-16, Improved Japanese			
Carrot	EC-178385, EC-187201, EC-27883 to EC-274886	Italy, USA, Nether-lands	High Vit.A High Vit.A Male sterile and restorer line, resistant to club rot	
Sem (Dolichos bean)	IC-16862, Pusa Bunch			

Collection Through Exploration

Emphasis was given for collection of cucurbitaceous vegetables from different parts of country in collaboration with NBPGR, State Agricultural Universities and ICAR Institutes. During the year, a number of crop specific explorations were undertaken to collect major and minor cucurbits. This includes bitter gourd, pointed gourd, cucumber, snap melon, kakrol, kartoli, pumpkin, sponge gourd and ridge gourd.

Table 5.5: Useful Indigenous Collections through Explorations

Tomato	HS-101, HS-102, HS-110, S-12, Punjab Kesari, Kuber, Kalyanpur T-1, KS-1, KS-2, Angurlata, CO-1, CO-2, CO-3
Brinjal	Pusa Purple Long, Pusa Purple Round, Pusa Purple Cluster, Pusa Bhairav, Arka Sheel, Arka Kusumkar, Arka Navneet, CO-1, MDU-1, Junagarh Long.
Hot pepper	NB-34 to NB-41, NP-46A to NP-51, Jawala, Pusa Red, Kalyanpur Red, Yellow Kalyanpur, Chaman, Pant C-1, Pant C-2, G-1, G-2, G-3, G-4, G-5

Explorations within India with International Collaboration

Following explorations were done within India with international collaboration and outside India for enhancement of vegetable germplasm (Table 5.6).

Explorations and Collections of Germplasm Outside India with International Collaboration

Germplasm collections are also made outside India under certain programme time to time (Table 5.7).

Status of Germplasm in India

The Indian gene center has rich diversity in vegetable crops. Main aim of NBPGR is collection, evaluation and maintenance of plant genetic resources of different crops. Not only this, it plays an important role in *ex-situ* conservation of crop genetic resources (National gene bank comprising long-term seed storage, in-vitro repository and clonal field gene banks). Conservation and sustainable management of plant genetic resources are cooperative endeavors and require active collaboration between NBPGR and the concerned national and international agencies. The National Bureau of Plant Genetic Resources, New Delhi from 1976 to 2005, has a total of 26,988 accessions of various vegetable germplasm collected from Indo-Gangetic plains, North-Eastern region, North-Western Himalayas, Eastern Uttar Pradesh, Bihar, parts of Maharashtra, West Bengal, Assam, peninsular region, tarai region in Uttar Pradesh, Chattishgarh, Madhya Pradesh, Vidarbha (MS) and tribal dominated belts in Central India, Orissa and adjoining Andhra Pradesh. The NBPGR New Delhi assembled systematically over 2900 germplasm lines of *Lycopersicon* including wild species from diverse agro-climatic zones of world. Major donor countries include USA, Philippines, U.K., Canada, Australia, Portugal, Italy, Argentina, Cyprus, Mauritius, Israel, Netherlands, Nepal, Tanzania, Turkey, Japan, Iraq, Chile, Yugoslavia, Denmark, Switzerland, China, Trinidad, South Africa, Czechoslovakia, Ghana, Germany, Hungary, Taiwan, Poland, Bulgaria, South America, Columbia, Hong Kong, Spain, United Arab Republic, Burma, Nigeria, New Zealand and Brazil. Realizing importance of *Solanum* species, explorations were undertaken in collaboration with IPGRI covering almost all brinjal growing areas of India and neighboring countries *viz.* Bangladesh, Nepal and Sri Lanka for collection of variability. Germplasm sampled from neighboring countries includes 175 accessions of eggplant and 27 other *Solanum* species from Bangladesh.

Table 5.6: Explorations (Multiple Crops) within India with International Collaboration

Collaboration/Scheme	*Areas Surveyed*	*Crops Collected*	*Collections*
PL-480 scheme	Different parts of india	Cluster bean	942
–do–		Legumes	8926
Indo-Canadian	Kumaon Himalayas	*Brassica* spp.	18
Indo-USAID collaborative explorations	Parts of Rajasthan, Madhya Pradesh and Uttaranchal	Cucumber	194
–do–		Snapmelon	236
–do–		*Cucumis* callosus	156
–do–		Snake cucumber (*C. melo* var. *utilissimus*)	24
–do–		Other cucurbits	48
IPGRI collaborative explorations	(i) North-eastern region) (ii) Parts of India, Nepal and Sri Lanka (iii) South Asia	Okra, eggplant and their wild relatives	4665

Table 5.7: Explorations and Collections of Germplasm Outside India

Country	*Crops Collected*
Central Asian Republics of the former USSR	Cassava, okra, winged bean, velvet bean, cucurbits, *Capsicum*, *Hibiscus* and tomato.
East Africa-Malawi and Zambia	Tuber crops, legumes, *Amaranths* and other vegetables
Kenya, Sudan and Ethiopia, Bangladesh	Eggplant

Table 5.8: Vegetable Germplasm Holding in National Gene Bank (as on 30-4- 2005)

Vegetables	*Number of Collection*	*Vegetables*	*Number of Collection*
Cucumber	638	Bitter gourd	973
Snap melon	697		
Musk melon	238	Kakrole	12
Kachri	348	Sponge gourd	852
Pumpkin	1280	Ridge gourd	628
Bottle gourd	1113	Satputia	19
Snake gourd	289	Ash gourd	439
Pointed gourd	324	Round gourd	64
Ivy gourd	154	Water melon	129
Musk melon	430	Kartoli	98
Tomato	1408	Brinjal	5029
Chillies	2573	French bean	1534
Section 1.03: C hardwickii	58	Jack bean	27
Rice bean	307	Lablab bean	904
Cluster bean	133	*Section 1.04: Canavalia*	18
Broad bean	123	*Section 1.05: Cowpea*	642
Garden pea	687	*Section 1.06: Onion*	697
Garlic	550	*Section 1.07: Colocasia*	874
Elephant foot yam	821	*Section 1.08: Potato*	247
Radish	175	*Section 1.09: Carrot*	32
Coriander	378	*Section 1.10: Fenugreek*	183
Cauliflower	141	*Section 1.11: Spinach*	133
Cabbage	20	*Section 1.12: Sweet potato*	27
Colocasia	440	*Section 1.13: Yams*	335

Procedure for Genetic Resources Exploration and Collection

A number of cultivated vegetables crops and their wild relatives possess enormous variability and genes resistant to biotic and abiotic stresses. It has a rich array of vegetable wealth both of indigenous and introduced types distributed in diverse ecological habitats and geographical regions. Due to long history of cultivation and continuous selection, several land races/primitive types suitable to different climatic and soil conditions are developed in various vegetable crops. Following steps are needed during exploration and collection of genetic resources.

Survey

In vegetables a large group of crops are annuals and propagated by seeds, while some are perennials and propagation is done by vegetative means. Principles for both groups of crops differ to some extent. Initially, survey is conducted for a particular crop in their region of diversity known as coarse grid survey and further intensive survey is undertaken from source locality for a particular genotype having desirable gene, which is known as fine grid survey. In coarse grid survey, a series of collections are made at a wide interval over the whole area. Further collection from the area, where interesting and intensive variation appear, the second expedition and intensive survey would then provide fine grid sampling by concentrating upon gene pool in this region. Hawkes (1976) is of the view that the above sampling strategy may be attempted wherever possible depending upon availability of time and resources. Before proceeding on expedition for gene pool sampling in vegetable crops a curator should have sound knowledge and understanding about the vegetable crops including their primitive/landraces/local types, their wild relatives, distribution of diversity, keen observation of variation in plants and environment, biotic and abiotic stress and their identifying key characteristics, understanding of gene pool concept, breeding system and population structure, coarse and fine grid sampling and propagation/seed multiplication technique.

Diversity Distribution and Variation in Adaptability Behavior

Evolution under domestication could be viewed at three different stages:

1. Domestication of wild species
2. Stabilizing semi-domesticated populations for production and productivity
3. Breeding for yield increase.

Plant domestication is induction of preferred taxa to, and their subsequent evolution in, man-made habitats. Most of the popular vegetable crop species have a long history of domestication yet wild species are till date exploited from their natural habitats particularly in tribal belts *e.g.* ferns, potherbs, roots and tuber crops. Adoption of some lesser-known vegetable crop species is also being taken up in different agro-ecological zones *e.g. Dioscorea* spp., *Colocasia* spp. and other yams in North-East India and coastal Kerala and *Houttuynia* spp. in Himalayan belt.

Basic changes that occur in a species on domestication could be due to selective harvesting and re-sowing as well as specific cultivation package/s in man made habitats. Harvesting automatically favours plants with the highest yield of economic plant part *e.g.* bigger raw fruit size, shape or colour of fruit/juice and leaves in different vegetable crops. Re-sowing with seed from specific plants automatically increases the fitness of these phenotypes. Selection favours uniformity in plant stand, increase in vigour particularly at seedling stage and loss of seed dormancy. Above all, there is a definite increase in both production and per unit area productivity due to farmer level selection in semi-domestcated types (de Wet, 1977).

Mechanism underlying increase in adaptedness, uniformity in crop population and increase in yield is associated with development of clusters of inter-related alleles at different loci and gradual amalgamation of cluster into large synergistically interacting complexes. Adaptedness of variable forms *i.e.* types of taxon and land races of primitive cultivars has been manifested as their superior ability to live and reproduce in respective environments.

Genetic variation and variation for adaptability of forms within species provide unlimited resources both for sustainable production and crop improvement. Recent theories on handling of plant genetic resources, including delineation of core subsets (Rana, 1992), give due weight age to both geographical and genetic background of accessions for their inclusion into reprehensive subset, out of large collection.

Table 5.9: Specific Adaptability of Various Vegetable Crop Species

Crops	*Area/region*	*Remarks*
Common bean	High altitudes of Himachal Pradesh and Western Arunachal Pradesh	Large variability
Lablab bean	Tripura	Large variability
Tree bean (*Parkia roxburgai*)	Mizoram, Manipur, Nagaland	Local preference
Horse-radish	Deccan Plateau	Local preference
Potato	Lahaul Valley	High yield Av. Tuber wt. 1 kg
Solanum gilo	North Eastern Region	Introduced from Africa and naturalized
Brinjal	Bundelkhand (U.P.) and Tripura	Primitive types available (Bholanath and Sheonath types)
Tomato	NEH (Meghalaya)	Cultivated in paddy lands from February to April
Bell pepper	Mezoram	Variability for fruit colour (white, red and Black)
Sweet Potato	Tripura	Large variability and Cross-4 highly successful
Radish	Meghalaya and Arunachal Pradesh	Radish Sel.1 highly successful
Red pepper	Assam	Baghi chillies, said to have high capsicine content (only medicinal use)
Chow-chow	Mizoram, Karnataka, Maharashtra	Highly naturalized
Bitter gourd	Tamil Nadu, Kerala, U.P. Bihar, West Bengal, Maharashtra	Highly specific adaptability
Kakrol	Mizoram, Tripura, West Bengal, Bihar, Vindhya Hills of U.P.	A specialty favourite with natives
Water melon	Rajasthan, Punjab, Haryana, Western U.P., Karnataka, M.P.	Near river beds only, develop more sweetness in arid zones
Musk melon	Rajasthan, Eastern U.P., Punjab	Develop more sweetness in arid zones
Buffallo gourd	Rajasthan	Develop more starch in roots and provide foliage, fruit (flesh contains more ware) for animals

Adaptability to Biotic and Abiotic Stresses

Wilsie (1962) reported distribution of climates, soil types, migration, introduction and selection as a few of the factors controlling plant distribution in post-domestication period. For example, although cowpeas are long day plants and beans are short day plants yet farmers' selections in different agro-climatic niches have stabilized various day neutral forms of both cowpeas and beans. Similarly, in onions long day types are available and relatively insensitive types are grown for bulbs in lowlands (Shanmugavelu, 1989).

Gableman and Gerloff (1978) reported occurrence of genetic variation for efficiency in micronutrient utilization in vegetable crops. Tolerance to salts and acidity or alkalinity also critically affect adaptability and evolution of forms, thereby making availability of diverse germplasm. Crops like garden beet, spinach, *Atriplex* spp. and asparagus have high salt tolerance whereas radish and beans have low tolerance to salts. Similarly, cucurbits prefer a soil pH between 6.0- 7.0 and no cucurbit can successfully grow below a pH of 5.0. Yet among the group, muskmelon is slightly tolerant to acidity while cucumber, pumpkin and squashes fall under moderate category.

Evolution of pathogen virulence has specific significance in relation to history of cultivation particularly mixed cropping where host forms are likely to have polygenic complexes for horizontal resistance. Thus screening and testing of diverse germplasm, including low yielding cultivars and landraces is important to get diverse genetic base particularly for resistance or tolerance to biotic and abiotic stresses.

Planning of Exploration

Broadly two kinds of explorations are formulated based on priority of crops, area/region to be surveyed. Both, crop specific and region specific/multi crop collection are advocated depending on situation. Crop specific explorations are undertaken as per needs of breeders. However, multi crop specific explorations are undertaken to cover probity areas, which hold rich crop genetic diversity in various vegetable crops, and maturity period, coincided at same time. Exploration is carried out by explorer, crop specialist/researcher, extension worker, and by botanist. Duration of exploration varies according to mission; within the country it is about 20 to 45 days whereas in foreign country larger duration is required, as otherwise it will be more expensive and less remunerative.

Selection of Site

Collection sites indicate the place of sample collection. This includes cultivated fields, natural habitats (mountains, valley, river sites, sea shore and forest), local village market, threshing floors and farmers seed store. In case of annual vegetable crops, the collection site is individual field or farm (if farmers use different seed stock) or a group of field or farms (if farmers use common seed stock). Further collection site may be reduced if crop is self pollinated and increased if the crop is cross-pollinated and showing high genetic diversity. In case of wild and weedy or weedy form relatives the selection of appropriate sampling site to collect diversity is more difficult. Species are often continuously distributed in nature or they follow island model or stepping stone model (Jain, 1975). In such cases, there are no artificial boundaries defining discrete, relatively homogenous populations as in cultivated crops. Thus Marshal and Brown (1975) suggested that plant explorer must decide where, in the continuum he/she will collect and total size of target area to be sampled (*e.g.* 1000, 10,000 or 100,00 m^2). Gene flow through pollen and seed dispersal is extremely limited in wild vegetable species. Natural population of wild species often shows marked genetic differences over small distances (Marshal and Brown, 1975).

Specific Area of Collection

Tribal dominated region inhabited by different ethnic groups provide another interesting situation for gene pool sampling of primitive types of vegetable germplasm. Such areas often possess rich diversity of various vegetable crops. The tribals hold primitive cultivars of different vegetable crops and usually do not exchange seeds between ethnic groups. In tribal areas particularly in North Eastern parts of Madhya Pradesh, Chhatisgarh, West Bengal, Bihar and Orissa, village market/haat are the most desirable place to collect diversity in a number of vegetable crops. Areas that remain to be covered include remote and less accessible areas, hilly terrain and tribal dominated belt.

Number of Sites

Number of sites depends on availability of variability and pollination nature (self and cross pollinated) of crop. However, each site provides opportunity for sampling different set of alleles, if no information on distribution of variation is available. Number of sites often depends on length of season, relative abundance of target species and roughness of terrain. Changes in eco-habitats/niches and farming system should also be taken into consideration for deciding number of sites. Optimum number of sites for sampling for a given species could be fifty.

Sample Size and Sampling

Optimum sample size per site would be number of plants required to obtain, with 95 per cent certainty that all alleles at a random locus occurring in target population with frequency greater than 0.05. Hawkes (1976) suggested that bulked seed samples from up to 50 per cent plants (minimum) should be collected from one site. However, this becomes difficult particularly in crops like brinjal and cucurbits, where availability of this much number of fruits is very rare at a site. Further in case of cross pollinated vegetables each sample size should be of 10,000 to 12,000 seeds and in case of self fertilized crop sample size should be of nearly 8000 seeds.

Random Sampling

Random sampling is usually practiced to collect maximum genetic variability of species, irrespective of relative frequency or rarity of any genes or linked genetic complexes. For this, at each collection site, single spike or a few pods/fruits are randomly collected at every second or third place along with a number of transacts through the crop at least at 50 places scattered over an area or samples are taken in clustered form. Hawkes (1976) suggested that bulked seed samples from up to 50 individuals and certainly not more than 100 should be collected randomly as multiple samples at one site. Sample should be clustered and cluster should be spread as evenly as possible throughout collecting area. Sampling technique may be appropriate in case of wild and weedy species.

Biased Sampling

Random sampling is not feasible in many of vegetables like brinjal and most of the cucurbits because farmers keep a few selected fruits only for seed purpose and many times ripened fruit is not available in field. Under this situation, explorer sticks to biased sampling. Harlan (1975) strongly pleaded about subjective, biased sampling. For biased sampling, crop specialist category of explorer is desirable because he has to be aware of morphological and physiological characters of the crop to be sampled. These characters include pattern, rate and habit of growth (Fordham, 1971) and seed source in relation to quality and germination requirement (Flint, 1970). Sampling of specific genotype forms a known source and also forms an important part in biased sampling.

Conservation Strategy of Vegetable Genetic Resources

Rich heritage and ethinic culture have favoured to preserve the richest diversity including rare landraces/primitive types of useful vegetables like eggplant, cucumber, ridge and sponge gourd and a number of root and tuber crop species. A number of vegetable crops were brought to India from other regions by travelers, invaders like Persian, Turkey, Moughals, Portugese, Dutch, French and British which acclimatized and developed good amount of diversity. Wild relatives/species of the important vegetable crops have commercial importance to a great extent. Most of these wild species grown in natural habitats possess genes resistant to biotic and abiotic stresses. However, diversity for valuable genetic resources is threatened in recent times. Therefore, conservation of vegetable genetic wealth particularly their wild relatives are thus essentially required for future utilization.

The conservation of PGR involves two basic strategies as per objective and priorities:

1. *In situ*
2. *Ex situ*

In situ

In situ conservation of PGR has to be carried out on farm, natural habitats/wild or landraces. Locally adapted materials are cultivated, utilized and conserved as part of traditional farming systems by paying due regard to natural ecosystems. Main aim of *in situ* conservation is to allow populations to maintain/perpetuate itself within the community environment, to which it is adapted so that it has potential for continued evolution (Anonymous, 1989 a,). Landraces/primitive cultivars developed in traditional agriculture system also need *in situ* conservation in their respective areas. These farming systems are of particular importance in maintaining local genetic diversity and providing food for local consumption and local markets. It will provide an option for further evolutionary changes likely to take place. However, for majority of the situation, *in situ* conservation is an ideal method of conserving wild plant genetic resources and perennial vegetables, which either do not set or set recalcitrant seed, do not produce plants true to type. *In situ* conservation of plant genetic resources has a number of advantages as compared to *ex situ* conservation (Anonymous, 1989 a).

1. It usually allows increased probabilities of conserving a large range of potentially interesting alleles.
2. It is specially adapted to species, which cannot establish or regenerated outside their natural habitats.
3. Allow natural evaluation to continue, a valuable option for conservation of diseases and pest resistant species, which can evolve with their parasites, providing, breeders with dynamic source of resistance.
4. It can serve several factors at once, since genepool of value to different sectors (crop breeding, production etc.) may often overlap and so can be maintained in same protected area.
5. It facilitates research on species in their natural habitats.
6. It assures protection of associated species of economic importance.

The 'National Man and Biosphere Committee' of Department of Environment has already identified sites as potential areas for biosphere reserve (Table 5.10). This covers all major biogeographic regions of the Indian subcontinent where flora and fauna can be conserved in natural habitats. These areas are identified based on their rich genetic diversity; floristic uniqueness endemic wealth of flora/ fauna and they are in totality representative of ecosystems occurring in different biogeographic regions.

Ex situ

Ex situ consevation requires collection and systematic storage of seeds/propagules outside natural habitats of species for short, medium and long term. *Ex situ* conservation PGR in the form of seeds in domesticated plant species for use in their genetic improvement was initiated with establishment of Imperial Agricultural Research Institute (IARI) in 1905 in village Pusa, Darbhanga district, Bengal (now in Bihar). In 1936, this institute was shifted to New Delhi. The conservation efforts of important PGR started in Division of Botany and later in the Section of Plant Introduction through frequent multiplication of seed materials. In 1976, Section of Plant Introduction was developed into National Bureau of Plant Genetic Introduction, which in 1977 was re-named as National Bureau of Plant Genetic Resources (NBPGR). Various possible approaches can be grouped in:

Plant Conservation

Botanical Gardens

India has more than 100 botanical gardens under different management systems located in different bio-geographic regions. Central and State Government manages 33 botanical gardens that maintain diversity in the form of plants or plant populations (MoEF, 1998).

Table 5.10: List of Biosphere Reserve, their Location and Demarcation of Area

Biosphere Reserve	*State (s) Location*	*Area (sq.km)*
Kerala	Tamil Nadu, Karnataka, Nilgiri	5,520
Namdapha	Arunachal Pradesh	4,500
Nanda Devi	Uttar Pradesh	2,000
Uttarakhand	Valley of Uttar Pradesh and Uttaranchal	3940
North Island of Andamans	The Andaman and Nicobar Island	1,375
Gulf of Mannar	Tamil Nadu	10,500
Kaziranga	Assam	7,600
Sunderbans	West Bengal	*
Thar Desert	Rajasthan	*
Manas	Assam	600
Kanha	Madhya Pradesh and Chattisgarh	*
Nokrek (Tura range)	Meghalaya	80
Rann of Kutch	Gujarat	5,000

* In the process of demarcation and delineation.

Source: Anonymous, 1987.

Arboreta

An arboretum generally refers to a place established for conservation of tree species or vegetatively propagated crops. Regional Plant Resource Centre at Bhubaneshwar established an arboretum with 1,430 species of tree, a palmeretum of 100 different types of palms, a bambastum with 61 collections of bamboo and an orchidarium housing 220 species of orchids. At the national level, interest in establishment of arboreta is very weak probably because of the cost involved.

Herbal Garden

This generally refers to gardens that predominantly conserve herbs, shrubs that are of medicinal and aromatic value. Concept of herbal garden has been picked up by non-governmental organization (NGOs) in India. Several non-governmental organizations in different parts of the country, particularly in tribal areas in Gujarat, Karnataka, Maharashtra, Madhya Pradesh, Chhattisgarh and Uttarakhand established herbal gardens with objective of conservation of local biodiversity in medicinal and aromatic plants and other economically important species. At community level, effort for such herbal gardens may increase income of local communities.

Field Genebank

Most institutions dealing with perennial or vegetatively propagated domesticated plants have field repositories for conservation of PGR.

Clonal Repositories

It includes clonally propagated crops.

Seed Conservation

Low Temperature Storage of Orthodox Seeds (Seed Gene Bank)

Seed genebank is responsible for conservation of seed accessions on long-term basis as base collections for posterity. It has 12 storage modules with a capacity to hold about one million accessions including vegetable at –20°C. The present base collection holding of vegetables in the National Gene Bank (NGB) is 15032.

Cryopreservation Storage of Orthodox Seeds and Recalcitrant (embryonic axes) Seeds

Cryopreservation is storage of biological samples in viable conditions at ultra low temperature of liquid nitrogen at –150 to –196°C. A total of 5328 accessions belonging to 502 species representing all major crops, have been cryopreserved at moisture content of 5-8 per cent in vapour phase of liquid nitrogen. Investigations are on to develop protocols for cryopreservation of more recalcitrant seed species and establish their base collections. As an alternative complementary method, attempts are also being made to cryopreserve pollen in case of trees or vegetatively propagated species.

In vitro Genebank Conservation

Includes (a) Conservation of cells, tissues, organs in glass or plastic containers under aseptic conditions through slow growth of cultures, (b) Cryopreservation of cultures (tissues, organs, pollen or cultures in liquid nitrogen at –150 to –196°C.). In India, to undertake conservation of vegetatively propagated species, a National Facility for Tissue Culture Repository was established at NBPGR in 1986 with financial assistance from Department of Biotechnology and since then the infrastructure has been further strengthened. The *in vitro* repository holds 1,396 accessions belonging to 124 vegetatively propagated plant species maintained for short term and medium term storage.

DNA Conservation

The basic objective in conservation of PGR is conservation of genetic diversity existing in the form of a functional unit called 'gene'. The whole genome in the form of genomic library or a sequence of DNA in the form of DNA library may be conserved following appropriate DNA conservation method.

Each technology is selected based on its merits in terms of utility, security, complementary and advantages over others.

National Active Germplasm Sites

The 40 National Active Germplasm Sites (NAGS) are holding active collections of relevant crop species. Eleven of them have medium term seed storage facilities in the form of cold storage modules maintained at 4°C and 35-40 per cent relative humidity. Active collections are basically for sustainable use in various crop improvement programmes. The NAGS are based at premier crop or crop group-based institutes. The crop-based institutes have multi-disciplinary team of scientists, better equipped for evaluation of germplasm for agronomic potentials and various yield reducing factors to identify accessions with desirable traits. This helps to generate information on potential value of various accessions to provide needed support for their utilization in breeding programmes. The National Active Germplasm Site (NAGS) for vegetable crops is Indian Institute of Vegetable Research, Varanasi (U.P.).

Germplasm Registration

The Indian Council of Agricultural Research (ICAR) has provided a mechanism for registering potentially valuable germplasm under 'Registration of Plant Germplasm' at NBPGR, New Delhi. A number of vegetable germplasm are being registered for their unique traits. The validity of registered germplasm is fixed in term of duration *i.e.* 15 and 18 years for clonally and seed multiplied vegetable crops, respectively.

Table 5.11: Registered Germplasm of Vegetables

Crop	*Line*		*Unique Traits*
Pointed gourd	IIVR PG-105	INGR-03035	Seedless fruits, obligate parthenocarpic, long duration fruiting (IIVR, Varanasi)
Bitter gourd	GY-63	INGR-03037	Gynoecious line with high yielding ability and attractive fruits (IIVR, Varanasi)
Water melon	RW-187-2	INGR-01037	High high yielding ability with yellow coloured flesh (ARS, Durgapura, Jaipur)
Water melon	RW-177-2	INGR-01038	Leaf mutant with simple unlobed leaves (ARS, Durgapura, Jaipur)
Bottle gourd	Androman-6	INGR-99009	Andromonoecious sex form of bottle gourd (NDUA&T, Faizabad)
Bottle Gourd	PBOG-54	INGR-99022	Segmented leaf genotype
Bottle Gourd	Cut Leaf		GB Pant Univ. of Agril. and Tech. Pantnagar
Potato	MP/99-322	INGR 04109	High starch and dry matter, low amylose and resistant to late blight
Chillies	MS1 and B line	INGR-04052	CGMS line with good GCA (IIHR, Bangalore)
Chillies	MS3 and B line	INGR-04053	CGMS line with good GCA (IIHR, Bangalore)
Chillies	IHR14 (PMR-14)	INGR-04054	Fertlity restorer line (IIHR, Bangalore)
Chillies	IHR-11-2 (IHR11#2)	INGR-04055	Fertlity restorer line (IIHR, Bangalore)
Chillies	MS-2A and 2B	INGR-03077	CMS line
Chillies	Capsicum Selection-2 (Nishat-1)	INGR-00001	High yielding with early maturity, export quality, superior shelf life and high nutritive value
Chillies	DRLT-1107	INGR-03092	High pungency
Brinjal	Hisar Jamuni (H-9)	INGR-99037	Intermediate, semi spreading, medium tall thornless, better ratoon for spring crop, fruit colour is retained for a long time less seed content, serves purpose of long and round types.
Brinjal	IIHR-3 (96-2-1)	INGR-03074	Resistance to bacterial wilt
Cucunber	AHC-2	INGR-98017	High yielding and long fruited types
Cucumber	AHC-13	INGR-98018	High yielding, unique, small fruited types, drought hardy, bear fruits even under high temperature in arid region with liomited irrigation
Kachari	Kachari AHK-119	INGR-98013	High yielding and drought hardy
Fenugreek (Methi)	Hissar Methi-346	INGR-01012	Downy mildew resistant genotype with quick germination, faster initial growth, long pod, bold seed with green tan seed coat colour and light green narrow leaves
Fenugreek (Methi)	Hissar Methi-350	INGR-01012	Powdery mildew resistant and Downy mildew tolerant genotype. Plant is tall erected; multi branched with whitish green stems and dark green narrow leaves.

Contd..

Table 5.11–Contd..

Crop	Line		Unique Traits
Peas	PMR-27	INGR-03078	Tendril type with resistant to powdery mildew
Peas	DPP-62	INGR-03079	Yellow wrinkled seeds with resistance to powdery mildew
Round melon	HT-10	INGR-99038	Intermediate semi spreading vine type roundmelon, tolerant to downy mildew and root rot wilt complex.
Snapmelon	AHS-10	INGR-98015	High yielding and drought hardy
Snapmelon	AHS-82	INGR-98016	High yielding and drought hardy
Tomato	Hisar Lalit	INGR-03036	Resistant to root knot nematode
Tomato	IIHR-2195	INGR-03076	Resistant to tomato leaf curl virus and bacterial wilt
Soybean	P-1366	INGR-01035	Vegetable type

Exchange of Germplasm

Ultimate goal of plant introduction to any region of world is use of germplasm to its fullest potential. In earlier days, agencies for plant introduction activities were travelers, pilgrims, invaders, explorers and naturalists. Movement of plants within old world countries was possible much earlier because of geographic contact compared to exchange of plants between new world and old world. Owing to these factors, the old world crops, *viz.* wheat and soybean were unknown in America about 400 years ago of which America is now a major exporter. Similarily, new world crops like chillies, and sweet potato introduced during the seventeeth century in India, show much more diversity due to better adaptations to Indian conditions.

After monumental work of N.I. Vavilov (1926), a Russian Botanist and explorer, concept of 'centres of origin' was analyzed, understood and recognized. Organized plant explorations and collections were undertaken jointly by different countries to these centers and the material collected was brought into their areas of collection and evaluated for various attributes. Spread of such findings created interest in plant breeders all over the world to acquire such materials for their utilization as primary or secondary introductions which further helped in the movements of crop plants (Dadlani, *et al.*, 1981). Classical example of introduction of new crops that helped other countries more than the place of origin of such plants are introduction of sunflower to USSR from central Mexico, USA, Chinese soybean to North America, Ethiopian coffee to Central and South America, Bahian cocoa to West Africa, Amazonian rubber to Malaysia, Asiatic yam into tropical America and Africa, wheat from Near East to USA; maize, tomato, potato and sweet potato from South America to Europe and Asia.

Procedure for Exchange of Germplasm

The NBPGR has brought out a brochure ' Guidelines for the exchange of seed/planting material' which is widely circulated amongst scientists in India. Guidelines for import of seed/planting material for research purpose are revised (Anon, 1989) in view of the enactment of new Seed Development Policy by Government. of India, which has also been Circulated amongst scientists. The Government has made the issuance of import permit mandatory for import of all seed/planting material for research purposes.

Import of Germplasm

All requests for indenting germplasm from abroad are to be made to NBPGR in prescribed application proforma giving specific details of required material stating sources/country as well as address of organization, so that the 'Import permit' is issued and sent to concerned scientist (s) for sending the same to the donor organization.

Concerned scientist/organization abroad is advised to take into consideration following requirements for mailing the materials to India.

Only healthy, viable and clean seed materials (free from soil, pests, pathogens and weeds) are to be forwarded without any seed treatment so as to facilitate proper quarantine examination. It may, however, be fumigated, if considered necessary.

The material is required to be accompanied with the 'Import permit' (which is to be sent to them alongwith our request letter) and phytosanitary certificate with additional declaration, if any, based on crop inspection certifying that the material is free from particular pathogen(s)/insects (s). It is to be further ensured that the package of seed/planting material must be addressed to the Director, NBPGR New Delhi who has to take delivery of the seed/planting material and conduct quarantine examination.

Export of Germplasm

Exchange of germplasm involves not only introductions but also supply of seed and other materials to collaborating scientist/organizations abroad. Export of vegetable germplasm is also made on basis of request recived by Bureau/ICAR Institutes/Agricultural Universities in India under various protocols/work plans/memorandum of understanding with different countries/CGIAR Institutions. This includes *Cucumis melo* (1), *Cucurbita pepo* (4), *Cucumis sativus* (1), *Luffa cylindrica* (1) to Italy; *Momordica charantia* (1) to Taiwan; *Sechium edule* (2) and *Triochosanthes dioica* (1) both to USA. During 2001-2003 a few cucurbitaceous germplasm were exported under phytosanitary certificate to other countries like cucumber (1), bitter gourd (1) Ridge gourd (1) and sponge gourd (1) to Iraq.

Table 5.12: Export of Germplasm

Crop	*Number of Samples*	*Country*
Cucumis melo	6	DPR Korea
Citrullus lanatus	1	DPR Korea
Cucumis sativus	1	DPR Korea
Momordica charantia	1	DPR Korea
Citrullus lanatus	1	Bangladesh
Cucumis sativus	1	Iraq
Lagenaria siceraria	1	Bangladesh, Iraq, Taiwan

Following guidelines are observed while responding to such requests:

1. Request for seed/planting materials received from concerned organizations/agencies are to be forwarded to NBPGR with relevant information so that prompt action on the supply of desired material could be taken.
2. Dispatch of seed/planting material is to be channelised through the Bureau so that prompt inspection of material could be done from quarantine angle and phytosanitary certificate issued.
3. No seed dressing with insecticides or fungicides is given while dispatching seed to the bureau.

Plant Quarantine

During introduction of germplasm, there is possibility of transfer of associated pests and diseases from one region/country to another as all types of planting materials are affected by diseases or injuries caused by one or more organisms *viz.* insects, nematodes, fungi, bacteria and viruses. It also results in distribution of several important pests and diseases from one country to another. Movement of germplasm has exposed the world to greater risks of introducing exotic pests and pathogens to 'clean areas'. If unhealthy looking plants are left out, latent or hidden infestation is likely to be collected, specially in case of vegetatively propagated organs like bulbs, tubers and rhizomes. Another danger is that some of the exotic pests and pathogens often find more favorable conditions or more suitable host plants for their growth and multiplication in new environment than are available in their original habitat. Problems of dissemination of pests and diseases through affected plant materials are further aggravated because the material can be sent across the world in a matter of a few hours, thus helping them to cover long distance in viable conditions.

Plant quarantine can be defined as "legal enforcement of measures aimed to prevent pests and pathogens from spreading, or to prevent them from multiplying further in case they have already found entry and have established in a new area". Though Plant quarantine measures may not guarantee an everlasting protection against the entry of exotic species but will certainly check or delay the introduction of these unwanted organisms and their subsequent establishment in hitherto clean areas.

Indian Plant Quarantine Regulations and Set-up

Realising importance of plant quarantine, Government of India legislated an Act in 1914, known as 'Destructive Insects and Pests Act of 1914' (DIP Act) under which various notifications were issued, restricting plants and plant materials from other countries as well as from one state to another state within the country (domestic quarantine). The latest regulation enacted under the DIP Act is the 'Plants, fruits and seeds Order, 1989 (PFS). This was necessitated to cater to needs of 'The New Policy on Seed Development' (NPSD) of Govt. of India which came into force on 1st October, 1988 with the objective to make avaible to Indian farmers the best genetic materials in the world to increase our agricultural productivity and to encourage the private sector seed industry in India not only to fulfill domestic requirements but also to develop export potential. While liberalizing import, care has been taken that there is absolutely no compromise on plant quarantine requirements. Although there are several requirements under the FPS Order, 1989, but for our purpose the most important requirements are:

1. No consignment shall be imported into India without a valid import permit issued by the competent authority. For importing germplasm Director, NBPGR is authorized by Govt. of India to issue import permits, both for Govt. institutions and private seed companies. For bulk consignments, the import permit is issued by Plant Protection Adviser to the Govt. of India.
2. No consignment shall be imported unless accompanied by an official phytosanitary certificate by an official of exporting country.
3. Seeds/planting materials requiring isolation growing under detention shall be grown in an approved post-entry quarantine facility.

   Import of soil, earth, sand, compost, and plant debris accompanying seeds/planting materials shall not be permitted.
4. Hay, straw or any other material of plant origins shall not be used as packing material.

Safe Movement of Vegetable Germplasm

There are glaring examples of losses caused by insect-pests and diseases introduced with import of planting materials like seed, propagules, fruits and plants. Dreaded golden nematode of potato introduced from U.K. has already infested entire Nilgiri hills. Of the several introduced exotic insect pests, spiraling whitefly, *Aleurodicus disperses* is quite recent (Russel, 1995) which has established successfully in Karnataka and found infesting 253 plant species. Introduction of exotic weeds such as *Lantana camara*, in early 19th century from Central America, *Parthenium hysterophorus* from Central and South America and *Phalaris minor* from Mexico in mid 20th century into India have become endemic sources of threat to our crop production and environment.

Table 5.13: Introduction of Insects/Nematodes

Insects/nematodes	*In Association with*	*From*
Acanthoscelides obtectus	*Cajanus cajan*	Brazil and Columbia
	Phaseolus vulgaris	Italy and Nigeria
Anthonomus grandis	*Gossypium* and *Vigna* species	USA
Ditylenchus dipsaci (nematode)	*Allium cepa*	UK
Heterodera schachtii (nematode)	*Beta vulgaris*	Germany, Denmark, Italy
Globodera rostochiensis (Golden nematode)	*Potato*	–

Table 5.14: Important Vegetable Diseases Believed to have been Introduced to India

Host	*Diseases*	*Pathogen*	*First Record in India*
Beans	*Fuscans* blight	*Xanthomonos phaseoli var. fuscans*	Pantnagar, Shimla, Solan, Delhi, 1972
Beans	Common blight	*Xanthomonos phaseoli*	Poona, 1972
Cabbage and cauliflower	Black rot	*Xanthomonos campestris*	Delhi, 1968
Chillies	Leaf spot	*Xanthomonos vasicatoria*	Maharashtra, 1950
Onion	Downy mildew	*Peronospora destructor*	Jammu and Kashmir, 1977
Potato	Late blight	*Phytophthora infestans*	1983
Potato	Wart	*Synchytrium endobioticum*	West Bengal, 1953

Evaluation, Promotion and Utilization of Vegetable Germplasm

Precise evaluation and documentation of germplasm are pre-requisites for their utilization. To promote use of promising germplasm/genetic stocks identified, evaluation of germplasm and conduct of field days should be organized besides publishing research articles and crop catalogues (printed and electronic). Development of core set of collection particularly in the crops having large germplasm collection can be a powerful tool for promoting utilization of germplasm. It is also important to accord recognition to those associated with development of improved/unique germplasm and genetic stocks, like plant breeders, farmer-breeders or other developers/innovators.

Characterization, preliminary evaluation and further evaluation of germplasm are prerequisites for utilization of plant genetic resources. Until characterization is not done, and attributes/traits are not known, germplasm has little practical utility. Germplasm evaluation, in broader sense, provides description of materials in a collection. Characterization and evaluation are normally the responsibility of crop curator, while further characterization and evaluation should be normally be carried out by plant breeders. Evaluation of genetic resources is a multi-dimensional endeavour; involving various disciplines like cytology, agronomy, genetics and biochemistry etc. The multidisciplinary participation to generate wide spectrum of information of germplasm collection would lead to its meaningful utilization. Various steps in plant genetic resource handling beginning from augmentation to seed increase, characterization, preliminary evaluation, detailed evaluation, utilization, regeneration and documentation need descriptions for practical purpose, as given below:

Characterization and Preliminary Evaluation

This consists of recording a limited number of additional traits, thought desirable by consensus of specialists/users of particular crops/species, which help in identifying useful germplasm. Evolution data refer to environmentally influenced characters. Important ones include site data; data on plant, leaf, flower, fruits, seeds and reaction to pests and diseases. Passport data are data collected by curator including origin of sample or it's known history. This is very important for identification and help in designating core collection, identifying duplicates and planning future exploration. Preliminary evaluation consists of recording data on a limited number of agronomic traits thought desirable by a consensus of research workers of particular crop. Traits, in general, have quantitative inheritance and, are influenced environmentally. Thus, to better assess their perfotrmance, the germplasm should be evaluated in the agro-ecological region from where the accessions are collected or in a similar environment. There is no prescribed limit for number of accessions in a trial planted for preliminary evaluation. Some trials may have many hundreds of accessions, each in small and perhaps single line plots. To minimize the inter-genotype competition, accessions should be arranged according to their height, growth habit, maturity and other such traits.

Large number of accessions in a trial usually needs a compromise between statistical and practical consideration. If these are ever in conflict, then practical consideration should take precedence over the statistical ones. In general, preliminary evaluation trials undertaken by curators have two features in common that set them apart from other field experiments (IPGRI, 2001).

Advanced Evaluation

Consists of recording a number of additional characters though to be useful in crop improvement. These include important agronomic traits, stress tolerance, disease and pest resistance and quality characters.

Observations are recorded on qualitative and quantitative traits. Qualitative traits include morphological, physiological and biochemical characters related to survival and scored using a number of checks to be determining variation within and among the traits. Quantitative traits are subject to environmental factors and are responsible for adaptation and productivity. These include productivity, quality components, resistance to diseases and pests as well as tolerance to adverse conditions or stress.

Much emphasis is presently given on multi-locational and multi-disciplinary approach of germplasm evaluation. For preliminary evaluation, locally adapted cultivars should be used as check and screening for specific diseases under controlled conditions or at hot spots should be carried out. Augmented design is invariably used due to large number of germplasm holdings under evaluation. Care should be taken to minimize natural cross-pollinated contamination and erroneous labeling. While regenerating, care should be taken to preserve original structure and productivity of accession. Main dangers are due to inappropriate handling of materials during sowing/planting, harvesting, threshing, cleaning, sub-sampling, packing and labeling. Frequent regeneration should also be avoided by producing sufficient seeds during initial seed increase and conservation in medium term storage in case of working collection and long term gene bank as base collection.

Table 5.14: Evaluation and Characterization of Some Vegetables

Crop	*Evaluating Center*	*Promising Accessions*
Bitter Gourd	NBPGR, New Delhi	High yielder-UB-22, UB-66, IC-74158, C-1621, K-3374
Bitter Gourd	IIVR, Varanasi	High yielder-IC-44428B, NIC-72285, IC-85604A, IC-85608B, IC-85611, IC-85636, V-89/0-103C, VRBT-29, VRBT-50
		High number of fruits/plant-Ic-45364A, VRBT-78, Ic-44426D, IC-44428B, IC-45350, NIC-72285, IC-85611
		CMV Resistance-IC-44428B, IC-72285, IC-85608B, V-89/0-103C, IC-85611, IC-85636, VRBT-50
		Fruit fly Resistance-IC-85641 and V-89/0-103C
Bottle Gourd	NBPGR,New Delhi	High yielder (Summer)–U-10-315, U-11-10, U-8-217, U-8-46, U-8-198, BDJ-864
		High yielder (Kharif)–U-11-216A, U-11-50, BDJ-53, BDJ-783, BDJ-896, EC-187247, BDJ-616
Bottle Gourd	IIVR, Varanasi	High yielder-U-10-316, IC-92363, VRBG-2, VRBG-4
		Red pumpkin beetle resistance-U-10-316, NIC-1225
		Leaf miner resistance-U-10-316, IC-92363 IC-92418
		Downy mildew resistance-U-10-316, IC-92362
Ivy Gourd	IGKVV, Raipur	High yielder-(3.20-21.21kg/plant)-Ac-59, 35, 54, 16, 6, 48, 10, 11, 14, 34, 52, 24, 17, 36, 2, 4, 31, 29
		Number of fruits/plant (209.33-1292.7)-Ac-35, 59, 11, 5, 16, 54, 24, 17, 10, 51, 61, 22, 58, 60, 53, 32, 31, 63, 37, 23
		Length of internodes (7.79-11.43cm)-Ac-17, 61, 5,14, 51, 6, 11, 15, 37, 1, 4, 29, 32, 36, 22, 2, 9, 55
		Diameter of stem (3.30-4.96)-Ac-4, 51, 63, 15, 36, 6, 11, 59, 61, 24, 62, 17, 16, 7, 32, 52, 58, 32, 2, 54, 5, 23, 4, 22, 9, 60, 29,
		Length of fruit (3.97-7.60)-Ac-6, 1, 62, 11, 33, 54, 61, 31, 59, Diameter of fruit (1.63-3.20)-Ac-59, 52, 31, 14, 35, 37
Ivy Gourd	IIVR, Varanasi	High yielder-VRK-05, VRK-10, VRK-20, VRK-35
		Diameter of stem-U-325/DA/DR/20, 325/DA/DR/36
		Length of internodes (CM)-U-35/DA/DR/48
		Resistance to mosaic virus-VRK-05, VRK-10, VRK-20, VRK-35
		Resistance to leaf miner-vrk-01, vrk-06, vrk-22, vrk-04, vrk-31, vrk-33 and vrk-55
Pointed Gourd	IIVR, Varanasi	High yielder-VRPG-13, VRPG-44, VRPG-70, VRPG-72, VRPG-93
		Number of fruits/plant-VRPG-13, VRPG-44, VRPG-70, VRPG-72, VRPG-88, VRPG-93, VRPG-113
		Heat tolerance-VRPG-72, VRPG-99, VRPG-19, U-34/BP/DR/07, VRPG-110, VRPG-110A, VRPG-110B
		Resistance to gall forming nematode-U-34/BP/DR/44, VRPG-72, VRPG-18, VRPG-96

Contd..

Table 5.14–Contd..

Crop	*Evaluating Center*	*Promising Accessions*
Ridge Gourd	NBPGR, New Delhi	Promising-IC-93399, U-324-138, U-20-202A, U-24-134, Nic-10216
Sponge Gourd	NBPGR, New Delhi	Promising-IC-92779, IC-92745 (very early), D-24-163, Nic-10237
Satputia	NBPGR, New Delhi	Promising-DCB-1363
Brinjal	NBPGR, New Delhi	IC-126879, NIC-5938, EC-304992 (Long deep purple); IC-90764, IC-112312, IC-90975, IC-126869, IC-144084 (Purple Long); IC-74209A, IC-136142, IC-144083, NIC-4573 (Green Long), IC-14405, NIC-13009 (Deep Purple Oblong); IC-90957, IC-127239, IC-113802, IC-112821 (Purple oblong); IC-127163 (white with purple stripe oblong), IC-137748, IC-144078 IC-136328 (Creamy white oblong)
Chillies	NBPGR, New Delhi	IC-119676, IC-119639, PSR-1775, JBT-12/83 MNCH-64, BD-119, BDS-2573
Tomato	NBPGR, New Delhi	EC-106285, EC-118292, EC-163683, EC-170662, EC-110578 (High yield); EC-41824, EC-164855, EC-169308, EC-122063, EC-126955 (Heat tolerant); EC-315487, EC-320578 (Moderate heat tolerant lines); EC-260639, EC-232429, EC-267729, EC-267726 (Suitable for fresh market); EC-161252, EC-100845, EC-110268 (Suitable gor transportation); EC-54628, EC-52062 (Paste type); EC-357833, EC-357827, EC-362956, EC-362940, EC-362944, EC-362947 (early maturing type)

Future Thrust

To meet needs of vegetable breeders in the country and elsewhere, some more elite germplasm will be introduced and indigenous variability will have to be collected. Considering the introductions already made and supplied to different centers in last 40-45 years and crop diversity build-up since 1940, following priorities need emphasis:

1. To assemble and inventorise all existing germplasm with different centers in various crops and characterization of same at multi-location using standard descriptors (already developed by NBPGR in collaboration with crop specialists) including diseases/pests resistance and quality attributes. Thus, status of field gene bank can be assessed.
2. Multiplication, seed increase and conservation of germplasm in the National Gene Bank at NBPGR with necessary passport and evaluation data.
3. To collect germplasm materials in different crops as per identified gaps and areas of germplasm availability including hot spot locations. Under NATP, collection of vegetable germplasm is given top priority. A massive programme through NBPGR, New Delhi and 10 Regional Stations (acting as Zonal centers) in collaboration with several CCPIs under National Agricultural Technology Project on Plant Biodiversity was undertaken. In this task, IIVR, Varanasi, IIHR, Bangalore, other ICAR institutes, SAUs and NGOs are helping a great deal.
4. The NBPGR imparts training in exploration/collection of germplasm. It has planned to conduct training on wild relatives of crop plants and plant identification, taxonomy/ biosystematics.

5. The wild germplasm assumes high priority for collection. It should be handled carefully for maintenance as these are sources of biotic/abiotic stresses, diseases/pests resistance.
6. Accessioning of all existing germplasm with different partners and registration of elite germplasm have become a necessary in the light of recent global development after signing of CBD and WTO.
7. Re-survey/collection in those areas from where useful genotypes have already been identified based on evaluation studies and biochemical/molecular linked desirable traits may be done.

References

Anonymous, 1987. Biosphere Reserves. Proc. 1st Nat. Symp. Sep.1986. Ministry of Environment and Forests, New Delhi p258.

Anonymous, 1989 a. Plant Genetic Resources: their conservation *in situ* for human use. IUCN-UNESCO/FAO, p 38.

Ariyo, O.J., 1993. Genetic diversity in West African okra (*Abelmoschus caillei*) Multivariate analysis of morphological and agronomical characteristics. *Gen. Resources Crop Evol.*, 40: 25-32.

Arora, R.K. 1985. Genetic resources of less known cultivated food plants. *NBPGR Sci. Monogr.* No. 9:125.

Bailey, L.H. 1950. *The Standard Encyclopedia of Horticulture*. The MacMillan Co., New York.

Bisht, I.S., Mahajan, R.K. and Rana, R.S., 1995. Genetic diversity in South Asian okra (*Abelmoschus esculentus*) germplasm collection. *Ann. Apl. Biol.*, 126: 539-550.

Bos, T., 1983. The optimum number of replication when testing lines or families on a fixed number of plots. *Euphytica* 32: 311-318.

Boswell, V.R. 1949. Our vegetable travelers. *Nat. Geogr. Mag.* 6: 145.

Chadha, M.L. and Tarsem, Lal, 1993. Improvement of cucurbits. (*In*) Advances in Horticulture, Vol. 5 Vegetable Crops. (Eds.) K.L. Chadha and G.Kalloo, Malhotra Publishing house, New Delhi.

Choudhury, B. 1967. Vegetables: National Book Trust, India.

Charrier, A., 1984. Genetic Resources of the Genus *Abelmoschus Med.* (okra, English translation). IBPGR, Rome.

Chevalier, A., 1940. L'origine, la culture et les usages de cinq Hibiscus de la section *Abelmoschus*. *Rev. Bot. Appl.*, 20: 319-328 and 419.

Dean, A. and Voss, D.1999. *Design and Analysis of experiments*. Springer Texts in statistics Springer, New York, USA, pp 67-97.

De Candolle, A. 1982. Origine des plantes cultivers.Germes Bailliere Paris France (English Translation, 1886, Kegan Paul).

Dadlani, S.A., Singh, B.P. and Singh, R.V. 1981. System of national and International exchange of germplasm and Methods of recording followed at NBPGR.(*In*) Mehra, K.L. Arora. R.K. and Wadhi, S.R. (Eds.) Plant Exploration and collections, Sci. Monog. No. 3. NBPGR, New Delhi.

De Wet, J.M.J. 1977. Increasing cereal yields: Evolution under domestication pp. 111-118. In D.S. Seigler (Eds) *Crop Resources*. Academic Press. New York.

Fedorov. 1974. Chromosome numbers of flowering plants. Atto Koetlz. Science Publishers, 624, Koring Stein, West Germany.

Federer, W.T. and Raghavarao, D. 1975. On Augmented Design. *Biometrics* 31: 29-35.

Federer, W.T. Nair, R.C and Raghavarao, D. 1975. Some augmented row-column designs. *Biometrics* 37: 361-374.

Fatokun, C.A. 1987. Wide hybridization in okra. *Theor. Appl. Gen.*, 74: 483-486.

Filov, A.I. and Vilmskaya, G.M. 1972. Cultivated cucurbits in different languages of the world. *Trudypo Prikladnoi Botanike Genetike L Selektsii*, 47: 138-150.

Flint, H.L. 1970. Importance of seed source ton propagation. Combined Proc. Int. Plant Propgator's Society, Wooster, Ohio 20: 470-475.

Fordham, A.J. 1971. Canadian hemlock variants and their propagation. Combined Proc. Int. *Plant Propagator's Society*, Wooster, Ohio 20: 171-179.

Ford-Lloyd, B.V. 2001. Genotyping in plant genetic resources. (In) Henry RJ (Ed) Plant genotyping-The DNA Fingerprinting of Plants. CABI Publishing, UK, 59-82, pp.

Gableman, and Gerloff 1978. Isolating plant germplasm with altered efficiency in mineral nutrition. *Hort. Sci.*13: 682-684.

Hamon, S., 1988. Organisation evolutive du genre *Abelmoschus* (Gombo): co-adaptation et evolution de deux species de gombo cultivees en afrique de l'ouest *Abelmoschus esculentus* et *Abelmoschus caillei.* (Eds.) ORSTOM, T.D.M. 46.

Hamon, S. and Hamon, P., 1991. Future prospects of the genetic integrity of two species of okra (*Abelmoschus esculentus* and *Abelmoschus caillei*) cultivated in West Africa. *Euphytica*, 58: 101-111.

Hamon, S. and Charrier, A., 1983. Large variation of okra collected in Benin and Togo. *Pl. Genet. Resources Newsletter*, 56: 52-58.

Hamon, S., Charrier, A., Koechlin, J. and van-Sloten, D.H., 1991. Potential improvement of okra (*Abelmoschus* spp.) through the study of its genetic resources (in French). *Pl. Genet. Resources Newsletter*, 86: 9-15.

Hamon, S. and Yapo, A., 1986. Perturbation induced within the genus *Abelmoschus* by the discovery of a second edible okra species in West Africa. *Acta Horticulturae*, 182: 133-143.

Harlan, J.R. 1975. Practical problem in exploration of seed crops. *In* Crop genetic resources for today and tomorrow. (Eds.) O.H. Frankel and J.G. Hawkes. Cambridge Univ. Press, London. 111-115 pp.

Hawkes, J.G. 1976. Sampling gene pools. Proc. NATO Conf. Conservation of threatened plants. Sec.I. *Ecology*. Plenum London.

IBPGR, 1991. Report on international workshop on okra genetic resources. Oct. 1990, IBPGR, Rome, pp. 2-3 and 8-12.

IPGRI, 2001. The Design and Analysis of evaluation Trials of Genetic Resources Collections. A guide for gene bank managers. IPGRI, Technical Bull. No. 4.International Plant genetic Resources Institute, Rome, Italy.

Jambhale, N.D. and Nerkar, Y.S., 1986. Parbhani Kranti- yellow vein mosaic resistant okra. *HortScience*, 21: 1470-1471.

Jambhale, N.D. and Nerkar, Y.S., 1990. Okra. *In*: *Vegetable Growing Handbook*: *Organic and Traditional Methods* (Eds) Splittstoesser, W.E. Melbourne, Thomas Nelson Australia,. 589-607 pp.

Jain, S.K. 1975. Population structure and the effect of breeding system. (*In*) Crop Genetic Resources for today and tomorrow. (Eds.) O.H. Frankel and J.G. Hawkes. Cambridge Univ. Press, London.pp 15-36.

Kempton R.A. and Fox, P.N.1997. Statistical Methods for Plant Variety Evaluation. Chapman and Hall, U.K.

Koechlin, J., 1991. African okras (*Abelmoschus* spp.): study of the diversity with regard to breeding (in French). In: Travaux et Documents Microfiches, 72. *Thesis*, Institut National Agronomiques Paris-Grignon, 180 pp.

Lin, C.S. and Poushinsky, G. 1983. A modified augmented design for an early stage of plant selection involving a large number of test lines without replication. *Biometrics* 39: 553-561.

Madeley, J. 1996. Yours for Food: Plant Genetic Resources and Food Security. Section 1: Plant Genetic Resources. People's Plan. BUKO Agrar, Nernstweg, Hamburg, Germany.

Marshal, D.R. and Brown, A.H.D. 1975. Optimum sampling strategies in genetic conservation. *In* Crop Genetic Resources for today and tomorrow. (eds.) O.H. Frankel and J.G. Hawkes. Cambridge Univ. Press, London. 37-52 pp.

Ministry of Environment and Forest (MoEF). 1998. Implementation of article 6 of the Convention on Biological Diversity in India. National Report (interim), Ministry of Environment and Forestry, Government of India.

Mohideen, M. K. and Irulappan, I. 1993. Improvement of Amaranthus. *In* Advances in Horticulture. Vol.5, Vegetable crops. 305-323. (eds) K.L. Chadha and G.Kalloo, Malhotra Printing House, New Delhi.

Munger, H.M. and Robinson, R.W. 1991. Nomenclature of *Cucumis melo* L. *Cucurbit Genet. Coop. Rep.* 14:33.

Pal, B.P., Singh, H.B. and Swarup, V., 1952. Taxonomic relationship and breeding possibilities of species of *Abelmoschus* related to okra (*Abelmoschus esculentus* L.). *Bot. Gaz.*, 113: 455-464.

PIER, 2001. " Invasive plant species: *Coccinia grandis*" Pacific Island Ecosystems at risk. United States Department of Agriculture, Natural Resources Conservation Services. The Maui Weekly Wrap. Maui News, p 10-18.

Rana, R.S. 1992. Core collection and the priorities of national programmes: Indian Perspective. Paper presented at the Intl. Workshop on Core Collection of plant Germplasm. Held at CENARGEN, Brasilia, Brazil. Sept.

Rana, R.S., Saxena, R.K. and Kochhar, Sudhir 1994. Descriptors and their descriptors status in relation to PGR conservation. *In* R.S.Rana *et al.*, Conservation and Management of Plant Genetic resources. NBPGR, New Delhi. pp. 192-201.

Seshadri, V.S. 1987. Genetic resources and their utilization in vegetable crops, pp. 335-343. in Plant Genetic Resources: Indian Perspective (Eds. R.S.Paroda, R.K. Arora and K.P.S. Chandel), NBPGR, New Delhi.

Sehgal, J.L., D.K. Mandal, C. Mandal and S. Vadivelu, 1992. Agro-ecological regions of India. Tech. Bull. NBSS Pub. No.24. NBSS and LUP (ICAR) N.Delhi.

Shanmugavelu, K.G. 1989. Production Technology of vegetable crops. Oxford and IBH Fransisco. 448 p.

Siemonsma, J.S., 1982a. La culture du gombo (*Abelmoschus* spp.) legume fruit tropical avec reference speciale a la Cote d'Ivoire. Thesis Univ. Wageningen, the Netherlands.

Siemonsma, J.S., 1982b. West African Okra, morphological and cytological indications for the existence of a natural amphidiploid of [*Abelmoschus esculentus* (L.) Moench. and *Abelmoschus* manihot (L.) Medikus. *Euphytica*, 31:241-252.

Shimotsuma. 1963. Cytogenetics and evolutionary studies in genus Citrullus, Seiken Zin Rept. Kihara Inst. Biol. Res. 15: 24-34.

Van Borssum Waalkes, J., 1966. Malesian *Malvaceae* revised. *Blumea*, 1: 1-251.

Vavilov, N.I.1926. Studies on the origin of cultivated Plants. Bull. Appl. Bot. 26 (2):1-248.

Vavilov, N.I.1951. Phyto-geographical basis of plant breeding. In: (Selected Writings of N.I Vavilov and translated by KS Chester) *The Origin, Variation, Immunity and Breeding of cultivated plants, Chronica Botanica* 13:364. Waltham Mass., USA.

Velayudhan, K.C., and Upadhyay, M.P., 1994. Collecting okra, eggplant and their wild relatives in Nepal. *Pl. Genet. Resources Newsletter*, 97: 55-57.

Whitaker, T.W. 1933. Bot. Gaz. 94: 780-790.

Wilsie, C.P. 1962. Crop adaptation and distribution. W.H. Freeman and Co. San Fransisco. 448 pp.

Zeven, A.C. and J.M.J. de Wet, 1982. *Dictionary of Cultivated Plants and Regions of Diversity*. Wageningen. 259 pp.

Zeven, A.C. and Zhukovsky, P.M. 1975. Dictionary of cultivated plants and their centers of diversity: excluding ornamentals, forest trees and lower plants. PUDOC, Wageningen. 219 pp.

Biodiversity in Horticultural Crops Vol. 1 *Pages* ***109–183***
Editors: **Professor K.V. Peter & Dr. Z. Abraham**
Published by: **DAYA PUBLISHING HOUSE**

Chapter 6

Tropical Vegetable Crops

K.R.M. Swamy and A.T. Sadashiva
Indian Institute of Horticultural Research,
P.O. Hessarghatta, Bangalore – 560 089

Agricultural biodiversity gives crop species including vegetables ability to adapt to changing environments, including new insect pests, diseases, soils and climatic conditions, besides providing new genetic material for breeding new varieties. These, in turn, provide a basis for more productive and resilient production systems, a basic for poverty alleviation, food and nutritional security. Since indigenous vegetables could help undercut the food insecurity at household level, there is urgent need to strengthen household food security by promoting their conservation, production, processing and utilization. Biodiversity of tropical vegetables, Dolanaceous fruits (Tomato, Hot pepper/Chilli and Sweet pepper/Bell pepper, Brinjal/Eggplant), Cucurbits (Watermelon, Muskmelon, Cucumber, Pumpkins and Squashes, Bottle gourd, Bitter gourd/Bitter cucumber, Ash gourd/Wax gourd, Round gourd/Round melon, Snake gourd and Pointed gourd, Sponge gourd and Ridge gourd, Chow-chow/Chayote/Guisquil, Ivy gourd, Stuffing cucumber/Korila), Malvaceous vegetable (Okra/Lady's finger), Leguminous vegetables (Common bean/French bean, Lima bean, Dolichos bean/Hyacinth bean/Indian bean, Cowpea/Yardlong bean, Sword bean, Winged bean), Alliums/Bulbous crops (Onion, Shallot, Garlic, Leek), other temperate vegetables (Carrot, Radish), Brassicas/Cole crops (Cabbage, White cabbage, Cauliflower and Broccoli), and Leaf vegetables (Amaranth, Spinach beet/Palak, Ceylon spinach) are reviewed. Biodiversity or biological diversity is defined as the variety and variability among living organisms and ecological complexes in which they occur. Genetic diversity refers to total amount of variation in a population or species. India, a treasure house of wide agricultural biodiversity, inhabits 260 million people below poverty line and 22 per cent of the world poor, who do not have access to quality food. The role of this diversity, especially that of vegetables, both cultivated and uncultivated could be significant in household nutrition, economy and food security. The trend, therefore, in new approaches, should be to increase food production without loss of on-farm biodiversity. Conventional approaches to crop conservation and improvement, to emphasize replacement of diversity with high yielding plant varieties which contributed to plant genetic erosion and have also expensive and ineffective for farmers in risk prone and socially marginal areas. Food insecurity, in turn, can be traced to breakdown of biodiversity-based production systems like mixed cropping, intercropping and crop rotation practices which provide diverse and seasonal food sources in difficult environments. Food security and social stability can be improved by enhancing capacity of communities to generate knowledge on biodiversity-based production systems and influence food policy.

Human food production has main stay in the husbandry of domesticated species and harnessing of wild plants and animals. 840 million people in the world do not have enough to eat and 500 million suffer from malnutrition and the population is still growing. This means that food production will have to increase 50 per cent by 2020. The solution, partly lies in biodiversity, in that it provides genetic information and materials used in plant and animal breeding. From millions of genes that serve as life's building blocks to thousands of plants and animals that inhabit earth to almost limitless combinations of organisms that make up natural ecosystems, biodiversity makes an essential contribution to feed the world. It will, therefore, be a key ally in fighting malnutrition and as such needs our protection (Pritam Kalia 2004).

Biodiversity Pillars

The greater part of the world's food supply depends on a very limited number of plant and animal species. About 7000 plants (2.6 per cent of all plant species) have been collected or cultivated for human consumption. Of these, a mere 200 have been domesticated and only a dozen, *viz.*, bananas, beans, cassava, maize, millets, potatoes, rice, sorghum, soya, sugarcane, sweet potatoes and wheat contribute about 75 per cent of the global intake of plant derived calories. Human food production, therefore, rests on a few species, leaving the majority of biodiversity little used and undomesticated (Pritam Kalia 2004).

Biodiversity and Nutrition

Quality of food, especially in terms of supplying essential vitamins and other nutrients, is central to achieve food security and avoid nutritional diseases/disorders. Although staple crops provide most protein and energy requirements, they are often deficient in minerals and vitamins. In rice consuming countries, for example, common nutritional deficiencies include: iron, vitamin A, iodine, thiamine, riboflavin, calcium, vitamin C and zinc. Vegetable diversity provides the basis for varied diets that contribute to balanced nutrition (Table 6.1). Alternatively, biodiversity represents variation in quality traits from which varieties with specific nutrient qualities can be selected. Scientists can be benefited from materials with high nutritional value for use in crop improvement. In addressing the global problem of poor nutritional quality and lack of diversity in diets, agricultural biodiversity at large and that of vegetable crops in particular can be of essential assistance (Pritam Kalia, 2004).

Genetic Resources, Erosion/Extinction

In recent past, genetic erosion resulted in reduction of existing varietal wealth in various crop species. Causes of this genetic erosion are many, but replacement of local varieties as a result of advancement is the most consistently cited reason. Food production is at risk from loss of agricultural biodiversity due to : (a) narrowing of future options through loss of genetic material that could be introduced into domesticated crops through breeding, (b) an increased susceptibility to diseases and pests because a fewer varieties and species are grown over large areas leading to dependency on pesticides and fertilizers, (c) destabilization of ecosystem through disrupting soil formation and predator prey cycles. It is estimated that there are roughly a quarter million plant varieties available for agriculture, but less than 3 per cent of these are in use today. With disuse comes neglect and possibly extinction, because:

1. Modern agriculture is concentrated on a small number of varieties designed for intensive farming. This has dramatically reduced diversity of plants available for research and development. This trend, and the increasing industrialization of agriculture are key factors in what is known as genetic erosion.
2. The world's food supply depends on about 150 plant species. Of those 150, just 12 provide three quarters of the world's food. More than half of the world's food energy comes from a limited number of varieties of three mega crops: rice, wheat and maize.
3. The world's poor depends on plants for as much as 90 per cent of their needs, *i.e.* food, fuel, medicine, shelter, transportation etc. Approximately 1.4 billion people, mostly resource poor farmers, use and improve their own crop seeds. This helps to maintain and enhance crop genetic diversity.
4. Millions of resource poor farmers eke out livelihoods on small pieces of marginal land in remote, mountainous or arid regions. Yet, it is estimated that they produce as much as 20 per cent of the world's food, largely without benefit of modern agricultural research (Pritam Kalia, 2004).

Table 6.1: Some Vegetables Grouped According to the Primary Type of Nutrient

Nutrient	*Vegetables*
Carbohydrates	White potato, beans, yam, sweet potato, cassava, taro (aroids).
Fat	Mature seeds of some legumes and cucurbits.
Protein	Beans, sweet corn, peas, leafy crucifers.
Carotene	Carrot, orange/yellow flesh sweet potato, squash, pepper, green peas, green leaf vegetables, green beans.
Vitamin C	Crucifers, tomato, pepper, melons, immature bean seed, bean sprouts, white potato, many leafy vegetables.
Minerals	Crucifers, most other leafy vegetables.

Source: Pritam Kalia, 2004.

Benefits of Biodiversity

1. Biodiversity gives species ability to adapt to changing environments, including new insect pests, diseases, soils and climatic conditions. Biodiversity also provides raw materials for breeding new varieties of crops including vegetables. These, in turn, provide a basis for more productive and resilient production systems. Use of crop diversity to produce stronger, more productive crops on farm, through field experimentation and/or gene transfer remains the best route to secure our food security and that of our children.
2. Farmers can improve their ability to meet food needs by growing diverse crops and their varieties. Different varieties, especially that of vegetables may have different tastes, mature at different times or have different cooking qualities. Some may grow well in rainfed areas while others in marshy land. Importantly, diversity contained in different varieties provides

farmers and scientists with options to develop, through selection and hybridization, new and more productive types that are resistant to pests and diseases resulting in a vast diversity.

3. Diversity can help to ensure an adequate and stable supply of food as well as enhancing its nutritional quality. It may also go a long way in safeguarding wiping out of an entire crop by pests, diseases or climatic conditions. In the Andes, farming communities use about 3000 different varieties of potatoes and in Java, farmers may plant more than 600 species in a single home garden, equipping them to secure their future.
4. Value of crop diversity goes well beyond its ability to support stable production systems in marginal environments. The world's population is still growing by an estimated 80 million people a year. This growth will bring with it a tremendous increase in food requirements, yet crop diversity is disappearing at unprecedented rates. It is anticipated that as much as 8 per cent of plant species could disappear in the next 25 years. Such reductions have serious implications for food security and crop diversity is the only vital component in meeting the food challenges we face now and forever. The wise use and management of crop diversity is perhaps one of the best ways to continue using agriculture sustainability that can go a long way in improving living standards of present and future generations (Pritam Kalia, 2004).

Poverty Alleviation through Biodiversity Conservation

Around 1.3 billion of the world's people live on less than one US dollar a day. Poverty affects both rural and urban areas, and it is associated with social problems such as malnutrition, disease, job insecurity and poor education. Food insecurity both now and in future is not only limited to developing countries, but also to 8 million people in industrialized and 26 million in countries in transition, who are under-nourished. However, close link between poverty and hunger means developing countries are the worst affected. Rural poor represents about 73 per cent of people living in poverty in developing countries. Social instability in South Africa and also in Asia including India is caused, in part, by food insecurity in rural communities leading to migration from rural to urban areas. Food security and social stability can be improved by enhancing capacity of communities to generate knowledge on biodiversity based production systems and influence food policy for giving weightage to food in wilderness, for instance: (a) many minor crops are important to millions of people in different parts of developing world especially the poor, *e.g.* cassava provides more than half plant food energy for people in Central Africa, (b) Wild species are an important source of vitamins, minerals and other nutrients especially in poor rural households in developing countries. These can be important resources for crop improvement as many of them contributed to evolution of present day cultivated types, for instance, wild cliff cabbage to present day cole crops and wild okra species to today's cultivated okra (Pritam Kalia, 2004).

Priority for Indigenous Vegetables

The indigenous vegetables, both cultivated and uncultivated have tremendous role in household nutrition, economy and food security. In many parts of the world, seasonality of food supplies is acute. Indigenous plants are often the best suited to the rigors of marginal environments, tolerating drought, poor soil fertility and local diseases and insect pests and serve as low-cost, reliable and culturally appropriate food sources in times of scarcity. There is an under-valued reservoir of diversity of Asian and African vegetables including that of leafy and starchy roots and tubers. These are extremely important for food security, nutrition and poverty alleviation. These indigenous vegetables are easier to grow, more resistant to pests, and acceptable to local tastes. Leaf vegetables are relatively high in quality protein and crude fiber (Table 6.2), whereas starchy vegetables are very important in supplying energy food needs of many populations in tropical and semi-tropical regions and to a large extent also those in temperate zones. India is endowed with an amazing number of different vegetables grown by farmers–175 different types in all including 82 leaf types and 41 root (tubers and bulb) crops. Potato, tomato, onion, cabbage and cauliflower account for 60 per cent of total production. Some of the indigenous vegetables are documented to have higher protein, phosphorus, iron, vitamins and carotene than the exotic vegetables. However, because of the introduction of exotic, high yielding food plants progressively a fewer wild plant species have been gathered or brought into cultivation in rural areas. To improve nutrition of rural communities and increase their household incomes, it is necessary to increase production, consumption and marketing of presently uncultivated indigenous vegetables. Healthy diversity through indigenous vegetables would be rewarding because they are hardy, nutritious, and storehouse of desirable traits. These so called indigenous vegetables, mostly leafy greens, roots and tubers etc., are often easier to grow, more resistant to pests and diseases, and acceptable to local tastes. They help diversify production systems, income, and diets for year round nutrition. Although, India made significant progress in increasing vegetable production since independence and continues to be second largest producer of vegetables in the world, yet we have not been able to ensure 300 g requirement of vegetables per capita per day to our countrymen recommended by dietitions. Possible way ahead could be to educate, enrich and empower our farmers, to take possible measures of population control and success in farmers economic development lies in diversifying their cereal-based production systems to include more remunerative crops like vegetables. Although, cereals are essential for sustaining livelihood of rural poor, but these alone cannot bail them out from poverty. A strong vegetable sector in India will lead to economic growth throughout the country as it generates more jobs. It will diversify and generate greater farm income (Pritam Kalia, 2004).

Table 6.2: Nutritive Composition of Some Selected Leaf Vegetables (per cent)

Vegetables	*Protein*	*Crude Fat*	*Crude Fibre*	*Ash*
Chard	15.6	3.9	9.8	15.6
Watercress	32.3	4.6	12.7	15.0
Amaranth	23.8	5.1	9.6	13.5
Crotalaria	34.2	3.9	9.8	7.3
Brussels sprouts	28.0	1.6	10.2	7.0
Spinach	20.9	1.6	4.6	15.6
Lettuce	21.4	2.1	10.7	8.6

Source: Pritam Kalia, 2004.

Biodiversity of Tropical Vegetable Crops

Solanaceous Vegetables

In this section, particular attention is given to tomatoes, sweet and hot peppers and brinjal/ eggplant, which are cultivated for their fleshy fruits. The three major food crops in this group thrive in tropical climates, although tomato and sweet pepper prefer a somewhat cooler climate than hot pepper and eggplant. All are very susceptible to low temperatures. Within this group, similarities occur regarding plant and seed characters, breeding behaviour, cultural practices and range of pests and diseases to which they are susceptible. Their nutritional value is rather low when compared to that of leaf vegetables. They supply minerals and vitamins in the diet and this contribution is significant, when large quantities are consumed. Tomatoes and hot peppers are indispensable flavouring ingredients for most tropical meals (Grubben, 1977). Solanaceous vegetables are an important group of crop plants. In India, the most widely cultivated solanaceous vegetables are tomato [*Lycopersicon esculentum* (L.) Mill.], eggplant or brinjal (*Solanum melongena* L.) and chilli (*Capsicum annuum* L.) and the cultivation of bell pepper or sweet pepper (*C.annuum* var. *grossum* L.) is also gaining importance. These vegetables are grown throughout the year in one or other part of the country (Kalloo *et al.*, 2005).

Tomato

Lycopersicon esculentum (L.) Mill. is an annual herb, variously called tomato, tomate, jitomate or faan ke'e, and love apple. The hermaphrodite flowers are self compatible and highly self-pollinating in cooler climates, but up to 24 per cent cross-pollination by insects is reported in tropical climate. At least 600 cultivars are described in seed catalogues. Cultivars are classified based on growth habit (determinate or indeterminate); fruit shape; cultivation methods (in the open or under glass); and whether the fruit is to be consumed fresh or processed.

Most of tomatoes grown in tropics for home use and local markets are primitive cultivars or local selections. Well-known cultivars which are grown in tropics for both fresh market and home use include the American cultivars: Bonny Best Campbell, Homestead, Indian River, Manalucie, Manapal, Marglobe, Floradel, Pritchard, Rutgers and BWN 21 hybrid. Cultivars suitable for processing include the Italian cultivars; Piacenza, Roma and San Marzono, and the American cultivar Heinz-1370. In the Far East, Best of All, Early Market and Marglobe are well-known cultivars.

Originating from Central America, tomato is now by far the most important world vegetable by production area, commercial value and consumed quantity. It has become a common vegetable throughout tropics. The plant requires moderately high day temperature (21-28 °C) and moderately cool nights (15-20 °C) for optimum growth and development. Fruit setting is reduced by temperatures which are either too low or too high and is inhibited below 13 °C and above 38 °C. Tomato is daylength neutral in respect of flowering, but in relatively long days, growth may be improved and dry matter content of fruits somewhat higher. Tomato cannot tolerate a badly-drained soil or a high relative humidity (80 per cent), both of which favour development of diseases and there is a high water requirement (8-10 mm/day) during period of fruit development. Both tall and dwarf cultivars are available, some are suited for lowlands and others for higher altitude cultivation. In temperate regions, most tomatoes are used in salads, but considerable amounts are also used in preparation of soups, sauces and beverages. In tropics, however, fruits are rarely eaten raw but are usually cooked for flavouring sauces. The average nutritional value (ANV) per 100 g of tomato is rather low, although better than that of eggplant. If level of consumption is high, tomatoes may contribute significant amounts of carotene, thiamin, niacin and vitamin C. In wet tropics, crop yields are much lower (3-10 tonnes/ha) than they are in the drier regions which have lower night temperatures (15-45 t/ha). This gave rise to an important regional trade in fresh tomatoes in these more favourable areas. Also, due to perishable nature of fruit and by its restriction to dry season production, use of tomato paste has become widespread; many countries have now established canning industries to reduce level of imports. Production techniques in tropics are generally rudimentary and many important production aids (fertilizers, pesticides and irrigation equipment) are lacking; consequently level of production is very low. An average production of about 20 tonnes/ha is obtained in mountainous regions and in cooler parts of tropics, but this is reduced to about 6 tonnes/ha in wet tropics. Adequate control of pests and diseases is extremely difficult, and in this respect, very a few cultivars which are adapted to the climatic conditions of wet tropics are available. Limited research on production techniques *e.g.* spacing, staking, pruning, mulching, control of pests and diseases, and application of growth regulators is carried out. The most important pests and diseases in wet tropics are: bacterial wilt (*Ralstonia*), leaf fungi (*Alternaria, Bortytis, Cladosporium*), root-knot nematodes and several virus diseases (tomato mosaic and streak, double virus streak, spotted wilt and cucumber mosaic). Many others have been recorded, which are locally or regionally serious (Grubben, 1977).

Origin and Distribution

Primary center of origin of *Lycopersicon lycopersicum* (L.) Karst (2n = 24) is situated in Mexico and in mountainous regions of west coast of South America. The progenitor is probably *L.esculentum* var. *cerasiforme* (Dun.) Alef. In Peru, Ecquador, Colombia and Chile, cultivated tomatoes exhibit spontaneous crossing with *L. pimpinellifolium* Mill., indicating a close genetic relationship. This and other wild species *e.g. L. cheesmanii* Riley, *L.chilense* Dun, *L.hirsutum* Humb and Bonpl. and *L.peruvianum* (L.) Mill. and a few *Solanum* species are now included in breeding programmes since they provide valuable genes for resistance to pests and diseases, such as *Fusarium, Cladosporium, Verticillium, Pseudomonas* and *Meloidogyne*. Several large germplasm collections, mainly of old and new cultivars selected in temperate regions, are being maintained for use of plant breeders. Primitive cultivars from tropical regions and wild or semi-wild South American tomato species and varieties are not well represented in these collections (Grubben, 1977).

Peru and Central Mexico were proposed as sites of origin and domestication of cultivated tomato. Evidences suggest that although, *Lycopersicon* spp. originated in Peru-Ecquador-Bolivia area of South America, domestication took place in Central Mexico, and maximum variability is present in Central Mexico and Coastal Peru. There are about 12 species in genus *Lycopersicon*. On the basis of fruit colour, all these species are classified into two sub-genera, namely, *Eulycopersicon* (characterized by red fruits with carotenoid pigmentation and annual growth habit) and *Eriopersicon* (characterized by green fruits with anthocyanin pigmentation). The cultivated species of tomato (*L.esculentum*) and one wild species [*L.pimpinellifolium* (Jusl.) Mill.] belong to *Eulycopersicon* and these two species are classified into five botanical varieties. One of these varieties, namely, *L.esculentum* var. *cerasifome* (cherry tomato) is the immediate ancestor of cultivated tomato (Kalloo, 1985). A characteristic feature of evolutionary process of tomato is that cultivated tomato is highly self-pollinated while there is considerable natural cross-pollination in cherry tomato (Kalloo *et al.*, 2005).

Genetic Diversity

Tomato belongs to one of the genetically best-known genera of plants and many genetic factors are investigated. A great deal of breeding work was carried out in developed countries (Europe, North America, Japan) and plant breeders are constantly developing new varieties. During the last ten years F_1 hybrids, showing heterosis and combining resistance to several diseases in addition to other desirable factors, became more popular. Cultivars are now available which have complete or partial resistance to *Fusarium* and *Verticillium* wilt, bacterial wilt (*Pseudomonas*), nail-head spot (*Alternaria*), leaf mould (*Cladosporium*), corky root (*Pyrenochaeta*), late blight (*Phytophthora*), bacterial canker (*Corynebacterium*), tobacco mosaic virus, spotted wilt virus and root-knot nematodes (*Meloidogyne*). Breeding work is currently in progress to limit damage by *Rhizoctonia* soil rot, *Stemphylium* leaf spot and several virus diseases. Breeding for resistance to various pathogens is the main task of tomato breeders. Improvement in quality (flavour, fruit shape, colour, vitamin C content, β-carotene content), earliness, and potential production are also under consideration. Tolerance to salt, resistance to cracking and good fruit setting at either high or low temperatures are locally important criteria. Many of the modern high-yielding cultivars are suitable for growing in mountainous regions of tropics but not in tropical, lowlands. Breeding work in tropics is mainly limited to cultivar screening and testing of cultivars from temperate regions; suitable cultivars adapted to the hottest season or for wet tropics generally are therefore rare or non-existent. In these regions, local selections are usually cultivated and these are fairly well adapted to high night temperatures and high humidities; they are also often less susceptible to diseases. In some countries (Guadeloupe, Mexico, Ghana, Nigeria, India, Hawaii) these cultivars were improved by selection and by crossing with exotic introductions. The IRAT station in Upper Volta developed a high yielding cultivar 'Farako-Ba' which has high level of resistance to Pseudomonas, by using Caribbean c.v. Floralou. Quality requirements of new tomato cultivars suitable for processing are strict concerning colour (bright red pulp), total soluble solids (>4.5 per cent), low acidity and a high content of reducing sugars (Grubben, 1977).

Over years, NBPGR, New Delhi evaluated a large number of germplasm accessions from diverse sources for various agro-botanical and economic characters (Srivastava *et al.*, 2000). Wide range of variability was observed for plant height (26-167 cm), branches per plant (3-19), clusters per plant (4-201), flowers per cluster (2-26), fruits per cluster (1-22), number of fruits per plant (8-790) and average weight of marketable fruit (30-167 g).

About 188 exotic accessions of tomato germplasm were evaluated at NBPGR Station, Hyderabad. Variability reported for some of the characters are as follows:

Seedling

Hypocotyl colour (green, one-fourth purple and half-purple from base, purple), its colour intensity, pubescence (present in all accessions), primary leaf length (7.0-32.4 mm), and width (3.0-9.4 mm).

Plant

Plant growth type (determinate, semi-determinate and indeterminate), plant size (small, intermediate like many improved varieties and large like many improved local varieties), vine length, stem pubescence density, stem internodal length, foliage density, number of leaves under first inflorescence (a few to many), leaf attitude, leaf type standard, (*L. peruvianum*, *L.pimpinellifolium* and others) besides variability in degree of leaf dissection and anthocyanin colouration of leaf veins. Vine size as a criterion of suitability of processing tomatoes differ from that for multi-hand-pick type or fresh market tomato. Processing types are designed for once over machine harvest or a maximum of 2 to 3 hand picks. Critical attributes of machine harvest type are determinate vine (small vine sizes) with good plant populations. Though in India, machine harvesting is not being practiced; in varietal breeding the self-pruning types with small plant size are most widely utilized.

Inflorescence

Type of inflorescence (uniparous, multiparous), corolla colour, blossom type, flower sterility type, petal length (6.973-15.236 mm), sepal length (3.82-15.2 mm), style position (inserted, same level as stamen, slightly exerted and highly exerted), style shape (simple, fasciated and divided), style hairiness, stamen length (5.322-9.73 mm) and dehiscence (poricidal like *L.pennellii* and longitudinal like many species).

Fruits

Exterior colour of immature fruits (greenish-white, light green, green, dark green, very dark green), presence and intensity of green backs, fruit pubescence (sparse like *L.esculentum*, intermediate like

L.pennellii and dense like *L.hirsutum*), predominant fruit shapes are flattened (oblate), slightly flattened, rounded, highly rounded, heart-shaped, cylindrical (long oblong), pyriforum, ellipsoid (plum-shaped) and other variability are fruit size homogeneity, fruit weight (1-200 g), fruit length (10.36-57.95 mm), fruit width (11.31-63.02 mm), fruit size very small (<3 cm), small (3-5 cm), intermediate (5.1-8 cm), large (8.1-10 cm), very large (>10 cm), exterior colour of mature fruits are green, yellow, orange, pink, red and its intensity, ribbing at calyx end (very weak, weak, intermediate, strong), ease/difficult of fruit to detach from the pedicel, fruit shoulder shape are flat, slightly depressed, moderately depressed, strongly depressed, pedicel length, presence of jointless pedicel, width of pedicel scar, size of corky area around pedicel scar, easiness of fruit wall to be peeled, skin colour of ripe fruit, thickness of fruit wall, thickness of pericarp, flesh colour of pericarp (green, yellow, orange, pink, red) and intensity, fruit cross sectional shape (round, angular, irregular), size of core, number of locules, shape of pistil scar (dot, stellate, linear, irregular), fruit blossom-end shape (indented, flat, pointed), fruit firmness. Besides, variability exists in days to flowering, days to maturity, ripening uniformity, number of inflorescences, number of flowers/inflorescence, fruitset/inflorescence, TSS and pH. There are many sources of plant material that have variations in earliness than presently available in commercial varieties. Earliness in tomato is period from seed germination to seedling emergence; seedling growth to last flowering; first flowering to fruit setting; fruit set to first ripe and first ripe fruit to peak production. Earliness is critical for economically desirable seasonal marketing demands.

For processing, tomato improvement, as it concerns growers, advances, extends and makes harvesting more manageable. Likewise, for processor it means more efficient in plant operation due to more even delivery of crop to the plant. In general, these types are of limited use due to low yield potential and deficiencies in quality–small fruit size, lack of firmness, crack susceptibility, low solids and poor colour. High solids in tomato fruits are associated with large, intermediate vines with low fruit to leaf ratio.

Seed

Seed shape (globular, ovate, triangular with pointed base), seed colour (light yellow, dark yellow, grey, brown, dark brown) and seed weight (<0.5 to 3 g) (Ram Srinivas and Varaprasad, 2000). At Indian Institute of Vegetable Research (IIVR), Varanasi, Uttar Pradesh, also a large number of accessions are evaluated including parthenocarpic lines introduced from the USA. A few accessions showed an ability to set fruit under high temperature (35-39 °C) conditions in April and May. EC130042, EC162935 and *L.cheesmanii* Riley, which set fruits at high temperatures, had stigma exertion of less than 1 mm, as compared to sensitive genotypes, which exerted longer stigma. Generally, heat tolerant lines set fruits but development of fruit was very slow and poor with the result that fruits remained smaller. However, heat tolerant accessions EC 168070, EC 130053-1 and EC 165393 had large fruit under Delhi conditions (Chandra and Thomas, 1993; Chandra and Gupta, 1994). Tomato is highly susceptible to a number of diseases and insect-pests. Sources of resistance to most diseases are identified. *Lycopersicon hirsutum* and *L.hirsutum f.glabratum* are generally resistant to numerous insect-pests. Farrer and Kennedy (1991) stated that resistance to 16 insect-pests species is reported in accessions of *L.hirsutum f.typicum* and *L.hirsutum f.glabratum*. Chaudhary *et al.* (1974) evaluated germplasm for resistance to nematodes and reported Anahu, Atkinson, Florida, Hawaii Cross, Hawaii-55, Nematex, Nemered, Pusa-20, VNF-8, 65 N-215-1 and 65 N-255-1 to be resistant.

Total soluble solids (TSS), sugars, acidity, lycopene and carotene are the main quality components in tomato fruit. Reducing sugars, namely, glucose and fructose are main components of soluble solids, whereas, protein, cellulose, pectic substances and hemicellulose are main insoluble solids. For processing, TSS is the most important trait, and in cultivated varieties, it ranges between 4 to 6 per cent. However, wild species, such as *L.chmiclewskii* and *L.cheesmanii* have up to 10 and 15 per cent TSS, respectively. For processing, acidity (less pH) is also important (pH should not exceed 4.4). *Lycopersicon pimpinellifolium* is a good source of high acidity. Tomato, particularly small-fruited varieties and wild species like *L.peruvianum* and *L.pimpinellifolium*, are good sources of ascorbic acid. Acidity also affects flavour, which depends on sugar:acid ratio.

Future Thrust

Germplasm of wild and hybrid types from Central and South America and of primitive cultivars present in various tropical and sub-tropical countries should be collected and conserved for breeding purpose. Loss of primitive cultivars is very real since these are gradually being superceded by commercial cultivars. International co-operation should be used to evaluate and document materials available in collections.

Breeding work aimed at producing cultivars which are well adapted to various ecological zones of wet tropics should be stimulated. Objectives should include breeding of true-to-seed or hybrid cultivars, tolerant to high night temperatures and a high level of humidity, producing good fruit set in hot weather with general resistance to major tomato diseases, particularly to bacterial wilt (*Ralstonia*), leaf fungi (*Alternaria, Cladosporium*), major virus diseases and root-knot nematodes (*Meloidogyne* spp.) (Grubben, 1977).

In tomato, cultivars are developed that possesses various desirable traits including resistance to tomato leaf curl virus (TLCV) transferred from *L.hirsutum* and *L.glabratum* (Kalloo and Banerjee, 2000). Development of pure lines possessing multiple resistance to major biotic stresses, for use as cultivar and parent of hybrids, needs to be strengthened. There is practically no research on breeding for resistance/tolerance to abiotic stress and quality traits, especially, temperature extremes. High β-carotene, high acidity and better shelf-life are other important traits. Thus, diverse germplasm should be assembled and evaluated for reaction to various biotic and abiotic stresses and other traits for utilization.

Hot Pepper/Chilli and Sweet Pepper/Bell Pepper

Capsicum annum L. var. *annuum* (Bell pepper, Cayenne, Chilli), is an annual herb with self-compatible hermaphrodite flowers and is widely cultivated in tropical areas. Most of hot peppers and all sweet peppers belong to this polymorphic cultigen and intercross freely, giving rise to a great range of types varying in fruit size, shape, colour and growth habit. *Capsicum frutescens* L. known as Tabasco or Bird Chili, is a perennial species occurring in a wild stage and is generally cultivated as a hot pepper. It does not, however, cross easily with other species. Several other wild or cultivated species of minor importance are described such as *C.pubescens*, Ruiz. and Pav., *C.chinense* Jacq. and *C.baccatum* var. *pendulum* Wild. (cultivated) and var. *baccatum* (wild). Leading sweet pepper cultivar is California Wonder, known for its large, smooth, attractive oblong-shaped fruits. Other popular North American cultivars, bred as pure lines, are: World Beater, Burlington, Yolo Wonder, Keystone, Resistant Giant and Ruby King, all of which have a certain degree of resistance to tobacco mosaic virus. A few cultivars of hot pepper are cultivated outside their regions of origin and nearly all those grown in tropics are local selections. Self-pollination in peppers is possible, but, for hot and bell peppers, insect pollination is the general rule. Between 2 per cent and 40 per cent of cross-pollination in areas of pepper cultivation are recorded, with a mean of 8 per cent, and it is therefore difficult to keep cultivars true to type. Hot peppers (chillies) are widely used as a flavouring ingredient (spice) and vegetable in tropical regions. Mild-sweet types of *Capsicum*, the so-called sweet or bell peppers, cultivated since the last century as a summer vegetable in Mediterranean area and in North America, are more popular both in tropics and in temperate regions (Grubben, 1977). Besides, chillies have medicinal and ornamental values too. Cultivated species of genus *Capsicum*, to which chilli belongs is *Capsicum annuum* Linn. with several varieties. Of these *C.annuum* var. *abbreviatum* is used as ornamental bush chilli (Erwin,1929). Temperature is particularly important and sweet peppers give optimal growth at dry temperatures in range 21-25 °C. Heat injury, retarded growth and poor fruit setting occur at temperatures above 30 °C. For hot peppers, optimum day temperatures are 24-30 °C; the limits for heat injury are also somewhat higher than they are for sweet peppers. Small-fruited cultivars are more tolerant to hot weather conditions than large-fruited ones. Water requirements of capsicums are similar to those of cotton (450 mm per annum), but supply must be regular, particularly for bell peppers.

Nutritional content of hot peppers is relatively high but intake per meal is generally low (104 g per caput). The ANV per 100 g of hot pepper is extremely high (27.92), due to the high dry matter content. Calculated on the basis of 100 g dry matter, ANV (80.7) is similar to that of bell pepper (82.6). Bell peppers can be consumed in much larger quantities, up to 20 g per meal and constitute a potentially important food source for nutritional improvement, especially with regard to their vitamin C content.

Hot pepper is one of the most important commercially grown vegetables in tropics, and is probably the most important, after tomato. It is also exported to temperate countries in a dried form, mainly from India, for use as a spice in flavouring sauces and canned products. However, greater part is produced for home consumption.

Sweet peppers are grown in some tropical and sub-tropical areas like Senegal, Kenya and North Africa during winter for export to Western Europe. Greater part of sweet pepper production is sold in local markets but general level of production and consumption in tropics are low. Hot pepper cultivation is hampered by many pests and diseases, although to a lesser degree than tomato. Tobacco mosaic virus is probably the most serious disease, but bacterial wilt (*Ralstonia*) and fruit flies are also common problems in pepper cultivation in hot climates. Bell peppers are more susceptible to diseases than hot peppers, especially wilt disease such as *Ralstonia*, *Fusarium* and *Sclerotium*. Reliable seeds of suitable hot pepper cultivars are either non-existent or not available. Good bell pepper seed may be found, but popular cultivars are relatively unsuitable for growing in wet tropics and during the hottest seasons (Grubben, 1977).

Genus of peppers, *Capsicum* (from box or case), consists of 90 or more species (according to classifications used), of which four to six are cultivated. Cultivated and wild species are described by Lippert *et al.* (1966) and by Heiser and Smith (1953). There are taxonomic controversies, but for practical purposes, most of desirable peppers are found in one species, *Capsicum annuum* L. Some forms of *C.frutescens* L. are used for their small and very pungent fruit in hot sauces but the species is of relatively minor importance. Two other species (*C.pubescens* Ruiz and Pav. and *C.pendulum* Willd.) are cultivated but are grown only on a small scale. There are many cultivated forms of C.annuum. Morphology of fruit and its pungency are often used to identify types of peppers, and this, sometimes led to considerable confusion, especially since all forms of intergrade can interbreed freely. Morphological forms should not be confused with botanical varieties. The most common forms are:

Long peppers–usually 20-30 cm long, yellow, or red in colour, tapering.

Sweet, bell, and paprika peppers–large, inflated, thick-fleshed fruits.

Chilli peppers–over 9 cm long, narrow, pointed.

Wrinkled peppers–less than 5 cm long, wrinkled appearance.

Cherry peppers–globose with firm flesh, yellow, red, or violet.

Cone or tabasco peppers–small conical, erect fruits.

Cluster peppers–fruits borne in clusters.

Common names for peppers usually consist of adjectives modifying the word pepper. Some reference to pungency (or lack of it) is frequently included. Common modifiers include bell, bird, capsicum, cayenne, chilli, paprika, pimento, red, sweet and wax (Martin *et al.*, 1979).

Origin and Distribution

Genus *Capsicum* (2n=24) originated in American tropics. It represents a diverse plant group having 25 wild and 5 cultivated species (Bosland and Votava, 2000). Based on gene flow through conventional hybridization, *Capsicum* species are grouped in three species complexes, namely, *C.annuum* complex, *C.baccatum* complex and *C.pubescens* complex. Among five cultivated species, *C.annuum* L., *C.frutescens* L. and *C.chinense* Jacq. belong to *annum* complex, while *C.baccatum* and *C.pubescens* Ruiz and Pav, belong to *baccatum* and *pubescens* complex, respectively. Among the five cultivated species (*C.annuum*, *C.baccatum* var. *pendulum*, *C.fruitescens*, *C.chinense* and *C.pubescens*), the most widely grown species is chilli (*C.annuum*) and sweet pepper (*C.annuum* var. *grossum*). *Capsicum annuum* was domesticated in highlands of Mexico and includes most of Mexican chilli peppers, most of hot peppers of Asia and Africa, and various cultivars of sweet peppers of temperate countries. However, due to non-adaptability of *C.annuum* to lowland tropics of Latin America, *C.frutescens* and *C.chinense* replaced its cultivation. *Capsicum frutescens* is also cultivated in Africa and India for either intact fruits or for oleoresin extraction. Cultivation of *C.baccatum* and *C.pubescens* is mostly restricted to Latin America (Pickersgill, 1997). Peppers are all of New World origin. Dried pods found in tombs in Peru suggest that peppers use for thousands of years. Center of common species, *C.annuum*, is Mexico. The minor species, *C.frutescens*, is widely distributed in tropical and sub-tropical America. Peppers were spread throughout Old World shortly after discovery of America. They were adopted widely and are now indispensable in India. Although the sweet or only slightly pungent forms are preferred in United States and Europe, very pungent peppers are the most widespread and economically important (Martin *et al.*, 1979). Areas of Meso-America and Andean zone are primary centers of *Capsicum annuum*. Since Columbian era, secondary centers developed in Southern and Central Europe, Africa and Asia. The wild variety *minimum* is probably the ancestor of cultivated variety *annuum*. Center of origin of *C.frutescens* L. (from *C.chinense* Jacq.) is also located in Meso-America and, as with *C.annuum*, secondary centers have arisen elsewhere.

Interspecific relationships in *Capsicum* and the relations between weedy and cultivated forms are studied. Variability existing in local populations of *C.annuum* is extensive. Many useful genes have already been identified in hot peppers and are being used in breeding work. Germplasm collections exist in many countries, mainly in agricultural stations where breeding work is being carried out. No systematic records of these collections are currently available (Grubben, 1977).

Genetic Diversity

Much breeding work was carried out on bell-peppers, mostly in USA, but also in France, Italy, Japan, Brazil, USSR, India and European countries such as Czechoslovakia, Bulgaria, Hungary. Important criteria for selection are high yield, resistance to virus diseases (tobacco and cucumber mosaic), good fruit shape with a sweet, non-pungent taste, a few seeds, and early production with first harvest approximately 60 days from transplanting. Cultivars such as Allbig give good fruit setting in hot weather. A few cultivars of Hungarian paprika, with elongated conical shape, were selected for high paprika-aroma content, thin skins facilitating drying, and for production of bright red powder on dehydration. More advanced techniques, such as application of haploidy, tetraploidy, cytoplasmic male sterility and heterosis breeding are used by plant breeders. For mechanized cultivation, types with compact habit, determinate growth and simultaneous ripening are required. Breeding of hot peppers is at a much less advanced stage. In USA, cultivars such as Anaheim and Chilli Cayene (both *C.annuum*) and types of Tabasco (*C.frutescens*) are widespread. Cultivars of *C.annuum* were selected for early yield, such as 120 days from transplanting to fruit harvest compared with 150 days for normal cultivars, high capsaicin content (the factor governing pungency), smooth shape for canning and thin walled fruits for drying. Several horticultural stations in tropical countries like Ethiopia, Ghana, Ivory Coast, Nigeria, Philippines, Mexico, Puerto Rico, India and Korea have programmes directed towards improvement of the local hot pepper cultivars, mostly by simple cultivar testing but also by hybridization. Improvement of disease resistance in locally preferred types and of types suitable for export as dried hot peppers would be the most important objectives. Donors for resistance to bacterial wilt (*Ralstonia*), virus diseases (tobacco mosaic, tomato etch, cucumber mosaic) and *Cercospora* are found, hot pepper cv. College No.9 Chilli possesses resistance to *Fusarium* wilt. Local West African (thick-walled) hot peppers are not frequently attacked by fruit flies. There exists an obvious resistance to *Sclerotium* (cv. World Beater) and *Phytophthora*; variation in susceptibility to nematodes and *Aphis* was noted in several cultivars (Grubben, 1977). The fact that all peppers are diploids (2x=24 chromosomes) and all can be grown as annuals has made considerable progress in genetics and breeding of these species possible (Lippert *et al.*, 1966). Progress was made on identifying individual chromosomes, and many of the simpler inherited characters that affect almost all distinguishable differences among varieties have been described. Trisomic races are being used to associate genes with particular chromosomes. Peppers are normally self-pollinated, but anthers do not dehisce until after anthesis. Cross-pollination by bees and other insects can reach 40 per cent but is normally less than 5 per cent.

India is the largest producer (0.8 million tones) of chillies in the world. It is consumed as a fresh vegetable and as a spice. Green chillies contribute 20 per cent of total export of fresh vegetables (excluding onion and garlic). *Capsicum*, a new world genus, has rich diversity. The word *Capsicum* originated from Greek word Kapso means to bite. In another version, it originated from the Latin Capsa meaning, a box which refers to shape of a fruit. *Capsicum* terminology is confusing. Various terms like pepper, Chili, Chile, chilli, aji, paprika and *Capsicum* are used interchangeably for plants in the genus *Capsicum*. The term chilli is used mostly in Britain, India, Africa and Eastern countries. Taxonomists use the term *Capsicum*, whereas in trade and scientific studies as pepper with various names bird pepper, chilli pepper, bell pepper and cayenne pepper. The International Standards Organization recognizes only 2 types: chillies and paprika to cover easily perceivable to strong pungent types and the big fleshy vegetable, sweet or just recognizable pungency types of *Capsicum*.

Capsicum is a small shrub of Solanaceae family. It is perennial in suitable climatic conditions, living for a decade or more but is more oftenly cultivated as an annual. It probably originated from an ancestral form in the Bolivia/Peru area. The *C.annuum* is distinguished by white flowers (rarely purple), anthers with anthocyanin, calyx with distinct points or teeth and longitudinal ridges. Flower 1/node (occasionally 2 at first flowering node), erect in wild form and pendant in many types. Seeds are tan. The *C. fruitescens* is characterized by waxy, pale greenish-yellow flowers, anthers with anthocyanin, calyx smooth, without longitudinal ridges. Flowers 2-5/node (usually 2), held stiffly erect with unopened flowers at an angle to peduncle. Seeds are tan. Cultivated types have less variability than *C. annuum*, but may have slender fruit up to 15 cm in length. Non-pungent forms are not known.

C. frutescens requires a tropical climate, hence is restricted in habitat. Flowers of *C. chinense* are white, anthers with anthocyanin, calyx ridged with short points and distinct annular constriction at base. Flowers 2-5/node (usually 3), peduncles variously curved in contrast to *C. frutescens*. Only commercial cultivar of the species is 'Habanero'. It does not differ greatly from *C. annuum* climatically. *C. baccatum* is distinguished by white flowers with discrete yellow spots, anthers white before dehiscence, brown afterwards. Calyx with ridges and distinct points or teeth, but without annular constrictions. Flowers in cultivated forms normally 1/node. Stems characteristically 4-sided. Seeds are tan. This is widely distributed in South America. The climatic requirements are similar to those of *C. annuum*. The *C. pubescens* requires a long cool growing season and is found only at higher elevations of tropics to sub-tropics. Flower is purple with a white basal zone, often with a conspicuous spot of yellow nectar. Anthers are purple. Seeds are black or nearly so. Plant is characteristically hirsute. Of the 5 cultivated species, *C. annuum*, *C. frutescens* and *C. chinense* consist of single gene pool, although some hybridization difficulties are there. The *C. baccatum* and *C. pubescens* neither hybridize with other 3 species nor with each other.

Horticultural classification of peppers grown in the USA is limited primarily due to more important forms of *C.annuum* in the USA and to cultivar 'Tabasco' of *C.frutescens*. Classification is based on pungency, colour, flower shape, size and use. There are 6 major groups under *C.annuum* and one group ('Tabasco') under *C.frutescens*. Under 6 major groups, there are many sub-groups bell, pimento, ancho, anaheima, cayenne, cuban, jalapeno, serrano, small bot, cherry, small wax and long wax. Of these bell and Anaheim are commercially important. Paprikas belong to anaheim group. Thus Capsicum is a highly polymorphic genus with much morphological genetic diversity both intra-specifically and inter-specifically. Intra-species and inter-species improvement is possible to evolve better types. The collection of chilli germplasm and its evaluation are in progress at the Kerala Agricultural University (KAU), Vellanikkara (Indira *et al.*, 1996).

Nine hundred and three germplasm accessions of chilli were evaluated for 42 traits at National Bureau of Plant Genetic Resources (NBPGR) Regional Stations at Hyderabad, Thrissur, Shillong, Bhowali and Amaravati. Promising chilli accessions identified at NBPGR are, for high fruit yield : BDJ22254, EC246007, IC18882, IC20901, IC92153, IC99912, IC92250, IC119556, IC147637, IC147719, IC99912, NIC19991, P2072; long fruit (>10 cm) : IC136061, IC136252, IC136266; PBNV/Mosaic resistance: EC121490, IC119611; drought tolerance: NIC23797, NIC23837 (Singh *et al.*, 2001). Chilli germplasm, 9852-173 and PDGI showed consistent tolerance/resistance against mite and thrips at IIVR Varanasi and 309-1-15, 300-1-5-1, S118, S635 and S565 showed consistent tolerance to thrips at IIHR. In addition, studies carried out at Regional Research Station, Lam, showed that Calaepin Red, Chamatkar, NP46-A, X1068, X743, X1047,BG4, X226, X230 and X233 were tolerant to thrips; LEC1, Kalyanpur Red, X1068, X204 and Goli Kalyanpur were tolerant to mites; and LEC28, LEC30, LEC34, Kalyanpur Red and X1068 were tolerant to aphids.

Fifty one lines of ornamental chillis (*C.annuum* var. *abbreviatum*) were collected through survey and exchange and were evaluated during 1988-89 at KAU, Thrissur. Beautiful plant types with dwarf stature and clustered fruiting habit are selected as ornamental types of chillies (Mini and Peter, 1994).

Future Thrust

Genetic material may be lost as primitive and older cultivars of sweet pepper are replaced by newer ones. An inventory should therefore be made of existing germplasm collections. There are a fewer risks of genetic erosion occurring in various hot pepper populations grown in tropics or in wild and semi-wild *Capsicum* species existing in Central and South America. However, collecting and testing of this germplasm source is advisable, to make it available to plant breeders in other tropical regions. Plant breeders should aim to produce bell pepper cultivars which are suitable for tropical climates; these should combine many types of disease resistance with a high vitamin C and pro-vitamin A content. Hot pepper cultivars should be selected for resistance to virus and to fruit flies and for high yield (Grubben, 1977).

Peppers are assured a successful future; they are already popular in diet and their use widespread. Excellent, disease-free varieties are already available, but they have not been universally adopted. Improvements constantly being made by plant breeders assure varieties that will be easier to grow with less disease and higher yields. But emphasis on compact, determinate varieties for machine harvesting should not hamper development of long-lived varieties for home-scale production. It should be simple to breed such varieties; necessary characteristics are already present in *C.annuum*. Considerable variation in carotene and vitamin C content also afford ample opportunity to select more nutritious varieties. Use of peppers as processed vegetables could increase. Pungent peppers are widely used in this fashion. And since peppers used as green are not as rich in vitamins as the yellow or red mature forms, peppers can contribute more to the diet if people merely change their usage habits (Martin *et al.*, 1979).

In chilli, there is ample scope of utilization of resistance sources available in *C. chinense*, *C. frutescens* and *C. baccatum*. All these species are cross compatible. *Capsicum chinense* and *C.baccatum* are potential sources for transfer of cytoplasmic male sterility (CMS) and resistance against several biotic stresses into *C. annuum*. There is need to identify more sources of resistance to thrips, mites and die back. Development of CMS lines particularly in chilli will be useful in hybrid breeding. Since most of sweet pepper lines lack restorer gene, more emphasis needs to be given on development and utilization of nuclear male sterility in this crop (Kalloo *et al.*, 2005).

Brinjal/Eggplant

Names eggplant, garden egg, aubergine, terong, aikwa and brinjal apply to several *Solanum* species with fleshy fruits, *S.melongena* L. being the most commonly cultivated species. *S.anomalum* Thonn., *S.macrocarpon* L., *S.aethiopicum* L., *S.incanum* L., *S.duplosinuatum* Klotsch. and several other minor species are cultivated in Africa and *S.quitoense* Lam., *S.muricatum* Ait., *S.topiro* Humbo. and Bonpl. in Central and South America. *S.aethiopicum and S.macrocarpon* are cultivated not only for fruits but also for their edible leaves. In South East Asia, small fruits of *S.ferox* L., and *S.torvum* Swartz. are eaten. *S.melongena* is self-compatible and highly self-pollinating. In hot, humid climate, up to 20 per cent cross-pollination from surrounding plants occurs, by insects or by wind.

Principal cultivars are: Black Beauty, Florida Market (resistant to Phomopsis fruit rot), Black Magic (an early hybrid, first harvest being produced after 72 days), Improved Long Purple, Peradeniya, Florida High Bush and Improved Muktakesu (resistant to *Cercospora*). In most countries, however, local cultivars are more common than high yielding exotic cultivars since they are more adapted to prevailing ecological conditions and are preferred by local population. Local cultivars exhibit a wide range of colour forms and shapes. Colour varies from white, yellow, green to purplish-black and shape from short to long and ovoid to sausage-shaped. Many of other species have much in common with *S.melongena* and are liable to same pests and diseases. Eggplants are cultivated in most tropical and sub-tropical areas, also in temperate areas which have a warm summer. Optimum day temperatures range from 25-35 °C and optimum night temperatures from 20-27 °C. Eggplants are more susceptible to low temperatures than tomatoes or peppers and cannot tolerate frost.

Eggplants grow well in wet tropics. They are less susceptible to diseases than either tomato or sweet pepper. As far as is known, primitive African species have same ecological requirements as *S.melongena* but South American species prefer a cooler climate. There is no reported daylength reaction. Fruits of eggplants have lower nutritional value than those of tomato. Intake of eggplant fruits per meal may be relatively high, up to 200 g/caput. Eggplant is an important vegetable in Central and South and South East Asia and also in a few African countries. In Ivory Coast, for example, *S.macrocarpon* and *S.incanum* are ranked second in quantity of vegetables consumed (about 4 kg/caput/year), and are placed after okra and before tomatoes and peppers. It is a common market vegetable in most tropical countries but most of production from home gardens or in mixed cropping with staple food crops is for home use. It is also an important vegetable in sub-tropics (India, North America, Central America) and in warm temperate regions (Mediterranean area, Central Asia, America). Eggplant is liable to various wilting diseases (*Ralstonia, Fusarium, Verticillium, Sclerotium*) and is susceptible to root-knot nematodes (*Meloidogyne*). Many other locally important pests and diseases are reported. Leaf curl is a serious virus disease in many parts of Africa and Asia (Grubben, 1977).

Origin and Distribution

Solanum melongena (2n = 24) probably originated in India and shows secondary diversity in China and South East Asia. *S.aethiopicum* (2n = 24), *S.macrocarpon* (2n = 36) and *S.incanum* (2n = 24) originated from tropical Africa. *S.quitoense* (2n =24), *S.muricatum* (2n = 24) and S.topiro (2n=?) all originated from South America and are mainly cultivated there although limited plantings of *S.muricatum* are observed in Ethiopia. About 100 old and modern cultivars of *S.melongena* were collected in seed banks in USA, USSR and in breeding institutes in Europe. A few collections of either numerous primitive or more advanced local cultivars were reported but interest was shown in these in Indonesia (Grubben, 1977). The genus *Solanum* originated mainly in Asia and Africa. The economically most important and best-known species, *S.melongena* was domesticated in the Indo-Myanmar region and has spread all over the world. Therefore, immense diversity of eggplant and its wild relatives exist in this region. Two African eggplants, the scarlet eggplant (*S.aethiopicum* L.) and the gboma eggplant (*S.macrocarpon* L.), are extensively cultivated in Africa and at very limited scale in other parts of world. There are many other species, for example, *S. nigrum* L. and *S. nodiflorum* Jacq., which are at times cultivated for (spontaneously occurring plants are looked after and harvested) in Africa and Asia (Daunay *et al.*, 2000). Many other species of *Solanum* are used as medicine (Bukenya and Carasco, 1999). *Solanum melongena* complex has three species, namely, *S. melongena, S. incanum* L. and *S. insanum* L. These are very closely related with each other but their taxonomic status is disputed. It is not clear whether *S.incanum* and *S.insanum* are progenitors or derivatives of *S.melongena*. Meiotic behaviour of some hybrids between these species was regular, suggesting close phylogenetic relationship among three species (Kalloo *et al.*, 2005).

Genetic Diversity

Much breeding work (pure line selection) was carried out, mostly in USA but also in India, Sri Lanka and elsewhere. The eggplant withstands inbreeding and shows a limited response to heterosis. Hybrid varieties were bred, applying male sterility, but these are not widely cultivated. In Hawaii, the high yielding cv. Waimanalo Long was selected. In India, interspecific hybrids with wild *Solanum* species are used in breeding for resistance to root-knot nematodes and to insect damage by *Leucinodes* (fruit and shoot borers). Resistance to leaf spot (*Alternaria*) has also been obtained. In Brazil, cultivars are selected which are tolerant to *Verticillium* wilt and anthracnose (*Colletotrichum*). In Japan, *S. integrifolium* (syn. *S. incanum*) is being used as a parent for introduction of disease resistance. The

Japanese cv. Nihon Nassu possesses resistance to bacterial wilt (*Pseudomonas*) and to anthracnose. In Guadeloupe, a breeding programme for bacterial wilt resistance is in an advanced stage. A cultivar Madinina, tolerant to bacterial wilt is already being commercially grown and a tolerant cultivar Bulacan, which is high yielding and contains a few seeds, was reported from Philippines. Breeding work on eggplant species other than *S. melongena* is either scarce or non-existent, except for some testing of local cultivars of *S. macorcarpon* and *S. incanum* in Ivory Coast and Ghana, screening of *S. macrocarpon* and *S. aethiopicum* to diseases in Nigeria, and testing of local cultivars of *S.quitoense* in South America (Grubben, 1977). At NBPGR, a total of 2,419 accessions including wild species were evaluated for 44 agro-botanical and economic characters, and a wide range of variability was recorded for plant height (40.3-96.5 cm), primary branches per plant (4.6-19.3), 75 per cent flowering after transplanting (23-82 days), horticultural maturity (55-87 days), flowers per cluster (2.2-9.30), fertile flowers per cluster (1.0-5.3), fruit length (4.3-36.7 cm), fruit circumference (6.5-46.3 cm), fruit weight (35-725 g) and the clusters per plant (8.4-35.5). In brinjal, *S. macrosperma* and *S.integrifolium* are resistant and tolerant to *Phomopsis* rot and drought, respectively. Both these traits can be transferred to cultivated *S.melongena* through conventional breeding methods, since there is no crossing barrier. Further, certain wild species possess resistance to biotic stresses, namely, resistance to shoot and fruit borer in *S.khasianum*, root knot nematode (RKN) in *S.sisymbrifolium*, and that to a number of root diseases in *S.torvum*. However, in these species crossability barrier with cultivated species needs to be overcome with the aid of biotechnolgoical approaches (Kalloo *et al.*, 2005).

Future Thrust

Primitive cultivars of *S.melongena* and of other eggplant species in all tropical countries are being rapidly replaced by more advanced, high yielding *S.melongena* cultivars. Much material is still present in Asian countries but there is noticeable erosion in Indonesia and the Philippines and some danger in Thailand and Malaysia. Collection of other eggplant species for use in breeding is recommended and material should be screened for presence of genes for disease and pest resistance. Resistance breeding to wilt diseases and root-knot nematodes should have first priority. Useful types for wider distribution may be found within local eggplant populations, for example, West African *S.macrocarpon*, which is cultivated for leaves and fruits in home gardens and in commercial plantings although *S.macrocarpon* is susceptible to *Cercospora* (Grubben, 1977).

Cucurbits

Cucurbits are an extremely important group of vegetables in both temperate and tropical regions. Nearly all the fleshy cucurbit fruits have a poor nutritional composition. Exceptions are yellow-fleshed pumpkins and melons which have high β-carotene content and bitter gourd which has a high iron and vitamin C content. The fleshy cucurbit fruit is appreciated in developed countries as a low calorie food. In tropics, daily intake of these fruits may be high, their contribution of vitamins and minerals should therefore not be underestimated. Pumpkins (*Cucurbita* spp.) may even serve as a source of carbohydrate. Cucumbers are eaten raw in salads, melons and watermelons consumed in a ripe state as dessert fruits and either immature or ripe fruits of many other vegetable cucurbits are cooked. Cucurbits form a distinct group of species with many similarities in botany, agronomy, ecological requirements and susceptibility to pests and diseases. Nearly all are monoecious plants with unisexual flowers, pollinated by insects. Cucurbits grow well at day temperatures between 20 and 35 °C; they cannot tolerate frost or strong winds. Low temperatures and a high relative humidity stimulate development of female flowers. Cucurbits show a slight photoperiodic reaction to short days. A few tropical cucurbits, like *Cucurbita moschata* Duch. require short days for flowering (Grubben, 1977).

Cucurbits constitute an important and large group of vegetable crops. These belong to family *Cucurbitaceae*. There are about 38 species of economic importance, which include a wide range of vegetables, namely, cucumber, gourds, melon, pumpkin and squashes. These are used as salad (cucumber, gherkin, long melon), as vegetable (all gourds), for pickling (bitter gourd, cucumber, gherkin, ivy gourd, pointed gourd), as dessert (muskmelon, snap melon, watermelon) or candied or preserved (ash gourd). A few cucurbits have other uses, such as, mature and dry fruit of bottle gourd is used for making vessels and musical instruments, fibrous material of *Luffa* is used as scouring and packing, and some species have ornamental and medicinal uses. Cucurbits are of considerable economic importance and are cultivated throughout the world from tropical to sub-temperate zones. China, Turkey, India and Iran are important cucurbit growing countries. Cucurbits are modest sources of calorie, minerals and vitamins. However, there are a few exceptions, for example, bitter gourd is rich in vitamin C and iron, pumpkin in carotene, kakrol in protein, and chow-chow has fairly high calcium (Sirohi *et al.*, 2005). There are about 28 genera and 86 species of Cucurbitaceae found in India, or roughly about 30 per cent of the genera and 10 per cent of the species of the entire family are indigenous to India. This represents a far greater number of species and genera than are found within the political boundaries of any other country. Moreover, there are more and different kinds of cucurbits cultivated in India than in any other political entity. There are at least fifteen species of cucurbits of economic importance in India, scattered among 10 genera, in contrast to USA where there are only 6 species of any considerable economic importance.

Cucurbitaceae has two sub-families, eight tribes representing varying degrees of circumscriptive cohesiveness and about 118 genera and 825 species (Jeffrey, 1990). Many cultivated and wild species which date back to pre-historic times, are associated with man's culture. Almost all the cucurbits are of tropical origin (Africa, America, Asia) and these are distributed all over the world in different ecological zones. Majority of cucurbits are monoecious annuals and propagated through seed but a few are vegetatively propagated (chow-chow, kakrol, pointed gourd) and a few are dioecious perennials (chow-chow, ivy gourd, kakrol, pointed gourd). Melons are mainly andromonoecious. Chow-chow is the only member of Cucurbitaceae that is viviparous, contains single seed and propagated by planting of whole fruit. Apart from cultivated species, enormous diversity occurs in the form of semi-domesticated and wild types. Semi-domesticated species and cucurbits of minor economic value are selected routinely by local tribes to meet their vegetable requirements and these play an important role in their life. Many of these species are fairly tolerant to biotic and abiotic stresses and can grow on soils, which are not suitable for other vegetables. Some of these species are nutritious and some have medicinal or other uses (Ram and Srivastava, 1999). Main species of cucurbits are described in this section.

Watermelon

Watermelon [*Citrullus lanatus* (Thunb.) Mansf] is a common summer crop and grown throughout India. There is a great deal of seed mixture in the named cultivars in India since little attempt is made to maintain cultivar purity. Usually these so-called cultivars are characteristic of single localities. Watermelon is a native of tropical and sub-tropical Africa. It reached India in pre-historic times. Diversity of fruits found in India is not nearly as great as one finds in countries bordering Mediterranean Basin where watermelon was presumably introduced at a much earlier date (Whitaker, 1973).

In *Citrullus*, three species are recognized. These are–*C.lanatus* (Thunb.) Matsum. and Nakai, *C.colocynthis* (L.) Schrad and *C.ecirrhosus* Cogn. Recently described species *C.rehmii* De Winter represents a valid fourth species (De Winter, 1990). *Citrullus lanatus* and *C. colocynthis* are found in India. In *C. lanatus*, domesticated watermelon is classified as var. *lanatus* whereas wild populations are generally classified as var. *citroides*, which include large watermelon. De Candolle (1882) considered that there was evidence of possibility of domesticated watermelon, *C.lanatus* being indigenous to tropical Africa specifically South Africa. It was probably introduced into cultivation in Egypt and spread to India in very early times. Cultivated forms have two centers of diversity, India and tropical and sub-tropical Africa. Cultivation of large watermelon (*C.lanatus* var. *citroides*) is comparatively recent and Soviet varieties grown today have shape of their African ancestor (*C.lanatus* var. *catter*). *Citrullus lanatus* ssp. *mucospermus* is supposed to be a wild form, and it has small fruit with white bitter flesh. *Citrullus colocynthis*, a potential crop of arid lands, grows wild in India exhibiting large variation in North-Western Plains and Central and South India and is an ancestor of watermelon (Shimotsuma, 1963). *Citrullus ecirrohosus* Cogn. is a perennial species without tendrils and with extremely bitter fruit. It is adapted to extreme climatic conditions of desert. Wild forms of several *Citrullus* species are found in dry sandy region of Kalahari Desert in South Africa.

Citrullus lanatus (Thunb) Mansf. (2n=22) is a short-lived trailing plant yielding 2-10 kg/annum of heavy, juicy and sweet fruits. It is widely cultivated in drier lowland regions of tropics. At least 300 local and modern cultivars are described. Charleston Gray, Mississippi, Sugar Baby, Chilian Black, Black Diamond, Congo and Crimson Sweet are cultivars which are particularly suitable for growing in tropics. Many local cultivars are known, most of them occurring in South East Asia and India.

In Africa, local cultivars with bitter fruits are widely cultivated for edible seeds. These seeds, called 'egusi' in West Africa, are an interesting product of traditional mixed cropping system, giving an additional crop with a high nutritional value. Very little information on this locally important crop is available. Fruits of several wild types are also used as a stewed vegetable. Watermelons require a relatively dry warm climate and a dry period is essential for fruit setting. It is a popular dessert fruit in near East, North Africa, Latin America and South East Asia. Commercial cultivation during dry season, under irrigation, is very common. No specific reaction to daylength is reported. Optimal day temperature range is between 25 and 30 °C. Nutritional value of watermelon is very low, lower even than that of cucumber. Seeds are particularly valuable and African cultivars contain 50 per cent oil and 35 per cent protein with high amounts of essential minerals (Fe, Ca, Zn, P, K, Mn). Amino acid composition compares with soybean but melon seeds have limited lysine and methionine.

Watermelons are highly commercialized and constitute an important product of city markets. Fruits are easily damaged and, in a warm and humid atmosphere, they can be stored for short periods only. This vegetable is not particularly suitable for intensive production and for home gardens since it requires a large area for full development.

Modern cultivars of watermelon are extremely susceptible to pests and diseases. Plants are often attacked by aphids, cucumber beetle and other insects. Ripe fruits are often seriously damaged by fruit flies and melon fly. Main diseases are: *Cercospora* leaf spot, anthracnose (*Colletotrichum*), *Fusarium* wilt, downy mildew (*Pseudoperonospora*), powdery mildew (*Sphaerotheca, Erysiphe*), and mosaic virus. The plant is also very susceptible to root-knot nematodes (Grubben, 1977). Dessert, beer, fresh juice or juice boiled down to heavy sweet syrup are various forms in which fruit is consumed. Rind is pickled or turned into a sugar candy (tuty-fruity). Seeds are protein-rich (25-32 per cent) with a high edible oil content (20-40 per cent). Roasted seeds are palatable (Dutta, 1992).

Origin and Distribution

The species originated in tropical Africa, where numerous wild types are present. It has been cultivated in Mediterranean area and in Asia for several thousand years where secondary diversity occurs (Grubben, 1977). It was introduced into India at an early date and reached China in AD 900-

1000. Widely cultivated throughout tropics and sub-tropics, watermelon is now grown largely in India, most parts of South East Asia, Africa, Caribbean and Southern States of USA. In India, river beds of Yamuna, Ganges, Narmada in the North and Pennar, Cauvery, Krishna, Godavari in the South are homes of the crop (Dutta, 1992).

Genetic Diversity

Spontaneous mutations and chromosomal rearrangements in watermelon are mainly responsible for genetic diversity expressed in plant morphology, physiology as well as their ability to survive under biotic and abiotic stresses. Hence, genetic diversity of land-races and wild relatives of crop are crucial to plant breeders in providing a diverse assortment to gene combinations. Excessive growth of vines of watermelon (2-6 m) often becomes a limiting factor in extensive cultivation in uplands. Dwarf mutants for example 'Bush Baby' help in restricting vine growth to 0.5-1 m and in controlling multiple branching from crown of plant. It helps in accommodating larger population (15,000-20,000 plants/ ha) in a unit area. Spontaneous mutations induced several economic variations in watermelon such as non-lobed leaves, glabrous stem and leaves and male sterility, which can be of use in hybrid seed production. Wide genetic diversity occurs in fruit size, weight, rind colour and pattern, rind toughness, flesh colour, texture and taste, seed size and seed colour of watermelon. Fruit is round to oval or oblong in shape, weighs 2-25 kg. The heaviest fruit weighing 117.93 kg (260 lb) was produced at Hope-Arkansas, USA. Choice of sizes varies among consumers and retailers. When cut and sold in pieces, larger fruits are preferred. Consumers in big cities prefer small fruit–an ice-box watermelon weighing 2-4 kg. Rind colour and pattern give the fruit an attractive appeal–colour varies from light green to dark green, dark blue and yellow, with or without bloom and the pattern from broad or narrow stripes, pencil linings and network lining to a mottled appearance. A rind lighter in colour is spotted by birds more conveniently unlike a deeper shade, a natural camouflage. When exposed to bright sunlight, the darker fruit gets sun-scald. Rind surface may be either smooth or with shallow furrows, thin or thick, tough or explosive. Thin but rough rind in watermelon is preferred to thick and explosive rind. Majority of Indian and American varieties of watermelon ('Peacock', 'Crimson Sweet', 'Arka Manik', 'Durgapura Meetha') have tough rind for long transportation. Watermelon varieties developed at Taiwan and Japan ('Yellow Baby' 'Carnival', 'Red Glory') have thin and explosive rinds difficult for long-distance transport in our climate. Flesh colour of fruit varies from white to yellow, canary yellow, orange, pink, red and crimson. In general, the bright red colour ('Giza' or deep crimson 'Arka Manik' and 'Crimson Sweet') are preferred in Indian markets. Yellow fleshed watermelon 'Yellow Baby' and 'Durgapura Kesar' do not gain with the Indian customer. Mouth appeal of fruit is attributed to sweetness, aroma and texture of flesh. Sweetness (TSS) in watermelon varies from 6 to 15 per cent. Ideally TSS value should be between 13 and 15 per cent. Sweetness in watermelon is mainly due to glucose, fructose, and sucrose; which increase with maturation, ripening and temperature. Ten to twelve per cent is approved by customers. Fructose and glucose are primary forms of carbohydrates transported in watermelon fruit tissue as photosynthetic products. These sugars combine in a 1:1 molecular ratio to form sucrose molecules. After anthesis (flower opening), reducing sugars (fructose and glucose) greatly increase in concentration from 20 to 36 days and decline thereafter. Non-reducing sugar (sucrose) starts accumulating 20 days after anthesis and thereafter increases linearly upto ripening. In general, total sugar concentration increases from 20 to 60 days after anthesis, the most rapid increase occurring between 20 and 36 days. Maximum flesh colour development in watermelon may not coincide with its sugar concentration. It is thus difficult to correlate the two with harvesting time. However, natural drying of tendril near fruit stalk, yellow ground spot on belly of fruit and dull sound of fruit when tapped with knuckles are fairly good indicators of a ripe watermelon ready for harvest.

Texture of flesh is crisp, granular, leathery or fibrous. A granular texture and delicate aroma are preferred by most consumers. Watermelon varieties–'Arka Manik', 'Crimson Sweet', 'Yellow Baby', 'Paradise', 'Sweet Marvel' and 'Sugar Delicata' have a good combination of high TSS, granular texture and fine aroma. Seeds per fruit, their size and colour also show remarkable variation. Number varies from 50 to 700 seeds in a fruit, length from 0.5 to1.3 cm and width from 0.1 to 0.5 cm. Triploid watermelons (seedless) have only a few hollow seeds. Seed coat possesses various colours like red, tan, white, black-grey, clump, white with tan or pink tip, tan with black tip and black-dotted pattern. Dark brown or black seeds of medium or small size embedded in yellow, red or deep crimson-coloured flesh appear very attractive and are preferred to large white, red or tan-coloured seeds. Watermelon variety 'Arka Manik' developed at IIHR, Bangalore, has triple diseases resistance to anthracnose, powdery mildew and downy mildew. Watermelon varieties 'Calhoun Gray', 'Summit', 'Smokylee', 'Dixilee', 'Sugarlee' and 'Texas W-5' are resistant to *Fusarium* wilt. Varieties 'Congo', 'Fairfax', 'Charlston Gray' and 'Crimson Sweet' are resistant to anthracnose.

The wild relatives of watermelon–*Citrullus colosynthis, C. ecirrhosus* and *C. naudinianus* have tolerance to drought and salt and are also cross-compatible with watermelon providing no restriction to gene flow. Behaviour of wild alleles in cultivated *C. lanatus* will not be known until it is transferred there. Tremendous genetic diversity exists in these wild species and needs intensive utilization through developmental breeding projects (Dutta, 1992).

Future Thrust

Both wild forms and primitive cultivars should be collected from tropical Africa, although genetic erosion in this African material is not yet widespread. Local cultivars grown in Asia, however, could easily be lost as they become replaced by high-yielding commercial cultivars. Selection of watermelon cultivars with horizontal resistance to many pests and diseases, with a view to successful cultivation in the wetter parts of the tropics, should be initiated. More attention must be paid to research and breeding programmes on the West African types which are grown for their seeds (*e.g.* cvs. Bara, Serewe and Sofin in Nigeria) (Grubben, 1977). Ecogeographical surveys of wild relatives and races of watermelon should be taken up.

Research on biochemical techniques to characterize genetic diversity should be taken up. Developmental breeding programmes for evolving watermelon varieties resistant to: (a) spotted-wilt virus, (b) Fusarium wilt, (c) anthracnose, (d) watermelon mosaic virus and (e) nematodes, should be taken up (Dutta, 1992).

Muskmelon

Muskmelon (*Cucumis melo* L.) is an important crop in Northern portion of Indian sub-continent. Cultivars grown are mostly primitive, and of poor quality characterized by relatively large soft fruits with smooth skin which cracks at maturity, and they have white or light pink flesh. Flesh has a musky, granular texture, and is low in soluble solids. Until recently, no developmental work was initiated to improve muskmelon in India, but this situation is changing. A few investigators claim that muskmelon is indigenous to India. This idea seems untenable because historical evidence suggests that muskmelon was a late arrival in India. Moreover, its nearest relatives are found in Africa, not India. Muskmelon, however, proliferated extensively under cultivation after being introduced into India, and at present there are well developed secondary centers of distribution in several areas of sub-continent (Whitaker, 1973).

Centre of origin for *C.melo* is still not clear although evidences point to Africa where wild species of *Cucumis* with same chromosome number (2n = 24) frequently occur. Domestication may have occurred independently in East Asia and South East Asia including India. Today, primary centre of diversity for this extremely polymorphic species is in South West and Central Asia, mainly Turkey, Syria, Iran, Afghanistan. North and Central India, Trans Caucasian, Turkmenistan, Tajikistan and Uzbekistan; and there is secondary centre of diversity in China and Republic of Korea. Munger and Robinson (1991) proposed seven horticulturally important groups of melons based on fruit characteristics and uses, rather than phylogeny.

1. *Cucumis melo* var. *cantaloupensis* (muskmelon): It is usually andromonoecious and has medium-sized fruit with netted, warty or scaly surface; usually orange but sometimes green flesh, and aromatic or musky flavour.
2. *C.melo* var. *flexuosus* syn. var. *utilissimus* (snake melon, long melon): It is monoecious and has long and slender fruit which is consumed immature.
3. *C.melo* var. *momordica* (snap melon): It is monoecious. It has ovoid to cylindrical fruit and white to pale orange pulp with low sugar content. Smooth surface of fruits cracks at maturity.
4. *C. melo* var. *inodorus* (winter melon): It is mostly andromonoecious and has white or green flesh that lacks musky odour. Shelf-life of fruit is longer than muskmelon and is indehiscent at maturity.
5. *C. melo* var. *dudaim* (mango melon, pomegranate melon): It is pickling type and has small, globular fruit, having white flesh with acid flavour.
6. *C. melo* var. *conomon* (pickling melon): It has fruit with smooth tender skin, white flesh, usually with a little sweetness and odour. It is andromonoecious, often pickled and also eaten fresh and cooked.
7. *C. melo var. agrestis* (kachri): It is mostly wild but is cultivated to a little extent. It has slender vine and small fruit.

Muskmelon, *Cucumis melo* L. (2n = 24) is a short-lived, twining, monoecious plant with predominantly unisexual flowers, individual plants yielding 0.5-2 kg of heavy, juicy fruits. Growth is less vigorous than that of watermelon. Many primitive and modern cultivars are known. In the USA, muskmelons, cantaloupes and winter melons are distinguished. Some popular American varieties, cultivated in many countries, are: Hale's Best, Honey Dew, Honey Rock and Smith's Perfect. Each production area has preferred cultivars. In India and South East Asia, approximately 100 local and commercial varieties are grown. *Cucumis melo* var. *conomon* (Thunb.) Greb, is a pickling melon, cultivated in China and South East Asia for ripe fruits which are used as a cooked vegetable. Melons are mainly grown during dry season in sub-tropics and are difficult to grow in tropics in low rainfall regions. They prefer somewhat lower temperatures than watermelons, the optimum growing temperature range being 20-35 °C. Nutritional value of white-green fleshed melons is higher than that of watermelons, owing to a greater vitamin C content. Deep yellow muskmelons or cantaloupes also have a much higher β-carotene content (1.7 mg/100 g) than the green-white fleshed melons (0-0.1 mg/100 g). Melons are a popular fruit vegetable in developed countries than watermelons, but they are much less important in tropical developing countries. Melons are affected by same pests and diseases as watermelons. They are very susceptible to shoot borers and fruit flies and cannot be grown successfully in wet climates. Melons may also suffer serious damage from downy mildew (*Pseudoperonospora*), powdery mildew (*Sphaerotheca*) and bacterial soft rot (*Erwinia*) (Grubben,1977).

Origin and Distribution

Muskmelon originated in tropical and sub-tropical Africa (South Africa), more specifically the Eastern region, South of Sahara Desert. This area has 40-odd wild species (2n = 24, 48,72) showing phylogenetically older characters, most being good sources of resistance to pests and diseases. Secondary centers of diversity are older cultivation areas in Asia (China, India, Iran, Southern Russia, Afghanistan, Pakistan). Working collections are maintained at several gene banks and at stations where breeding work is carried out. Muskmelon was introduced into India, probably during Mughal invasion around 14th century from Central Asia region. Since then, it spread to different parts of the country as far down to Andhra Pradesh and Karnataka. Genes for resistance to powdery mildew and many primitive characters have all been located in the indigenous material (Grubben, 1977; Tarsem Lal and Dhaliwal, 1992).

Genetic Diversity

Root system of muskmelon is extensive but shallow. Plants develop a strong taproot. Leaves are alternate, simple, palmately five-lobed and somewhat angled when young, becoming subcordate at maturity. Tendrils are borne in the axils of leaves and are simple. Cultivated muskmelons are mostly andromonoecious (bisexual flowers and staminate flowers: on separate nodes). Only a few are monoecious. Staminate flowers are borne in axillay clusters of 3-5. Muskmelon includes fruits popularly known as cantaloupes, honey dews, cassabas and Persians. Fruit is an inferior berry or pepo; varying in size with the cultivar; in shape (oblate, spherical, round, oval, or elongate). In most cultivars, fruit has 10 sutures. Towards maturity, fruit surface may be smooth or netted. Edible portion is mostly pericarp with very little mesocarp. Surface of immature fruit is usually green, becoming yellow or brownish upon maturity. Fruits may or may not slip on maturity. Flesh colour may be green, white or salmon-orange. Fruits are primarily used fresh when ripe, unripe may be cooked as vegetable by village folk. Flesh provides a wholesome food, rich in vitamins A, B and C and minerals calcium, phosphorous and iron in available form. Seed kernels are edible, tasty and nutritious. They contain a high percentage of oils, and are a rich source of energy.

Plants flourish well under warm climate and cannot tolerate frost. Optimum temperature for germination of seeds is 35°C. With increase in temperature, plants complete their vegetative growth earlier. Stormy weather particularly dust storm during flowering reduces fruit setting. Dry weather with clear sunshine during ripening ensures a high sugar content, better flavour, quality and a high percentage of marketable fruits. High humidity increases incidence of diseases, particularly those affecting foliage. (Tarsem Lal and Dhaliwal, 1992).

A great deal of breeding work on melons is initiated, mostly in the USA. Hybrid varieties with resistance to *Fusarium* wilt and to powdery and downy mildews are developed. Advanced cultivars combine high yield and quality with resistance to several diseases (Grubben, 1977).

Future Thrust

Primitive American and Mediterranean cultivars have been collected. There is, however, a danger that local Asian cultivars will be lost. If breeding work for tropics are considered, increasing carotene content should be an important objective although production of cultivars with high degree of pest and disease resistance should have the first priority (Grubben, 1977). To strengthen national resources in the crop, countries in Central Asia, the Gulf Mediterranean region, tropical East and South Africa (the major muskmelon growing areas) may serve for collection of germplasm through intensive surveys. Local types should continue to be preserved. Breeding for resistance against downy mildew, powdery mildew, viral disease complex, and insect pests like red pumpkin beetle, fruit fly and red spider mite are priority areas.

Crop production would be stabilized and use of chemicals restricted, since muskmelon is a dessert fruit. For heterosis breeding male sterility and gynoecism in the crop need to be exploited. F_1 hybrid seed production would be efficient and economical. Riverbed cultivation is more than 80 per cent of total area planted under melons in the country. It requires scientific approach for commercial success (Tarsem Lal and Dhaliwal, 1992).

Cucumber

In India, cucumber (*Cucumis sativus* L.) is an important summer vegetable. Cucumbers are used raw, pickled, and oddly enough, as a cooked vegetable. Cooked cucumbers make a delicious dish. There is much evidence to suggest that the cucumber is indigenous to Himalayan region of Northern India. Cucumbers of India show enormous variation, and in the foothills of the Himalayas, a wild form of cucumber, *C.hardwickii* Roy is found. *Cucumis hardwickii* crosses readily with different cultivars of *C. sativus* without loss of fertility in the F_1 or F_2 generations. Cucumber and *C.hardwickii* have 7 pairs of chromosomes, contrasted with other species of *Cucumis* which are mostly African and have 12 pairs of chromosomes. Morphologically cucumber is a much different plant from the African species of *Cucumis* and should probably be placed in a separate genus to properly express its relationship with African members of the genus. A taxonomist proposing such a change, however, would surely be shouted down by his fellows as a disturber of status quo, and bitterly denounced as a splitter. Cucumbers of India are conveniently separated into 4 categories: (1) the exotic group of European and American cultivars; (2) cultivars from Western part of India that are xerophytic in nature; (3) Chinese cultivars with long fruits and glossy skin; and (4) Himalayan types, mostly with rusty skin and some with perfect flowers. It is among this fourth group that we find the most primitive types of cucumbers (Whitaker, 1973).

Cucumis has more than 30 species of which cucumber (*Cucumis sativus*) and melon (*C.melo*) have major economic importance. *Cucumis sativus* originated in India where its close relatives with same chromosome number (2n = 14) occur. Among all these species, *C.sativus* and to a limited extent *C.callosus* (Rottle.) Cogn. are cultivated while *C.hardwickii, C.hystrix, C.prophetarum* L. and *C.setosus* Cogn. are wild (Chakravarty, 1982; Arora and Nayar, 1994). Rich variability is found in foothills of North West Himalayas and Southern Hills for *C.hardwickii*; North East Himalayas (Meghalaya, Mizoram, Assam) for *C.hystrix*; Rajasthan (Sirohi district) and parts of Gujarat, Maharashtra and Tamil Nadu for *C.prophetarum* and parts of Maharashtra for *C.setosus*. Cucumber is cultivated for at least 3,000 years in India and 2,000 years in China. China is a secondary center of diversity where it occupies the second largest area among vegetables after Chinese cabbage. Two other species of importance in *Cucumis* are *C.anguria* L. (West Indian gherkin) and *C.metulliferus* E.Mey. (African horned cucumber). *Cucumis anguria* produces extraordinarily large number of fruits and has potential for wider use in gherkin production (for example, as small pickled fruit). For many years, it was thought to be of American origin, but sexual compatibility and electrophoretic studies showed that it is a cultigen, which descended from a non-bitter mutant of a bitter African species *C.longipes* Hook. f. *Cucumis metuliferus* is also of African origin and has commercial potential (Sirohi *et al.*, 2005).

Cucumber *Cucumis sativus* L. (2n = 24) is a monoecious, annual climber or creeper. Fruits are used in an unripe state–in temperate regions they are usually eaten raw in salads or pickled, but they are also stewed in tropical regions. Many cultivars of this ancient cultigen exist, often with varying ecological requirements. Cultivars which perform well under tropical conditions include: Tax Long,

Ashley, Fengshan Green and Poinsett, the latter having high resistance to powdery and downy mildews. 'Eprix' pickling cucumber has resistance to nematodes. Southern Cross, Market King and Pixie are recommended for Pacific Islands. Cucumber is an old cultigen from Asia and the Mediterranean area, cultivated for centuries. It is a typical vegetable of warm temperate and cool tropical areas. The optimum range of day temperature is between 22 and 30°C. The pickling cucumber cultivars, also called gherkins, prefer a somewhat cooler climate than cucumbers.

The calorific and nutritional value of cucumbers is very low. It is a popular vegetable in developed countries where there is an excessive calorie intake. Owing to the very high yields obtained, nutrient production per ha is reasonable. Cucumber after tomato is the second most important vegetable species in Western Europe, where it is cultivated mainly in glasshouses. In sub-tropics and tropics, however, cucumber is much less important. It is rarely cultivated in tropical Africa and America, but is more popular in Asia, particularly in cooler areas. Export of cucumbers and pickling cucumbers to temperate areas could become economically important for some regions.

Diseases of watermelons may also affect cucumbers, but to a more serious extent. During wet season, many leaf diseases, including mildews, cause a serious loss of production (Grubben, 1977).

Origin and Distribution

Cucumber probably originated from Asia (Himalayas, Burma and North India) where wild and weedy types of *Cucumis* are present *e.g. C.hardwickii* Royle. Related wild *Cucumis* species occur in Africa. Secondary centres of cucumber are in China and Near East. Many collections for plant breeders are available, comprising both primitive races and more advanced cultivars (Grubben, 1977).

Genetic Diversity

A great deal of work on breeding and genetics of cucumber has been done. Genetically, it is one of the best known vegetable species. Parthenocarpic and female cultivars are developed, and hybrids exhibiting heterosis are produced from female lines. Resistance to many diseases was obtained by hybridization, the most important being resistance to powdery and downy mildews, anthracnose, scab (*Cladosporium*), bacterial soft rot (*Erwinia*) and mosaic virus. Interspecific hybridization may open new perspectives (Grubben, 1977).

Future Thrust

With increase in breeding programmes and commercial seed production, old cultivars from India, China and other parts of Asia should be collected systematically for conservation, especially since material from secondary centers of diversity are useful in breeding for resistance to diseases. Selection work should aim at producing suitable cultivars for hot and wet tropics but no priority should be given to cucumber in these areas, as economic importance is limited and nutritional value very low (Grubben, 1977).

Pumpkins and Squashes

Cucurbita genus is important and better understood amongst cucurbits because of extensive investigations on phylogeny (Whitaker and Davis, 1962). It is comparatively a closed group or system of about 27 cultivated and wild species concentrated in tropical regions of Central and South America; and all the species have chromosome number, 2n = 40. This genus includes three species of commercial importance, namely pumpkin (*Cucurbita moschata*), summer squash (*C.pepo*) and winter squash (*C.maxima*); and two lesser-known species, namely, fig leaf (*C.ficifolia*) and *C.mixta* (Andres, 1990). Robinson and Decker-Walters (1999) renamed *C.mixta* as *C.argyrosperma*. Cultivated *Cucurbita* originated in two centers in new world. Central Mexico is center of origin for *C. ficifolia*, *C. mixta*, *C. moschata* and *C. pepo*, while Southern Peru, Bolivia and Northern Argentina are centers of origin for *C. maxima*. Areas of diversity are Northern Mexico for *C.pepo*, from Central Mexico and the Yucatan Peninsula to Coasta Rica for *C. mixta*, Central Mexico through the high plateau of Central America and along Andes to Central Chile (1000-2000 m elevation) for cold tolerant *C. ficifolia*, Southwards from Mexico City through Central America into Northern Columbia and Venezuela for *C. moschata*, in Northern Argentina, Bolivia, Southern Peru and Northern Chile for *C. maxima*. Whitaker and Bemis (1975) assigned a key role to *C. moschata* as the nearest putative ancestor of the genus. They suggested *C. moschata* as a link between wild and cultivated species and that all the five cultivated species are independently domesticated. In India, *C.moschata* is more widely distributed than other species. According to Decker (1988), archaeological evidences indicate occurrence of wild populations of *C.pepo* in Mexico and Eastern USA around 10,000 and 30,000 years ago, respectively. Domestication of this species took place independently in these two areas. *Cucurbita argyrosperma C.Huber* was apparently domesticated in Southern Mexico where archaelogical evidences of its domestication date back to 5200 BC. Wild population of sub-species *sororia* that occurs in Mexico and Central America, gave rise to domesticated sub-species *argyrosperma*. Land races of sub-species *argyrosperma* var. *argyrosperma* may have been selected first, the other two group of cultivars, namely, var. *callicarpa* and var. *stenosperma* probably evolved from Northern and Southern land races of var. *argyrosperma*, respectively (Merrick, 1990). *Cucurbita ficifolia* was cultivated in Andes Mountains of Peru in pre-historic times. As species spread northwards, a large number of perennial xerophytic types evolved adapted to desert and plateau regions of Northern Mexico and South Western USA. Attempts are made to domesticate another xerophytic species, namely, *C.foetidissima*. This species is perennial and grows wild in arid deserts of South Western USA and Mexico. Its ability to withstand drought has attracted interest in its commercialization. It produces high yield of seed that is rich in protein and oil, and that of fleshy tuberous roots. In addition to xerophytic types, species adapted to humid forest conditions also migrated to North, and one such species *C. okeechobeensis* is found as far as North Florida. Amongst wild species, *C. lundelliana* Bailey, *C. martinezii* Bailey and *C. okeechobeensis* Bailey are resistant to powdery mildew disease. *Cucurbita martinezii* also possesses resistance to squash mosaic and watermelon mosaic virus whereas *C. foetidissima* Kunth is resistant to drought. *Cucurbita pedatifolia* Bailey is a possible bridge species between xeropytic and mexophytic species. The Indian sub-continent is the center of origin for a number of wild and cultivated cucurbits.

Fruits of *Cucurbita* spp. are called pumpkins, winter squashes, summer squashes or vegetable marrows, but these names give no indication of relevant botanical species. The most commonly cultivated species are *C.pepo* L., *C.moschata* Duch, and *C.maxima* Duch. Fruits of *Cucurbita* spp. with a hard outer skin, suitable for long storage and used when ripe, are called 'winter squash' in USA. This latter term also includes the 'pumpkin'; this type is the one most commonly grown in tropics. Pumpkins are yellow-fleshed and are normally described as belonging to species *C.maxima*. A few important pumpkin cultivars are: Sugar, Cheese and Cushaw and Baby Delica, Butternut, Blue Kierj are recommended in the Pacific Islands. Summer squash or vegetable marrow (mostly *C.pepo*) is the name given to *Cucurbita* spp. consumed when immature. Well-known cultivars are: Long White, Cocozelle and Zucchini, all bush types of Italian origin. Other cultivars have a vine-like growth habit, such as Crookneck and Straightneck. Winter squash cultivars include: Table Queen and Hubbard. Many local cultivars of pumpkins, marrows and (winter) squash are known in sub-tropical and tropical areas, principally in South East Asia, and Meso America, less so in Africa. Tropical lowland pumpkins are generally referred to as *C.moschata*.

Pumpkins, and to a lesser extent marrows, are popular vegetables in Asia and Latin America, but are less widely grown in Africa. Pumpkins are cultivated in home gardens, on refuse heaps, and also as intercrops in fields *e.g* together with maize. In addition to ripe fruits, flowers and young leaves are also eaten. Bushy types of marrow are cultivated by commercial vegetable growers. Performance is better in wet areas than in dry and they may be grown during rainy season when melons and watermelons are less successful. Pumpkins and squashes are grown from sea level up to altitudes of 2000 m amsl (Grubben, 1977).

Generally cucurbits are low in nutriive value. Pumpkin has high stored carbohydrate (energy), vitamin C and minerals. Yellow- and orange-fleshed fruits are very high in carotene, a precursor of vitamin A. Pumpkin seeds contain protein and edible oils, and constitute a dietary item for Latin Americans, people of Southern Asia and Central Africa. In fact, pumpkin has a place of high value due to its very high productivity per unit area, good storability, availability for a long period, more consumption, better suited for distant transportation and extensive cultivation in India. Pumpkins are marvels of vegetable world due to their many unusual and extravagant characters. Pumpkin fruits are used as vegetable both in immature and mature stages. They are consumed as vegetable, processed food and stock feed. Flesh is delicious when stewed, boiled or baked. Fruit is sweetish when fully mature and may be used in sweets, pie, jams or candied or fermented to make a beverage. Male flowers and tender twigs of it are also consumed (Sirohi, 1993). Nutritional value of *Cucurbita* species with yellow or red flesh is higher than that of cucumber, due to high carotene content. In unripe state, nutritional value of fruits of marrow is much lower. However, their consumption is less widespread than that of pumpkins or squashes, which are generally consumed in ripe state. Pumpkin leaves are a popular leaf vegetable in Asia (Philippines) and West Africa. Pumpkins are sold in tropical markets as a cheap form of produce which is of minor importance. Cultivation is relatively easy and level of production is high. Pumpkins store well over long periods, especially hard-rind winter squashes. Pumpkins are subject to same diseases as watermelon, but to a lesser extent. In hot, wet tropical climates, many fruits may be lost due to *Choanephora* wet rot, and *Alternaria* leaf spot may cause moderate infections (Grubben, 1977).

Origin and Distribution

Cucurbita spp. are native to Southern part of North America, Mexico and Central America, South to North Argentina. *C.lundelliana* Bailey is probably the progenitor of cultivated species and they are related to several other wild or weedy species. *C.maxima* shows secondary diversity in India. In parts of South East Asia there are very old introductions of *Cucurbita* spp. Breeding stations maintain working collections (Grubben, 1977). Genus *Cucurbita* gives pumpkin, squashes or vegetable marrows. It is the most economically valuable genus with 27 species, five of which are cultivated all over tropical and sub-tropical world. It is indigenous to New World (America). From archaeological remains, it is possible to trace man's association with these species backward in time for a maximum of 10,000 years. Among cultivated species, *Cucurbita moschata* Duch. ex Poir, known as pumpkin, is the most important and extensively cultivated in India (Sirohi, 1993).

Genetic Diversity

In pumpkin, high genetic variability in respect of both qualitative and quantitative plant and fruit characters are found. Very little improvement work is done on this vegetable crop. At Indian Agricultural Research Institute (IARI), New Delhi, a large germplasm of this cucurbit (indigenous and exotic) was collected, purified and assessed to develop high yielding varieties and F_1 hybrids with desirable fruit characters. All these lines are highly susceptible to low temperature (below 8 °C) and die below 5 °C.

Fruit setting is adversely affected, if temperature rise above 38-40 °C. The lines are significantly different in vine growth (1.5-6.0 m). Very little difference is observed in stem characters of different lines except in surface of stem which varies from soft hairy, harsh to somewhat prickly. Leaf characters could be classified into: leaf shape (ovate, orbicular, reniform, retuse), leaf size (small, intermediate, large), leaf margin (smooth, dented), leaf lobes (absent, shallow, intermediate, deep), and leaf spots (white or silver, absent). Two genotypes were developed with light yellow spots on surface of leaves and inheritance of two colours white and yellow, studied. These two spot colours differed by a single pair of genes. This trait assumes importance and can be effectively used as a 'gene marker' in the commercial hybrid seed production of pumpkin. Pumpkin is monoecious in sex behaviour, flowers are unisexual solitary in the leaf axils. Generally stalks of the male flowers are long (3-10 cm) and thinner (0.50-0.75 cm) than the female flowers. The stalk of the female flowers is distinctly angular and longitudinally furrowed, 2-6 cm long and 0.75-1.0 cm thick. Like other cucurbits, male flowers appear before female flowers in this crop. However, few lines showed a reverse trend by producing the female flowers first. There was much variation in the ratio of male:female flowers (5 male : 1 female to 20 male : 1 female).Variation in days to male flower opening was 55 to 70 days, and for female flower opening it was 60 to 75 days.

Anthesis of flowers takes place before sunrise. If pollination are attended before 8.00 AM, fruit-setting could be increased. Honeybees are the chief insect pollinators for good crop. The flowers of pumpkin should be handled or touched as less as possible for better fruit-set in case of hand-pollination because they are very sensitive. Very wide genetic variability in fruit characters is available. That the fruit shape (globular, round, flattened, dumbbell, disc, long, cylindrical, oval, acorn or heart-shaped, pyriform, crooked neck, elongate form), fruit size index (small, medium, large and very large), flesh thickness (thin, medium, thick, very thick), flesh colour (cream, yellowish green, yellow, orange, salmon), fruit skin colour at maturity (green, blackish green, cream, yellow, orange, brown), fruit skin (smooth or no ribs, superficial, intermediate, deep), flesh texture (smooth firm, soft spongy, fibrous, grainy) and peduncle separation from the fruit (easy, intermediate, difficult). The number of fruits per plant varies from 1 to 6 and average fruit weight from 1.5 to 40.0 kg. However, the largest pumpkin fruit may weigh more than 180 kg. Maturity of fruits or number of days from sowing to harvesting is quite long (100-150 days). Variations among genotypes for storing their mature fruits at room temperature could be divided into: poor (one week), low (one month), intermediate (2 months), high (3-4 months) and very high (above four months). Variations among genotypes for seed characters (seed size, colour and seed weight) has also been observed (Sirohi, 1993).

Breeding problems in marrows (*C.pepo*) have received much attention in the USA and South Europe. High yielding hybrid cultivars are now becoming popular. Amphidiploidy and embryo cultures are useful techniques for creating inter species hybrids. In Japan and the USA, commercial cultivars of hybrids of the three major cultivated species were developed. Little work has been done on tropical pumpkins (Grubben, 1977). Among cultivated species of *Cucurbita* (*C. moschata, C. pepo Linn., C.maxima* Duch, *C. mixta* and *C. ficifolia* Bouche), certain degree of cross compatibility was demonstrated through hand-pollination. Genetic variability within the species for interspecific compatibility was observed. Most compatible crosses produce only slightly fertile F_1 hybrids and self-fertile F_2 progeny. A wild species *Cucurbita lundelliana*, resistant to powdery mildew, is called a 'Bridge species' of *Cucurbita*. It is cross-compatible with all cultivated species of genus *Cucurbita* to the extent of producing fruits and seeds. This is an important species for plant breeders to evolve better varieties of cultivated species for quality, disease resistance and adaptability (Sirohi, 1993).

Future Thrust

Ancient tropical cultivars are in danger of disappearing with introduction of modern commercial cultivars, but, to date, danger appears to be slight. Screening of various tropical cultivars of yellow-fleshed pumpkins and further improvement by hybridization is highly desirable. The yield potential should also be improved (Grubben, 1977).

Bottle Gourd

Bottle gourd [*Lagenaria siceraria* Moll. (Standl.)], is extensively grown in India. Immature fruits are used as vegetable. Mature fruits with their characteristically hard shells are used as containers, domestic utensils, musical instruments, floats for fishing nets, and many other purposes. There is much variation in size and shape of *Lagenaria* fruits in India. Indians developed cultivars especially adapted to rainy season. In fact, many localities have cultivars with local names and are adapted to needs of specific locality in which they are grown.

Latest research suggests that Bottle gourd is native to tropical and sub-tropical Africa, the species of which it is most closely allied are from this area. It was probably transported to Americas at an early date without assistance of man (Whitaker, 1973). *Lagenaria*, previously considered a monotypic genus, has six species which are of African origin. *Lagenaria siceraria* (bottle gourd) is the only cultivated species. It is cultivated throughout tropical and sub-tropical regions in Asia, Africa and South America. Based on morphological differences, cultivars and land races of African and New World origin are recognized as *L.siceraria* ssp. *siceraria* and those of Asia as *L.siceraria* ssp. *asiatic*. Other three species, namely, *L.abyssinica* Hook. f., *L.breviflora* Robert. and *L.sphaerica* (Sond.) Naudin, are wild perennials (Sirohi *et al.*, 2005).

The bottle gourd *Lagenaria siceraria* (Mol.) Standl (2n = 22) is apparently the only species of its genus. The name is synonymous with *L.vulgaris* Ser. and *L.leucantha* (Duch.) Rusby. It is also called 'trumpet gourd', the calabash gourd, and the white-flowered gourd. The name "bottle gourd" is especially appropriate, because this plant species is one of the few from which useful and lasting containers can be made (Martin, 1977).

Bottle gourd is a vigorous creeping or climbing annual plant. It is mainly cultivated for the hard fruit skin, which is used in manufacture of domestic utensils. Other cultivars are grown as vegetables, young fruits and, occasionally, ripe seeds, being eaten. Many local cultivars with various shapes are grown, but commercial varieties are also available. Bottle gourd is a popular vegetable in China, India, South East Asia and Sri Lanka where fruits are selected which are less bitter. It is grown for market and for domestic use. Crop is well adapted to a hot, dry climate. Fruits have a low nutritional value (ANV =1.61), comparable with that of cucumber. Bottle gourd rarely suffers from diseases and pests although plants are susceptible to anthracnose during rainy season in India. Production level is low, compared with that of cucumber. Storage life is very limited (Grubben, 1977).

Bottle gourd may be the only cucurbit plant known in both the New and Old Worlds in early pre-historic times. Archeological evidences support man's association with bottle gourd in Peru from 11000 to 13000 years B C. The genus *Lagenaria* considered monotypic for a long time was later reported to have 5 dioecious and perennial species. Out of all, only *Lagenaria siceraria* was monoecious, annual and widely cultivated throughout tropics and sub-tropics. In addition to its use as vegetable, tender fruits are used for making sweet meats and hard shells of mature fruits for making water jugs, domestic utensils, musical instruments and floats for fishing nets. Bottle gourd fruits contain small quantities of certain minerals and vitamins and its flesh, tender stem and leaves have medicinal values (Sirohi and Sivakami, 1991).

Origin and Distribution

Several wild species of genus *Lagenaria* occur in tropical Africa, which may be a center of diversity. It spread throughout tropics at an early date and secondary centers are present in tropical America, Brazil, India and South East Asia, where the vegetable types are found. A few materials are available in national institutes, particularly in India (Grubben, 1977). Bottle gourd traveled to India from Africa where it evolved into numerous local varieties, and from India to China and to Indonesia and as far as New Zealand. Archeological remains show that bottle gourd was used in Egypt about 3500 to 3300 BC. First records in China are of the 1st century AD and in New Zealand by 12th century AD. *Lagenaria* also traveled to New World. Dried gourds survived in sea water for at least 224 days, and seeds remained viable (Whitaker and Carter, 1954). Fruits might have travelled to New World by sea. Remains found in Mexico date from 7000 to 5500 BC and in Peru from about 10000 BC (Richardson, 1972). Bottle gourd is thus an ancient crop, widespread and well used, from warm parts of Temperate Zone throughout dry and wet Tropics. It is the only crop known to have been cultivated in pre-Columbian times in both Old World and New World. Evidence from Heiser's study (Heiser, 1973) of the penis gourd in New Guinea supports the idea of cultural diffusion across the Pacific. Gourds of New Guinea are more like those of South America than those of Africa.

Genetic Diversity

Cultivation of this cross-pollinated crop over thousands of years exposed a large variation in several of its qualitative and quantitative characters. In an intensive breeding programme at Indian Agricultural Research Institute (IARI), New Delhi, more than 300 genotypes (lines) were developed by repeated inbreeding and through individual plant selection. In general, it was observed that bottle gourd loves hot and humid climate for good vine growth and high yield and therefore productivity per unit area is more in *kharif* than when grown in summer. Majority of the lines are highly susceptible to low temperature (below 8 °C) and usually die if temperature drops below 5 °C. Variability is observed in certain lines in adverse conditions and possibility of breeding off-season cultivars in cucurbit may thus be feasible. Fruit-setting is affected adversely in very hot summer (around or above 40 °C) due either to a very poor or no activity of insect-pollinators. Hand-pollination gives good results and can effectively be carried out in kitchen gardens and in breeding plots to ensure fruit-set.

Though very little differences were perceived in morphological attributes of stem, leaf and flower, in bottle gourd a peculiar leafy calyx is present in male flowers of a few lines. Vine may grow from 1.5 m to 8 m long. All genotypes of cultivated bottle gourd are strictly monoecious bearing male and female flowers separately on same plant. Generally, male flowers appear first followed by both female and male flowers. A few lines showed a reverse trend by putting forth female flowers first. There was much variation in ratio of male and female flowers–from 6:1 to 25:1 though genetically controlled, the sex ratio is influenced to some extent by temperature, humidity, light, soil fertility and soil moisture. A few workers reported the modification of sex ratio in cucurbit through application of certain chemicals and plant regulators capable of increasing femaleness, thereby yield. Their results suggest temporary variations in sex ratio may be created to advantage in desired lines (Sirohi and Sivakami, 1991). Bottle gourd is a vigorous, annual, running or climbing vine with large leaves and a lush appearance. It is fast growing and begins to flower only 2 months after seeding. Thick stem is furrowed longitudinally. Vine is much branched and climbs by means of branched tendrils arising from stem with leaf. Foliage is covered with soft hairs and has a musky or foul smell when crushed. Leaves of bottle gourd are up to 40 cm broad, circular in overall shape, with a cordate base and smooth margins, or with a few broad lobes, or with undulate margins. Leaves have a velvety texture because of fine hairs, especially on undersurface. Bottle gourd is monoecious. Flowers are borne singly on axils of leaves, the males on long peduncles and females on short peduncles. Flowers are white and attractive, up to 10 cm in diameter, with spreading petals. Ovary is inferior and in the shape of future fruit. Otherwise, male and female flowers are similar in appearance. Anthers are borne on short filaments grouped at center of flower. Stigmas are short, thickened and branched. In bottle gourd, a number of male flowers are produced before first female flower appears. More female flowers are produced in winter than in summer, and more by young plants than by old plants. Flowers, opening in late afternoon, are pollinated by diurnal and nocturnal insects. In India, the most important of these are *Carpophilus dimidiatus*, the corn sap beetle; *Planidia interpumetella* and *Pygaronia* spp., moths; and *Apis florae*, a small bee. Pollen is abundant and usually fertile. Stigmas are receptive to pollen germination from 6 hours before to 36 hours after anthesis (Singh and Singh, 1970). Fruit size is related to vigor of plant, number of fruits previously set, and time of year. This relationship seems to be no more than competition of fruits for available nutrients. Large fruits are obtained by pruning away newly set young fruits. Fruits of bottle gourd show an immense amount of variation in shape, ranging from flattened or discoidal forms to bottle like forms with one to three swollen regions, to long club shapes, sometimes over 1 meter. Furthermore, natural forms are further varied by artificially restricting growth with bands to develop special forms. Brownish seeds are numerous in a whitish-green pulp. The most spectacular variation in bottle gourds is with respect to fruits. Background colour is either light green or dark green. Dark green colour can be distributed as a solid colour, as regular or irregular strips, and as an irregular blotch. Size of fruit varies from 5 to 35 cm in diameter and from 10 to 120 cm in length. Shape of fruit is variable. Fruit can have a sterile (seedless) neck, which varies from a few to 40 cm in length and 2 to 4 cm in width. Wider necks usually contain seeds and neck may have a seed-containing bulge. Seed containing portion of fruit varies from discoidal to globose, to ellipsoidal, cylindrical, club shaped, or long and narrow. Long, narrow forms are the best for vegetables, and globose types serve as containers. Large, almost tube like fruits are useful for storage of seeds, flour, and so forth. Cultivars of Lagenaria are known throughout West Africa, where bottle gourd is grown for containers, but these cultivars are largely unrecorded in terms of name and characteristics. Probably no region has as much diversity and so many named varieties of bottle gourd as India (Martin, 1979).

Fruit shapes are very distinct, broadly classified into long, oblong-round and miscellaneous types. Genotypic expression of long bottle gourd fruits are long-straight club with or without 'neck', long–curved club with or without 'neck' and long-straight cylindrical. Oblong fruits were oblong-club, oblong pear or oblong-cylindrical. Round-fruited were seen as disc, round-flat, almost round, round-oblate, round-deep oblate, globe, round-pear and ovate. Top or acorn, pot or pitcher, dumbbell and bulb formed miscellaneous group. Blossom end of fruit was rounded, pointed, nippled, flattened or depressed. Blossom-end scar on fruit was variously observed as a spot of small, medium or large size, sometimes undesirably extra large, flared and tough. Fruit stalk (peduncle) was 8-25 cm long. An unusual trait, fruit 'slipping' from stalk even with a slight jerk to fruit, seen in one of the genotypes, is an undesirable trait. It would not only make post-harvest handling difficult, but lessen market quality of fruit. Most of the genotypes produce light green fruits at edible stage. Some variation was also observed, a few lines producing either green, dark green or greenish-white fruits. Some lines had striped or mottled fruit surface while majority had nonstriped fruit skin. Flesh colour was either white, cream or greenish-white, and texture soft-firm, soft non-firm or slightly tough or even leathery. In bottle gourd, yield is directly affected by important quantitative traits: fruit size, length and girth; number of fruits to a plant and fruit weight. Ample diversity for all the characters is present in the germplasm. Among long-fruited lines, fruit girth varies from 10-40 cm grouping into small, medium, large and extra-large. Fruits per plant vary from 2 to 20 and average fruit weight 250-12,000 g. Direct correlation was observed between fruits per plant, fruit size and weight. Total yield per plant is between 0.5-25 kg. First harvest could be taken between 55 and 120 days during summer and *kharif*. Variations among genotypes for certain mature seed characters have also been recorded. Seeds per fruit 200-1,500 and weight 50-500 g; colour off-white, light brown or dark brown. Rich variability available in bottle gourd germplasm, specially for fruit characters combined with monoecious sex behaviour, narrow sex ratio of male to female flowers, solitary nature and medium size of flowers (facilitating easy handling) and many-seeded nature of fruit, offers much scope for developing desirable types in gourd vegetable through careful selection and planned hybridization (Sirohi and Sivakami, 1991).

Future Thrust

Bottle gourds, for example, yield one of the few containers that can be produced in garden. They are made from a renewable resource, and their use does not interfere with ecological relations or destroy a fossil fuel. As they can be grown on fences, they are one of the crops which convert this portion of a small lot to food production. Another possible use of these species, still almost unexplored, is the use of seeds for their protein and oil contents. Varieties with edible seeds, free of toxic substances, might be found. *Lagenaria* could become important, easy-to-grow sources of protein and oil for hot, humid tropics (Martin, 1979). Although not generally in direct danger of genetic erosion (although this is clearly evident in Indonesia) an evaluation of local cultivars could be useful.

Bitter Gourd or Bitter Cucumber

Bitter gourd (*Momordia charantia* L) is extensively grown over all of the Indian sub-continent. The species is probably indigenous to India as it is found wild in many areas of the country. There is considerable variability in fruits. Warty, green fruits are about 15-20 cm long. There is also a form with short fruits, about one-half the length of those with longer fruits. Forms with short fruits are usually grown in dry season, long fruited forms in rainy season. Fruits are most commonly used with meat and fish dishes and in curries (Whitaker, 1973). The genus *Momordica* has 45 species, native to tropical Africa. Of these, seven are found in India but only four species (*M.balasamina, M.charantia, M.cochinchinensis, M.diocia*) are commonly grown. Balsam apple (*Momordica balsamina*) occurs in semi-dry North-Western Plains, Eastern Ghats and Western Ghats. Bitter gourd (*Momordica charantia*), a very popular vegetable in India, is grown all over tropical and sub-tropical regions. It has large diversity in India, China and South East Asia. It is most likely that its cultivation started in India, although a number of related wild forms exist in Africa. Sweet gourd (*Momordica cochinchinensis*) is popular in Tripura, Assam and West Bengal, whereas, kakrol (*M.dioica*) is cultivated in West Bengal, Assam, parts of Bihar and adjoining areas. There are several other *Momordica* species of minor importance, which include *M.foetida* Schumach. and Thonn., a traditional medicinal plant in West Africa, and *M.grosvenori* Swingle, which has potential for extraction of sweet glycoside component (Sirohi *et al.*, 2005).

Bitter gourd, *Momordica charantia* L. (2n = 22) is a slender, climbing plant. The bitter young fruits are used as a vegetable and leaves and tender shoots are also consumed. Synonyms of bitter gourd are bitter cucumber and balsam pear. Thick fleshed and less bitter cultivars grown in Taiwan, yield 20 tonnes/ha in 100 days. Bitter gourd is a very popular vegetable in India, China and South East Asia and is well adapted to hot and humid climates. It is grown in kitchen and market gardens. Nutritional value of green, small-fruited types of bitter gourd is surprisingly high for a cucurbit fruit. It has an ANV of 4.10, due to high iron and vitamin C content. Fruits are an important product in local markets in South East Asia and India. Storage possibilities are limited, no serious diseases have been reported, with the exception of fruit flies (Grubben, 1977). In West Indies, tropical Mexico and South America, bitter gourd is one of the most commonly used medicinal herbs. In United States, England and France, it is grown as an ornamental annual. Fruits of this vegetable are rich in several minerals and vitamins specially iron and ascorbic acid. They are a tonic, stomachic, carminative and produce a cooling effect. Fruits, leaves and roots have long been used as folk medicine against diabetes. Fruits of this vegetable are cooked in many ways–fried, boiled, stuffed, dried and pickled. Tender shoots and leaves of plant are consumed in some parts of the country (Sirohi, 1991).

Origin and Distribution

Origin is probably India but secondary diversity is seen in China and South East Asia. Wild species of *Momordica* are present in many tropical areas. Local cultivars of *M.charantia* are many in Asia and rare, but not unknown in Latin America (Grubben, 1977). It originated in tropics of Old World and is widely distributed in China, Malaya, India, tropical Africa and certain other countries. India is an important center of genetic diversity for bitter gourd (Sirohi, 1991).

Genetic Diversity

Selections are made by seed growers in India, Japan and China. In Bangalore (India), dwarf cv. Arka Harit is bred, combining high quality with a high yielding capacity (Grubben, 1977). Though a wide variability for plant and fruit characters is available in the country for this crop, very little has been done in improving existing genotypes. At IARI, New Delhi, most of the lines characterized are for susceptibility to low and very high temperatures; fruit setting severely affected if temperature drop below 10 °C or shoot up to 40 °C or more. A few lines set normal fruits in both summer and *kharif*, while a few set good fruit either in summer or *kharif*. Lines are different to some extent in morphology of stem and their leaf-size, shape, surface, number and depth of lobing and hairiness. Bitter gourd is a monecious, climbing and cross-pollinated annual. There is marked difference in male and female sex ratio among the lines of this crop–5 male : 1 female up to 40 male : 1 female. Generally male flowers appear first in plant, but in stray cases a reverse trend may also be observed. Distinct variability is recorded in fruit size (2.0-25 cm length and 2.0-15 cm girth), fruit shape (pendulous, fusiform, oval, round, spirally oblong, clubs), fruit surface (ribbed with broken or unbroken tubercles, smooth,soft spring), fruit colour (green, pale green, dark green, glossy green, white, creamy white), meatiness of fruit flesh (2-15 mm), seeds per fruit (4-50), seed size (5-15 mm), seed colour (white, brown, dark brown, tan), maturity (45-80 days), fruits per plant (6-85), fruit weight (10-200 g) and yield per plant (0.4 to 6.0kg). Even though a distinct difference in the content of certain minerals is established in different cultivars, no correlation is found between cultivars nor between size and colour of fruits and their nutritive value. The general belief that small and green bitter gourds are richer in nutritive values than longer and white ones is totally wrong (Sirohi, 1991).

Future Thrust

Local cultivars of bitter gourd may vanish as they become replaced by commercial cultivars. Measures for conserving wild cultivars are recommended especially in Indonesia and China. More information are required on distribution, variability and use of *M.balsamina* L. and *M.cochinchinenisis* Spreng. Bitter gourd offers good possibilities of breeding for higher productivity. Its economic importance justifies a high priority among the cucurbits, equivalent to that given to melons and pumpkins (Grubben, 1977).

Ash Gourd/Wax Gourd

Ash gourd, (*Benincasa hispida* Cogn.) is a minor cucurbit crop in India, but it is widely grown, mostly in home gardens throughout the country. Immature fruit is cooked and served as a vegetable.

It is also used in making candies. Thick flesh serves as a vehicle for impregnation with sugar. Wax gourd is highly prized for its medicinal value, presumably for curing patients with diseases of nervous system. Fruits are large, mostly 30 cm or more in diameter, and covered with a bluish, ash-gray bloom (Whitaker,1973). Ash gourd belonging to a monotypic genus, is a native of Indonesia (Java) and Japan. It was domesticated in India during pre-historic times. It is widely grown all over the country in tropical and sub-tropical regions and possesses large variability for morphological and quality characters. Its fruits have long storability. It is a potential crop of tropical Africa and America (Sirohi *et al.*, 2005).

Benincasa hispida (Thumg) Cogn. is commonly called ash gourd, hairy melon, wax gourd, winter melon, ash pumpkin and white pumpkin. *Benincasa hispida* is a vigorous, annual, climbing or creeping plant. At least ten different local cultivars are known in South East Asia, differing mainly in fruit shape. Fruit weight varies from 1 to 40 kg. A few crops are grown for production of immature fruits, but ripe fruits, which can be stored in a cool, dry place for at least six months, protected by their waxy outer layer, are more widely used.

Benincasa hispida is an important vegetable in Southern China, India and all countries of South East Asia, being grown from sea-level up to altitude of 1,500 m. above msl. It was introduced into Cuba and is found in a few regions of America, especially the Southern USA, where there are Asiatic communities. Plant requires a hot and moderately humid climate. In North India, it was demonstrated that the highest production (20 tonnes/ha) is obtained from sowings from June to August, flowering takes place in October-November; both cool weather and short days during these months stimulate development of female flowers. Although young leaves, flowers and both immature and mature seeds may be consumed, main product of wax gourd is mature fleshy fruit, seeds of which are removed before cooking. Food value is very low and is comparable with that of cucumber. It contains no pro-vitamin A (Grubbben, 1977).

Medium dry, low-lying tropical areas are chosen spots of ash gourd. It is a warm season vegetable cucurbit grown for succulent hairy fruits much used in confectionary and Ayurvedic medicinal preparations. An enzyme extracted from ash gourd juice is successfully used in the production of cheddar cheese instead of calf rennet. *B.hispida* is a potential fodder crop too. The sweet delicacy petha is made of ash gourd. The ayurvedic preparation 'Kooshmanda rasayan' is prepared from fruit. Potential of this crop as a nutritious cattle feed needs to be explored. Seeds could be a potent source of vegetable oil and protein (Peter *et al.*, 1991). Wax gourd is cultivated in home gardens with supports but is also grown as a cash crop in the field without supports. It is a highly commercialized vegetable due to appreciable long periods of storages without deterioration. The wax gourd is not very susceptible to pests and diseases. Fruit flies may damage young fruits and cater pillars sometimes feed on leaves. Serious fungal diseases, have not been reported. The production level is low, compared with that of the cucumber (Grubben, 1977).

Origin and Distribution

Although cultivated throughout tropical Asia, origin of this cultigen is probably Indonesia. No wild relatives or ancestors are known. Small working collections are available at horticultural stations in South East Asia and India (Grubben,1977).

Since existence of wild populations of *Benincasa* are unconfirmed, center of origin too remains uncertain. May be ash gourd is a native of Indo-Malaysia, whose cultivated form may have originated in South-East Asia. Chinese scriptures of ADV and VI refer to ash gourd. Grown extensively in tropical regions of the Old World, *Benincasa* is less common to New World (Peter *et al.*, 1991). A few selection work on local cultivars is reported from India and Japan, where wax gourd appears in seed catalogues.

Genetic Diversity

Four major categories are recognized: (1) Unridged winter melon group; (2) Ridged winter melon group; (3) Fussy-gourd group and (4) Wax gourd group. The wax gourd group is prominent with 3 cultivars:

1. Fruits nearly round and essentially hairless;
2. Fruits nearly round and hairy; and
3. Fruits oblong and hairy.

Ash gourd is a robust, hispid and monecious annual vine with stout prominently furrowed stems and bifid tendrils. The five-parted yellow flowers are solitary in leaf axils, measuring 4-6 cm in diameter and 3-5 cm in length excluding height of ovary. Staminate flowers have long peduncle (5-15 cm); peduncles of carpellate flowers are much shorter (1-3.8 cm). Sepals are foliaceous. Fruits are covered with white silky hairs. Flower opening takes place in early hours (04.30 to 07.30) with anther dehiscence between 03 and 05 hours before anthesis until eighteen hours after anthesis. Pollen fertility is maximum (95 per cent) at anthesis (Peter *et al.*,1991).

Future Thrust

Collection of primitive cultivars for plant breeders is recommended but there appears to be no danger of losing such cultivars at present. Selections of highly productive cultivars of good quality, within the available material through variety testing and inbreeding, is advised (Grubben,1977). Selection for bushy type of ash gourd and prospects of breeding for higher protein lines are feasible, owing to presence of considerable variability in the crop. Ash gourd is graft-compatible with a muskmelon-scion. Such a scion could rescue muskmelon crop in areas prone to wilt caused by *Fusarium oxysporum* ssp. *melonis*. Ash gourd lines are resistant to wilt (Peter *et al.*, 1991).

Round Gourd/Tinda/Round Melon

Round gourd *Praecitrullus fistulosus* Pang. or *Citrullus lanatus* var. *fistulosus*, is one of the most widely grown summer vegetables in Northern India. It is used for food in immature state, normally cooked with other vegetables before seeds become hard. It is also used for pickles and preserves. *Praecitrullus fistulosus* is a peculiar species of much systematic interest. To casual observer, vine and flowers are almost identical with watermelon. Fruits, however are much smaller than those of watermelon *i.e.*, about size of a large turnip. Unlike watermelon with 11 pairs of chromosomes, tinda has 12 pairs, same as African species of *Cucumis*. Considerable effort is made in attempting crosses between tinda x watermelon, and tinda x muskmelon (*C.melo*). An occasional fruit with the tinda x watermelon cross, and a few viable seeds are produced, but F_1 plants are not viable. The cross with muskmelon failed to set fruit. This species needs much more investigation to establish its relationship with *Citrullus* and other closely related genera (Whitaker, 1973). *Praecitrullus fistulosus* (round melon), once included in genus *Citrullus*, is its closely related species. It is originated in India. Genetic variability in fruit morphology is concentrated in North-Western India (Western Uttar Pradesh, Haryana, Punjab), but the variability is rather limited suggesting its partial or recent domestication (Sirohi *et al.*, 2005).

Snake Gourd and Pointed Gourd

In *Trichosanthes*, 41 species are reported, and 21 of these originated in India, parts of tropical Asia or Indo-Malayan region. Among these, snake gourd (*T.cucumerina* var. *anguina* syn. *T.anguina*) and pointed gourd (*T.dioica*) are cultivated throughout India. Major diversity is distributed in North-Central and North-Eastern India for *T.dioica*; whereas *T.anguina* is distributed in warm humid areas throughout the country but rich diversity is observed in North-Eastern region, West Bengal, Malabar Coast of Karnataka and Kerala, and in lower and mid-hills of Eastern Ghats (Sirohi *et al.*, 2005).

Snake gourd (*Trichosanthes anguina* L.) is widely cultivated in South India particularly during rainy season, and is highly prized as a summer vegetable. Plants are either staked or trained to grow on a trellis. Young fruits are harvested and cooked in immature states. Flowers are unusual, like the showy flowers of certain Passifloraceae. They are white, large, and fringed (Whitaker, 1973).

The pointed gourd/Parwal (*Trichosanthes dioica* Roxb.) is dioecious and perennial. It is extensively cultivated during rainy season, and is the only cultivated cucurbit, commonly propagated by cuttings. There are a few good clones available with attractive fruit characters. Cuttings are taken from female plant, and when planted in field are interspersed at random with a few male plants for pollination. This species is indigenous to Bengal-Assam region, and requires a hot, humid climate for the best development. It is easily digestible, can be used as a diuretic, a laxative, and is useful for correction of circulatory disorders (Whitaker, 1973).

Sponge Gourd/Ridge Gourd

Luffa has nine species out of which seven (*L.acutangula* (L.) Roxb., *L.cylindrica* M.Roem, *L.echinata* Roxb., *L.graveolens*, *L.tuberosa* Roxb., *L.umbellata*) are found in India. This genus has a long history of cultivation in the tropical Asia and Africa. The probable center of origin and the primary gene center of *Luffa* is India. *Luffa acutangula* (ridge gourd)and *L.cylindrica* (sponge gourd) are grown throughout India in tropical and sub-tropical climates. *Luffa acutangula* has three varieties: var. *acutangula* is grown in South East Asia and other tropical areas; var. *amara*, a wild form is confined to Peninsular India, while var. *forskallii* (Harms.) Heiser and Schilling, another wild form, is confined to Yemen. *Luffa echinata* grows in natural habitat in Western Himalayas, Central India and Gangetic Plains. *Luffa graveolens* is a wild species distributed in parts of North-Central India, and *L.hermaphrodita* (*satputia*), that bears fruits in clusters, is cultivated in Eastern Uttar Pradesh and Bihar (Sirohi *et al.*, 2005).

The sponge gourd, *Luffa cylindrica*, is a common vegetable in India. The young fruits are used for food, and the dry, inner fibrous tissue of the mature fruit is used as sponge, in bathing, scouring, and as filters. Ridge gourd or angled *luffa* (*L.acutangula*) is also used for food, usually cooked with meats, but more commonly used as an ingredient of curries. Young, tender fruits are used for this purpose. Fruits of both species are bitter. These two species of *Luffa* are indigenous to tropical Asia, more specifically India, and have become feral in many areas (Whitaker, 1973).

Of the genus *Luffa*, two species are commonly cultivated. *L.acutangula* (L.) Roxb ($2n = 26$) is widely known as ridge gourd/angled loofah, *L.cylindrica* (L.) Roem (*L.aegyptiaca* Muell.) ($2n = 26$) is known as smooth gourd, smooth loofah, sponge gourd or dishrag gourd. There are at least two uncultivated species in the genus, *L.echinata* and *L.graveolens* (Dutt and Roy, 1971).

Sponge gourd or rag gourd, is a very robust climbing annual cucurbit, widely grown for its ripe fruits which yield a fibrous, spongy material. The young fruits are often used as a cooked vegetable. Special vegetable cultivars of loofah are grown specifically for this purpose and seed is commercially available. Ridge gourd, is cultivated in Far East to a lesser extent than *L.aegyptiaca*, but for the same purposes. Loofah is widespread in humid tropics, but is only important as a vegetable in India and South East Asia. It thrives well in tropical lowlands and in areas with day temperatures over 25 °C. Nutritional value is rather low (ANV = 2.04) comparable with that of waxgourd. It has a short storage life. Loofah is grown mainly in kitchen gardens, where, plants are supported by trellises or fences; it thrives well on poorer sands. It is a cheap vegetable, sold in markets in many regions of Asia (Grubben, 1977).

Origin and Distribution

Primary gene centers occur in tropical India. A few other *Luffa* species also occur here. *Luffa* species originated in India. Wild forms of *L.acutangula* are still found in North West India. *L. cylindrica* originated from *L.acutangula*, but the two species differ so much in morphology and in time of flowering that such an origin does not seem likely. Furthermore, *L.cylindrica* is more widespread than *L.acutangula* and tends to go wild in many parts of the Tropics. It has evolved into many forms, and the best varieties for fiber (sponges) are now found in Japan. Both species are introduced to all parts of the Tropics (Martin, 1979).

Genetic Diversity

Smooth loofah is a vigorous, annual, climbing vine with medium-size leaves and a cucumber-vine appearance, and it has a fetid odour when foliage is crushed. Stem is five-angled or almost smooth, with a fine, sparse pubescence. Leaves are more or less ovate in overall shape, or at times broadly kidney shaped, cordate at base, and acute or acuminate at tip. They are characterized by three to seven angles or lobes, and surface is slightly rough. The species is monecious, and flowers open during day. Male flowers are bright yellow, up to 10 cm in diameter, rotate, with short stamens and large, clustered anthers. Flowers occur in axillary racemes, sometimes grouped in a tight cluster, of which one or two flowers open daily. Female flowers are usually solitary, arising on axil, or sometimes a female flower develops at tip of a male inflorescence. Ovary is inferior, 2 to 10 cm long, smooth, and glabrous. Fruits are similar in shape to a cucumber but usually much larger, green and sometimes mottled, smooth but with longitudinal lines. Mature fruit has a network of fibers inside and a large number of flat seeds a little more than 1 cm long and has less than 1 cm wide. These have a fine wing and are smooth and black or whitish. Angled loofah is similar in appearance to smooth loofah but is readily distinguished in that its flowers open in late afternoon and stay open through night. Stems are slender and hairy, as is foliage in general. Leaves are ovate in general outline or cordate with acuminate tip, or even with a tail. Surfaces of leaves are covered with short and stiff hairs. Species is monoecious, although hermaphroditic varieties are known. Male flowers are similar to those of preceding species but smaller, up to 8 cm in diameter, and pale yellow. Female flowers rise from stem between two male inflorescences. Hermaphrodite variety bears clusters of flowers, each of which produces a small fruit. Ovary is inferior, spindle shaped, and densely pubescent and has 10 longitudinal ribs. Fruits are cylindrical to club shaped and reach 70 cm in length and 12 cm in width. They are strongly ribbed, green, and slightly scabrous to touch. Seeds are slightly longer and slightly narrower than those of *L.cylindrica* and are black with a pitted appearance (Martin,1979).

Among loofahs, sexual system is much more complex. In addition to common monecious condition, variety 'Satputiya' produces only hermaphroditic flowers, and dioecious varieties are found. Most study was made in common monoecious type. As with *Lagenaria*, *Luffa* spp. produce principally male flowers at first. Once female flowers appear, a ratio of about 25 male flowers to 1 female flower is established in *L.cylindrica*, and 43 to 1 in *L.acutangula* (Katiyar, 1969).This ratio can be changed by chemical treatments. Indole acetic acid doubles percentage of female flowers (Katiyar, 1969), whereas morphactin and gibberellic acid increase vegetative growth and percentage of male flowers (Krishnamoorthy, 1971 and 1972). The stigmas of *Luffa acutangula* are receptive to pollinization from 2 hours before to 36 hours after anthesis, and the stigmas of *L.cylindrica* from 4 hours before to 60 hours after anthesis (Katiyar, 1971). *Luffas* are cross-pollinated by a wide number of insect species, including *Anthophora* spp., *Apis florea*, *Ceratina voridissima*, *Crocisa histerio*, *Eristalis transpositus*, *Megachile monticola*, and *Xylocopa* spp. (Siddiqui and Mecci, 1974). The 4 *Luffa* species are diploids with 26 chromosomes. Their interspecific crossing relationships are studied extensively by Dutt and Roy (1971).Chromosome pairing data were used to determine degree of relationship between species, and phylogenetic relationships were established. *L.graveolens*, a monoecious species, seems to be the ancestor of other species. *L.echinata*, a dioecious species, arose directly from *L.graveolens*. In a separate evolutionary path, *L.cylindrica* evolved, and from it the monecious form of *L.acutangula*. From the latter developed the hermaphroditic form, sometimes called *L.hermaphrodita*. The fairly fertile crosses among the species suggest the ease of interbreeding these species to obtain desirable combinations of characters. Loofah gourds probably vary as much as *Lagenaria* in physiological traits. In Puerto Rico, great differences were seen in earliness, longevity, and fruit production. In both species, fruits vary in overall shape, from short and compact to long, narrow, or clublike cylinders. In *L.cylindrica*, varieties differ in bitterness of fruit. Only sweet ones are used as vegetables. Sponge gourds differ in size and thickness of fibrous mass used as sponge and in quality characteristics of this fiber (Martin, 1979).

Future Thrust

There is no imminent danger of genetic erosion of *Luffa* spp. This vegetable is of minor economic importance and has a low nutritional value. In general, no high priority for breeding work is recommended, although collection and use in breeding of old cultivars in Malaysia and the Philippines are recognized in these countries (Grubben, 1977). *Luffa* could become important an easy-to-grow source of protein and oil for hot and humid tropics. Sponge gourd yields another renewable resource of bio-diesel that is likely to be useful for ages to come. Each species is unique in its own way and valuable (Martin, 1979).

Chow-chow/Chayote/Guisquil

Sechium, formerly a monotypic genus, has now eight species. Chow-chow (*Sechium edule*) is native of Mexico and Guatemala and is under cultivation for centuries. Wild forms of this species and its close relative, *S.compositum*, are found in that region (Sirohi *et al.*, 2005).

Sechium edule (Jacq.) Swartz (2n= 24) is a vigorous perennial climber. The fleshy fruits contain only one large seed which does not withstand drying and consequently entire fruit must be planted. In addition to fruits, young leaves and fleshy roots may also be consumed. Chayote is a popular vegetable in many areas of Central America, India and South East Asia. Chayote is grown commonly in Sierra Leone, Ghana and Malawi but it could be grown more widely in other areas. It is cultivated in tropical lowlands but higher yields are obtained at altitudes from 500 to 1,500 m above msl. and where nights are cooler. It does not withstand drought. *Sechium* produces the best in light shade (Grubben, 1977). Almost all parts of chow-chow are edible. Young leaves, shoots and tendrils are cooked as greens. Fruits are either fried or cooked like bottle gourd. Rich in Calcium (17-140 mg/100 g), fruits of chow-chow are a good supplement in a tropical diet. Starchy roots resemble sweet-potato in taste. Starch obtained from roots can be used in textile industry and in making biodegradable plastics. Mature stem gives quality fiber. Stems, fruits and leaves can also be used as fodder (Dutta, 1994). Chayote is a very cheap vegetable with a low nutritional value (ANV = 1.81). Since its production capacity is high and leaves are edible, it is a very suitable vegetable for home garden production. Consumption of chayote leaves is very common in South East Asia as too is the cooked tuberous starchy root. Chayotes are commonly found in Latin American markets. The city of Sao Paulo in Brazil consumes 25 million kg per annum. Chayote is also important as a market vegetable in India and

South East Asia. Chayotes are rarely damaged by pests and diseases. High costs are incurred for supports, if crop is grown on a commercial scale (Grubben, 1977).

Origin and Distribution

Origin of this cultigen is South Mexico and Central America. Local cultivars are recorded here and in South East Asia but rarely named since they are recognized as white or green, pointed or blunt, spinous or smooth, round or long. Since chayote cannot be propagated from dried seeds, no commercial cultivars are available. No breeding work has been recorded (Grubben, 1977). Recently eight wild species were collected from Central America, the center possessing maximum genetic diversity. Genetic diversity is conserved *in situ* in natural habitat, where over a period of time, evolutionary processes allow plants to adapt to environmental changes. This period is followed by on-farm cultivation and conservation by farmers. *Sechium edule* is now cultivated in South East Asia (India, Malaysia, Indonesia and the Philippines), Central and Western Africa, Central and South America (Brazil) and most other tropical regions. In India, its local cultivars are commonly grown in coastal and low hills (1,000–2,000 m above msl.) of Maharashtra, Karnataka, Tamil Nadu and North East regions. The *ex situ* conservation of genetic diversity of chow-chow in gene banks is scattered and needs immediate attention (Dutta, 1994).

Genetic Diversity

Unlike other seeded cucurbits, chow-chow has only one seed in fruit with viviparous germination (seed germinates inside fruit). Seed is large (3-5 cm long, 2-2.5 cm wide and 1-1.5 cm thick) and embedded near broader blossom end of fruit. Plant is thus propagated by planting whole fruit enclosing sprouted seed. However, to get true-to-type plants, propagation is done through stem-cuttings (15-20 cm long).

Chow-chow is a semi-perennial plant remaining economically viable up to 5 years. It produces vigorous vines 12-15 m long, requiring support–trellises or training on to a 'Pandal'. In Maharashtra, vines are pruned during December-May to a length of 120-150 cm. In cooler regions, vines are killed due to frost. Leaves of chow-chow are ovate with pointed apex and rough surface. They are 10-15 cm long. Flowering occurs 100-120 days after planting in tropical and sub-tropical regions having12 to 12 ½ hour day-length. Flowers are monoecious. Male flowers are borne in clusters in a 1.5-4.0 cm long peduncle. Female flowers are solitary. Usually both male and female flowers are borne in same axil. Creamy-white flower possesses large nectaries. Pollination is through bees. Fruit shape varies from oblong to obovoid to pyriform (pear-shaped) with longitudinal furrows deeply emerged at apex. Fruit colour varies from green to yellowish-green or creamy-white. Fruit is either smooth or thinly pubescent (soft spines), 7-20 cm long, 5-8 cm in diameter, and weights 100-450 g. Fruit stalk is 2-3 cm long. Flesh is thick (2-3 cm), succulent, crisp white or greenish-white. Fruits are ready for harvest 120 days after planting. It takes 25-30 days after flowering to attain marketable size. Harvesting of tender fruits lasts for 2-3 months in a season and total yield is 50-60 tonnes/ha. A single plant bears 70-100 fruits (30-50 kg) during first year and 500-600 fruits during second to third year. Plants bear profusely during first 3 years and moderately during fourth and fifth years. In Karnataka, 3 crops may be raised in a year. Fruits can be stored for a month at 10-15 °C and 85-90 per cent relative humidity. Roots are tuberous in nature which helps plant to survive under unfavourable conditions. Roots are starchy in nature and become ready for harvest after 3-4 years of planting (Dutta, 1994).

Future Thrust

The rapid decline of plant genetic diversity is mainly due to quick deforestation of natural forest ecosystem by human interference. Of immediate concern, therefore, is collection, conservation and sustainable use of genetically diverse resources on global basis. Collections made as a result of international collaboration should not become property of any single nation but should be held in a trust for easy access to present and future users, world over. Policy-makers should not impose any form of intellectual property protection on plant genetic resources, a common heritage of mankind (Dutta, 1994). Local cultivars should be collected for variety testing. There is erosion of old cultivars in Philippines and Indonesia. Although fruits have a rather low nutritional value, this vegetable is relatively easy to grow in home gardens and has a very high yielding capacity. It is recommended for selection (Grubben, 1977).

Ivy Gourd/Coccinia

Coccinia genus has nearly 30 species, confined to tropical Africa except for ivy gourd (*C.grandis*), which occurs in cultivated and wild forms in South Eat Asia in addition to Africa. It is also grown in Latin America, Australia and other tropical areas. In India, it is cultivated in warm humid areas (Sirohi *et al.*, 2005).

Ivy gourd/Coccinia (*Coccinia indica* (Naud.) Wight and Ern. Syn. *C.grandis*) is dioecious. Fleshy fruits are 3-4 cm long, smooth and grown on trellis. Young shoots are tied in bundles, and sold in markets for use as a cooked vegetable. Green fruits after cooking are used in sauces and curries. The entire plant, leaves, stems, and roots, purported to have medicinal uses. This species is indigenous to tropical Asia, and possibly to tropics of Northern Africa (Whitaker, 1973).

Stuffing Cucumber/Korila

Cyclanthera has about 30 species and all of these are native to tropics of South America. Stuffing cucumber (*Cyclanthera pedata* var. *edulis*) is cultivated in tropical America and Asia. It is a potential crop for wide cultivation in high altitude areas. Its closely related species *C.brachystachia* (Ser.) Cogn. is also cultivated for edible fruits in Nepal and South Taiwan (Sirohi *et al.*, 2005).

Malvaceous Vegetable

Okra/Lady's Finger

Okra, *Abelmoschus esculentus* (L.) Moench or *Hibiscus esculentus* L., is called gumbo, gombo, or lady's finger in English-speaking countries, bhindi in India, and bamyah or bamiat in Arabic-speaking countries. Preferred generic name *Abelmoschus* is suggestive of musky odour of seed. Genus includes a few other species: *A.ficulneus* (L.) Wight et Arn.; *A.manihot* (L.) Medic., sunset hibiscus, used for its edible leaves; and *A.glutinotextile Kagaava*, a cross between *A.esculentus* and *A.manihot* (Martin and Ruberte, 1978). Okra is an annual erect, poorly or non-branching herb, cultivated for its young fleshy fruits. Clemson Spineless, Lousiana Green Velvet, Ladyfinger, Perkins Spineless and Emerald are well-known American cultivars of short-duration type. Pusa Sawani is a popular Indian cultivar. In USA, 7 tonnes/ha of young fruits are harvested *i.e.*, 5-10 fruits per stalk. Okra crop in humid tropics generally yields less than 2-3 tonnes/ha. Ripe pods contain 30-80 seeds; the weight of 1000 seeds is 50 g. One ha of okra produces 1500 kg of seed; but 400 kg of viable seed is a good average yield in tropics. Soaking hard-skinned, fatty seeds improves germination. Many local cultivars occur in Africa and India, differing from each other in growth habit (branching or non-branching, large or dwarf, late or early, hairy or glabrous, light-green or dark-green or red) and fruit characteristics (upright or pendulous, slender or wider, with 5 to 10 ridges). Skin may be soft or hard, varying in specific gravity and dry matter content. Okra is often cultivated on marshy land at end of rainy season, in monoculture or in mixed cropping with maize, sorghum or upland rice. Okra is a hot weather, tropical lowland crop, susceptible to drought and low night temperatures. It grows well at day temperatures between 25 and 40 °C with night temperatures over 22 °C. Most tropical cultivars are short-day plants. Okra has a good nutritional value. It is a very popular, tasty, gelatinous vegetable, used in a number of local dishes. Extremely high calcium content is particularly valuable. Okra is widely grown in the Sahel area as a food crop for dry season. Fruits, cut into slices and dried in sun, can be stored for long periods. Leaves are also used as a vegetable and may be dried. Okra is a typical market vegetable. After harvesting, it can be easily transported in bulk and stored for several days without much loss of quality. Considerable differences in keeping quality exist among cultivars. Okra is not as cheap as leaf vegetables because yield potential is low, and consecutive harvests are time-consuming. Dried okra slices are an important market product in Sahel zone. Fresh okra is important in Brazil, West Africa, India and in many other regions in tropics. It is also a popular vegetable in Near East and southern part of USA (Grubben, 1977). Okra is well distributed in both Temperate Zone and Tropics. It is especially important in North East Brazil, India, and West Africa and is grown on a small scale almost everywhere. Dried okra is popular in Turkey, and oil from seed is used in Greece. Okra has probably been bred more systematically in southern United States than in any other region, but it has also been improved systematically in India and Brazil. In West Africa, improved varieties were introduced from other countries (Martin and Ruberte, 1978).

Okra is very susceptible to powdery mildew (*Oidium abelmoschi, Erysiphe cichoracearum*), *Verticillium* and *Fusarium* wilts, yellow vein mosaic virus, root-knot nematodes and many insect pests (*Heliothis* caterpillars, leaf beetles, bugs).

Origin and Distribution

Hibiscus esculentus L. (Syn. *Abelmoschus esculentus* (L.) Moench, is a cultigen originated in tropical Asia. It was present in Africa and India as a polyphyletic species (Joshi and Hardas, 1976). A semi-wild ancestor, *A. tuberculatus*, occurs in India. Collections are available in gene banks in USA and India (Grubben, 1977). Genus *Abelmoschus* is of Asiatic origin, but exact ancestral home of cultivated *A. esculentus* is disputed. It is domesticated in Ethiopian region or in West Africa (Joshi *et al.*, 1974). A late 19th century record shows its occurrence in wild state in White Nile in Sudan (Singh *et al.*, 1975). South-East Asia as a center of diversity for the genus *Abelmoschus* was considered by van Borssum-Waalkes (1966). Indian origin of *A.esculentus* was also proposed by Zeven and Zhukovsky (1975), and Zeven and de Wet (1982). Vredebregt (1990) observed that rich diversity in *A.ficulneus* in India combined with that of *A.tuberculatus* may be an argument for an Asian origin of *A.esculentus*. *Abelmoschus ficulneus* contributed to second unknown genome of cultivated okra, one genome being from *A.tuberculatus*. The ancient Sanskrit literature in India also mentions *Tindisha* and *Gandhamula* designating Bhindi (okra). Centers of diversity of *Abelmoschus* include West Africa (Benin, Togo, Guinea), India, and South East Asia (Burma, Indochina, Indonesia and Thailand) (Chevalier 1940, van Borssum Waalkes 1966; Hamon and Hamon 1991; Bisht *et al.*, 1995). Two species, *A. esculentus* and *A. caillei*, are cultivated for their fresh fruits. To avoid confusion, the name okra will be used only for *A. esculentus* which is an annual cultigen in low altitude regions of tropics and subtropics, with an extension to temperate climates in Mediterranean basin. *Abelmoschus caillei*, a relatively new species, is most probably indigenous to West Africa (Siemonsma 1982; Hamon *et al.*, 1986). Geographical distribution is from North West Guinea extending to South East Africa. It is cultivated in most humid parts of Guinea, mostly in an intercropped system with *A.esculentus* (Hamon and Charrier, 1983; Hamon 1988). *A. moschatus* and *A. manihot* are semi wild, and show a greater diversity than the cultivated forms (Hamon and Hamon, 1991). *A. moschatus* is morphologically and genetically the most different from other species (Hamon and Yapo 1986) with probably the widest geographical distribution (Hamon and Charrier 1983);

which extends from South Pacific Islands through Indochina and India and over to Central and Western parts of Africa. The cultivated form (*A.moschatus* subsp. *moschatus* var. *moschatus*), better known as "musk mallow", "Jew's mallow", or "ambrette" in Africa, Asia, and America, has fragrant seeds used for perfumery (Hamon and Charrier 1983). Subspecies *biakensis* is endemic to Papua New Guinea (Charrier 1984). Cultivation of *A.manihot* is mainly in South East Asia and is known under vernacular name *"aibika"*. It is also found in India, Northern Australia, and less frequently in American continent and tropical West Africa. Cultivation is exclusively for leaf consumption. The species *A. tetraphyllus* (IBPGR 1991), *A. ficulneus, A. crinitus, A. angulosus* (van Borssum Waalkes 1966; Bates 1968), and *A.tuberculatus* are truly wild. *A. ficulneus* is found from Northern parts of Australia to South East Asia and from India to most parts of Africa. *A. tetraphyllus, A. crinitus, A. angulosus*, and *A. tuberculatus* are exclusively of Asian origin. *A. tetraphyllus* var. *tetraphyllus* and var. *pungens* are endemic to South East Asia with extension of former to Papua New Guinea and New Ireland. *A.crinitus* is found at low altitudes in India and South East Asia, and *A.angulosus* at high altitudes in India, Sri Lanka and Indochina, and Indonesia. *A.tuberculatus* is endemic to Northern and Western parts of India (Charrier 1984; Velayudhan and Upadhyay 1994).

Discovery of *A.caillei*, the second edible okra species, in West Africa (Chevalier 1940) and later evaluations (Hamon and Yapo 1986; Hamon 1988; Jambhale and Nerkar 1983, 1989) received much attention in last decade. It is distinguished from *A.esculentus* by shape and number of epicalyx segments and fruit shape and orientation (Siemonsma 1982), and shows great variability (Hamon and Charrier 1983; Hamon *et al.*, 1991; Ariyo 1993). *A. caillei* is prized for its potentially prolific yield, vigorous growth and tolerance to negative environments and serving as a source of many desirable characteristics for okra (Fatokun 1987; Ariyo 1993). It is a short day plant, photoperiod sensitive, and requires selection for photoperiod insensitivity to make it more adaptable (Ariyo 1993).

Genetic Diversity

Flowers of okra are self-fertile but usually up to 20 per cent cross-pollination occurs by insects. Varying chromosome numbers between 2n = 72 and 144 are reported. Hybrids between Indian and African cultivars are partially sterile. These are being studied in India for resistance to yellow vein mosaic virus. A popular Indian early cultivar with very long fruits, Pusa Sawani, has some tolerance to virus. Sources of tolerance to *Fusarium, Verticillium* and root-knot nematodes are reported. American seed companies selected attractive cultivars like Clemson Spineless, Louisiana and Emerald for fresh market and special short-fruited cultivars like Goldcoast for canning. National Research Institutes in Ivory Coast, Ghana and Benin carried out a few testing of local cultivars. Hybrid varieties of okra are now available. In India, it was shown that crossing inbred lines gives rise to heterosis and male sterility could be obtained by chemical induction. In Japan, amphidiploids were obtained from interspecific crossings with related wild species (Grubben, 1977). Ten *Abelmoschus* species, namely, *A.angulosus* (2n = 56), *A.tuberculatus* (2n = 58), *A.manihot* (2n = 66), *A. moschatus* and *A. ficulneus* (2n = 72), *A. esculentus* (2n = 130), *A. tetraphyllus* var. tetraphyllus, *A. tetraphyllus* var. *pungens* and *A. crinitus* (2n = 138), and *A. cailei* (2n = 196) occur in India. Of these, only *A. esculentus* is only cultivated in India. *Abelmoschus moschatus* occurs wild though it is cultivated for its aromatic seeds.

Maximum variation was observed in cultivated species *A.esculentus*. The countries with the largest collection or diversity are Turkey and India, and to a lesser extent Afghanistan, Iran, Pakistan, Bangaldesh and erstwhile Yugoslavia. In cultivated okra, genetic diversity was observed for absence or presence/number of ridges on fruits, extent of hairiness and pigmentation on fruit and other plant parts, length : width (diameter) ratio of fruits, fruit attachment, lobbing of leaves, branching and plant height. A few of these characters are highly influenced by environment. In hilly areas (warmer parts of temperate region), smooth-fruited variety develops hard hair and sometimes non-pigmented ones develop partial pigmentation (Dhankar *et al.*,2005). Commercial okra cultivars are erect annuals, becoming woody at maturity. Main stem branches at base with usually 2 to 5 but as many as 20 predominant branches/plant. Orthotropic stemmed types are not exclusive (Hamon and Sloten 1989; Ariyo 1990; Ariyo and Odulaja 1991). The plant often reaches 60 to 180 cm in height (Hamon and Sloten 1989; Ariyo 1990), but an extremely tall type in central parts of Nepal reaches up to 4 m (Velayudhan and Upadhyay 1994). Leaves are present at each node, solitary flowers are axillay, and fruits are born upright. Simple leaves are palmatifid to palmatisect (Charrier 1984; Ariyo and Odulaja 1991) with a symmetrical lamina of chartaceous texture. Base is cordate, margins toothed (serate), and apex acute (Bhat *et al.*, 1988). Effective methods for leaf area determinations are proposed by Srinivas (1982) and Abdullahi and Jasdanwala (1991). Red pigmentation occurs in stems, petioles, leaf veins, pedicel, petal bases and fruits, which gives the plant ornamental value (Martin *et al.*, 1981; More and Vibhute 1983). The plant carries unicellular trichomes on almost all parts of plant, providing protection from pests such as leafhoppers (Uthamasamy 1985). Seeds are relatively large and heavy (60 g/100 seeds) and have a light green to gray colour (Martin and Rhodes 1983). Shape is round to reniform, and their surface is rather glabrous or sometimes hirsute or scattered, having simple short hairs (trichomes) on surface (Wyatt 1985). Fruits consist of 55 to 62 per cent pericarp, 30 to 40 per cent seeds, and 7 to 11 per cent axil (Singh and Agarwal 1988). Detailed characterization of 241 accessions of wild species germplasm comprising *A.ficulneus* (68 accessions), *A.manihot* ssp. *tetraphyllus* (81 accessions), *A.moschatus* (7 accessions), and *A.tuberculatus* (85 accessions) was also undertaken at NBPGR and diversity in various qualitative and quantitative traits studied (Bisht *et al.*, 1995). Diversity for a range of morphological characters, namely, epicalyx segments, shape, size and persistence; petal colour and petal blotch; stem and fruit pubescence; pigmentation of various plant parts; days to flower; plant height and fruit yield/plant were important components of variation and contributed significantly to total variation in germplasm accessions of above four species studied. In *A.ficulneus*, much of the within species variation was recorded for traits such as epicalyx segments (4.0-12.0), epicalyx length (5.0-15.0 mm), fruit colour, position of fruit on main stem, fruit pubescence, plant height (12.0-75.0 cm), days to maturity (85-173) and fruits per plant (39-131). Much of the variants in *A. manihot* ssp. *tetraphyllus* accessions were recorded for plant height (30.0-109.0 cm), days to maturity (103-193) and fruits per plant (190-221). Wide variation was recorded in *A.tuberculatus* accessions with regard to plant height (31.0-135.0 cm), stem pubescence, stem colour, epicalyx shape, epicalyx persistence, epicalyx segments (5-12), epicalyx length (4.0-15.0 mm), number of internodes (6-32), days to maturity (86-151), fruit colour, fruit pubescence, fruits per plant (10-98) and seeds per fruit (16-41). Not much variation was recorded within species for *A.moschatus* accessions studied (Bisht *et al.*, 1995).

Relationships and genetic integrity among cultivated and wild forms of *Abelmoschus* species were studied by Babu and Dutta (1990) and Hamon and Hamon (1991). In most cases, interspecific crosses among *Abelmoschus* species are not readily obtained, and F_1s are often sterile (Siemonsma 1982, Charrier 1984). However, *A.esculentus* crosses relatively easily with *A.caillei* (Nerkar and Jambhale 1985; Hamon and Yapo 1986; Fatokun 1987; Jambhale and Nerkar 1989) and may produce partially fertile plants. *A.tetraphyllus* x *A.esculentus* hybrids are also possible (Sharma and Dhillon 1983; Hamon and Yapo 1986; IBPGR 1991) and have been stabilized for breeding purposes. Charrier (1984) suggested three ploidy levels; where *A.moschatus* (n=36), *A.ficulneus* (n=36), *A.tuberculatus* (n=29), *A.esculentus* (n=36), *A.manihot* (n=30-34) are at ploidy level 1; *A.esculentus* (n=62-65), *A.tetraphyllus* (n=69) are at ploidy level 2 and *A.caillei* (n=99) is at ploidy level 3. Fourth ploidy level is also possible. The genus, therefore, may be a "polyspecies complex" where continuous interchange of genes throughout its evolution has complicated determination of relationships.

Future Thrust

Collections of local Asiatic and African cultivars should be made both for direct breeding purposes as well as for conservation of germplasm. Danger of genetic erosion is great as a consequence of growing popularity of early, high yielding Indian and American cultivars. African farmers need long-duration cultivars, suitable for their traditional cultivation methods. For fresh market, however, growers require early-maturity cultivars possessing a high level of disease resistance, with special emphasis on mildew and virus. Yield potential of okra is far too low but screening of many local cultivars would be the first step to improve okra crop (Grubben, 1977). Okra will continue to be utilized principally as a welcome and productive vegetable throughout the tropics. Because of its adaptability, dependability, and resistance to hot and humid weather, okra will always be appreciated. It merits a place in home gardens and on subsistence farms, but it should never replace vegetables that are more nutritious but harder to produce. Potential for okra seems to be as an oil and protein crop (Karokoltsides and Constantinides 1975). Simple techniques are developed to produce okra seed meal at household level, and such meal can be substituted for a part of wheat flour, and the oil used in baking. A technique to combine uses might be to grow okra as a vegetable crop during rainy seasons, permitting pods to mature during dry season as a source of oil and meal. Breeding work is desirable to incorporate resistance to diseases and insects in both garden and commercial varieties for hot and humid Tropics (Martin and Ruberte, 1978). Many of the limitations in okra production can be overcome by advances in genetic improvement. Genetic resistance is greatly needed in many parts of the world because okra is extremely sensitive to a large number of pests and diseases. Resistance to area-specific viruses, such as Yellow Vine Mosaic Virus in Asia and Okra Mosaic Virus in Africa; soil-borne fungi, such as *Fusarium* spp. and *Verticillium* spp.; and root-knot nematodes may be the first priority of breeders throughout the world. Biotechnological techniques could have practical relevance in okra. Progress in breeding would be facilitated by increased germplasm exchange. Publication of world germplasm collections would greatly facilitate use of exotic germplasm. Techniques to overcome interspecific barriers would expand available gene pool. In this connection, research on *in vitro* technology such as embryo rescue technique needs to be explored in okra. Furthermore, addition of photoperiod insensitivity to *A.caillei* would allow further exploration in this species by itself or by introgression into *A.esculentus* (Duzyaman, 1997).

Legume Vegetables

Distinction between grain legumes, cultivated as field crops, and vegetable legumes (leguminous vegetables), is not always clear. The shoots or young leaves of many legumes are occasionally consumed, but this is of minor importance. Composition of leguminous vegetables varies. Dry beans have a very high ANV, due to their dry matter content, but the ANV calculated per 100 g dry matter is low. Fresh seeds and young pods have a fairly high ANV, sprouted beans a much lower one. Young seeds, at harvest, weigh approximately twice that of dried seeds. Ability of plants of Leguminoceae to fix nitrogen by means of *Rhizobia* bacteria is extremely important for field grown pulses, but is less so for intensive production, since such vegetables are generally well provided with soil organic material. Long-lasting or perennial, climbing, deep-rooting tropical leguminous vegetables such as Lima bean, winged bean, hyacinth bean and yardlong bean, are valuable plants for home garden production since they are less susceptible to diseases and are often more drought resistant. Climbing types, planted against fences and hedges, make full use of available space and light. Climbing types are often used in mixed cropping using dried stalks of maize, sorghum or castor as a support, or they may possibly be planted against stems of shrubs remaining in field after the bush has been cleared. Only a few climbing legumes *e.g* common French bean, Lima bean and Winged bean, yield a sufficiently high quantity of seed to justify their cultivation as grain legumes in small gardens (Grubben, 1977).

Common Bean/French Bean

Common bean or French bean, *Phaseolus vulgaris* L. (2n=22) is a species of which many cultivars are grown for pods, green seeds or ripe, dry seeds. Climbing cultivars (pole, snap or French beans) are popular in home gardens and for commercial production of young pods in cooler parts of tropics. At least 200 commercial cultivars of this vegetable are known. Distinction between those grown as vegetables and those grown for dry seeds is not clear and the same cultivars may be used for production of pods, young seeds and mature dry seeds. Contender and Canadian Wonder are often used as bush type of snap beans in the tropics, yielding up to 15 tonnes/ha. Canfreezer and Florigreen are suitable cultivars of pole type. Many local cultivars are grown which are inferior in quality to these cultivars but which have a greater degree of horizontal disease resistance. Western seed companies produce seeds of bush type of French bean in upland areas of East Africa which have a suitable warm-temperate climate, where occurrence of halo blight (*Pseudomonas*) is minimal. Common bean is a warm-temperate plant, suitable for cultivation in tropics at elevations above 1,000 m above msl. To obtain a good seed set, mean day temperature should not exceed 30 °C. Plant is indifferent to day length. In a humid lowland climate, flowering and fruit set are poor, leading to low yields of dry beans and fresh pods. Nevertheless, French beans are widely grown for supplying markets of large cities, due to extremely high prices paid by public. In areas of high humidity, plants are subject to several diseases and pests, main diseases being: dieback and root-rot caused by *Pythium*, *Sclerotium*, *Fusarium*, anthracnose (*Colletotrichum*), *Pseudomonas* and *Xanthomonas* common (bacterial) blight, *Cercospora* leaf spot and mosaic virus. Important pests are: spider mites, bean flies (*Melanagromyza phaseoli*), aphids and pod borers.

Nutritional value of dry beans is very high, due to low water content. As a fresh vegetable, nutritional composition is fairly good; fresh pods are rich in calcium and vitamin C and young seeds also contain reasonable amounts of calcium, iron and vitamin C. Common bean is the principal grain legume of Central and South America and of East Africa. As a vegetable, cultivated for fresh pods (Fresh beans, snap beans) and, to a lesser extent, the young seeds, it is extremely important in Western Europe and USA, but less so in tropics. Climbing cultivars of French beans are popular in home gardens since they produce pods over a longer period than bushy cultivars which are more generally grown on a commercial scale. Cultivation methods used in growing of snap beans, such as staking, density and control of diseases and pests are relatively primitive. Fruit setting is often poor and many cultivars are susceptible to drought (Grubben, 1977).

Origin and Distribution

Centre of origin is Central America. Wild *P.vulgaris* and the closely related *P.coccineus* occur in this region as well as hybrids and *P.darwinianus* (Hdz. X and Miranda) Smartt. Many local primitive cultivars are present in cultivation areas. The IBPGR committee on *Phaseolus* germplasm has reviewed areas of diversity and needs for collection. Large collections of bean types are amassed as an urgent necessity for use of plant breeders and as a protection against genetic erosion. Germplasm banks in the USA and also at CIAT in Colombia possess large collections. The International Bureau of Plant Genetic Resources (IBPGR) designated CIAT to hold a world collection. Other important material is held at INIA and Chapingo College, Mexico (Grubben, 1977).

Genetic Diversity

A great deal of advanced breeding work was carried out in the USA and Western Europe on *P.vulgaris*, both as a pulse and vegetable. Some work has been reported from Central and South America. Cultivated types hybridize easily with wild forms and with the related species *P.coccineus*. The latter species has already been used by plant breeders as a donor for many types of disease resistance. Genetic base of bean cultivars, both in Europe and in USA is narrow and useful combinations may be expected from hybridization between wild and related species, also with primitive tropical cultivars. Screening and testing of low-yielding cultivars grown in tropics in America, as well as Africa and Asia, may give rise to selected types with valuable characters like resistance to disease and drought and also tolerance to high temperatures. Many local bean cultivars grown in Brazil are more or less tolerant to *Xanthomonas* (Grubben, 1977).

Future Thrust

Local cultivars in the tropics disappear as they become replaced by high-yielding modern cultivars. Collection of these cultivars is therefore recommended. Selection of high-yielding cultivars of snap beans adapted to tropical conditions and with a high level of pest and disease resistance is advisable, due to their potentially high economic value (Grubben, 1977).

Lima Bean

Lima bean, *Phaseolus lunatus* (2n=22) is a very variable species. Green-white cultivars are often grown for fresh consumption and commercial dwarf cultivars with large white-green seeds are popular in USA as canned vegetable. A few perennial, drought and disease resistant types are cultivated in tropics but annual types which are of better quality and produce higher yields are more popular. Small-seeded, often red-coloured types are sometimes called sieva beans and are used as a pulse. Lima beans are cultivated as pole beans in home gardens; they are supported by bushes or maize and sorghum stalks when grown as a field crop. *P.lunatus* grows well in areas with a day temperature range of 22-30 °C. Above 30 °C, flower shedding occurs. Lima bean is grown at altitudes up to 1,500 m, but is more commonly cultivated in lowland areas where it produces well during dry season. Most of local cultivars are short-day plants, whereas many improved cultivars are day-neutral. It is a popular vegetable in most of the tropical lowland areas of Africa, Asia and Latin America. The dry Lima bean has a high nutritional value (compared to hyacinth bean). It has a high protein, fibre, calcium and iron content. Fresh seeds are rich in calcium, iron and vitamin C but may contain HCN which can be adequately removed by cooking. The ANV for dry matter is low. Lima beans, as green beans, are a very important vegetable for canning in USA but the crop is of less importance as a pulse (dry seeds) or as a vegetable (young seeds) in tropics where they are mainly grown for home consumption. Coloured, ripe seeds may contain HCN which is removed during cooking. The yield potential of Lima bean in tropics is very low. Many destructive diseases affect the crop in USA *e.g. Pseudomonas* bacterial spots, *Diaporthe* pod-blight, *Phytophthora* downy mildew and *Colletotrichum* stem anthracnose. These diseases are rarely serious in tropics, due partly to higher temperatures which are unsuitable for development of certain fungi, but also to higher disease resistance of some primitive, heterogenous local populations. Damage by pod insects and poor fertilization are causes of low yields obtained during rainy season. Virus and spider mites are sometimes locally important (Grubben, 1977).

Origin and Distribution

Many primitive cultivars are grown in tropical countries, mostly in West Africa, Malagasy and South East Asia. These can be considered as secondary centres of diversity and escaped types are sometimes found in the natural vegetation. The primary centre is situated in Central America, from where different cultivar groups spread to North and South America, in pre-Columbian times. Wild types are present in these areas. The largest germplasm collections are available in USA but CIAT will also hold material at request of IBPGR. Other material is held by INIA and at Chapingo College, Mexico (Grubben, 1977).

Genetic Diversity

Lima bean is a self-pollinating plant, but up to 20 per cent cross-breeding occurs. Primitive cultivars have proved to be useful in breeding programmes; a nematode-resistant bush cultivar Nemagreen was selected in the USA after crossing with a semi-wild pole type. Fordhook 242, Henderson, Ventura and several other cultivars are popular bush beans in the USA. Climbing cultivars such as Carolina, Easy Shell and Florida Speckled Butter are popular in home gardens. The American high-yielding Lima beans are now becoming widely cultivated in tropics. Lima beans are less susceptible to pests and diseases than are common or French beans; they are, however, often attacked by cucumber virus. Lines with resistance to Mexican bean beetle were selected in Mexico (Grubben, 1977).

Future Thrust

Risk of loss of germplasm is not yet serious but may become so in next a few decades, when more productive cultivars become available. Collection of local cultivars is not urgent, but is needed for establishment of breeding programmes.

Evaluation of agronomic characters in hundreds of local types of Lima beans existing in tropics, and of their wild relatives, will provide a broad base for future breeding work (Grubben, 1977).

Dolichos Bean/Hyacinth Bean/Indian Bean

Dolichos bean, *Dolichos lablab* L. (syn. *D.purpureus* L., *Lablab niger* Medik.) is either a vigorously growing annual or perennial pole bean. Plant habit resembles that of climbing Lima bean (*P.lunatus*). Seeds have a white, protruding hilum approximately 10 mm long, and varies in colour from white to red and black, according to cultivar. Plants require strong supports, due to their prolific leaf and stem production. Many local cultivars are known in Africa and Asia. In tropical Africa, they are grown in home gardens and are supported by remaining shrubs and trees when grown as field crop. Commercial cultivars are available, especially in India and Sudan. In Caribbean area, crop is known as Seim or Waby salad bean. Cultivars Darkness and Daylight are well-known. Dwarf determinate, early-maturing cultivars, for cultivation as a pulse, are known in Egypt, Sudan and Australia. The Hyacinth bean is grown as fodder crop in India.

Hyacinth bean is grown commonly in Africa and Asia in tropical lowlands up to altitudes of 2,000 m above msl. Local cultivars are mostly short-day or day-length neutral plants, but long-day cultivars exist in India. The plant is fairly drought resistant. Young pods, as well as young and ripe seeds, are eaten. Pods and young seeds are fairly rich in calcium and vitamin C. Hyacinth beans, as dry beans, young pods and unripe seeds, are typically products for home consumption. They are sold in local markets but do not have much commercial value. Leaves and young pods are also eaten. Hyacinth bean is generally resistant to pests and diseases but damage caused by *Agromyza* stem flies, black bean aphids, pulse beetles and *Xanthomonas* bacterial blight are reported. Local cultivars are often too prolific vegetatively and give very low yields (Grubben, 1977).

Origin and Distribution

Origin of *D.lablab* is probably tropical Africa, with Hindustani Centre and parts of South East

Asia as secondary centre of diversity. Several wild types are described but no large collections of germplasm are known (Grubben, 1977).

Genetic Diversity

Hyacinth bean is mainly self-pollinated, but up to 5 per cent cross-pollination by insects may occur. Selection work has been carried out in Trinidad, where the Waby salad bean has been developed. Dwarf, determinate, early-maturing cultivars have been selected in Sudan, Egypt, and Australia. Most of breeding work on *Dolichos* as a vegetable has been carried out in India. Resistance to stem fly, aphids and bacterial blight has been found (Grubben, 1977).

Future Thrust

Danger of genetic erosion is limited, at present, *i.e.*, as long as no superior cultivars are introduced. Collection of local and advanced cultivars for breeding purposes would be useful. Yield of hyacinth beans should be improved. The crop shows a high potential, both as a field-grown pulse and as a vegetable for home gardens (Grubben, 1977).

Cowpea/Yardlong Bean

Vigna unguiculata (L.) Walp. cultivar group *sesquipedalis* (2n=22, 24) the yardlong bean, string bean or asparagus bean is a type of cowpea, used as a vegetable. It is cultivated for very long and tender pods, but unripe or ripe seeds are also eaten. Some climbing cultivars of *Vigna unguiculata* with short, stiff pods, are also cultivated as a vegetable crop in small gardens and leaves and pods as well as young or ripe seeds are used. These cowpea cultivars are common in tropical Africa and are often supported by shrubs and maize or sorghum stalks when grown with other food crops. Bush cowpeas are grown in USA as green beans and as fodder. A few commercial cultivars of yardlong bean are available, although local cultivars are most widely grown in Africa. Cowpea is mainly cultivated in regions which have a pronounced dry season during which pods are harvested, whereas, yardlong bean is more frequently grown in wet tropics in areas which do not have a significant dry season. Yardlong bean is liable to same pests and diseases as cowpea, but to a lesser extent. However, it is more susceptible than cowpea to extremely high temperatures (>35 °C) and drought and requires more water, due to large leaf surface area produced. All cultivars are short-day plants or day-neutral. Important diseases are: anthracnose (*Colletotrichum*), *Sclerotium* footrot, *Fusarium* wilt, *Cercospora* and other leaf diseases and cowpea curculio (*Chalcodermus*) and bean fly (*Melanagromyza*). Cowpea is the most important pulse of West Africa and it is also widely cultivated in Central America, where it is second in importance to common bean; it is also grown, to a lesser extent, in Asia (India). Yardlong bean is mainly grown in India and South East Asia, but Asiatic immigrants introduced it into Caribbean area. It is rarely found in other tropical regions. Nutritional value of yardlong beans and of young cowpea pods is fairly high, approximately equal to that of snap beans. Food value of unripe and ripe seeds is comparable with those of common and Lima beans. Yardlong bean is commonly seen in local markets in Asia but is produced mainly for home consumption. Both cowpea and yardlong bean are very susceptible to pests and diseases (Grubben, 1977).

Origin and Distribution

The primary centre is West Africa, with a secondary centre in India. Many local cultivars are developed in South East Asian countries, where humid conditions promote a certain degree of cross-pollination by insects. The largest collection of cowpea cultivars can be found at IITA in Ibadan, Nigeria where a germplasm catalogue is available. Other collections are established in India and USA and at national institutes in the tropics. Local cultivars of yardlong beans are given limited attention (Grubben, 1977).

Genetic Diversity

Much research and breeding work on cowpea are carried out in USA, Nigeria and, to a lesser extent, in some other countries (India). Many valuable genes are identified *e.g.* for male sterility, high protein and methionine content, resistance to thrips, pod borers, *Meloidogyne, anthranose, Xanthomonas, Cercospora, Uromyces* and cowpea mosaic virus. Less attention has been given to improvement of yardlong bean. Hybridization with local strains has been reported from India and Philippines. Bush types which do not need trellis supports are developed from climbing local cultivars, *e.g.* Los Banos Bush Sitao no.1. Well-known cultivars include Singapore Light Green and Bulacan Dark Purple. Yardlong bean crosses readily with cowpea and intermediate types are common (Grubben, 1977). *Genus Vigna* family Leguminoceae, sub-family Fabaceae, is a genus having about 170 species. Of them, 120 are in Africa, 22 in India and South-East Asia, and a few in America and Australia. Two species are wild, while 3 are cultivated for pods or forage crops. Cultivated cowpea originated in Central Africa from where it spread in early times through Egypt or Arabia to Asia and Mediterranean regions. It was brought to West Indies in sixteenth century by Spaniards and was introduced to America in about 1700 AD.

Cowpeas are now widely distributed throughout tropics and sub-tropics. Of the total world production, about 80 per cent comes from Nigeria alone. Other major cowpea producers are Upper Volta, Uganda and USA. It is also grown on a limited scale in Mediterranean region, South Africa and Australia. In India, it is mainly grown in central and peninsular regions for vegetable, fodder, hay, ensilage, dry seeds, green manure and cover crops. Major cowpea producing states are Uttar Pradesh, Punjab, Delhi, Haryana, Bihar, Andhra Pradesh, Karnataka and Gujarat. Botanical varieties of cultivated cowpea are as follows:

V. unguiculata (common Cowpea)

Plants are viny, sometimes erect with short pendent, 20-30 cm long pods with small seeds. This includes a large number of varieties distributed mostly in Africa compared to tropical and sub-tropical Asia.

V. cylindrica (catjang Bean)

Plants are more erect than those of *V.unguiculata*, pods small, 7-13 cm long, more or less erect. Seeds small, 5-6 cm long, kidney-shaped, thick as broad bean.

V. sesquipedalis (Asparagus or Yardlong Bean)

Plants are trailing or climbing, pods pendent, 30-90 cm long, fleshy and inflated, tending to shrink when dry. Seeds elongated, kidney-shaped and 8-12 mm long. Cultivars belonging to this group are found in Indonesia, Philippines, Sri Lanka and India. Cowpea utilized for green pods (vegetable) or dry seed (pulse), forms an important component of farming systems from arid to humid tropics. Cowpea cultivars grown for immature green pods, used as a vegetable are variously known as Asparagus bean, snake bean and yardlong bean. When grown for dry or immature seeds, they are known as black-eye pea, kiffir pea, China pea and southern bean.

Cowpea is well-adapted to stress with excellent nutritional qualities containing about 24.6 per cent crude protein. It is a key dietary staple for the poorest sector in many developing countries. As a crop of all round utilization, cowpea is grown for dry seeds, immature seeds, immature green pods, green leaves and even roots. Cowpea is a vigorously growing, bushy or trialing, annual herb with a strong tap root. The leaves are trifoliate with long petiole. Leaflets are large, hairy, dark green, cordate and subtended by inconspicuous stipules. Flowers are shortly pedicelled, borne on axillary racemes,

2-3 flowers to each peduncle, carrying 3 bracts at base, calyx glabrous or less placidly wrinkled with small lobes, triangular and acute, standard whitish to violet with violet wings and truncate at base, white, light pink or purple or light blue in colour, with their keel incurved and arched upward. Pods 20-30 cm (variable in size) long, cylindrical, smooth and slightly curved with a thick obtuse beak, pendent and slightly constricted between seeds. Seeds are globular to kidney-shaped, 5-12 mm long, ranging from white, green, yellow, red, brown to black, sometimes speckled or blotched. Seeds are characteristic in having a marked hilum surrounded by a dark ring. Chromosome complement of cowpea, 2n = 2x = 22, is typical of tribe Phaseoleae (Mishra and Mishra, 2001).

Future Thrust

Primitive yardlong bean cultivars should be collected since they are subject to genetic erosion especially in South East Asia. Yield potential, as well as pest and disease resistance, are important aims for breeding programmes on yardlong beans. These should preferably be in close collaboration with existing cowpea breeding programme (Grubben, 1977).

Sword Bean

Sword bean, *Canavalia gladiata* (Jacq.), DC (2n = 22, 44) is a widespread legume in tropical Asia, cultivated for young tender pods. It is a deep rooted, drought resistant, perennial climber, sown against fences in home gardens and close to tree stumps and shrubs when grown as a field crop. Dry beans are coarse-skinned and are not very appetizing, but they have a high nutritional value, similar to that of other dry beans. Unripe seeds are more palatable. Seeds generally contain HCN which is removed by cooking (Grubben, 1977).

Origin and Distribution

Sword bean originated from Indochina-Indonesia centre with probably a secondary centre in tropical Africa. The wild *C.gladiolata* Sauer of Burma is probably its ancestor and *C.polystacha Schweinf* of China is closely related (Grubben, 1977).

Genetic Diversity

Hybridization between sword bean, which is partly cross-pollinating, and South American jack bean *C.ensiformis* (L.), DC (2n = 22), a forage and cover crop of minor importance, may be possible. No important pests or diseases are recorded, other than spider mites. In wet regions, production is low and many young pods are heavily damaged by pod borers (Grubben, 1977).

Future Thrust

Degree of variation which exists in various local cultivars is not fully known although there are distinct botanical varieties. No breeding work is reported. There is danger of genetic erosion and the crop should be studied further. This drought resistant plant could become a valuable vegetable legume or pulse in hot dry areas (Grubben, 1977).

Winged Bean

In the hot, humid tropics, a single legume stands out as easy to grow, resistant to pests and diseases, productive, appealing to the eye and to the palate, and highly nutritious; the winged bean, *Psophocarpus tetragonolobus* (L.) DC (2n = 26). It is similar to soybean in its potential for Tropics. Although well known and widely distributed in South East Asia, it is little known throughout Tropics.

Psophocarpus tetragonolobus DC also known as the Goa bean is a prolific, climbing, perennial or annual plant. It is a minor vegetable in South East Asia and India and is cultivated mainly in home gardens for young pods. In Java, it is cultivated as a pulse on dykes between rice fields. In Burma, the fleshy roots (yielding 3 tonnes/ha) are sold as a delicacy. It is now becoming more widely known in many parts of tropical Africa. Quantity of ripe seeds produced is high and this can be partly explained by long duration of growth period. Plants require strong supporting stakes. Winged bean is noted for high level of *Rhizobium* activity in root system. Seeds are not easily separated from pod (Grubben, 1977). Winged bean is most commonly used as a green vegetable harvested when pods are about half grown. Leaves are also edible as a pot herb, as are leaves of many legumes. Tuberous roots, when they occur, are also eaten. Dried seeds do not cook well, except with a special technique. They are parched before eating, or a curd is prepared from expressed juices of soaked, ground beans, as with soybean. The greatest potential for winged bean is in home garden or in small-scale, high-intensity agriculture, because vines need staking, bear over a long period of time, and require regular harvesting. Potential of high-protein dried seed has scarcely been evaluated (Martin and Delpin, 1978). The winged bean is a tropical short-day plant, well adapted to lowland conditions. It is not drought-resistant and has a high water requirement (8 mm/day) for adequate production. It is remarkably free from pests and diseases. Winged bean has been introduced to Caribbean by Asian immigrants. In Africa, it is not widely cultivated, although it is introduced into a few West African countries. Composition of pods is similar to that of common bean and yardlong bean. Ripe seeds have a high protein content, comparable with that of soybean, but higher than that of common bean, Lima bean, mung bean, hyacinth bean and pigeon pea. The cooked dry bean is suitable as a protein-rich food for young children, since testa of large seeds can be removed very easily. Winged beans are commonly cultivated in home gardens for immature pods and there is a growing interest in its cultivation as a pulse in tropical lowland areas. Seeds, roots and leaves are also used, in addition to pods. Roots leaves are also very rich in nutrients (compared to cowpea leaves) (Grubben, 1977). Genus *Psophocarpus* (from Greek for "noise" and "fruit" in reference to explosive opening of fruit when dried in sun) includes wild, semi-cultivated and cultivated species. Taxonomic relationships have not been clarified, however. The most common species is winged bean itself, *P.tetragonolobus*, but other species have special value in limited areas. *Psophocarpus palustris* Desv., the swamp winged bean, is a cultivated and wild species of tropical Africa, introduced to a few other regions. Occasionally its leaves, young pods, and tuberous roots are eaten, and at times its seeds, which are parched before eating. It is also suitable as a cover crop, but has little other significance. A common synonym is *P.scandens* (Endl.) Verd. The species *P.lancifolius* Harms, *P.monophyllus* Harms, and possibly others are found wild in Africa (Martin and Delpin, 1978). Some of the many common names of *P.tetragonolobus* are shared with other bean species: asparagus bean, four-angled bean, Goa bean, Manila bean, Mauritius bean, princess pea, and winged bean. The term "winged bean" is becoming the most popular of the English common names (Hymowitz and Boyd, 1977).

Origin and Distribution

Origin of winged bean is probably tropical Asia and recent work indicted that Papua New Guinea may also be a centre of diversity. Local cultivars are available in Papua New Guinea and Indonesia (Grubben, 1977). Geographical origin of winged bean is still uncertain. Burkil (1935) believes that historical and nomenclatorial evidence point to an African origin, perhaps, Madagascar. But centre of variation now seems to be Papua New Guinea (Khan, 1976). In Indonesia and Papua New Guinea, hundreds of varieties exist with significant variation in all principal characteristics. It is difficult to believe that such variation developed after introduction of one or a few varieties. Burmese varieties, grown chiefly for their tuberous roots, are somewhat different from varieties of Papua New Guinea and may in fact represent more highly evolved and selected types. Winged bean was well distributed in southern Asia from India to Islands of Pacific in pre-Columbian times. It is an important vegetable in India, Ceylon, Burma, Thailand, Vietnam, the Malay Peninsula, and the Philippines. It is especially important in highlands of Papua New Guinea. Although found throughout tropical Africa, it is not especially important there. It was introduced to New World, but found only sporadically through Caribbean and American Tropics (Martin and Delpin, 1978).

Genetic Diversity

Evaluation of local cultivars was reported from India and South East Asian countries. Local cultivars are very uniform and are probably self-pollinating to a high degree. Large and small-podded cultivars are known (Grubben, 1977). Pod of winged bean is highly variable. Details of variations are given by Khan (1975), but in some cases the ranges he reports are inadequate. Lengths range from 6 to 40 cm and widths from 1.2 to 4 cm. Pod may be square or rectangular (flattened) in cross section. It is often coloured dark red or purple by anthocyanins. Wings vary in size and shape. They may be relatively narrow, with entire margins, or wide and thick, with undulate margins. Fiber within the pod varies. In better varieties, wings as well as the fiber are reduced. Number of seeds within pod varies from 5 to 40. Seeds are spherical, weigh from 0.05 to 0.45 g, and vary in colour from white to tan to dark brown or almost black. Seed colour is influenced by anthocyanin and tannin contents and by environment during maturation. Winged bean is a genetically diversified species that exhibits a wide variety of easily classified or continuous characters. Among first 26 accessions introduced at Mayaguez Institute of Tropical Agriculture (MITA), 64 types were found. However, many of these differed only in simple characteristics, such as flower or seed colour, that might be found in any segregating population (Martin and Delpin, 1978).

Principal variation found so far is in following characteristics:

Seed colour, seed size, ring around hilum, hard seed (poor germination without scarification), vigour of vegetative growth, dwarf or normal stature, response to long days in Puerto Rico, tuberous root production, degree of nodulation of roots, anthocyanin of stem, size of leaves and leaflets, flower colours, flower colour distribution, pod colours, pod colour distribution, pod length, pod width, smoothness of pod surface, seeds per pod, cross sectional shape of pod, development of wings, undulation of wings, dry pod weight, total yield, perennial versus facultative annual growth habit, drought resistance, hardiness of seeds when cooked, bitterness of cooked seeds, protein content and oil content (Martin and Delpin, 1978).

Future Thrust

There is some risk of genetic erosion and local cultivars should be collected for plant breeding purposes especially in South East Asia. Considerable international interest has recently been shown in winged bean. It is apparent that it is likely to gain importance as a vegetable and pulse crop in lowland tropics. The breeding objectives can be defined after a study of available cultivars and cultivation techniques (Grubben, 1977). Winged bean, is an excellent vegetable for production in the home garden and for the market. Winged bean has immense worth, and efforts should be made to popularize it and to distribute seeds widely. Although especially adapted to hot, rainy season and coastal regions, it should be possible to select varieties for year-round production and for adaptation to specific environments, including tropical highlands.

Because it requires staking and harvest by hand, winged bean is a labour-intensive crop. This is not a disadvantage in many parts of Tropics. Needs of plant are simple, and so it is suited to a low level of technology. Year-round production, at least of some varieties, is a marked advantage of crop for small-scale production and use. Processing potentials of winged bean, particularly as a snap bean substitute, need investigation. Beans of appropriate varieties should be useful either canned or frozen (Martin and Delpin, 1978).

Alliums/Bulbous Crops

Allium species (onion, shallot, garlic, leek, etc.) make up a closely related group both taxonomically and in cultivation methods. *Allium cepa* L. the dry or bulb onion, is the most important species of Allium group. It can be regarded as being the most important single vegetable species of tropics, after tomato. Nutritional value of onions is very low, nevertheless, they are important for flavouring. Their medical properties are often stressed, garlic, onion and other species aid digestion and purify blood (Grubben, 1977). Bulbous crops have an underground stem modified in form of tubers, corms, rhizomes or bulbs, which not only store food material, but also serve as organs of perennation for plants. Genus *Allium* belongs to family Alliaceae, has bulb as an underground storage organ, which is also used for vegetative propagation. Alliums represent a major group among bulbous crops and these are cultivated for food, medicine or religious purposes since early times. Most of these crops are strongly flavoured and are characterized by sulphur-containing compounds which impart them a distinctive smell and pungency (Rabinowitch and Currah, 2002). Characteristic flavour and odour are produced when plant tissue is bruised, cut or macerated. Use of onions and garlic in medicine in India in sixth century BC is mentioned in ancient Indian literature.

India occupies second position in onion and third in garlic production in the world. Fresh and processed produce of these crops are exported and it is an important foreign exchange earner. Onion is known to lower blood sugar and promote bile production. It is cooked as a vegetable or consumed raw, as a pickle or as dehydrated product in the form of powder, flakes and bulbs. Dehydrated products of onion are in great demand in world market. India accounts for more than 70 per cent share in export of fresh onion. Garlic is consumed green or mostly as a spice, and is also used in processed form like pickles, flakes, granules etc. Garlic is carminative and gastric stimulant. It helps in digestion of food, reduces cholesterol levels in human blood and also lowers blood sugar. It has higher nutritive value compared to other cultivated alliums (Pandey *et al.*, 2005).

Genus *Allium* was earlier classified under family *Liliaceae* for a long time, later under family *Amaryllidaceae*, and presently it is placed under family Alliaceae (Purseglove, 1972). *Allium* is a very complex genus accounting for over 700 species of bulbous herbs. According to Bujisen (1990), the genus comprises of 1000 species, which are distributed throughout world. Onion has the primary center of origin in Central Asia and secondary centre in Near East (McCollum, 1976). It was domesticated at several places and occupied vast area in Western Asia, extending from Palestine to India. Most of the botanists believed that it had its origin in Iran, Pakistan, Afghanistan, Kazakhistan, Russia and adjoining regions. Shallot is indigenous to Palestine and owes its name to town of Ascalon (Giles, 1943). It is treated as a separate species by some authorities. In India, it is grown in a few parts of Himalayas and also in Peninsular India. Welsh onion is native to Central and North Western China and cultivated to some extent in India, particularly in North-Eastern region and Western Himalayas. Chinese chives are been domesticated in the Orient and it has changed little from wild. For garlic, Central Asia is the primary center and Mediterranean region is the secondary center of origin (McCollum, 1976). It is grown in India and China. It was known to Egyptians before 3000 BC and also to ancient Greeks and Romans. Europeans carried it to Western Hemisphere. Leek has its center of origin in the Eastern Mediterranean region from where it spread to Europe by Middle Ages. In India, it grows well in Western Himalayas. Chives have a Mediterranean origin. Other domesticated species occur in Turkey, Iran, Iraq, Pakistan and the erstwhile USSR. Remaining species occurring in wild, were classified primarily, based on leaves or floral morphology under different sections and groups, although, their taxonomic classification, correct nomenclature and synonymy continue to be vague, complex and unresolved (Stearn, 1946). Regions of high species diversity exist in Commonwealth of Independent States (CIS), Turkey, Iran, North Iraq, Afghanistan, Pakistan and hilly tracts of North and North-eastern India. According to Bujisen (1990), all taxa of genus *Allium* cultivated in South-East Asia are introduced either from China, central Asia or Europe. Besides, wild species are generally restricted to Northern Hemisphere and are totally absent in tropical parts of South-East Asia. Species cultivated in South-East Asia which have their origin and secondary centre of diversity in Sino-Japanese centre, include welsh onion, Rakkyo, Chinese chives etc. Above species are also found in Indian gene centre, some of them still in wild state. Jones and Mann (1963) discussed many aspects of onion and its allies. Domesticated *Allium* species come from Near East, Central or Eastern Asia.

In India, genus *Allium* is widely distributed in temperate and alpine regions. About 30 species are reported to occur in Indian gene center (Babu, 1977). Majority of wild species are new records. Negi and Pant (1992) reported occurrence of lesser-known wild species in mountainous regions and discussed information on their habitat, local use etc. Prominent wild species included in the study were *A.carolinianum* DC. (syn. *A.blandum* Wall.), *A.chinense* G. Don (syn. *A.bakeri* Regel.), *A.consanguineum* Kunth, *A.humile* Kunth (syn.*A.govanianum* Wall. ex Baker), *A.przewalskianum* Regel. (syn. *A.jacquemontii* Regel., *A.stolczkii* Regel.), *A.stracheyi* Baker, *A.victorialis* L. and *A.wallichii* Kunth. According to Hooker (1973), the most widely cultivated Allium species in the Indian sub-continent include, onion, garlic, shallot and leek. Genetic diversity in cultivated alliums was reported from different parts of the country. Prominent areas for genetic diversity are Jammu and Kashmir, Himachal Pradesh, Uttaranchal, North-East region, Rajasthan, Gujarat, Maharashtra, Andhra Pradesh, Karnataka and Tamil Nadu (Pandey *et al.*, 2005).

Onion

Allium cepa var. *cepa* L. (2n = 16), the dry or bulb onion, is one of the most ancient vegetables in the world. Many commercial cultivars are selected either for eating raw or boiled, or for dehydration. These cultivars differ in yield potential, quality (shape, colour, pungency, storability), disease resistance and climatic requirements. On the American market, three types are distinguished.

1. Bermuda-Ganex-Grano short-day type, flat to conical-shaped, mild, early maturing.
2. Late maturing type, long-day globe onions, mild, *e.g.* White Globe, Sweet Spanish.
3. Creole type, with very strong flavour, *e.g.* Ebenezer and White Portugal.

First group is the most suitable for cultivation in tropics. Onions are sometimes cultivated for use as green bunching onions, harvested when young, with or without a small bulb. Well-known cultivars for this purpose are Evergreen White Bunching and Evergreen. They can be grown successfully in most areas of tropics which have short days. In general, most of dry onions cultivated in tropical areas of Asia and West Africa are local selections of fairly primitive cultivars. Most of onions grown in Latin America and East Africa, however, are exotic cultivars such as Red Creole, Granex, Bombay Red. In West Africa, Red Kano, Red Patna, Texas Grano, Yellow Bermuda and Early Yellow Cape Flat are well-known cultivars. Onions are a cool season crop, growing well in day temperatures ranging from 13 to 24 °C. They are not particularly susceptible to frost. For bulb initiation, a critical minimum daylength must be reached. European and American summer onions are "long day" cultivars, not suitable for tropics but a few early-maturing American cultivars are available in international seed trade. Many local onion cultivars, however, grow well at high temperatures up to 30 °C and form bulbs in short days (11-12 hours). Short day cultivars are not short day plants but cultivars which bulb earlier. Bulbing is also stimulated by high temperatures. Flowering normally occurs only if bulb formation are followed by a cool period; onion in temperate regions is therefore regarded as a biennial plant. In tropical regions with cool nights, bolting of a percentage of young plants of local cultivars, in which bulb formation has started, is a common occurrence, resulting in considerable yield reduction. If the emerging flower stalk are pinched out, however, bulb formation may continue. Due to its special ecological requirements, onion is normally cultivated only during cool season in sub-tropics and tropics, but rarely between 10 °N and 10 °S, except in tropical highlands above 1,500 m amsl.

In general, nutritional value of dry onions is low. The ANV is only 2.05. Onions are an indispensable part of many diets as a flavouring, and they are consumed regularly in small quantities, being comparable with hot pepper. If immature bulbs and tops are consumed, the nutritional value will be much higher and the ANV may reach 3.96, with a high vitamin C and calcium content. When used raw in salads, nutritional value is somewhat higher *e.g.* for onion 2.30 and green onions and tops 4.69 ANV. In wet tropics and in a large part of India and South East Asia (Indonesia) onion is cultivated and consumed to a lesser extent than it is in other countries. In general, onion is the second most important world vegetable species (after tomato) and good storage and bulk transport properties of dry onions facilitated development of a world-wide trade. They are always in demand even in the smallest local markets. Primitive local cultivars which are grown in many countries (India, Nigeria, Mexico) often show tendency to bolting and have a poor keeping quality. High losses due to pests and diseases are reported, the most important being thrips, onion fly (*Hylemia antigua*) and nematodes (*Ditylenchus dipsaci*), purple blotch (*Alternaria porri*), downy mildew (*Peronospora*), pink root (*Pyrenochaeta*) and *Colletotrichum* smudge. White rot (*Sclerotium cepivorum*) is a serious disease in some sub-tropical areas (Brazil). Technical information concerning onion cultivation in tropics (density, fertilizing, disease and pest control, storage and processing) is very limited (Grubben, 1977).

Origin and Distribution

Allium cepa is a cultigen without wild types, but in its primary centre of diversity (Central Asia) several related wild species are found, *e.g.* *A.vavilovii*, Popov and Vved. *A.oschaninii*, Fedtsch. and *A.pekemense*, Fedt. Hundreds of locally grown, primitive cultivars exist, mostly in Near East (which, with the Mediterranean, represents a secondary center of diversity) are in central Asia, but also in all traditional onion growing regions of the tropics. Large collections of onion cultivars and of wild types are available in Europe and USA (Grubben, 1977).

Genetic Diversity

Allium cepa var. *cepa* was subjected to selection for several thousands of years especially in India and Near East. In China selection of other species of *Allium* occurred. Breeding programmes resulted in production of high-yielding and high quality cultivars such as Early Yellow G1 which are widely distributed in temperate regions. Breeding work on this crop is relatively easy; the onion plant is a self-compatible out-breeder, pollinated by insects. It withstands certain degree of inbreeding, giving rise to heterosis when strains of good combining ability are intercrossed. Application of male sterility has made commercial seed production of hybrids possible and, at present, numerous hybrid forms (F_1) are cultivated, combining uniformity and good quality characters with resistance to several diseases. Modern short-day cultivars frequently grown in tropics are: Texas Early Grano no.502, Early Yellow Cape Flat, Yellow Bermuda, Red Creole, Tropicana, El Toro and Bombay Red. In a cultivar testing and breeding programme in Northern Nigeria, Texas Grano showed a high yield capacity but a low of dry matter (DM) content (50 tonnes/ha x 6 per cent DM = 3 tonnes DM) and Bombay Red a lower yield capacity (30 tonnes/ha) but a high DM content (15 per cent) giving 4.5 tonnes DM/ha. Texas Grano, with brown scale leaves, is a good export quality onion but had a short storage life, whereas Bombay Red had excellent storage quality but red colour, popular in local markets, made it less suitable for export. However, the local poor quality cultivars are proved superior to internationally grown normal and hybrid cultivars in yield, resistance to diseases and sometimes also in storage quality. Good results are reported also from IRAT, Nigeria. A selection of a local cultivar, 'Blanc de Galmi', produced up to 60 tonnes/ha, with 15 per cent DM and had excellent storage qualities. Similar results are reported from several countries with a tradition of onion cultivation (Taiwan, Mexico and India). Local and modern cultivars show important differences in their susceptibility to pests and diseases. A certain degree of resistance or tolerance to thrips, onion fly and nematodes is reported. Italian Red and local Brazilian cultivars are less susceptible to purple-blotch, a serious disease under wet conditions. A few cultivars are less susceptible to downy mildew. Commercial cultivars Granex and Excel L35 show resistance to pink root. No data are available on resistance to *Colletotrichum* smudge. Apart from selecting for yield, storage potential and disease-resistance several onion cultivars are selected for use in food industries *i.e.*, for canning or dehydration. A few of these cultivars contain about 20 per cent dry matter. Wild *Allium* species have not yet been used in onion breeding. Up to the present, interspecific crosses have failed, except with *A.fistulosum*. Gene translocation to *A.cepa* might be feasible (Grubben, 1977).

Future Thrust

Primitive onion strains existing in Asian countries, Africa and Central and South America should be collected, since they are gradually being replaced by modern cultivars. Breeding work on this economically important crop should be accelerated. Requirements for future selection of onions for tropics include following: a high yield ability, combined with a high DM content, good quality for both local and export markets, good storage properties, minimal bolting and a combination of resistance to many diseases and pests. Sterile male lines should be selected for production of hybrid seed. If further improvement by crossing between local strains and modern temperate cultivars become limited *e.g.* for resistance to diseases, pests and drought, crosses with other *Allium* species *e.g. A.fistolosum* should be considered (Grubben, 1977).

Shallot

Shallot, *Allium ascalonicum* L., usually referred to in this way but more correctly as a form of *A.cepa* (2n = 16) is an onion-type, propagated by cloves. Many clones are known in temperate and in tropical regions of Asia and Africa. Clones which flower readily as well as clones which seldom or never flower are known; both are related to onions. Shallots for planting are used locally, but are rarely commercially sold. Seeds are not used for propagation since they usually result in production of undesirable types. Shallots are better adapted than onions to cultivation in the tropics. They are widely cultivated in West Africa (Ivory Coast, Ghana), in Caribbean countries, and elsewhere in tropical lowlands of South East Asia and Central and East Africa but thrive the best in drier climates at medium altitudes. Shallots have a relatively low nutritional value, similar to that of dry onions. Frequently, the leaves and young bulbs are consumed, and value is thereby increased making it comparable to that of green onions. Shallots are often considered a luxury, since they have more pungent flavour than onions. In market, shallots command a better price than onions. In the forest zone of South Ivory Coast, South Ghana and Caribbean, they are an important commercial vegetable, often replacing onions. Since the crop is labour-intensive and has a lower yield potential than onions, shallot is not normally cultivated in cool and dry areas where onions can be grown. Main problem for shallot growers is conservation of suitable planting material. In humid regions, this degenerates quickly probably due to facts such as fungus diseases, nematodes, spider mites and virus diseases and has to be renewed after 3-5 generations with material from drier regions. Shallots suffer from same diseases as onions and are particularly susceptible to *Sclerotium rolfsii* (Grubben, 1977).

Origin, Distribution and Genetic Diversity

As shallot is a botanical form of dry onion, centre of diversity is identical. No breeding work on shallots in tropics is reported and no collection of shallot cultivars is known (Grubben, 1977).

Future Thrust

Shallot cultivars could well disappear if they become replaced by onions. Local tropical cultivars should be collected and stored in a germplasm conservation centre for vegetatively propagated vegetables. Testing of local cultivars and improvement by breeding are justified for those regions where shallots are economically important *e.g.* West Africa (Ghana, Ivory Coast) (Grubben, 1977).

Garlic

Garlic, *Allium sativum* L. (2n = 16) is a very pungent species propagated by cloves. Many local clones are known, a few are well adapted to tropical conditions *e.g.* high temperature up to 30 °C. Garlic is extensively grown in East Asia, Mediterranean and in Latin America, but is relatively unknown in Africa although it could be grown much more widely in drier regions. It is unsuitable for growing in wet humid areas. Nutritional value is approximately similar to that of dry onions. It is more widely used as spice than a vegetable, and intake per caput per meal is very low. Plant is very susceptible to *Alternaria* and other pests and diseases which attack onions. Conservation of cloves as planting material for following year is difficult since they do not store well (Grubben, 1977).

Origin, Distribution and Genetic Diversity

Ancestor of old cultigen is possibly *A.longicuspis* Regel (2n =16) which occurs in Central Asia. Reports on breeding work are very scarce and no international trade in planting material of superior clones is known (Grubben, 1977).

Future Thrust

There appear to be little genetic erosion in local strains. These should, however, be collected for future breeding. Breeding work on this economically important vegetable spice, through cultivar testing, should be carried out, aiming at higher yields and improved disease resistance (Grubben, 1977).

Leek

Leek, *Allium porrum* L., is a tetraploid with 2n = 32 and is a temperate leaf vegetable, showing retarded growth and low yield capacity at day temperatures above 24 °C. Leek cultivation in cool season and in tropical mountain regions is possible. Seed production in such areas may be successful after a cool season but is not economically justified. American Flag, Musselburgh and Italian Giant are well-known cultivars. Leek suffers from same pests and diseases as onions, but to a much lesser degree. Nutritional value of leek is better than that of all other *Allium* species. It has an ANV value of 6.63 due to relatively high calcium, carotene, iron and vitamin C contents. Although highly regarded by Europeans living in tropics, this species is of minor importance to local populations (Grubben, 1977).

Origin, Distribution and Genetic Diversity

Neither the breeding of leeks specifically for tropical conditions nor existence of local strains of leek in tropical or sub-tropical countries was reported except in Far East. Wild and cultivated *A.ampeloprasum* L. types related to *A.porrum*, occur in Near East (Grubben, 1977).

Future Thrust

At present no action for collection of germplasm or breeding of leeks for tropical regions appears necessary (Grubben, 1977).

Other Temperate Vegetables

Carrot and radish are typical temperate vegetables which are of regional importance in tropical areas.

Carrot

Carrot, *Daucus carota* (L.) (2n = 18) is a biennial temperate vegetable and many cultivars exist. Some of them, Danvers Half Long and Intermediate are suitable for cultivation in upland regions of tropics, and others *e.g.* Chanteney, Early Nantes and Cape Market are suitable for low to medium altitudes. Imperial Long, Scarlet, New Kuroda, Chanteney and Egmond Gold are recommended for Pacific Islands. Seed sown at rate of 3-4 kg/ha will give a density of 200 plants/m^2. After 100 to 150 days, crop yields between 5 tonnes/ha (low) to 30 tonnes/ha (high). Carrots are usually cultivated as a cool season crop. A high day temperature, above 25 °C, causes slow growth rate and blanched yellow-white roots of poor quality are produced. Flowering and seed setting are successful only in cool regions with a mean day temperature below 20 °C (Brazil); seed yields of about 200 kg per ha may be obtained. The ANV rating is 6.48; the very high value is due to β-carotene content of 4.2 mg/100 g. Carotene content of carrots in temperate regions is assessed as 6.00 mg/100 g, normally resulting in an ANV of 9.00. It is known that high temperatures and short-day conditions decrease β-carotene content. Carrots are important market vegetables in many areas in tropics, especially Brazil, Venezuela, Nigeria and Indonesia. *Alternaria* leaf blight and root-knot nematodes are major cultural problems in tropics. Carrot seedlings are easily washed out by heavy rains and many fungi or bacteria cause root-rot *e.g. Erwinia, Stemphylium, Sclerotium, Fusairum* (Grubben, 1977).

Origin, Distribution and Genetic Diversity

Carrots have origin in South West Asia from where they spread to China and Mediterranean area. Several wild annual forms occur in these regions. Much breeding work was undertaken in West Europe (the Netherlands, France), Japan and USA. Present cultivars are selections from old landraces, selected for carotene content, colour, shape, earliness, yield, resistance to cracking and decreased bolting. Canning and storage quality, which are related to dry matter content, are also important. Inbreeding induces reduction of vigour. Hybrid cultivars, however, are not yet in general cultivation. Local carrot cultivars from Brazil include: Kuroda and Piracicaba (Grubben, 1977).

Future Thrust

Preservation of old local strains from temperate regions should be ensured by developed countries. Risk of disappearance of wild germplasm is not imminent. To promote cultivation of this very nutritious vegetable, cultivars should be bred which are suitable for tropical conditions (Grubben, 1977).

Radish

Radish, *Raphanus sativus* L. (2n=18) is a short-lived annual, grown throughout tropics. Seed production of temperate and Asiatic sub-tropical types are successful in cool areas. Japanese radish (daikon) forms large, long or round, white roots instead of small red, round European type. Seeds sown at a rate of 40 kg/ha yield 10-40 tonnes of fresh roots after 30-70 days with a seed yield of 800 kg/ha. Radishes are fairly resistant to pests and diseases. Blackrot (Aphanomyces) and Alternaria are mentioned. Nutritional value of radish is low (ANV=3.22) (Grubben, 1977).

Origin, Distribution and Genetic Diversity

Much breeding work on radish was done in developed countries, but only a few reports, mainly from India, are available on results of breeding in tropical countries. Germplasm of wild radish types and species are abundant in Europe and Mediterranean (Grubben, 1977).

Future Thrust

No high priority to germplasm collection and breeding need be given, but more information is needed on diversity in Chinese or Japanese radish (Grubben, 1977).

Brassica/Cole Crops

Genus *Brassica* includes a number of popular vegetables. Most of these species are biennial, cold-resistant, temperate plants, requiring a cold period for flowering and seed production. Some Brassicas are indigenous to tropics. Taxonomy, especially of cultivated types, is very difficult and includes many synonyms; clear descriptions of species, sub-species and botanical varieties are rarely available. More important Brassicas grown in tropical areas as vegetables are, White cabbage (*B.oleracea* L. var. *capitata* L.), cauliflower (*B.oleracea* var. botrytis L.) and broccoli (*B.oleracea* var. *italica* Plenck.). Although the genus *Brassica* is very polymorphic, all its vegetable species have similar cultivation and ecological requirements and are susceptible to same types of pests and diseases. All species are out-breeders, pollinated by insects. Inbred lines, if crossed with each other, give rise to heterosis. Hybrid seed is relatively easy to produce with aid of self-incompatible mother plants. Cauliflower is an exception, in that a high percentage of selfing is normal and inbreeding does not lead to genetic depression. Many interspecific crosses, amphidiploids and allopolyploids are recorded and are used in plant breeding programmes. Commercial cultivars have usually been selected from old local cultivars by mass selections, but are gradually being replaced by hybrid cultivars (Grubben, 1977).

Cabbage/White Cabbage

Cabbage/White cabbage, *Brassica oleracea* var. *capitata* (2n= 18) is a temperate biennial plant, one of the most ancient and widely cultivated European vegetables. Two hundred and eighty cultivars are grown in Western Europe, differing in shape and size of head, shape and colour of wrapper leaves, resistance to diseases, earliness and storage quality. Late types (90-120 days from transplanting) such as Glory of Enkhuizen produce large heads of 2-5 kg whereas early types such as Golden Acre and Copenhagen Market produce small heads of 1-2 kg in 60-90 days. Early types are more popular in tropics than are late types. A local, vegetatively propagated, white cabbage is grown in Indonesia, at medium elevations. After heads are harvested, young lateral shoots which arise from remaining stem are used as new planting material. Due to its poor quality, it is gradually being superseded by cultivars propagated from seed. White cabbage is commonly cultivated in all tropical upland regions. Its optimum day temperature requirement is low, between 15 and 20 °C. Good heads may be obtained at temperatures up to 28 °C, but a longer growing season is necessary. In recent years, a few Japanese early cultivars are introduced; these perform well under tropical lowland conditions. Cabbage is not daylength sensitive and flower initiation is mainly determined by temperature. Bolting occurs after a period of a few weeks if the temperature falls below 10 °C. Nearly all cabbage seed used in tropics is imported from temperate regions, but cultivars are developed which will flower at higher temperatures, giving rise to local cabbage seed production industries (Brazil, Venezuela). Bolting in cabbage cultivars also occurs in Indonesia. White cabbage has a fairly good nutritional value with an ANV of 3.52. It cannot, however, compete with green, leaf vegetables. Calcium and iron content of cooked cabbage can be readily adsorbed by human body. Importance of cabbage in tropics is increasing rapidly, one of main reasons being that it is a high prestige vegetable which can fairly be produced in considerable quantities. It is not liable to damage during transport and can be stored for several weeks. It is in demand for city markets, where it often replaces green leaf vegetables used in traditional dishes. In contrast to this development, importance of white cabbage in developed countries is gradually decreasing since it is being replaced by cauliflower and other vegetables. In Philippines, 10 per cent of budget spent on vegetables is used in production of cabbage which is a very popular vegetable. Cabbage is also important in upland regions of Indonesia and East Africa (Kenya, Tanzania), in Northern India, and in Meso and South America. Apart from high temperature, main limiting factor is its great susceptibility to pests and diseases. Black rot (*Xanthomonas campestris*), clubroot (*Plasmodiophora brassicae*), bacterial soft rot (*Erwinia*) and *Fusarium* and *Sclerotium* wilts appear to be the most important diseases. Caterpillars such as *Plutella* and *Pieris* cause severe damage in many areas and many other pests and diseases are also reported. *Alternaria*, anthracnose (*Colletotrichum*) and downy mildew (*Peronospora*) are occasionally serious (Grubben, 1977).

Origin and Distribution

Origin of white cabbage is either Western part of Mediterranean area or Asia Minor. Cultivated species *B.oleracea* have several wild ancestors. Heading types may have arisen, by selection, in Western Europe in early Middle Ages. Many working collections and germplasm collections of wild and primitive types of white cabbage and other Brassicas useful for breeding purpose are already available (Grubben, 1977).

Genetic Diversity

White cabbage is one of the most intensively studied vegetables regarding its genetic and breeding behaviour. Most of the older cultivars are result of mass selection. Modern cultivars are selected from crosses involving high quality parents. Hybrid cultivars which combine a high degree of uniformity, a high yield capacity and other desirable factors are becoming increasingly popular. The most important development in this field is selection of tropical types by Japanese breeders, such as K-Y Cross and K-K Cross; these are suitable for hot climates and mature in 60 days. In wet tropical climates, level of infection by bacterial and fungal diseases is very high (from 30 to 70 per cent). These diseases are difficult to control by cultivation methods but good results have already been obtained in breeding programmes aimed at introducing resistance to some pathogens. In Indonesia, hybrids between Glory of Enkhuzen x Pudjon are resistant to blackrot. In Brazil, a local cultivar Louho was selected, which showed a high level of resistance to blackrot and gave good seed production at high temperatures. Many cabbage cultivars are resistant to different physiological races of clubroot (Grubben, 1977).

Future Thrust

Collection of germplasm in Mediterranean area and in South West Asia has attracted attention of plant breeding institutions. Danger of losing wild Brassica types is not great and there appears to be little need for immediate action. Existing collections of local Asiatic white cabbage cultivars should be revised. High priority should be given to breeding cabbage cultivars which give high yields in hot climates, produce fertile seed and combine many kinds of disease resistance. Attention should also be paid to those types which have a high nutritional value (Grubben, 1977).

Cauliflower and Broccoli

Cauliflower (*Brassica oleracea* var. *botrytis*) and broccoli (*B.oleracea* var. *italica*) are closely related annual plants with very similar characteristics. Main difference between the two varieties is that white cauliflower is harvested only once whereas small green broccoli heads may be cut from same plant over an extended period. Cauliflower heads should be compact and white, whereas green broccoli inflorescences or curds have a loose structure. Mechanized harvesting is more practical with cauliflower than with broccoli. In tropics, main cauliflower cultivars are late-maturing European ones, adapted to short daylengths but those are being replaced by Indian and Japanese cultivars such as Early Patna and Snow Queen, which are adapted to warm climates. Tropical cultivar Pua Kea was selected in Hawaii. Calabrese is perhaps the most common of the many broccoli selections. Cauliflowers and broccoli are both cool weather crops. Cauliflower, in particular, is more difficult to grow in warm climates than white cabbage but broccoli is better adapted to high temperatures. No satisfactory curd formation takes place in cauliflower at temperatures higher than 25 °C and heat injury or scald is a frequent phenomenon. Seed production of cauliflower in warm climates is either non-existent or very poor but broccoli gives a good seed yield at higher elevations. Nutritional value of cauliflower is good (ANV=4.12), due to a high vitamin C content, Broccoli has very high value (ANV=6.5) due to high calcium, carotene and vitamin C content. Popularity of these vegetables in city markets is increasing. They are already important in cooler parts of Latin America, India and Far East. Both cauliflower and broccoli are susceptible to same pests and diseases as white cabbage, but to a lesser degree. Main limiting factor is poor formation of curds in cauliflowers grown in high temperatures (Grubben, 1977).

Origin and Distribution

Origin and main centre of diversity are in Mediterranean. Working collections of old and modern cultivars are available at breeding stations (Grubben, 1977).

Genetic Diversity

Indian selection Patna and, within the past few years, the Japanese hybrid cultivar Snow Queen are early tropical cauliflowers developed by modern plant breeding techniques. Genetic constitution and breeding methods used with both cauliflowers and broccoli hybrids were developed both in USA and Japan. In Campinas (Brazil), local cauliflower cultivar Piracicaba, which produced seed, was selected (Grubben, 1977).

Future Thrust

Collection of germplasm has no high priority for tropics. Resistance to serious diseases is a much more important requirement. More attention should be paid to broccoli, which is easier to cultivate and has a better nutritional value than cauliflower (Grubben, 1977).

Leaf Vegetables

In tropics, numerous wild and cultivated plants are in use as leaf vegetables. About 1,000 species with edible leaves, belonging to some 125 different plant families are reported. All dark-green vegetables are more or less similar in their nutritive value and methods of consumption and preparation. Approximately 80 per cent of more popular dark-green leaf vegetables are wild plants and are mainly gathered for home consumption. These 600-800 species constitute only a very small part, probably far less than 10 per cent of total quantity of leaf vegetables consumed. Interest in them is declining, with increase in population density, changes in wild vegetation and increase in urbanization associated with economic development. Amaranth is the most important annual leaf vegetable in Benin and Sierra Leone, *Celosia* is the most important one in South Nigeria, *Solanum* in South Togo, taro and *Solanum* in Ghana and *Corchorus* in South Ivory Coast and Liberia. This illustrates great diversity of locally important leaf vegetables which can be found within a limited area. In tropical lowlands, very expensive European type vegetables are consumed only to a limited extent by more affluent sector of population and are not consumed in rural areas. In cooler tropical regions, more European type vegetables may be consumed. Dark-green leaf vegetables have high nutritional value, especially their content of calcium, iron, β-carotene and vitamin C. These vegetables are also rich in other protective nutrients such as phosphorus, vitamin B complex and folic acid. From details of nutritional values of 24 leaf vegetables, following conclusions may be drawn:

1. Dry matter content of fresh product varies considerably between different species. It has been proved that this content also varies within each vegetable species. In Amaranth leaves, dry matter content ranges from 8 to 21 per cent depending on cultivar, climatic conditions, soil type, fertilizers applied, irrigation, degree of maturity, method of harvesting, transporting and handling. It is better to consider nutritional value of dry matter of vegetables than that of fresh product. Moreover, leaves with a low dry matter content lose water during cooking, whereas, leaves with a high dry matter content take up water. Unfortunately, data on nutritional value of cooked vegetables are very scarce.
2. Proteins (29.5 per cent of dry matter) are generally high in sulphur containing amino acids and consequently they improve biological value of main protein-supplying food types (cereals, tubers, pulses).
3. Records of nutritional values show great variation in vitamin and mineral contents. This may be due to different methods of analysis, but growing conditions of vegetables may also influence quality. For example, it was proved that a high level N-fertilization increased β-carotene content of amaranth from 17.0 to 40.0 mg per 100 g dry matter. Available data on nutritional values should therefore be interpreted with some caution.

Leaf vegetables generally have a low economic value and social prestige and are more commonly consumed by less affluent sectors of society. Many leaf vegetables are produced for domestic consumption and leaves are easy to grow without much attention. Successful market products among leaf vegetables, are those which are easy to grow in gardens, give high yields with a short growth period, are easy to harvest with a minimum of labour and have a widely acceptable taste. Most important are amaranth and two European type vegetables *i.e.* spinach beet and lettuce.

Several of these species suffer seriously from pests and diseases and seeds or planting materials of good cultivars are seldom or never available. Main limitation, however, is lack of any knowledge of nutritional value of leaf vegetables among public in general. Short storage capacity of fresh leaves, which demands a well-organized marketing system, is also a serious handicap (Grubben, 1977).

Amaranth

Amaranthus spp. are widely distributed, short-lived weeds, leaves of which are frequently used as pot-herbs. Some species are cultivated in home gardens and for marketing, *i.e. A.tricolor* L. (2n = 34) mostly in East Asia, *A.cruentus* (L.) Sauer (2n = 32) mostly in Africa and *A.dubius* Mart. ex Thell. (2n =

64) in Caribbean area. These are probably the most important leaf vegetable of lowland tropics of Africa and Asia, but are scarcely known in South America. Nutritional value is high, although high oxalic acid content slightly decreases ANV rating. A considerable amount of research was carried out on amaranth in those regions where it occurred as a serious weed, and it was included as a test plant in physiological research. Amaranthus is a plant with a C-4 cycle photosynthetic reaction which makes it an efficient producer of carbohydrate. Some Indian cultivars are markedly short-day plants; market growers plant them in beginning of summer and harvest them over several months by repeated cutting until beginning of flowering at end of the season. Cultivation method is either by sowing directly, or transplanting seedlings raised in a nursery. When plants are pricked out, either dense spacing (10 x 10 cm)with one single harvest by uprooting, or wide spacing (25 x 25 cm) with several harvests by repeated cutting is practiced. Amaranth is a suitable plant in a crop rotation since it is not affected by more common soil diseases and pests (nematodes, *Fusarium*, *Sclerotium*, bacterial wilt). Serious pests and diseases are: damping off (*Pythium*), wet-rot (*Choanephora*), caterpillars (*Hymenia*) and stem borers (*Lixus*). Early flowering occurs as a consequence of short day length or may result from a short period of water stress (Grubben, 1977).

Origin, Distribution and Genetic Diversity

Amaranth species show centers of diversity in Central and South America, India, South East Asia and secondary diversity in West Africa and East Africa. Many local cultivars and wild forms occur especially in Africa. Some cytogenetic work was carried out in India. Crossing with related wild species or with grain amaranths is possible, but not between three vegetable species mentioned. All are monoecious plants with wind-pollinated, unisexual flowers. However, abundant pollen production leads to a high degree of self-fertilization. Commercial cultivars exist in India, Taiwan, the Caribbean and USA; these are remarkably uniform. Considerable differences exist between three main species. African cultivars of *A.cruentus* were originally grain-amaranths; they have long stems, a high DM-content in leaves and bear large inflorescences. Cultivars of *A.dubius* and *A.tricolor* have a much lower seed production; their habit is similar to spinach, with short stems and succulent leaves. Wide variation between local cultivars within each species in growth habit, disease resistance, taste and quality implies possibilities for future breeding. Small collections of local cultivars are kept at AVRDC in Taiwan, TNAU, Coimbatore and other institutes in India, and also at national research institutes (Grubben, 1977).

Future Thrust

Local cultivars should be collected since they are apt to disappear rapidly with the introduction of modern cultivars in India and South East Asia. Breeding work should be stimulated, with a view to selection of high yielding cultivars with resistance to wetrot and stem borers, with a good cooking quality and low oxalate content (Grubben, 1977).

Spinach Beet/Palak

Swiss chard, spinach beet or leaf beet, (*Beta vulgaris* L.var. *bengalensis* Roxb.) (2n = 18) is a temperate annual vegetable widely cultivated in Mediterranean area, but becoming increasingly popular in cool tropical areas. Stalk type, the real 'Swiss chard' which is popular in Mediterranean area, has a much lower nutritional value than leafy types. Dark green Lucullus, Fordhook Giant, Large Rib and Rhubarb Giant are well-known leaf cultivars. Since only the large, outer leaves are harvested, harvest period is prolonged and level of production high. Swiss chard is a very important vegetable in North India and, to a large extent in cooler parts of South America. Diseases and pests are identical with those of sugarbeet *i.e.* nematodes, aphids, yellow mosaic virus and *Cercospora* leaf spots (Grubben, 1977).

Origin, Distribution and Genetic Diversity

Breeding work on Swiss chard is undertaken in India, resulting in introduction of new commercial cultivars. These, together with local cultivars from India, Hongkong and South China, yield better than cultivars introduced from temperate regions and produce viable seed in tropics. Collections of both cultivated and wild *Beta* species and cultivars from Centre of diversity, *i.e.*, East Mediterranean area and secondary centers in Asia and North Africa are kept by sugarbeet breeders in West Europe and North America but are inadequate for improvement of tropical spinach beets (Grubben, 1977).

Future Thrust

Genus *Beta* has a high priority for IBPGR since the local, South West Asian cultivars and wild species are rapidly disappearing. Materials collected for breeding work with sugarbeet can also be used for spinach beet programmes. This vegetable merits more attention as a suitable market vegetable for cooler parts of tropics (Grubben, 1977).

Ceylon Spinach

Ceylon spinach, *Basella alba* L. (2n = 48) is a climbing perennial plant, mostly cultivated as an annual vegetable against supports in home gardens but in a few areas as a vine-like market vegetable which is not staked. It is a short-day plant, with a C4-cycle photosynthetic pathway similar to that of amaranth. Flowering is precluded at a day length of more than 13 hours but water stress promotes early flowering. *Basella* is known as a minor vegetable in all tropical lowland areas. It is often called Ceylon spinach or Malabar nightshade. Three main types are in cultivation. The most common has dark-green, oval or nearly round leaves; a less popular type, often referred to as *B.rubra*, has red oval round leaves and red stems and a third, less commonly grown type has cordate-shaped, dark-green leaves. Ceylon spinach is succulent and slightly mucilaginous vegetable. It is very susceptible to root-knot nematodes but is seldom or never attacked by insects, due to its very thick leaf cuticle. Red spots on leaves and stems, caused by *Cercospora* and *Acrothecium*, are rarely serious. Market growers in Asia normally use cuttings as propagating materials whereas, in Africa, seeds are usually sown (Grubben, 1977).

Origin, Distribution and Genetic Diversity

Basella has its origin in South Asia but is widely cultivated in tropical Asia, Africa and New World. No reference to breeding work has been encountered. Indian seed catalogues offer *Basella* seeds. No collections of germplasm are known (Grubben, 1977).

Future Thrust

Ceylon Spanish is a very productive leaf vegetable, suitable for both home gardens and market production in the low humid tropics. It shows a remarkable resistance to pests and diseases. There is no danger of genetic erosion. Collecting local types and screening for high yielding capacity are advisable but are of low priority (Grubben, 1977).

Remedies for Food Insecurity

Home Garden and Seed Exchange Network

Seed collection and exchange programmes with people's participation help in retaining variety and richness of our traditional food habits. A few small seeds can feed a family, a fistful of seeds the whole community'. Nature has blessed every region with a variety of grains and vegetables. These varieties, which have sustained for long under changing environmental conditions, are fast disappearing. Recent fusion of technology with agriculture resulting in hybrid seeds pushed local varieties to background. This threat to food diversity demands creation of awareness about seed diversity in India and this little effort can make available wide variety of semi-wild and cultivated plant species. Home gardens provide access to food (and thus food security) and nutrition and provide an additional source of income. Even the poor are a pat of this small scale production system. Importantly, home gardens are immense repositories of genetic resources and wealth.

Participatory Programme

This programme with participation of village women folk especially, can go a long way in collection and documentation of home garden species and ensure good quality, healthy organic and open pollinated seeds. Home garden and seed exchange effort are testament to the idea that participatory approaches to food security are the best begun with saving seed diversity and control over local varieties with communities.

Diversification of Production Systems

Two production systems, namely vegetables for cereal-based cropping system and intensive year round vegetable production comprising periurban production and homestead production should be concentrated upon depending on the importance of vegetables in an area *i.e.* local niche, ecological limitations, marketing opportunities etc.

Vegetables in cereal-based cropping system that maintain soil fertility while serving the market will ensure sustained income for current and potential vegetable growers and ensure supply of essential micronutrients. There are several other systems based on intensive production of vegetables year round on same piece of land. Mainly, the following production systems should be stressed upon as these have potential.

Periurban

It supplies fresh perishable vegetables, generates jobs and allows economical use of even small pieces of land, can adjust quickly to market needs, and can recycle urban waste. More importantly, periurban production increases vegetables supply and income for urban poor.

Homestead Production

It is important to family diets and especially to children's health.

Consumption Awareness

Increasing vegetable consumption becomes a global priority as most people's diets still fall short of mark. Some of the world's widest spread and debilitating nutritional disorders, including birth defects, mental and physical retardation, weakened immune systems, blindness and even death are caused by diets lacking in vitamins and minerals. Low vegetable intake is a major contributing factor to such deficiencies. Over two billion persons in developing world suffer from a lack of vegetables in their diets. Encouraging people to eat more vegetables should, therefore, be an awareness drive. There are various reasons why different populations tend to shy away from vegetables–cost, convenience, taste and stigma, to name a few. Different agencies are working together to address these obstacles and find ways to boost vegetable consumption around the globe. These are so indispensable because they are full of health protective substances, which serve an array of important functions in the body. To increase utilization, focus should be on marketing and increasing public awareness in addition to promoting postharvest technology (Pritam Kalia, 2004).

Future Strategies

1. Programmes of collection and characterization of indigenous vegetable crop varieties and land races should be supported and extended, paying particular attention to their ability to yield under low input conditions. In conjunction with this, economic incentives and institutional barriers to maintaining vegetable crops and biodiversity rich farming systems must be encouraged.
2. Monitoring of biodiversity, inventorying of indigenous vegetable plant resources and identifying losses in biodiversity as a basis for introducing more effective biodiversity conservation measures, need enforcement.
3. Vegetable breeding needs to be relooked and efforts made to include local needs and constraints into criteria for selection of new varieties. This will reduce risk of imposing high input varieties on farmers who do not have resources to pay for them.
4. Determining optimum sizes of ecosystems for *in situ* conservation and development of improved methods of *ex situ* conservation of sexual seeds and vegetative materials of indigenous vegetable crops and other beneficial plants.
5. All introductions of alien species and varieties especially from the continents, should be subjected to increased vigilance through risk and impact assessments to ensure environmentally sound and sustainable food production and pose no threats to human health.
6. Genetic improvement and increased utilization of indigenous vegetable crops including roots, tuber, rhizomes, leafy and others should be encouraged.
7. Investing household willingness to invest in and manage protected plant habitats containing indigenous vegetable crops and other useful plants that generate products desired. This information can be used to encourage local community investment in preservation of biodiversity.
8. Exploring possibilities of using biotechnology for indigenous vegetable biodiversity to promote food and nutritional security in India (Pritam Kalia, 2004).

Conclusions

Average daily per caput consumption of vegetables in the developing countries is about 100 g *i.e.* only 46 per cent of the quantity consumed in developed countries. From nutritional surveys, it can also be concluded that nutritional needs in hot zones differ from those in temperate zones. Special emphasis should be given to vitamin A deficiency, hence need for leaf vegetables in particular. Although large regional differences exist in nutrient requirements from vegetables, in addition to basic food, it can be concluded that an average daily per caput consumption of 150 g of vegetables would be sufficient, if one-third of this quantity consisted of leaf vegetables.

Principal conclusions concerning different crops are as follows:

Tomato

Research and breeding programmes should be directed towards cultivation of this crop in the tropical lowlands. Collection of germplasm has already been given much attention, tomato being a main project subject of Asian Vegetable Research and Development Centre (AVRDC), Taiwan.

Peppers

Sweet pepper is an important vegetable in both tropical and subtropical areas, whereas hot pepper is more important in the tropics. Both have high vitamin C and carotene content. Breeding of hot pepper for higher yields must have first priority.

Watermelon

In tropical Asia, especially India, many valuable local cultivars are being replaced by commercial cultivars. Collection of germplasm therefore deserves a high priority.

Pumpkins, Melons, Bitter Gourd

Pumpkins and yellow or red melons merit attention due to their high carotene content and bitter gourd for its high vitamin C and iron content.

Wax Gourd, Chayote/Chow-chow

A high priority for research should be given to wax gourd, since it stores well, and also to chayote, because it is very suitable for cultivation in home gardens in humid areas at medium altitudes and its leaves also are eaten.

Onion

Primitive onion strains in Asian countries (India), West Africa and South America are gradually being replaced by commercial modern cultivars. They should therefore be collected.

Shallots, Garlic

A high priority for research should be given to shallots and garlic which are both locally very important. Shallots are particularly important for cultivation in tropical lowlands, as a substitute for onions.

Hyacinth Bean, Lima Bean, Winged Bean, Yardlong Bean

A first priority for research and breeding should be given to hyacinth bean, Lima bean, winged bean and yardlong bean. Climbing types of these four species are suitable for cultivation in home gardens in high rainfall areas of tropical lowlands.

Common Bean

Common or French bean is a main subject of research in national institutions in temperate countries and at CIAT in Columbia since it is widely grown as a vegetable or as a pulse.

White Cabbage

Production of white cabbage is rapidly increasing, especially for supplying to markets, where it is replacing traditionally grown, local leaf vegetables. Underlying reasons are that this crop is relatively easy to grow. It is also high yielding, transports well and stores for some time. Availability of Japanese heat-resistant cultivars supports this trend.

Tropical Leaf Vegetables

Species of leaf vegetables should have special attention since they are so important in diets in less affluent areas. They are early, highly productive and inexpensive vegetables, mainly grown in compounds and in small family gardens for home consumption. There are more than 1,000 tropical plants with edible leaves but only about 10 per cent of these are currently cultivated. A first priority for research and breeding has been allocated to annual crops such as amaranth and to shrubs or trees such as, drumstick tree. The first one is a widely cultivated annual vegetable, suitable for both home gardens and market production. Trees and shrubs are available plants for home gardens in wet tropics.

Lettuce, Spinach Beet

Consumption of lettuce and spinach beet in tropics is increasing, especially in urban areas. Research and breeding of lettuce and spinach beet (cultivars suitable for hot climates) should be given high priority due to their economic and nutritional importance.

Okra

Okra is one of the most important tropical vegetables grown in Africa, America, South West and South Asia. Primitive cultivars are in danger of being lost, as high-yielding commercial cultivars become popular.

Carrots

Carrots are becoming popular in many tropical upland regions. Primitive and wild types occur in South West Asia. Carrots are potentially important for their high carotene content; for this reason they deserve a high priority in research and breeding programmes.

Other Vegetable Species

Many other tropical vegetables (mushrooms, bamboo shoots, palm cabbage) and some temperate vegetables such as radish and sweet corn are locally of importance. No high priority for research and breeding should be given to these crops (Grubben, 1977).

References

Abdullahi, M.U. and Jasdanwala, R.T., 1991. Enlargement quotient to estimate leaf area in two cultivars of okra [*Abelmoschus esculentus* (L.) Moench.]. *Journal of Agronomy and Crop Science*, 167: 167–169.

Andres, T.C., 1990. Biosystematics, theories on the origin and breeding potential of *Cucurbita ficifolia*. In: *Biology and Utilization of the Cucurbitaceae*, (Eds.) Bates, D.M., Rabinowitch, R.W. and Jeffrey, C. Cormell University Press, Ithaca, New York, USA, pp. 102–119.

Ariyo, O.J., 1990. Measurement and classification of genetic diversity in okra (*Abelmoschus esculentus*). *Annals of Applied Biology* 116: 335–341.

Ariyo, O.J., 1993. Genetic diversity in West African Okra (*Abelmoschus caillei*) (*A.chev.*) Stevels—Multivariate analysis of morphological and agronomical characteristics. *Genetic Resources and Crop Evolution*, 40: 25–32.

Ariyo, O.J. and Odulaja, A., 1991. Numerical analysis of variation among accessions of okra [*Abelmoschus esculentus* (L.) Moench.]. *Malvaceae*. *Annals of Botany*, 67: 527–531.

Arora, R.K. and Nayar, E.R., 1994. *Wild Relatives of Crop Plants in India*. Science Monograph No. 7, National Bureau of Plant Genetic Resources, New Delhi, India, 90 pp.

Babu, C.R., 1977. *Herbaceous Flora of Dehradun*. Publication and Information Department, Council of Scientific and Industrial Research, New Delhi, India. 721 pp.

Babu, K.V.S. and Dutta, O.P., 1990. Pollen fertility studies in *Abelmoschus* spp. *South Indian Horticulture*, 38: 109.

Bates, D.M., 1968. Notes on the cultivated Malvaceae. 2. *Abelmoschus. Baileya*, 3: 99–112.

Bhat, R.B., Inamdar, J.A. and Weber, D.J., 1988. Leaf architecture of some Malvaceae. *Angewandte Botanik*, 62: 135–146.

Bisht, I.S., Mahajan, R.K. and Rana, R.S., 1995. Genetic diversity in South Asian Okra (*Abelmoschus esculentus*) germplasm collection. *Annals of Applied Biology*, 126: 539–550.

Bosland, P.W. and Votava, E.J., 2000. *Peppers : Vegetable and Spice Capsicums*. Centre for Agriculture and Biosciences International Publication, Commonwealth Agricultural Bureaux International, Wallingford, UK.

Bujisen, J.R.M., 1990. *Taxonomic Survey of Allium Species cultivated in South-East Asia*. Rijks Herbarium, Rapenburg, The Netherlands, 47 pp.

Bukenya, Z.R. and Carasco, J.F., 1999. Ethnobotanical aspects of *Solanum* L. (Solanaceous) in Uganda. In: *Solanaceae IV: Advances in Biology and Utilization*, (Eds.) Nee, M., Symon, D.E., Lester, R.N. and Jessop, J.P. Royal Botanic Gardens, Kew, UK.

Burkill, I.H., 1935. *A Dictionary of the Economic Products of the Malay Peninsula, 2 Vols.* Oxford University Press, Oxford, UK, 2402 pp.

Chakravarty, H.L., 1982. *Fascicles of Flora of India, Fascicle II, Cucurbitaceae*. Botanical Survey of India, Howrah, India.

Chandra, U. and Gupta, P.N., 1994. Evaluation of tomato germplasm adaptable to abiotic stress conditions of northern India. *Indian Journal of Plant Genetic Resources*, 7: 165–172.

Chandra, U. and Thomas, T.A., 1993. Sources of tomato germplasm adaptable to hot summer season in India. In: *Adaptation of Food Crops to Temperature and Water Stress*, (Ed.) George Kuo C. Asian Vegetable Research and Development Center, Taiwan, Publication No. 93, 410, 501–502.

Chaudhary, B., Tiwari, R., Mathai, P.J., Joshi, G.C., Kesavan, V., Chakravarti, A.K., Kalda, T.S. and Sivakami, S., 1974. Breeding for disease resistance in some tropical and sub-tropical vegetable crops. *Indian Journal of Genetics*, 34A: 328.

Charrier, A., 1984. Genetic resources of the genus *Abelmoschus* Med. (Okra) (English translation). IBPGR, Rome, 61 pp.

Chevalier, A., 1940. L'origine, la culture et les usages de cinq Hibiscus die la section *Abelmoschus*. *Review of Applied Botany*, 20: 319–328, 402–419.

Daunay, M.C., Lester, R.N., Hennart, J.W. and Duranton, C., 2000. Egg plants: Present and future. *Capsicum Eggplant Newsletter*, 19: 11–18.

Dhankar, B.S., Mishra, J.P. and Bisht, I.S., 2005. Okra. In: *Plant Genetic Resources: Horticultural Crops*, (Eds.) Dhillon, B.S., Tyagi, R.K., Saxena, S. and Randhawa, G.J. Narosa Publishing House, New Delhi, India, pp. 59–76.

De Candolle, A., 1882. *Origine des Plantes cultivies*. Germes Bailliere, Paris, France, 377 pp.

Decker, W.D.S., 1988. Origin, evolution and systematics of *Cucurbita pepo* (Cucurbitaceae). *Economic Botany*, 42: 4–15.

De Winter, B., 1990. A new species of *Citrullus* (Benincaseae) from the Namib Desert, Namibia. *Bothalia* 20: 209–211.

Dutt, B. and Roy, R.P., 1971. Cytogenetic investigations in Cucurbitaceae. I Interspecific hybridization in *Luffa. Genetica* (The Hague), 42: 139–156.

Dutta, O.P., 1992. Genetic diversity in cucurbits... watermelon. *Indian Horticulture*, 37: 35–36.

Dutta, O.P., 1994. Genetic diversity in cucurbits ... Chow-chow. *Indian Horticulture*, 38(4): 36.

Duzyaman, E., 1997. Okra: Botany and horticulture. In: *Horticultural Reviews*, Vol. 21: 42–72., (Ed.) Jules Janick. John Wiley and Sons, Inc., New York.

Erwin, A.T., 1929. A systematic study of the peppers. *Proceedings of the American Society of Horticultural Science*, 26: 128–131.

Farrer, R.R. and Kennedy, G.G., 1991. Insect and mite resistance in tomato. In: *Genetic Improvement of Tomato, Monographs on Theoritical and Applied Genetics*, Vol. 14, (Ed.) Kalloo, G. Springer-Verlag, Berlin, Heidelberg, Germany, 121 pp.

Fatokun, C.A., 1987. Wide hybridization in okra. *Theoritical and Applied Genetics*, 74: 483–486.

Giles, W.F., 1943. Onions and other edible *Alliums*. *Journal of Royal Horticultural Society*, 68: 193–200.

Grubben, G.J.H., 1977. *Tropical Vegetables and their Genetic Resources*. International Board for Plant Genetic Resources, Rome, Italy, 191 pp.

Hamon, S., 1988. Organisation evotutive du genre *Abelmoschus* (Gombo): Co-adaptation et evolution de deux species de gombo cultivees en Afrique de l'quest *A. esculentus* et *A.caillei*. Edit. ORSTOM TDM 46.

Hamon, S. and Charrier, A., 1983. Large variation of okra collected in Benin and Togo. *Plant Genetic Resources Newsletter*, 56: 52–58.

Hamon, S., Charrier, A., Koechlin, J. and Sloten, D.H.V., 1991. Potential improvement of okra (*Abelmoschus* spp.) through the study of its genetic resources (in French). *Plant Genetic Resources Newsletter*, 86: 9–15.

Hamon, S., Clement, J.C., Leblanc, J.M. and de Kochko, A., 1986. The cultivated okra in West Africa: collecting missions in Guinea. *Plant Genetic Resources Newsletter*, 65: 34–37.

Hamon, S. and Hamon, P., 1991. Future prospects of the genetic integrity of two species of okra (*Abelmoschus esculentus* and *A.caillei*) cultivated in West Africa. *Euphytica*, 58: 101–111.

Hamon, S. and Sloten, D.H.V., 1989. Characterization and evaluation of okra. In: *The Use of Plant Genetic Resources*, (Eds.) Brown, A.H.D., Frankel, O.H., Marshall, D.R. and Williams, J.T. Cambridge University Press, Cambridge, UK, pp. 173–196.

Hamon, S. and Yapo, A., 1986. Perturbation induced within the genus *Abelmoschus* by the discovery of a second edible okra species in West Africa. *Acta Horticulturae*, 182: 133–143.

Heiser, C.B.Jr., 1973. The Penis gourd of New Guinea. *Annual Association of American Geography*, 63: 312–318.

Heiser, C.B. and Smith, P.G., 1953. The cultivated *Capsicum* peppers. *Economic Botany*, 7: 214–227.

Hooker, J.D., 1973. Liliaceae *Allium* L. *Flora of British India*. Vol. VI (Reprinted by Singh and Singh). M/s Periodical Experts, New Delhi, India, pp. 337–345.

Hymowitz, T. and Boyd, J., 1977. Origin, ethnobotany and agricultural potential of the winged bean–*Psophocarpus tetragonolobus*. *Economic Botany*, 31: 180–188.

IBPGR, 1991. Report on international workshop on okra genetic resources (October, 1990) pp. 2–3, 8–12. IBPGR, Rome.

Indira, P., Rajan, S. and Peter, K.V., 1996. Genetic diversity in capsicum. *Indian Horticulture*, 40: 23.

Jambhale, N.D. and Nerkar, Y.S., 1983. Screening of okra cultivars, related species and interspecific hybrid derivatives for resistance to powdery mildew. In: *National Seminar on Breeding Crop Plants for Resistance to Pests and Diseases*. Tamil Nadu Agricultural University, Coimbatore, (Abstr.).

Jambhale, N.D. and Nerkar, Y.S., 1989. Evolution of Guinean okra cultivars from West Africa. *Journal of Maharashtra Agricultural Universities*, 14: 215–216.

Jeffrey, C., 1990. Systematics of the cucurbitaceae: an overview. In: *Biology and Utilization of Cucurbitaceae*, (Eds.) Bates, D.M., Robinowitch, R.W. and Jeffrey, C. Cormell University Press, Ithaca, New York, USA, pp. 3–9.

Jones, H.A. and Mann, L.K., 1963. *Onions and their Allies*. London and New York, Leonard Hill (Books) Limited, 286 pp.

Joshi, A.B., Gadwal, V.R. and Hardas, M.W., 1974. In: *Evolutionary Studies in World Crops*, (Ed.) Hutchinson, J.B. Cambridge University Press, London, UK, pp. 99–105.

Joshi, A.B. and Hardas, M.W., 1976. Okra. In: *Evolution of Crop Plants*, (Ed.) Simmonds, N.W. Longmans, London, UK, pp. 194–195.

Kalloo, G., 1985. *Tomato*. Allied Publishers Pvt Ltd, New Delhi, India.

Kalloo, G., 1992. Utilization of wild species. In: *Distant Hybridization in Crop Plants*, (Eds.) Kalloo, G. and Chaudahry, J.B. Springer-Verlag, New York, USA, pp. 149–167.

Kalloo, G., 1993. Eggplant (*Solanum melongena* L.). In: *Genetic Improvement of Vegetable Crops*, (Eds.) , G. and Bergh, B.O. Pergmon Press, Oxford, UK, pp. 587–604.

Kalloo, G. and Banerjee, M.K., 2000. H-24: Moderately leaf curl resistant variety of tomato (*Lycopersicon esculentum* Mill.). *Vegetable Science*, 27: 117–120.

Kalloo, G., Umesh Srivastava, Major Singh and Sanjeet Kumar, 2005. Solanaceous vegetables. In: *Plant Genetic Resources: Horticultural Crops*, (Eds.) Dhillon, B.S., Tyagi, R.K., Saxena, S. and Randhawa, G.J. Narosa Publishing House, New Delhi, pp. 19–33.

Karakoltsides, P.A. and Constantinides, S.M., 1975. Okra seeds: A new protein source. *Journal of Agriculture and Food Chemistry*, 23: 1204–1207.

Kataria, A.S. and Gill, S.S., 1988. *Catalogue on tomato*. All India Co-ordinated Research Project on Vegetable Crops, Division of Vegetable Crops, Indian Agricultural Research Institute, New Delhi, India.

Katiyar, G.P., 1969. Effect of IAA on sex expression and pollen fertility in *Luffa* sps. *Allahabad Farmer*, 43(3): 215–217.

Katiyar, G.P., 1971. Stigma receptivity in *Luffa* species. *Farm Journal* (India) August 1971, pp 24–25.

Khan, T.N., 1975. Variation in winged bean in Papua New Guinea. *Proceedings of a Symposium on South East Asian Plant Genetic Resources*, Bogor, Indonesia, pp. 152–155.

Khan, T.N., 1976. Papua New Guinea: A centre of genetic diversity in winged bean [*Psophocarpus tetragonolobus* (L.).]. *Euphytica*, 25: 693–706.

Krishnamoorthy, H.N., 1971. Effect of morphactin on growth and sex expression of *Luffa acutangula*. *Z Pflanzenphysiology* Bd., 65: 88–91.

Krishnamoorthy, H.N., 1972. Effect of GA_{39}, GA_{4-79}, GA_5 and GA_9 on the sex expression of *Luffa acutangula* var. H-2. *Plant Cell Physiology*, 13: 381–383.

Lippert, L.F., Smith, P. and Bergh, B.O., 1966. Cytogenetics of the vegetable crops. Garden pepper, *Capsicum* spp. *Botanical Review*, 32: 24–55.

Malik, S.S., Srivastava, U., Tomar, J.B., Bhandari, D.C., Pandey, A., Hore, D.K. and Dikshit, N., 2001. Plant exploration and germplasm collection. In: *National Bureau of Plant Genetic Resources: A Compendium of Achievements*, (Eds.) Dhillon, B.S., Varaprasand, K.S., Kalyani Srinivasan, Mahendra Singh, Sunil Archak, Umesh Srivastava and Sharma, G.D. National Bureau of Plant Genetic Resources, New Delhi, India, pp. 31–68.

Martin, F.W., 1979. *Vegetables for the hot humid tropics part 4. Sponge and bottle gourds, Luffa and Lagenaria.* Agricultural Research (Southern Region), Science and Education Administration, USDA, New Orleans, Lousiana. pp 19.

Martin, F.W. and Delpin, H., 1978. *Vegetables for the Hot, Humid Tropics, Part 1: The Winged Bean, Psophocarpus tetragonolobus*. Southern Region Agricultural Research Service, USDA, New Orleans, Louisiana. 22 pp.

Martin, F.W. and Rhodes, A.M., 1983. Seed characteristics of okra and related *Abelmoschus* species. Qualitas Plantarum. *Plant Foods and Human Nutrition*, 33: 41–49.

Martin, F.W., Rhodes, A.M., Ortiz, M. and Daz, F., 1981. Variation in okra. *Euphytica*, 30: 697–705.

Martin, F.W. and Ruberte, R., 1978. *Vegetables for the Hot, Humid Tropics, Part 2: Okra, Abelmoschus esculentus*. Office of the Regional Administrator for Federal Research (Southern Region), Science and Education Administration, US Department of Agriculture, New Orlean, Louisiana 70153 (September) pp 1–22.

Martin, F.W., Santiago, J. and Cook, A.A., 1979. *Vegetables for the Hot, Humid Tropics, Part 7: The Peppers, Capsicum species*. Agricultural Research (Southern Region) Science and Education Administration, USDA, New Orleans, Louisiana.

McCollum, G.D., 1976. Onion and allies. In: *Evolution of Crop Plants*, (Ed.) Simmonds, N.W. Longman Group Limited, London, UK, 339 pp.

Merrick, L.C., 1990. Systematics and evolution of a domesticated squash. *Cucurbita argyrosperma* and its wild and weedy relatives. In: *Biology and Utilization of the Cucurbitaceae*, (Eds) Bates, D.M., Robinowitch, R.H. and Jeffrey, C. Cormell University Press, Ithaca, New York, pp. 77–95.

Mini, C. and Peter, K.V., 1994. Ornamental chillies. *Indian Horticulture*, 38(4) : 30–31.

Mishra, S.K. and Mishra, S.K., 2001. Genetic diversity.... Cowpea [*Vigna unguiculata* (L.) Walp.]. *Indian Horticulture*, 45: 36.

More, D.C. and Vibute, U.D., 1983. Inheritance of pigmentation in okra. *Journal of Maharashtra Agricultural Universities*, 8: 137–140.

Munger, H.M. and Robinson, R.W., 1991. Nomenclature of *Cucumis melo* L. *Cucurbit Genetics Cooperative Report*, 14: 43.

Negi, K.S. and Pant, K.C., 1992. Less known endemic wild species of *Allium* (Amaryllidaceae). *Economic Botany*, 46: 112–116.

Nerkar, Y.S. and Jambhale, N.D., 1985. Transfer of resistance to yellow vein mosaic from related species into okra [*Abelmoschus esculentus* (L.) Moench]. *Indian Journal of Genetics and Plant Breeding*, 45: 261–270.

Pal, B.P., Singh, H.B. and Swarup, V., 1952. Taxonomic relationships and breeding possibilities of species of *Abelmoschus* related to okra (*A.esculentus*). *Botanical Gazzett*, 113: 455–464.

Pandey, U.B., Ashok Kumar, Ruchira Pandey and Kamala Venkateshwaran, 2005. Bulbous crops: Cultivated Alliums. In: *Plant Genetic Resources: Horticultural Crops*, (Eds.) Dhillon, B.S., Tyagi, R.K., Saxena, S. and Randhawa, G.J. Narosa Publishing House, New Delhi, India, pp. 108–120.

Peter, K.V., Sadhan Kumar, P.G., Mini Raj and Prasanna, K.P., 1991. Genetic diversity in cucurbits ... Ash gourd. *Indian Horticulture*, 36: 35.

Pickersgill, B., 1997. Genetic resources and breeding of *capsicum* spp. *Euphytica*, 96: 129–133.

Pritam Kalia, 2004. Vegetable biodiversity for food and nutritional security. *Indian Farming*, 51: 19–26.

Purseglove, J.W., 1972. Alliaceae. In: *Tropical Crops–Monocotyledons*, (Ed.) Purseglove, J.W. Longman Group Limited, Harlow, U.K., 607 pp.

Rabinowitch, H.D. and Currah, L., 2002. *Allium Crop Science: Recent Advances*. Commonwealth Agriculture Bureaux. International Publishing House, UK, 515 pp.

Rai, M. and Gupta, P.N., 1995. Genetic resources of Solanaceous vegetable. In: *Genetic Resources of Vegetable Crops: Management, Conservation and Utilization*, (Eds.) Rana, R.S., Gupta, P.N., Mathura Rai and Kochhar, S. National Bureau of Plant Genetic Resources, New Delhi, India, pp. 91–110.

Ram, D. and Srivastava, U., 1999. Some lesser known minor cucurbitaceous vegetables: their distribution, diversity and uses. *Indian Journal of Plant Genetic Resources*, 12: 307–316.

Ram Srinivas, T. and Varaprasad, K.S., 2000. Genetic diversity ... tomato. *Indian Horticulture*, 44: 25.

Rana, R.S., 1991. Structure and distribution of plant trichomes in relation to taxonomy: *Hibiscus* L. *Feddes Repertorium*, 102: 335–344.

Richardson, J.B.III., 1972. The pre-columbian distribution of the bottle gourd: A re-evaluation. *Economic Botany*, 26: 265–273.

Robinson, R.W. and Decker-Walters, D.S., 1999. *Cucurbits*. Commonwealth Agricultural Bureaux. International University Press, Cambridge, UK, 226 pp.

Saxena, R.K., 1981. *Catalogue on Tomato* (*Lycopersicon esculentum* Mill.) *Germplasm*. National Bureau of Plant Genetic Resources, New Delhi, India.

Sharma, B.R. and Dhillon, T.S., 1983. Genetics of resistance to yellow vein mosaic virus in interspecific crosses of okra (*Abelmoschus* species). *Genetica Agraria*, 37: 267–275.

Shimotsuma, 1963. *Cytogenetics and evolutionary studies in genus Citrullus, Seiken Zin Report Kihara Institute of Biology R S*, 15: 24–34.

Siddiqui, O. and Mecci, A.K., 1974. Insect complex in the pollination of sponge gourd, *Luffa aegyptiaca*. *Proceedings of the Pakistan Science Conference*, 25(3–A): 52.

Siemonsma, J.S., 1982. West African okra-morphological and cytological indications for the existence of a natural amphidiploid of *Abelmoschus esculentus* (L.). Moench and *A.manihot* (L.). Medikus. *Euphytica*, 31: 241–252.

Singh, Ram and Agarwal, R.A., 1988. Okra fruit components and their proportions in okra genotypes. *Research and Development Report*, 5: 99–100.

Singh, Major, Chakraborty, S., Kirti Srivastava and Kalloo, G., 2001. Biotechnolgoical advances in vegetable crops. In: *Emerging Scenario in Vegetable Research and Development*, (Eds.) Kalloo, G. and Kirti Singh. Research Periodicals and Book Publishing House, New Delhi, India, pp. 58–84.

Singh, C.B. and Singh, S.N., 1970. Studies in the floral biology of white bottle gourd [*Lagenaria sicedraria* (Mol.) Standl.]. *Allahabad Farmer*, 44(3): 133–138.

Singh, H.B., Swarup, V. and Singh, B., 1975. *Three Decades of Vegetable Research in India*. Indian Council of Agricultural Research Technical Bulletin. New Delhi, India.

Sirohi, P.S., 1991. Genetic diversity in cucurbits.... Bitter gourd. *Indian Horticulture*, 35(4): 36–37.

Sirohi, P.S., 1993. Genetic diversity in cucurbits.... Pumpkin (*Cucurbita moschata* Duch. Ex Poir.). *Indian Horticulture*, 38: 35 and 37.

Sirohi, P.S., Munshi, A.D., Gunjeet Kumar and Behera, T.K., 2005. Cucurbits. In: *Plant Genetic Resources: Horticultural Crops*, (Eds.) Dhillon, B.S., Tyagi, R.K., Saxena, S. and Randhawa, G.J. Narosa Publishing House, New Delhi, pp. 34–38.

Sirohi, P.S. and Sivakami, N., 1991. Genetic diversity in Cucurbits ... Bottle gourd. *Indian Horticulture*, 36: 44–45.

Srinivas, K., 1982. Leaf area determination in okra (*Abelmoschus esculentus* L.). *Progressive Horticulture*, 14: 186–187.

Srivastava, U., and Sharma, G.D., 2000. *Evaluation Studies in World Germplasm of Tomato* (*Lycopersicon esculentum* Mill.) *and Developing Core Set*. National Bureau of Plant Genetic Resources, New Delhi, India, 304 pp.

Stearm, W.T., 1946. Nomenclature and synonymy of *Allium odorum* and *A. tuberosum*. *Herbertea*, 11: 227–245.

Tarsem Lal and Dhaliwal, M.S., 1992. Genetic diversity in cucurbits....Muskmelon. *Indian Horticulture*, 37: 55–56.

Uttamasamy, S., 1985. Influence of leaf hairiness on resistance of bhendi or lady's finger, *Abelmoschus esculentus* (L.) Moench to the leaf hopper, *Amrasea devastans* (Dist.). *Tropical Pest Management*, 31: 294–295, 341, 345.

Van Borssum Waalkes, J., 1966. Malesian Malvaceae revised. *Blumea*, 1: 1–251.

Velayudhan, K.C. and Upadhyay, M.P., 1994. Collecting okra, eggplant and their wild relatives in Nepal. *Plant Genetic Resources Newsletter*, 97: 55–57.

Vredebregt, J.H., 1990. Taxonomic and ecological observations on species of *Abelmoschus* Medik. In: *Report of an International Workshop on Okra Genetic Resources*, National Bureau of Plant Genetic Resources. New Delhi, India, pp. 69–76.

Whitaker, T.W., 1973. The cucurbits of India. *Proceedings of the Tropical Region American Society for Horticultural Science,* 17: 255–259.

Whitaker, T.W. and Bemis, W.P., 1975. Origin and evolution of the cultivated *Cucurbita. Bulletin Torrey Botanical Club,* 102: 362–368.

Whitaker, T.W. and Carter, G., 1954. Oceanic drift of gourds–experimental observations. *American Journal of Botany,* 4: 697–700.

Whitaker, T.W. and Davis, G.N., 1962. *Cucurbits. Botany, Cultivation and Utilization.* Interscience Publishers, Inc. New York, USA. 250 pp.

Wyatt, J.E., 1985. Inheritance of photoperiod sensitivity, hirsute seed and albinism in okra. *Journal of American Society for Horticultural Science,* 110: 74–78.

Zeven, A.C. and de Wet, J.M.J., 1982. *Dictionary of Cultivated Plants and their Regions of Diversity.* Center of Agricultural Publishing and Documentation. Wageningen, The Netherlands, 259 pp.

Zeven, A.C. and Zhukovsky, P.M., 1975. *Dictionay of Cultivated Plants and their Centers of diversity.* Centre of Agricultural Publishing and Documentation. Wageningen, The Netherlands, 219 pp.

Biodiversity in Horticultural Crops Vol. 1 *Pages* ***184–196***
Editors: **Professor K.V. Peter & Dr. Z. Abraham**
Published by: **DAYA PUBLISHING HOUSE**

Chapter 7

Tropical Tuber Crops

M.S. Palaniswami and Shirly Raichal Anil
Central Tuber Crops Research Institute,
Sreekaryam, Trivandrum – 695 017

Plant biodiversity, in particular, is arguably the single greatest resource that humankind garnered from nature during our long cultural development. Scientists described more than 2,50,000 species of mosses, ferns, conifers, and flowering plants, and more than 50,000 plant species are yet to be documented, primarily in the remote, little-studied reaches of tropical forests

Nowhere is the value of biodiversity more evident than in our food supply. Roughly one third of all plant species have edible fruits, tubers, nuts, seeds, leaves, roots, or stems. In the 1920s, the legendary Russian plant explorer Nikolai Vavilov identified geographic centres of crop diversity, including Mesoamerica, Central Andes, Mediterranean Basin, Near East, highland Ethiopia and Eastern China. He also inferred correctly that most centres correspond to where crops were first domesticated. People everywhere rely on plants for staying healthy and extending quality and length of their lives. The World Health Organization estimates that 3.5 billion people in developing countries rely on plant-based medicine for their primary health care, in which tuber crops also play a major role(Tuxill, 1999).

There are 25 hot spots of biodiversity in the world of which Western Ghats and Eastern Himalayan are the most important one. Seven per cent of world flora and 6.5 per cent of world's fauna are in India. Indo-Burma region is one of the eight hottest hot spots of the world, and is the centre of origin of tuber crops like yams and taro, *Amorphophallus* and *Alocasia*.

Tropical root crops with a world production of around 350 million metric tonnes have great potential in meeting the basic food and energy needs of developing world as food, feed and for industries. Burgeoning population, food shortages, ever-increasing need to explore new frontiers to alleviate world hunger and relative rise in prices of fossil-based energy contributed a great deal to the increasing importance of root and tuber crops as source of food, feed and energy. They grow well in a wide range of soil types so long as there is adequate rainfall.

Tropical tuber crops form an important group of subsistence crops for millions of people (about one fifth of world population), especially to those living in tropics and sub-tropical zone. Tuber crops are the third important food crops after cereals and grain legumes (Chadha and Nayar, 1994). The carbohydrates synthesised in leaves are stored in the form of starch in the underground tubers and roots, which not only form food but also a source of raw material in certain industries.

Among the many tuber crops available, only a few are domesticated and widely cultivated for edible and industrial purpose. Major tuber crops are Cassava, Sweet potato, Asiatic yam, African yam, Lesser yam, Elephant foot yam, Taro/bunda, Tannia and Country potato. They are widely cultivated and used for edible and industrial uses. In addition, there are minor tuber crops like Arrow root, Winged bean, Yam bean, *Canna,* Safed musli, Jerusalem artichoke etc. (Table 7.1) (AICRPTC, 2005).

Table 7.1: Major Tropical Root and Tuber Crops

Crop	*Scientific Name*	*Family*
Cassava	*Manihot esculenta* Crantz	Euphorbiaceae
Sweet potato	*Ipomoea batatas* (L.) Lam	Convolvulaceae
Greater yam	*Dioscorea alata* L.	Dioscoreaceae
White yam	*Dioscorea rotundata* Poir	Dioscoreaceae
Lesser yam	*Dioscorea esculenta* (Lour.) Burls	Dioscoreaceae
Aerial yam	*Dioscorea bulbifera* L.	Dioscoreaceae
Taro	*Colocasia esculenta* (L.) Schott	Araceae
Tannia	*Xanthosoma sagittifolium* (L.) Schott	Araceae
Elephant foot yam	*Amorphophallus paeoniifolius* (Dennst.) Nicolson	Araceae
Giant taro	*Alocasia macrorrhiza* (L.) Schott	Araceae
Giant swamp taro	*Cyrtosperma chamissionis* (Schott) Merr.	Araceae
Country potato	*Solenostemon rotundifolius* (Poir) J.K. Morton	Labiatae
Arrowroot	*Maranta arundinacea* L	Marantaceae
Canna	*Canna edulis* (Ker-Gawler)	Cannaceae
Winged bean	*Psophocarpus tetragonalobus* (L.) DC	Leguminoceae
Yam bean	*Pachyrrhizus erosus* (L.) Urban	Leguminoceae

Cassava

About ninety eight species of *Manihot* are recognized. All of them are native to New world and are concentrated in four regions of diversity in Brazil and Central America. All the *Manihot* species so far examined have 2n=36 chromosomes. *M. esculenta* or cassava is cultivated for its edible tubers. It is an important food crop for people in tropics and subtropics. Its wild relatives are sources of many useful characters like high protein content (Nassar and Dorea, 1982), tolerance to drought (Nassar, 1978, 1991), and apomixis (Nassar *et al.,* 1998). *M. glaziovii* Muell. Arg and *M. piauhyense* were tried for rubber production (Purseglove, 1968). *M. anomala* is a tuber bearing species with high leaf retention. *M. caerulescens* is resistant to drought and excessive soil toxicity, *M. tristi* is used in breeding programme for improving protein content (Jos and Sreekumari, 1994). *M. grahami* and *M. anisophylla* have tolerance to seasonal low temperature. *M. brachyloba* is a Mexican species found in Brazil (Ghosh *et al.,*1988).

A large number of hybrids showed extremely important economic value. For example hybrid of cassava with *M. oligantha* showed high protein content (4 per cent compared to 1.5 per cent in the cultivated clones), low HCN content and high production of root (Nassar and Dorea, 1982). Another example is the hybrid of cassava with *M. neusana* which exhibited for the first time the crop apomixis and led to production of apomictic clones. Hybrids of cassava with *M. glaziovii* and *M. pseudoglazziovii* are very tolerant to drought and had high productivity of roots (Nassar, 2003).

Interspecific hybrids between cassava and its wild relatives show relatively regular meiosis, and further generations can be obtained. Electrophoresis shows affinity among species of different sections, and between a few of them and cassava. Both polyploidy and apomixis may have contributed to speciation in this genus. Polyploidy offered genetic variability while apomixis is responsible for perpetuating new hybrid types adapted to different environments. Cassava may have originated by hybridization between two wild Manihot species followed by vegetative reproduction of the hybrid (Nassar, 2001). Studies on the degree of genetic relatedness using RAPD and AFLP molecular markers suggest that the species *M. flabellifolia* and *M. peruviana* gave rise to cultivated species (Colombo *et al.*, 2000).

Interspecific hybridization was also carried out in as early as 1943 by Prof. Abraham *et al.*, between *Manihot esculenta* and *M. glaziovii*. Both are very closely related species, chromosome number in both being 2n=26. This close relationship between two species and the apparent wild nature of ceara rubber plant (*M. glaziovii*) gave the idea of crossing it with *M. esculenta* to introduce wild genes into the make up of cultivated cassava plant. (Abraham *et al.*1964). Central Brazil was identified as one of the four centres of diversity of *Manihot* species. Twenty six of 98 recognized species of *Manihot* were recorded in this region. These include *M. divergens* (*M. violecea* var. *divergens*), *M. mossamedensis, M. sparsifolia, M. alutacea, M. peltata, M. gracilis, M. oligantha, M. fluticulosa, M. paviaefolia, M. procumbens, M. falcata* (*M. hirariana*), *M. reptans, M. violacea, M. anomala and M. tripartita.* (Nassar, 2001).

Table 7.2: Species Diversity of *Manihot*

Species	Remarks
Manihot esculenta	Native of Brazil, cultivated for edible tubers
M. glaziovii Muell. Arg.	Ceara rubber–yields (Purseglove 1968) rubber of good quality but latex coagulates immediately on exposure
M. dichotome Ule	From Brazil
M. piauhyense Ule	From Brazil tried for rubber production
M. anomala	Tuber bearing and high leaf retention, Allen 1979
M. caerulescens	Resistance to drought and excessive soil toxicity
M. tristis	Wild species used in breeding programme for improving protein content.
M. grahami	Having tolerance to seasonal low temperature.
M. anisophylla	Having tolerance to seasonal low temperature.
M. brachyloba	A Mexican species found in Brazil.

Sweet Potato

Sweet potato is one of the world's highest yielding crops/unit area/unit time. It is a native of tropical America. A few varieties have carotenoids as equivalent to that of carrot. Among related species of *Ipomoea*, sweet potato is the only species producing edible tubers and is a hexaploid. It is supposed to be the product of cross between diploid *I. leucantha* (BB) and tetraploid *I. littoralis* (BBBB) and *I. trifida* is the wild ancestor of sweet potato. A few other species in the section have thickened roots and they are diploid annuals or perennials or tetraploid perennials (Ghosh *et al.*, 1988). Sweet potato belongs to family Convolvulaceae which consists of about 1000 species of which *I. batatas* is of economic importance as food. Though centre of origin is South America, wild relatives were found in Malabar coast. Van Rheedes (1692) reported a number of *Ipomoea* spp. like *I. mauritiana, I. alba, I. aquatica, I. marginata* and *I. campanulata* most of which are medicinal from Malabar coast in his *Hortus Malabaricus*.

Table 7.3: Species Diversity of *Ipomoea*

Species	*Remarks*
Ipomoea mauritiana Jacq.	Medicinal plant common on coastal plain of Malabar Hort. Malabaricus 11: 101-102, t.49-1692
I. alba L.	Found in most region of coastal Malabar Hort. Malabaricus 11: 101-102, t.50-1692.
I. aquatica Forssk.	Medicinal–found in stagnant fresh water ponds throughout Malabar
I. marginata (Desr.) Verdc.	Leaflets crushed and used as shampoo for bath. Found in uplands throughout Malabar.
I. campanulata	Uplands in moist shaded places in Malabar–antidote for snake poison. Hort. Malabaricus 11: 115, t.56-1692.
I. marginata (Desr.) Verdc	Leaflets are crushed and used as shampoo and occur in uplands in Malabar. Hort. Malabaricus 11: 109-110, t.53-1692.
I. pescaprae (L.)	Common in beaches of Malabar Hort. Malabaricus 11: 117, t.57-1692.
I. asarifolia (Desr.) Roem and Schutt	Found in uplands throughout Malabar.
I. pes-tigridis (L.)	Occurs throughout Malabar Hort. Malabaricus 11: 121, t.59-1692
I. quamoclit (L.)	Cultivated as a garden plant in Malabar, with beautiful red flowers.

Yams

Indo-Burma region is one of the centres of origin of yams (Coursey, 1976). *D. oppositfolia, D. wallichii, D. spicatum, D. whittei* and *D. tomentosa* are a few of spp. of *Dioscorea* found in Western Ghats, Kerala, Maharashtra (CTCRI 1998). *D. esculenta* var. *spinosa* (CTCRI, 1998) was also collected from Western Ghats. *D. rotundata* is a native of West Aftrica. *D. cayenensis* (Lam) having yellow flesh and rich in catotenoids is a native of West Africa. *D. mexicana* is a non edible spp. used in pharmaceutical industry. *D. bulbifera, D. hispida, D. pentaphylla* and *D. alata* were reported by Van Rheedes (1688) in *Hortus Malabaricus.* Cinnamon yam (*D. batatas*), a native of China is the hardiest of yams and is a perennial climber with heart-shaped leaves and long vertical tubers, up to 1 m (3 feet) deep. It is also known as Chinese yam. The botanical name, *D. batatas* is now replaced by *D. opposite.*

Wild relatives of yams are seen in Western Ghats, especially in Southern parts of Western Ghats (Agasthyamalai ranges) in almost all the reserve forests (Ramanathan, 1978). *D. wallichii, D. pentaphylla, D. oppositifolia, D. tomentosa, D. spicata, D. intermedia, D. hispida, D. bulbifera,* and *D. alata* were distributed throughout this range (Prain and Burkill, 1938). In deciduous forest region of Agasthyamalai range, wild yams especially *D. oppositifolia, D. wallichii,* and *D. pentaphylla* are present (Henry *et al.*,1984). Distributions of the above yams are more or less uniform in nature at the ranges (Chandra Babu *et al.*, 2000). *Dioscorea* species are distributed nearly throughout India except in the dry north-western regions.

They grow at elevations of 2500-5000 m in Himalayas. Genetic diversity and resulting heterotic groupings in 19 varietal groups of cultivated Guinea yams (*D. rotundata* and *D. cayenensis*) and their wild relative *D. praehensilis* were assessed using amplified fragment length polymorphism (AFLP) genetic markers (Mignouna, *et al.*, 1998). High levels of genetic diversity were found in natural populations of *Dioscorea tokoro*. Genetic diversity was measured by investigating: (1) allozyme allele frequencies ; (2) nucleotide differences in haplotypes of the Pgi [phosphoglucose isomerase (glucose-6-phosphate isomerase)] locus; and (3) microsatellite variation. Most of the genetic diversity resided within each population and the diversity caused by population differentiation appeared to be small (Terauchi, 1995).

Degras *et al.* (1996) reported complexity of speciation in *Dioscorea* spp. He indicated *D. cayenensis–rotundata* and a wild species *D. praehensilis*. Martin and Rhodes (1978) established how the so-called *D. cayenensis* and *D. rotundata* could not be maintained as true good species. Hamon and Toure (1990) gave a comprehensive view of the complex with the definition of nine classes where seven wild species are associated to some, from numerical taxonomy based on morphological and starch–gel electrophoresis data. Yam improvement has normally being done by introduction and clonal selection resulting in considerately reduced genetic variability (Coursey, 1976).

Table 7.4: Species Diversity of *Dioscorea*

Species	*Remarks*
D. dumetorum (Kunth) Pax	Bitter yam, trifoliate leaf and stem pubescent. Tubers large and coarse tubers are bitter and sometimes poisonous. Native: Tropical Africa
D. rotundata (Poir)	White yam, stem twines to right, wingless and circular in cross section. Tuber cylindrical, smooth skinned, white firm fleshed. Native : West Africa
D. alata L.	Greater yam or water yam, winged stem, stellate in cross section twining to the right, tuber shape extremely variable but mostly cylindrical, tuber flesh is white or purple and watery in texture. Native : South East Asia (Burma)
D. cayenensis Lam. (yellow yam)	The tuber flesh is yellowish in colour due to presence of carotenoids stem twine to left, possess thorns. Native : West Africa
D. esculenta (Lour) Burk (lesser yam/chinese yam)	Stem twines to left, stem thin often pubescent and cylindrical, tubers small, characteristically borne in clusters, tubers are white fleshed and sweet in taste when cooked. Native : Indo China
D. bulbifera L.	Aerial yam, characterised by production of large number of bulbils (aerial tubers) at the axil of a leaf. Tubers are generally single, round and unbranched and covered with roots. Grows wild in forest areas in Kerala. Native: Asia and Africa
D. trifida L. (Cush-cush yam)	Vines twine to left, produce small tubers. Stem is winged rectangular in cross section. The tuber flesh is white, yellow, pink or purple and highly palatable. Native : North of South America
D. opposita Thumb. (Cinnamon yam/Chinese yam) (*D. batatas* syn.)	Tolerate colder climate, cylindrical and twines to right, tuber is thin and very long and grows downward to soil. Bulbils occur in leaf axils. It is the hardiest of yam. Native : China
D. japonica	As food and medicine. Native : Japan
D. hispida	Grows wild in Ponmudi forest areas in Kerala. Tubers are single, spherical, irregularly branched and covered with roots. Tubers irritate the throat and are not edible.
D. mexicana	Used for extraction of pharmaceuticals (diosgenin) on commercial basis

Contd..

Table 7.4–Contd..

Species	*Remarks*
D. composita	Used for extraction of pharmaceuticals (diosgenin) on commercial basis
D. floribunda	Used for extraction of pharmaceuticals (diosgenin) on commercial basis
D. wallichii Hook f.	Is a wild species growing in South Wyanad. This plant produces many tubers in a cluster. Tubers are long and narrow and grow parallel to ground. Tuber flesh is fibrous and cream in colour.
D. pentaphylla	Is a wild species producing elongated bulbils in abundance on vines. Underground tuber is elongated and narrow. The flesh is fibrous, but smooth at bottom and this portion is eaten by hill tribes.
D. tomentosa	Is a wild species which does not produce bulbils. Under-ground tubers are one to many, 1 to 2 metres long and narrow with fibrous flesh. Trace of sapogenin content is reported in this species (Barua *et al.*, 1956).
D. oppositifolia	Is a wild species of *Dioscorea* which is eaten by hill tribes.

Aroids

Colocasia and *Xanthosoma* are together called cocoyams. *Colocasia esculenta* (taro) may be classified as distinct varieties like *C. esculenta* var. *esculenta* and *C. esculenta* var. *antiquorum* depending on arrangement of corms and cormels. Taro originated in South Central Asia, probably in India or Malaysia. Wild types of *Colocasia* were found during exploration of Western Ghats. *Colocasia esculenta* also has a type which produces stolons rather than corms/cormels.It is swamp taro grown for its edible stolons in different parts of West Bengal and Assam (AICRPTC, 2005). Pandey *et al.* (1993) recognised a diverse group of *C. esculenta* from North Eastern India with many groups showing differences in altitudinal adaptations. Lekhanpaul *et al.* (2003) from analysis of genetic diversity in Indian Taro, indicated presence of a very close gene pool of wild and cultivated forms with extremes levels of phenotypic and genotypic variation.

Xanthosoma sagittifolium or tannia originated in tropical America, but related species like *X. violaceum* (pink fleshed) was reported from Western Ghats and Uttar Pradesh (CTCRI, 1998). *X. violaceum* is a species of *Xanthosoma* possessing violet petioles and leaves and cultivated for its edible tubers. Foliage is also used as a vegetable. Other edible species are *X. artovirens* (yellow fleshed), *X. caracu* (white fleshed) and *X. brasiliense*, grown not for tuber but for its leaves to be used as salad.

Alocasia or giant taro is cultivated to a lesser extent and includes species like *A. macrorrhiza*, *A. denudata* Engler, *A.indica*, *A.cucullata*, *A.formicula* (Onwueme, 1978). Ancient polynesians (Hawaiians) call *A. macrorrihiza* 'Ape'. By a few, this plant is referred as "wild" Taro. This variety of *Alocasia* is sometimes referred to as elephant ears or upright elephant ears. Upright is used as a description of leaves standing straight up pointing skyward. It can easily grow to a height well over 2 m and is a native of Sri Lanka. The corm and leaf juice are used for medicinal purpose in India. Another species *A. cucullata* (Chinese taro) is a small elephant ear with thick, shiny green, heart-shaped leaves resembling little elf hats each with curlicue, twisted tips and upturned leaf margins. Plant forms a nice, sturdy, 1 m tall x 1 m wide clump with leaf blades about 4 m long and 2 m wide on upright, roundish green petioles (www.plantdelights.com/ catalog/spring/pge5html). The flower spathe is hooded, *i.e.* cucullate, with the whole flower structure hiding below the leaf. *A. cucullata* is just perfect where an elephant ear form and texture are needed without the large size.

Table 7.5: The Species Diversity of *Alocasia*

Species	*Remarks*
Alocasia cucullata (Lour) G. Don	Chinese Ape, Chinese Taro-Leaves and tubers used as vegetable
A. lowii Hook.	Leaves olive green and purple. Cultivated as ornamental
A. sanderiana hort.	These species have been recorded as being irritant (Burkill 1935).
A. denudata Engl.	This species is recorded as being an irritant (Burkill 1935). The species is used as an irritant in dart poison and is also a contact poison. If the powdered plant is mixed with bamboo hair and rubbed over the body, severe itching and sometimes death may occur.
A. indica Schott	Indo Malayan Alocasia–This species was recorded as being irritant. The rhizome may be used as a rubefacient.
A. macrorrhiza G. Don. (syns. *Arum macrorrhizum* L., *Alocasia macrorrhiza* Schott, *Colocasia macrorrhiza* Schott)	Giant Alocasia, Giant Elephant's Ears, Giant Taro, Taro, Cunjevoi, Cunjerboy, Ape, Spoon Lily, Australian Cabbage The whole plant is irritant; the sap of leaves and petioles will blister the lips and is powerfully and sometimes destructively irritant to eyes. The sliced tubers and leaves may be used as a rubefacient, and the acrid juice is used to combat caustic effect of sap of *Melanorrhoea laccifera* Pierre
A. portei Becc. and Engl.	Reported as having irritant juice
A. metallica Hook.	Has rich bronze and purple leaves, cultivated as ornamental
A. acuminata Schott.	Cultivated species
A. montana (Roxb.) Schott.	Roots of this plant are used for poisoning tigers.
A. fornicata (Roxb.) Schott.	Medicinal

With 12 species, *Cyrtosperma* is a small genus. All species with exception of *Cyrtosperma chamissionis* (now *C. merkusii*) are confined to New Guinea and associated islands (Thompson,1982), the Philippines, Borneo, Sumatra, Java, and Oceania. *C. chamissionis* or Giant swamp taro is a terrestrial, usually grown in a tall pit gardens and in freshwater swamp or marsh area and is the largest of taros. Originated in South-East Asia, perhaps in Indonesia, it spread throughout Malayan Archipelago and Pacific territories. It is a minor crop in Melanesia and parts of Polynesia, but a major food in Micronesia (Plucknett, 1976). The giant swamp taro has traditionally served Micronesians in many different ways. Leaves are used as food wrapings or plate; whereas, the stem, tuber, is cooked and eaten in many ways. It also serves as a "famine food" in time of food scarcity (www.comfsm. fm/~dleeling/ botany/1998/vhp/araceae.html). Subfossil leaf fragments of giant swamp taro (*C. chamissonis*) were recovered from archaeological contexts dating as early as A.D. 1489 (mean date) on Henderson Island (22° 22' S, 128° 18' W), Pitcairn group—a raised limestone (*makatea*) island isolated at extreme margin of South Eastern Polynesia and the Indo-West Pacific biotic province (Hather and Weisler, 2000). Species that are potentially excellent horticultural subjects, both for their long lasting inflorescence and spectacular foliage, are *C. cuspidispathum, C. macrotum, C. carii*, and *C. beccarianum* (www.aroid.org/ genera/cyrtosperma/ Cyrtospermalist.html).

Amorphophallus paeonifolius (Dennst.) Nicolson or elephant foot yam has its centre of origin in India. Many wild spp. were explored from Western Ghats like *A. dubius* which is medicinal, *A. wightii* and *A. sylvaticus*, Kunth. A mention of *A. paeonifolius* (Dennst) Nicols. var. paeonifolius which is a thorny type is made in *Hortus Malabaricus* (Van Rheedes, 1692). Other species are *A. oncophyllum, A. variabilis* and *A. rivieri* (Ghosh *et al.*, 1988) which produce small tubers which are not edible. Corms of

A. paeonifolius vary, one type with conspicuous corms and less or no cormels and the other type with one corm and more cormels.

Table 7.6: The Species Diversity of *Amorphophallus*

Species	*Remarks*
Amorphophallus paeonifolius (Dennst.) Nicolson	Cultivated species
A. prainii Hook. f.	This species irritate the oral cavity when eaten (Burkill 1935).
A. rivieri Durand (syn. *A. konjac* K. Koch)	Devil's Tongue, Leopard Palm, Umbrella Arum, Snake Pal Sap of flower produces marked itching
A. sylvaticus Kunth Wild Suran	Fruit causes a lasting burning of tongue and lips together with salivation and numbness. In Indian indigenous medicine, crushed seeds are applied externally to bruises and also into cavities in teeth as an analgesic
A. variabilis Blume	Tubers of this species are inedible except in Java where they are eaten after suitable preparation (Burkill 1935)
A. commutatus (Schott) Engl.	Appendages eaten as vegetables but flowers emit a good deal of stench
A. dubius Blume. Purple stalked dragon	Tuber is administered in piles and apoplexy
A. sylvaticus (Roxb) Kunth.	Found in deccan peninsula and suspected to be poisonous to human beings and livestock. Seeds are used for tooth ache and bruises and in treatment of glandular enlargement

Minor Tuber Crops

Maranta arundinacea L. or West Indian Arrowroot, a herbaceous, tropical perennial plant, with a creeping rhizome, indigenous to tropical America to Northern South America and the Lesser Antilles, where it was known to the Indians during pre-Columbian days. It has fleshy cylindrical tubers with rings of scars which are leftovers of large thin scales. The stem reaches a height of 2 m and has creamy white flowers at the end. It has numerous, ovate leaves, 6 to 30 cm, with long sheaths that often envelope the stem. Arrowroot grows for 6–12 months before harvest. The plant bears fruit, little berries or capsules. Maranta starch can be used as a thickener for sauces and gravies. It is also used for making clear glazes for fruit pies. Starch is extracted from rhizomes not older than a year. In Suriname, starch is used by Amerindians as babypowder. A very light weight white powder, when applied to skin, gives a soft, silky smooth feel. Tuber is rich in beta-carotene, niacin, riboflavin and thiamin. The genus consists of about 25 spp. of perennial herbs in tropical America. *M. bicolor* Ker. and *M. leuconeura* Morr. are cultivated as ornamentals for their attractive variegated leaves (Purseglove, 1972). *M. leuconeura* or Prayer Plant is a native of Brazil. Leaves fold upwards at night, hence the common name.

Curcuma angustifolia or Indian arrowroot occurs wild in India and used in the same way as true arrowroot *Maranta arundinacea. Curcuma zeodaria,* a native of North Eastern India, is known for its yellow musk smelling odour with a slight smell of camphor and is used in medicine and perfumery (Purseglove, 1972).

Coleus parviflorus Benth. or *Solenostemon rotundifolius* Poir J.K. Morton (Chinese potato or country potato) is an important minor tuber crop grown extensively as a vegetable in most of homestead gardens in Kerala, Tamil Nadu and Karnataka. Belonging to family Labiatae the genus consists of 200 species. Country potato is a native of Africa. The plant is propagated vegetatively and seed set is completely absent as a result of which there is little variability within the species (Vimala, 1994). *Coleus amboinicus* Lour commonly known as wild oregano, is used in the local folk medicine for whooping cough and anti-inflammatory treatment (Rejesus, *et al.*,1999).

Yam bean (*Pachyrrzhizus erosus* (L) Urban, a native of Mexico and parts of Central America is grown for its fleshy edible tubers. In India, it is popularly grown in parts of West Bengal, Bihar, Orissa and Assam (Vimala, 1994) for its tuber and seed. The genus consists of six species of perennial herbs with trifoliate leaves and tuberous roots. *P. erosus* and *P. tuberosus* are cultivated for their edible tubers. *P. ahipa* (Wedd.) Perodi, a non-climbing species, is cultivated for its small fusiform tubers (Purseglove, 1968).

Winged bean (*Psophocarpus tetragonolobus* (L.) DC is known as Goa bean or princess pea. The genus consists of 5 species of twinning and tuberous-rooted herbs. *P. tetragonolobus* is cultivated in parts of tropics and probable centre of origin is tropical Asia where it is now mainly cultivated. (Purseglove, 1968). *P. palustris* Desv. grows wild throughout tropical Africa. Winged bean is grown for its immature pods, which are cooked and eaten like French beans and tuberous roots, eaten raw or cooked. Young leaves, shoots or flowers are also eaten as vegetable. It is a green manure crop also, because of its excessive nodulation, being a Leguminaceous crop. It is a native of either India or Mauritius or originated in African region, where all the 4-5 species of the genus are found wild.

Canna genus contains 55 species of herbs widely distributed in the tropics and sub-tropics particularly in Western Hemisphere. *Canna edulis* or purple arrowroot, a native of tropical America is a perennial herb grown for its edible rhizomes. *C. edulis* is grown commercially in Australia for production of starch (Purseglove, 1968). The tuber and tops of plant are used as livestock feed and starch is easy to be digested and hence used as food for children and invalids. Young rhizhomes are eaten as vegetable. The shoot and petiole are used as fodder(Vimala,1994).

Safed musli (*Chlorophytum borivillianum*) is an important Ayurvedic medicinal minor tuber crop. Tubers are white, and hence called as safed musli. Tubers are of great medicinal value containing the steroid sapoginine (1-2 per cent), proteins (10-12 per cent) and calcium. Tubers are fat free. It is useful on *pitta* and *vata*. It is effective against on *fatigueness, daaha* and in blood purification. It is useful against renal calculus, dhupani, sangrahani, leucorrhoea and diabetes. It is a basic ingredient in *chyanprash*.

Jerusalem Artichoke (*Helianthus tuberosus L.*) is a hairy, tuber-bearing perennial from sunflower family (Compositae). It is a native of North America where is was cultivated by Indians. It grows from 1.5 to 3.0 m high. Edible tubers are somewhat potato-like but carbohydrate present is inulin rather than starch and flavour resembles that of artichokes. Tubers may be eaten raw or boiled, pickled, made into chips or ground into flour and is equal to potato in food value(Vimala,1994).

Conservation

During this century, professional plant breeders used this rich gene pool to create high-yielding crop varieties responsible for much of the enormous productivity of modern farming. Gene banks focused almost exclusively on storing seeds of crop varieties and their immediate wild relatives.

Work on genetic resources of tropical tubers is under way since 1978. Till this date, a large amount of tuber crops germplasm belonging to various crops and their wild relatives are collected, characterised, evaluated, documented and conserved either in field gene bank or in *in vitro.* Currently 4219 accessions of different species of tuber crops are maintained under AICRPTC in ten State Agricultural Universities and two ICAR Institutes (ICAR Research Complex, Shillong and CARI, Port

Blair) serving as National Active Germplasm sites spread over 11 states and one union territory. Germplasm maintained at different centres are indicated in Table 11.7 (AICRPTC, 2005).Apart from this, CTCRI has a strong gene pool collection at Trivandrum and Bhubaneswar (CTCRI, 2004).

Table 11.7: Collection and Maintenance of Germplasm of Different Tropical Tuber Crops Under AICRPTC Centres

Crops/Centre	*Dholi*	*Raje-ndra nagar*	*Coim-ba-tore*	*Jorhat*	*Dapoli*	*Kal-yani*	*Faiza-bad*	*Ranchi*	*Jag-dal-pur*	*Shil-long*	*Nava-sari*	*Port Blair*	*Total*
Cassava	22	46	316	26	22	9	12	14	59	7	5	6	544
Sweet Potato	1166	156	86	30	82	138	83	75	120	19	9	41	2005
Colocasia esculenta var. antiquorum	25	99	0	72	53	86	82	34	89	35	16	0	591
Colocasia esculenta var. esculenta	75	0	8	0	0	15	53	40	25	0	0	1	217
Alocasia macrorrhiza	0	0	0	4	1	0	0	0	0	0	0	0	5
Swamp taro	0	0	0	24	0	45	0	0	1	0	0	0	70
Elephant foot yam	18	32	5	11	8	18	49	4	33	1	3	2	184
Tannia	0	0	2	10	11	9	0	3	7	0	0	0	42
Giant Swamp taro	0	0	0	0	0	9	0	0	0	0	0	0	9
D. alata	12	15	6	33	28	20	10	14	31	3	4	3	179
D. esculenta	10	0	5	4	6	2	0	0	0	0	0	5	32
D. rotundata	2	0	2	2	0	4	0	0	2	0	0	5	17
D. bulbifera	0	0	0	0	7	0	0	0	13	0	0	0	20
Yambean	137	6	0	1	12	2	32	0	14	0	0	0	204
Country potato	1	0	1	1	2	0	1	3	6	0	0	1	16
Arrowroot	0	0	1	2	3	0	1	2	9	0	0	0	18
Wingedbean	20	0	0	0	0	0	0	0	0	0	0	0	20
Others	1	0	1	0	9	0	0	6	27	0	2	0	46
Total	**1489**	**354**	**433**	**220**	**244**	**357**	**323**	**195**	**436**	**65**	**39**	**64**	**4219**

Field gene banks are useful in maintenance of active or working collections. But in storage of base collections, field gene banks become unmanageable. Furthermore, plants are at risk from pests and diseases and are subject to environmental constraints, such as drought or flooding (Hanson, 1985). Hence *in vitro* conservation is increasingly being considered as the safer, more practical option for crops that produce recalcitrant seeds or that are propagated clonally like tuber crops. Cassava had been chosen by IBPGR and CIAT for a pilot project to develop and test *in vitro* active gene bank management procedures (Withers *et al.*, 1991). The *in vitro* methods developed initially for propagation and genetic manipulation of plants species, are found useful and are currently being adopted for germplasm conservation of tuber crops in major gene banks at IITA (Nigeria), CATIE (Costa Rica), CIAT (Columbia) and INRA (Guateloupe) (Plucknett *et al.*, 1983). At CTCRI, Thiruvananthapuram *in vitro* methods for conservation of tuber crops like cassava, sweet potato, *Dioscorea*, *Colocasia* and *Coleus* are being standardized and more than 848 accessions of different tuber crops are being conserved in this manner (Unnikrishnan *et al.*, 1992; CTCRI, 2004).

For most of this century, plant diversity was treated as "common heritage" of humankind, freely available to anyone who can use it, with proprietary ownership only granted via patent law to individuals who demonstrate trade secrets or uniquely improve a crop variety or other plant.

Formal agreements for sharing benefits of plant diversity are negotiated extensively in search for new pharmaceuticals from plants in biodiversity-rich developing countries. First such "bioprospecting" agreement was announced in 1991 between Merck Pharmaceuticals and Costa Rica's nongovernmental National Institute of Biodiversity (InBio).

For all of human history, we depended on plants and the rest of biodiversity for our soul and subsistence. Now roles are reversed, and biodiversity's fate depends squarely on how we shape our own future. From reducing overexploitation of wild plants to establishing traditional resource rights for biodiversity stewards, many options are available for developing cultural links that support plant diversity rather than diminish it.

References

Abraham, A., Panicker, P.K.S and Mathew, P.M. 1964. Polyploidy in relation to breeding in tuber crops. *J. Indian Bot. Soc.* Vol. XLIII (2) 278-282.

AICRPTC, 2005. Annual Report of All India Coordinated Research Project on Tuber Crops 2004-2005. CTCRI, Thiruvananthapuram, 150 pp.

Barua, A.K. 1954. Steroid sapogenins from Indian Dioscorea plants. I. *J. Indian Chem. Soc.* 31: 172-177.

Burkill, I.H. 1935. A dictionary of the economic products of the Malay Peninsula Crown Agents, London.

Chadha, K.L. and Nayar, G.G. 1994. History of Tuber Crops Research in India. In: Advance in Horticulture vol. 8-Tuber Crops. (Eds.) Chadha, K.L. and Nayar, G.G. Malhotra Publishing House, New Delhi.pp. 1-16.

Chandrababu, S., Sukumaran, S and Raj. A.D.S. 2000. Distribution of wild varieties of yams, amorphophallus and their mycorrhizal associations observed from Kuttiar region of lower Kodayar ranges. *Proc: Mano Research Forum.* M.S. University, Tirunelveli. B. 24-25.

Colombo, C., Second, G., Charrier, A. 2000. Genetic relatedness between cassava (Manihot esculenta Crantz) and M. flabellifolia and M. peruviana based on both RAPD and AFLP markers. *Genetics-and-Molecular-Biology*. 23(2): 417-423.

Coursey, D.G. 1976. *Yams. Dioscorea* spp. (Dioscoreaceae) *In:* Evolution of Crop Plants. pp. 70-74. (ed.) N.W. Simmonds Longman Group Limited.

CTCRI, 1998. Annual Report 1997-1998. CTCRI, Thiruvananthapuram.

CTCRI, 2004. Annual Report 2003-2004. CTCRI, Thiruvananthapuram.

Degras, L., Pierre, F. and Arnolin, R. 1996. Some Breeding Aspects of Yam (*Dioscorea* spp.). *In:* Tropical Tuber Crops (Eds.) Kurup, G.T., Palaniswami, M.S., Potty, V.P., Padmaja, G., Kabeerathuma, S and Pillai, S.V. Oxford IBH Publishing Co Pvt Ltd., New Delhi. pp. 146-159.

Ghosh, S.P., Ramanujan, T., Jos, J.S, Moorthy, S.N. and Nair, R.G.1988.Tuber Crops. Oxford IBH Publishing Co Pvt Ltd., New Delhi.

Hamon, P. and Toure, B. 1990. The classification of the cultivated yams (*Dioscorea cayenensis-rotundata* complex) of West Africa. *Euphytica,* 47 179-188.

Hanson, J. 1985. Methods of storing tropical root crops germplasm with special reference to yam. *FAO/ IBPGR Plant Genetic Resources. Newsletter* 64: 24-32.

Henry, A.N., Chandra Bose, M., Swaminathan, M.S and Nair, N.C. 1984. Agasthyamalai and its environs: A potential area for a biosphere reserve. *J. Bombay Nat. Hist. Soc.* 81 : 282-290.

Hather, J.G. and Weisler, M.I. 2000. Prehistoric Giant Swamp Taro (*Cryptosperma chamissionis*) from Henderson Island, Southeast Polynesia. Pacific-Science. 54(2): 149-156.

Jos, J. S and Sreekumari, M.T. 1994. Genetic Resources in cassava. In: Advance in Horticulture vol. 8- Tuber Crops. (Eds.) Chadha, K.L. and Nayar, G.G. Malhotra Publishing House, New Delhi. India. pp. 17-33.

Lekhanpaul, S., Velayudhan, K.C. and Bhat, K.V. 2003. Analysis of genetic diversity in Indian taro [*Colocasia esculenta* (L.) Schott] using Random Amplified Polymorphic DNA (RAPD) markers. *Genetic Resources and Crop Evolution.* 50(6):60 3-609.

Martin, F.W and Rhodes, A.M. 1978. The relationship of *Dioscorea cayenensis* and *D. rotundata. Trop. Agric.* 55: 193-206.

Mignouna, H.D., Ellis, N.T.H., Knox, M.R., Asiedu, R. and Ng-QN. Tropical root crops: staples for sustainable food security into the next millennium. Proceedings of the 11th International Society for Tropical Root Crops Symposium, St. Augustine, Trinidad, 20-28 October 1997. *Trop. Agric.* 75(1-2): 224-229.

Nassar, N.M.A. 1978c. Some further species of *Manihot* with potential value to cassava breeding. *Can. J. Pl. Sc.,* 58: 915-916.

Nassar, N.M.A. 1991. Production of triploid cassava, *Manihot esculenta* Crantz by hybrid diploid gametes. *Field Crops Res.,* 13: 173-182.

Nassar, N.M.A. 1997. Development of cassava interspecific hybrids for savanna (cerrado) conditions. *J. Root Crops.* 22: 9-17.

Nassar, N.M.A. 2001. Cassava, Manihot esculenta Crantz and wild relatives: their relationships and evolution. Genetic-Resources-and-Crop-Evolution. 48(5): 429-436.

Nassar, N.M.A. 2003. Keeping options alive and threat of extinction: A survey of wild cassava survival in its natural habitat. *J. Root Crops.* 29(2): 17-20.

Nassar, N.M.A. and Dorea, G. 1982. Protein contents of cassava cultivars and its hybrid with *Manihot* spp. *Turialba* 32(4): 429-432.

Nassar, N.M.A., Vieira, M.A., Vicira, C. and Grattapaglia, D. 1998. Molecular and embryonic evidence of apomixes in cassava interspecific hybrids (*Manihot* spp.) *Can. J. Pl. Sc.,* 78: 348-352.

Onwueme, I.C. 1978. Tropical Tuber Crops. John Wiley and Sons, New York.

Pandey, G., Sharma, D.D. and Hore, D.K. 1993. Genetic diversity of Arum (*C. esculenta* germplasm in North Eastern India) *Ind. J. Agl. Sciences.* Vol. 63: 665-667.

Prain, D and Burkill, I.H. 1938, An account of the genus *Dioscorea* in the east. Part II *The species which twine to the right:* with addenda to part I and summary.Printed at Bengal Govt. Press.

Plucknett, D.L. 1976. *Edible aroids. Evolution of Crop plants* (Simmonds, N.W., Ed.). Longman Group Ltd., England.

Plucknett, D.L., Smith, N.J.H., Williams, J.T. and Murthi N. Anishetty. 1983. Crop germplasm conservation and developing countries. *Science* 220 : 163-169.

Purseglove, J.W. 1968. Tropical Crops : Dicotyledons. London: Longman.

Purseglove, J.W. 1972. Tropical Crops : Monocotyledons. 2. London: Longman.

Ramanathan, S. 1978. Management plan for the Kalakkad wild life sanctuary, Govt of Tamil Nadu, Forest dept.

Rejesus, B.M., Punzalan, E.G., Vasquez, E.A. and Kraus, W. Philippine. 1999. Biological activities of the compound from *Coleus amboinicus.J. Crop.Sci.* (Philippines), 24(1): 81.

Terauchi, R. 1995. Induced mutations and molecular techniques for crop improvement. Proceedings, Vienna, Austria, 19-23 June 1995, pp. 301-308.

Thompson, S.A., 1982. *Cyrtosperma chamissonis* (Araceae): ecology, distribution and economic importance in the South Pacific. (now *C. merkusii*). *J. Agric. Trad. Bot. Appl.* 29(2): 185–203.

Tuxill, J. 1999. Appreciating the benefits of plant biodiversity. In : "State of the World, 1999". Worldwatch Institute, 1776 Massachusetts Avenue, N.W., Washington, D.C.

Unnikrishnan M., Nair, N.G., Nayar, G.G. 1992. Preliminary studies on the conservation of germplasm of tuber crops through *in vitro* cultures. *In:* New Trends in Biotechnology. Oxford IBH pp. 51-55.

Van Rheedes. 1688. *Hortus Malabaricus.* University of Kerala Thiruvananthapuram. Vol. 7 pp.119-187. Reprinted in 2003.

Van Rheedes. 1692. *Hortus Malabaricus.* University of Kerala Thiruvananthapuram. Vol. 11 pp. 54-158. Reprinted in 2003.

Vimala, B. 1994. Genetic Resources and Varietal Improvement in Minor Tuber Crops. In: *Advance in Horticulture Vol. 8: Tuber Crops.* (Eds.) Chadha, K.L. and Nayar, G.G. Malhotra Publishing House, New Delhi-110064. India. pp. 139-153.

Withers, L.A., Zakri, A.H., Normah, M.N., Karim, A.G.A., Senawi, M.T. 1991. Conservation of plant genetic resources through *in vitro* methods. Proceedings of the MNCPGR-CSC international workshop on tissue culture for the conservation of biodiversity and plant genetic resources held in Kuala Lumpur, Malaysia, 28-31 May 1990. pp. 1-18.

www.aroid.org/genera/cyrtosperma/ Cyrtospermalist.html.

www.comfsm.fm/~dleeling/botany/1998/vhp/araceae.html.

www.plantdelights.com/ catalog/spring/pge5html.

Biodiversity in Horticultural Crops Vol. 1 *Pages* ***197–209***
Editors: **Professor K.V. Peter & Dr. Z. Abraham**
Published by: **DAYA PUBLISHING HOUSE**

Chapter 8

Orchids of Western Ghats, India

C. Sathish Kumar and S. Ganeshan
Tropical Botanic Garden and Research Institute,
Pacha Palode, Trivandrum – 695 562, India

The 1600-km long mountain chain called Western Ghats which lie on the western side of the Indian peninsula is not only known for its spices and condiments and timbers, but also for its orchids-unique species of *Aerides* (*A. crispa* Lindl., *A. maculosa* Lindl.), *Paphiopedilum druryi* (Bedd.) Stein, *Vanda*s and leafless *Vanilla*s. Alteration of the pristine habitats that harbour the last remaining wild populations of orchids for various developmental purposes has severely affected the survival of many a species. The region has 302 species, 3 subspecies and 4 varieties of orchids in 80 genera. Of this, three genera (*Aenhenrya, Smithsonia* and *Xenikophyton*) and 133 species are endemic (44 per cent). A synopsis of the orchid flora is provided with emphasis on the showy and promising ones. Orchid Researches currently underway at Tropical Botanic Garden and Research Institute, situated at the lap of the Western Ghats are also highlighted.

Indian orchids played their historic role in the early periods of orchid breeding and trade. Arrival of Indian orchids from Assam jungles and the east caused great sensation in the British Auction Halls in the 19th century. There are many stories of adventurous enthusiasts who set out to India in search of orchids and ended up in tiger dens! In those days of no forestry operations virgin forests of India were beamed with wildlife. Several mountain chains with such forests ran across the country. Western Ghats is one such mountain chain.

The Western Ghats lie between 8° 20'–20° 40' N latitudes and 73°-77° E longitudes on the west coast of India extending from Gujarat in the north to Tamil Nadu in the south traversing the states of Maharashtra, Goa, Karnataka and Kerala at a length of 1600 km. Whereas the Himalayas form the youngest mountain chain in India, Western Ghats is the oldest with characteristic archaean crystalline hard rocks. A mosaic of different vegetations types such as evergreen, semi-evergreen, moist deciduous and dry deciduous forests with unique shola forests and extensive high altitude grasslands occur here. The lives of the people of this region are intimately woven with the Western Ghats. The great rivers of Godavari, Bheema, Krishna, Tungabhadra and Kaveri and umpteen small rivers originate in this Ghats. More than 200 dams are located in these rivers and their tributaries. More than 60 per cent of the electricity consumed by these states is generated here. Too many developmental projects, fast expanding plantations, mining activities and consequent deforestations have caused irreparable damage to the Western Ghats and its biodiversity.

Historical Background

Historically, the Arabs, the Europeans and the Chinese traders and travellers visited this land of spices and condiments during the 15th and 16th centuries. Van Rheede (1636-1691), the Dutch Commander of Malabar was quite convinced of the botanical marvels when he brought out his *Hortus Malabaricus* in 12 volumes between 1678 and 1693. This was the first work, which depicted the orchids of Malabar and brought to the notice of the world. Among the 690 species described and illustrated, Rheede included 17 species of orchids now referable to *Acampe, Bulbophyllum, Cleisostoma, Cymbidium, Dendrobium, Geodorum, Liparis, Rhynchostylis, Rhytionanthos, Seidenfia* and *Taprobanea*. For the next 150 years this was the only work available on Malabar flora. The accuracy of the drawings and the authenticity and depth of the descriptions were matchless when *Hortus Malabaricus* was published. Linnaeus admitted that *Hortus Malabaricus* gave him many insights for preparation of his *Species Plantarum* (1753). John Graham (1805-1839), George Samuel Perrottet (1793-1870) and Achilles Richard (1769-1852), Robert Wight (1798-1872), Nicholas A. Dalzell (1817-1878) and Alexander Gibson (1800-1867), Richard Henry Beddome (1830-1911), Thomas Caverhill Jerdon (1808-1863), Joseph Dalton Hooker (1817-1911), Theodore Cooke (1836-1910), Cecil Ernst Claude Fischer (1874-1950) and Blatter and McCann were prominent in bringing out treatises on Orchids of Western Ghats region.

With the reorganization of the Botanical Survey of India in 1956 and opening of two regional circles in Western Ghats- one at Pune near Mumbai (Bombay) and the other at Coimbatore- collection activities were revamped and discoveries of orchids gradually came to light. Santapau and Kapadia (1966) brought out a critical assessment of the *Orchids of Bombay* containing pithy descriptions and critical notes. Abraham and Vatsala (1981) in their *Introduction to Orchids* described 150 native orchids of South India. Joseph (1982; 1987) gave an excellent report on the *Orchids of Nilgiris*.

The following summary gives the number of species, subspecies and varieties available in different genera of orchids in Western Ghats: 302 species, 3 subspecies and 4 varieties in 80 genera. Of these, 3 genera (*Aenhenrya, Smithsonia* and *Xenikophyton*), 138 species, 2 subspecies and 4 varieties are endemic (shown in brackets).

1	*Acampe*–4 (1)	11	*Cheirostylis*–3 (1)
2	*Acanthephippium* 1(0)	12	*Chiloschista*–2 (1)
3	*Aenhenrya*–1(1)	13	*Chrysoglossum*–1(0)
4	*Aerides*–4 (1)	14	*Cirrhopetalum*–2(1)
5	*Anoectochilus*–2(2)	15	*Cleisostoma*–1(0)
6	*Aphyllorchis*–1(0)	16	*Coelogyne*–6 +2 var. (2 + 2 var.)
7	*Arundina*–1 (0)	17	*Corymborkis*–1 (1)
8	*Brachycorythis*–3(2)	18	*Cottonia*–1 (1)
9	*Bulbophyllum*–18(12)	19	*Crepidium*–2(0)
10	*Calanthe*–3 (0)	20	*Cymbidium*–2+ 1 subsp. (0)

Figure 8.1: ***Aenhenrya rotundifolia*** **(Blatt.) Sathish & F. Rasm.**

Figure 8.2: ***Arundina graminifolia*** **(D. Don) Hochr.**

Figure 8.3(a): ***Brachycorythis splendida*** **Summerh.** **(b):** ***B. wightii*** **Summerh**

Figure 8.4: ***Chiloschista fasciata*** **(F. v. Muell.) Ormerod & Seidenf.**

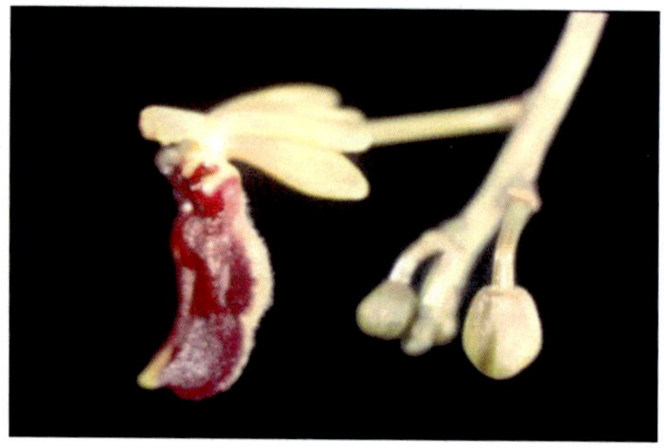

Figure 8.5: ***Cottonia peduncularis*** **(Lindl.) Rchb. f.**

Figure 8.6a: ***Dendrobium heterocarpum*** **Lindl. (Manipur)**

Figure 8.7: ***Diplocentrum congestum*** **Wight**

Figure 8.6(b): ***D. heterocarpum*** **Lindl. (local)**

Figure 8.8: *Disperis neilgherrensis* Wight

Figure 8.9: *D. zeylanica* Trimen

Figure 8.10: *Spathoglottis plicata* Bl. x

Figure 8.11: *Ipsea malabarica* (Rchb. f.) Hook. f.

Figures 12 & 13: Two Clones of the Hybrid Ipseglottis Prof. A. Abraham

Figure 8.14: ***Luisia macrantha*** **Blatt. & McCann**

Figure 8.15: ***Paphiopedilum druryi*** **(Bedd.)Stein**

Figure 8.16: *Pecteilis gigantea* (L.) Rafin

Figure 8.17: *Phaius luridus* Gard, ex Thw.

Figure 8.18: *Rhytionanthos indicum* Sathish & Garay

Figure 8.19: *Seidenfadeniella filiforme* (Rchb. f.) Christ.& Ormd.

Figure 8.20: ***S. rosea*** **(Wight) Sathish**

Figure 8.21: ***Smithsonia maculata*** **(Dalz.) Saldanha**

Figure 8.22: ***Taprobanea spathulata*** **(L.) Christ.**

Figure 8.23: ***Trias bonaccordensis*** **Sathish**

Figure 8.24: ***Vanda thwaitesii*** **Hook. f.**

Figure 8.25: ***Vanda wightii*** **Rchb. f.**

Figure 8.26: *Vanilla wightiana* Lindl. ex Hook. f.

21	*Dendrobium*–18+2 subsp. (11+2 subsp.)	40	*Liparis*–13(4)
22	*Dienia*–1(0)	41	*Luisia*–6(3)
23	*Didymoplexis*–2(1)	42	*Nervilia*–6(1)
24	*Diplocentrum*–2(1)	43	*Oberonia*–28(15)
25	*Diploprora*–1(0)	44	*Pachystoma*–2(1)
26	*Disperis*–2(1)	45	*Paphiopedilum*–1(1)
27	*Epipactis*–1(0)	46	*Papilionanthe*–1(0)
28	*Epipogium*–1(0)	47	*Pecteilis*–1(0)
29	*Eria*–12+1 var.(10+1 var.)	48	*Peristylus*–13(8)
30	*Eulophia*–10(2)	49	*Phaius*–1(0)
31	*Flickingeria*–2(2)	50	*Pholidota*–1(0)
32	*Gastrochilus*–3(1)	51	*Phreatia*–1(0)
33	*Gastrodia*–1(0)	52	*Podochilus*–1(1)
34	*Geodorum*–1(0)	53	*Polystachya*–1(0)
35	*Goodyera*–2 (0)	54	*Pomatocalpa*–1(0)
36	*Habenaria*–39 (22)	55	*Porpax*–2(1)
37	*Hetaeria*–2(1)	56	*Pteroceras*–2(1)
38	*Ipsea*–1(1)	57	*Rhynchostylis*–1(0)
39	*Kingidium*–3(2)	58	*Rhytionanthos*–3(2)

59	*Robiquetia*–3(1)	70	*Taprobanea*–1(0)
60	*Saccolabiopsis*–1(0)	71	*Thelasis*–1(0)
61	*Satyrium*–1(0)	72	*Thrixspermum*–3+1 var. (1 var.)
62	*Schoenorchis*–(2)	73	*Thunia*–1(0)
63	*Seidenfadeniella*–2(1)	74	*Trias*–2 (2)
64	*Seidenfia*–6 (4)	75	*Trichoglottis*–1(0)
65	*Sirhookera*–2(0)	76	*Tropidia*–4(0)
66	*Smithsonia*–3(3)	77	*Vanda*–4(1)
67	*Spiranthes*–1(0)	78	*Vanilla*–2(1)
68	*Taeniophyllum*–2(1)	79	*Xenikophyton*–2(2)
69	*Tainia*–1(0)	80	*Zeuxine*–6(1)

Endemism in Orchids

Of the 302 species, 3 subspecies and 4 varieties of orchids known in 80 genera from Western Ghats, 138 species, 2 subspecies and all the four varieties are endemic (45.7 per cent). Nowhere in India, Orchid species have this much endemism. Three genera *viz. Aenhenrya* (1 species), *Smithsonia* (3 species) and *Xenikophyton* (two species) are also endemic.

Notes on a few orchids are given below.

Aenhenrya Gopalan

This belongs to the sub tribe Goodyerinae and is represented by *A. rotundifolia* (Blatt.) Sathish and F. Rasm. It is obviously related to *Anoectochilus* and *Odontochilus* and has green leaves with silvery veins and a large solitary flower. This is a good addition to the group of jewel orchids (Figure 8.1).

Arundina Bl.

Populations of *A. graminifolia* (D. Don) Hochr. in Western Ghats markedly differ from the Himalayan counter part in being short-stature, bearing deep pink flowers, which are not fully opening (Figure 8.2). Because of the heavy demand we have micro-propagated the selfed pods and seedlings were distributed.

Brachycorythis Lindl.

Of the three species in the Western Ghats, two are endemic and are very showy (Figure 8.3).

Chiloschista Bl.

Of the two species available, *C. fasciata* has very beautiful flowers (Figure 8.4).

Cottonia Wight

This monotypic genus represented by *C. peduncularis* (Lindl.)Rchb.f. belongs to Aeridinae and has very long peduncles bearing curiously shaped flowers (Figure 8.5). Our cross with *Acampe praemorsa* (Roxb.) Blatt. and McCann produced short peduncle racemes in the first generation.

Dendrobium Sw.

Of the 18 species and 2 subspecies found in the Western Ghats, 11 species and 2 subspecies are endemic. South Indian populations of *Dendrobium heterocaprum* Wall. ex Lindl. are in many respects different from the Himalayan populations (Figure 8.6a & b for comparison).

Diplocentrum Lindl.

This genus belongs to Aeridinae and has 2 species- *D. congestum* Wight and *D. recurvum* Lindl. distributed in South India with the latter species extending to Sri Lanka. The lip in both cases has two collateral spurs (Figure 8.7).

Disperis Sw.

This predominantly African genus, has two species in the Western Ghats- *D. neilgherrensis* Wight and *D. zeylanica* Trimen (Figures 8.8 and 8.9).

Ipsea Lindl.

This is a small genus having 3 species allied to *Pachystoma* and *Spathoglottis* found in the Western Ghats (*I. malabarica* (Rchb.f.) J. D. Hook.), Sri Lanka (*I. speciosa* Lindl.) and Thailand (*I. thailandica* Seidenf.). Indian species is popularly known as the Malabar Daffodil Orchid. We crossed this with *Spathoglottis plicata* Bl. and the first generation shows very unusual flowers (Figures 8.10–8.12).

Luisia Gaudich.

Some of the finest species (*L. birchea* Bl., *L. evangelinae* Blatt. and McCann and *L. macrantha* Blatt. and McCann) are available here (Figures 8.13 and 8.14).

Paphiopedilum Pfitz.

Western Ghats is the western limit for the distribution of the genus *Paphiopedilum* and has a single species, *P. druryi* (Bedd.) Stein. Despite many times designated as extinct in wild, this species still occurs at Agastyamala (Figure 8.15).

Pecteilis Rafin.

Represented by *P. gigantea* (L.) Rafin., often confused with *P. susannae*(L.)Rafin., but very distinct with deeply laciniate petals (Figure 8.16).

Phaius Lour.

This Old World genus is represented by only one species, *P. luridus* Gard. ex Thw. *P. pictus* Hunt from Australia does not seem to differ from this species (Figure 8.17).

Rhytionanthos Garay *et al.*

This genus has 3 species in Western Ghats- *R. indicum* Sathish and Garay, *R. nodosum* (Rolfe) Garay *et al.*, and *R. rheedei* (Manilal and Sathish) Garay *et al* (Figure 8.18).

Seidenfadeniella Sathish

This genus occurs in Western Ghats and Sri Lanka and has only two species- *S. filiformis* (Rchb.f.) E. A. Christ. and Ormerod and *S. rosea* (Wight) Sathish. Both are pendulous, flimsy and terete-leaved epiphytes with decurved and graceful flowers (Figures 8.19 and 8.20).

Smithsonia Saldanha

This endemic genus has 3 species–*S. maculata* (Dalz.)Saldanha, *S. straminea* Saldanha and *S. viridiflora* (Dalz.) Saldanha (Figure 8.21).

Taprobanea Christ.

This is a segregate from *Vanda* in which it was till recently placed, *Taprobanea spathulata* (L.) Christ. is a good breeding material and under cultivation it is quite floriferous (Figure 8.22).

Trias Lindl.

Two species are available- *T. bonaccordensis* Sathish and *T. stocksii* Benth. ex J. D. Hook (Figure 8.23).

Vanda Jones ex R. Br.

Four species are found in Western Ghats- *V. tessellata* (Roxb.) Hook. ex G. Don, *V. testacea* (Lindl.) Rchb.f., *V. thwaitesii* J. D. Hook. and *V. wightii* Rchb.f., the last two were till recently believed extinct (Figures 8.24 and 8.25).

Vanilla Plumier ex P. Miller

This is a pantropical genus with two leafless species in Western Ghats- *V. walkeriae* Grah. and *V. wightiana* Lindl. ex J. D. Hook (Figure 8.26).

Xenikophyton Garay

This endemic genus has two species here- *X. seidenfadenianum* M. Kumar, S. Seq. and J. J. Wood and *X. smeeanum* (Rchb.f.) Garay.

Orchid Researches at Tropical Botanic Garden and Research Institute (TBGRI)

Orchids have a special place of pride in the mandate of Tropical Botanic Garden and Research Institute (TBGRI). Prof. A. Abraham (1914-1994), the founder director of TBGRI was a great orchid enthusiast and lover. He introduced many exotic hybrids into Kerala and dreamt of a day when orchid hobby would lead to a good industry as happened in Singapore, Thailand and Malaysia. Prof. Abraham could live up to see his dream fulfilled. His encouragement enriched the orchid collections of TBGRI. We now have over 600 species of orchids from Western Ghats, Eastern Himalayas and the Andaman and Nicobar Islands, grown in 5 orchidaria and more than 150 hybrids have been brought from Singapore and Thailand. Apart from the taxonomic studies on Indian orchids, we are seriously pursuing reproductive biology studies particularly of endangered species and breeding works with Indian species. Micro-propagation studies of different genera of orchids were done and protocols perfected. Seedlings raised out of this process are being distributed to the public at reduced rate. TBGRI is now recognized as the Coordinating Centre of an all India project on orchid taxonomy supported by the Ministry of Environment and Forests, Government of India.

References

Abraham, A. and P. Vatsala, 1981. *Introduction to Orchids.* Tropical Botanic Garden and Research Institute.

Sathish Kumar, C., 2003. *Orchid Works at TBGRI.* Tropical Botanic Garden and Research Institute, Trivandrum.

Sathish Kumar, C. and K. S. Manilal, 1994. *A Catalogue of Indian Orchids.* Bishen Singh Mahendra Pal Singh, Dehra Dun.

Biodiversity in Horticultural Crops Vol. 1 *Pages* ***210–233***
Editors: **Professor K.V. Peter & Dr. Z. Abraham**
Published by: **DAYA PUBLISHING HOUSE**

Chapter 9

Conservation of Spices Genetic Resources through *in vitro* Conservation and Cryopreservation

*K. Nirmal Babu, S.P. Geetha, D. Minoo, G. Yamuna, K. Praveen, P.N. Ravindran and K.V. Peter**

Indian Institute of Spices Research, P O Marykunnu – 673 012, Calicut
**Kerala Agricultural University, P.O. KAU, Trichur – 680 656,*

Plant genetic resources, a subset of biodiversity, contain the genetic material, which hold key to food security and sustainable agricultural development (Iwananga, 1994). Plant genetic resources constitute genotypes or populations of cultivars (landraces, advance/improved cultivars), genetic stocks, wild and weedy species, which are maintained in the form of plants, seeds and tissues. Genetic resources are non-renewable and are among the most essential of the world's natural resources. Plant breeders require reservoir of genetic variation for crop improvement. Due to spread of high yielding varieties and selection pressure, genetic variability is gradually getting eroded resulting in large-scale depletion of variability. This situation thus demands priority action to conserve such germplasm and it is essential that these be well conserved, be it at species, genepool or ecosystem level, for present and future generations (Frankel, 1975).

Spices and herbs are aromatic plants–fresh or dried plant parts like foliage, young shoots, roots, bark, buds, seeds, berries and other fruits of which are mainly used to flavour our culinary preparations, confectionary. They are also major ingradients in indigenous medicine and perfumery. Spices and herbs are grown throughout the world–different plant species in different regions. Peninsular India is a rich repository of spices and over 100 species of spices and herbs are grown. The other major spice growing countries are Brazil, China, Guatemala, Indonesia, Madagascar, Nigeria, West Indies, Malaysia, Sri Lanka, Spain, Turkey, Medeterranion region and the Central America. Black pepper, cardamom, ginger, turmeric, vanilla, capsicum, cinnamon, clove, nutmeg, tamarind, coriander, cumin, fennel, fenugreek, dill, caraway, anise and herbs like saffron, lavender, thyme, oregano, celery, anise, sage and basil are important as spices. India being the native home of many spices, their conservation and characterization are one of the priority programmes. Deforestation, habitat degradation and over-exploitation caused considerable loss of diversity in spices.

In many spices, conventional seed storage can satisfy most of the conservation requirements. But in crops with recalcitrant seeds and those having conservation needs cannot be satisfied by seed storage, have to be stored in vitro. Most field gene banks are prone to high labour cost, vulnerable to hazards like natural disasters, pests and pathogens attack (especially viruses and systemic pathogens), to which they are continuously exposed and require large areas of space (Chandel and Pandey, 1991). This supports *in vitro* conservation. In addition, other resources like continuous supply of standard stock cultures for experiments to examine physiological and biochemical processes, cell and callus lines developed for *in vitro* synthesis of valuable secondary products, flavours and other important compounds will benefit strongly from *in vitro* cultures. Most of the spice crops are either vegetatively propagated or have recalcitrant seeds. The spices germplasm is mostly conserved in field gene banks. Most of the spices are plagued by destructive and epidemic diseases caused by viruses, bacteria and fungi. This makes germplasm conservation in field gene bank risky. Thus *in vitro* storage system becomes important in the overall strategy of conserving genepool. Each technology should be chosen on the basis of utility, security and complementarity to other components of strategy. A balance needs to be struck between seed, field genebank, *in vitro* conservation of propagules, tissues, pollen, cell lines and DNA storage for overall objective of conserving gene pool (Withers, 1991).

Micropropagation

Plant regeneration and successful cloning of genetically stable plantlets in tissue culture are important pre-requisites in any *in vitro* conservation effort. Micropropagation protocols for stable cloning of elite genotypes of spice crops were standardised. Protocols are available for black pepper and its related species, cardamom, large cardamom, ginger, turmeric and related genera, large cardamom, kasturi turmeric, mango ginger, *Kaempferia galanga, K. rotunda, Alpinia* spp, vanilla and related species, cinnamon, camphor, cassia and herbal spices like lavender, celery, thyme, mint, anise, savory, spearmint and oregano (Nirmal Babu *et al.*, 1997, 2005, Minoo 2002). These techniques form the base for establishing tissue cultures and developing *in vitro* technology for conservation. The basal media used are MS (Murshige and Skoog, 1962) for crops like cardamom, ginger, turmeric, kasturi turmeric, mango ginger, large cardamom, *Kaempferia*, *Vanilla* spp. seed and herbal spices and WPM-Woody Plant Medium (Mc Cown and Amos, 1979) for black pepper and its related species, cinnamon, camphor and cassia. Simultaneously these tissue-cultured plants are being evaluated for their morphological and genetic stability in culture (Luckose *et al.*, 1993, Chandrappa *et al.*, 1997, Nirmal Babu *et al.*, 2003, Madhusoodanan *et al.*, 2005). Though micropropagation protocols were standardized using growth regulators, all the *in vitro* storage experiments were carried out using growth regulators free media to reduce rate of multiplication which in turn will reduce extent of variation.

Protocols are available for micropropagation and multiplication of many endangerd species like *Piper hapnium, P. silentvallyensis, P.schmidtii, P. wightii, P. barberi* (Figure 9.1), *Vaniilla aphylla* (Figure 9.2) *V. pilifera, V. walkyrie, V. wightiana, Kaempferia rotunda and Alpinia galanga* are available (Peter *et al.*, 2002, Minoo 2002, Nirmal Babu *et al.*, 1999, 2005).

Bertaccini *et al.* (2004) and Du *et al.* (2004) reported micropropagation and establishment of mite-brone virus-free garlic.

In vitro Conservation

In vitro conservation involves maintenance of explants in a sterile, pathogen-free environment and is widely used for conservation of species that produce recalcitrant seeds, or do not produce seeds (Engelmann, 1997).

Various *in vitro* conservation methods are used. For short-and medium-term storage, the aim is to increase intervals between subcultures by reducing growth. This is achieved by modifying environmental conditions, including culture medium to realize slow-growth conservation. The most widely applied technique is temperature reduction (varying from 0-5°C for cold tolerant species to 9-18°C for tropical species) that can be combined with a decrease in light intensity or storage in dark (Engelmann, 1997) and adjustment of growth medium. Alternatives to standard slow-growth conservation include modification of gaseous environment of cultures, desiccation and encapsulation of explants.

Principle of slow growth storage is that the safety of *in vitro* culture be ensured without disadvantages of frequent subculturing. Thus the risk of contamination at each transfer interval, inputs in terms of labour and consumables are reduced. Several methods, such as temperature reduction, medium modification, use of osmoticums, etc. reduce rate of growth of tissue cultures, so that it can be kept unattended for moderate length of time (Withers, 1980; 1987; 1991; Withers and Williams, 1986).

Factors like temperature, culture medium, use of osmoticum, and physiological state of ex-plant, culture vessel, minimization of evaporation loss, encapsulation and desiccation influenced slow growth. Growth reduction can generally be achieved by lowering culture temperature, but scope for temperature reduction depends upon species to be conserved. Tissue cultures of cardamom could not be stored under low temperatures (Nirmal Babu *et al.*, 1994). Normal culture vessel allowed comparatively better gaseous exchange but there is faster rate of moisture depletion in culture media and drying up of cultures between 120 and 180 days depending upon species. Use of screw caps, polypropylene caps or aluminum foil as vessel closures minimized moisture loss, helped in increased longevity of cultures and the subculture period could be prolonged upto 30 days (Nirmal Babu *et al.*, 1994, Geetha *et al.*, 1995, Geetha 2002). Sealing culture tube with parafilm helped in reducing the chance of evaporation and consequent dehydration. Longevity of culture was reported earlier in ginger and turmeric (Balachandran *et al.*, 1990). Normal culture room temperatures of 22 ± 2 °C are suitable for storage of spices germplasm (Dekkers *et al.*, 1991, Nirmal Babu *et al.*, 1994; Geetha *et al.*, 1995).

In vitro Conservation of Spices

Conservation of pepper, cardamom, herbal spices, vanilla and ginger germplasm in *in vitro* gene bank by slow growth was reported (Balachandran *et al.*, 1990, Dekkers *et al.*, 1991, Geetha *et al.*, 1995, Tyagi *et al.*, 1996, Nirmal Babu *et al.*, 1996, 1997, 1998, 1999, Peter *et al.*, 2001; Geetha 2002; Ravindran *et al.*, 2004). Highlights of present status of information on *in vitro* and cryo conservation of important spices are given in Table 9.1.

Black Pepper and Related Species

Protocols for *in vitro* conservation by slow growth of black pepper and its related species *viz.*, *P.barberi*, *P.colubrinum*, *P.betle* and *P.longum* were standardized (Geetha *et al.*, 1995) by maintaining cultures at reduced temperatures, in presence of osmotic inhibitors, at reduced nutrient levels, or by minimizing evaporation loss by using closed containers. Black pepper cultures could be maintained in half strength WPM supplemented with 15 gl^{-1} each of sucrose and mannitol for one year with 85 per cent survival. In *P.barberi* full strength WPM with 25 gl^{-1} sucrose and 5 gl^{-1} mannitol was suitable for storage of cultures upto one year with 80 per cent survival. Shoot tips of *P.longum* and *P.colubrinum* could be stored upto one year in full strength WPM with 20 gl^{-1} sucrose and 10 gl^{-1} mannitol with 75 per cent and 70 per cent survival respectively. *P.betle* cultures could be stored in half strength WPM supplemented with 20 gl^{-1} sucrose for one year. Conserved materials of all these species showed normal rate of multiplication when transferred to multiplication medium after storage. Normal sized plantlets when transferred to soil established with over 80 per cent success. They developed into normal plants without any deformities and were morphologically similar to mother plants.

Table 9.1: Present State of Information on *In vitro* and Cryo Conservation of Spices

Application	*Technique*	*Reference*
Black pepper and related species	Meristem culture	Philip *et al.*, 1992
Disease eradication propagation	Shoot Culture Leaf/root	Broome and Zimmerman, 1978
Slow growth storage	*In vitro* plantlets at 22ºC	Nirmal Babu *et al.*, 1999
Cryopreservation	Seeds	Chaudhury and Chandel 1994
	Synseeds	Ravindran *et al.*, 2004
Slow growth storage	Shoot cultures	Tyagi *et al.*, 1996, Geetha *et al.*, 1995, Peter *et al.*, 2002
Slow growth storage	Plantlets	Nirmal Babu *et al.*, 1997
Slow growth storage and cryopreservation	Plantlets and shoot tips	Ravindran *et al.*, 2004
***Allium* spp**		
Disease eradication	Meristem culture and themotherapy	Conci and Nome 1991
Propagation	Microbulbets	Novak and Havel 1981
Cryopreservation	Shoot culture, microbullbelts	Keller 1991
Slow growth storage	Apical meristem	El-Gizawy and Ford-Lioyd 1987
Cryopreservation	Shoot tips	Niwata 1995
Cardamom		
Propagation	Shoot buds, callus regeneration	Rao *et al.*, 1982, Nirmal Babu *et al.*, 1997
Slow growth storage	*In vitro* plantlets at 22ºC	Nirmal Babu *et al.*, 1997, Geetha *et al.*, 1995, Peter *et al.*, 2002
Cryopreservation	Seeds	Chaudhury and Chandel 1995
Slow growth storage and cryopreservation	Plantlets and shoot tips	Ravindran *et al.*, 2004
***Zingiber* spp.**		
Disease eradication	Shoot cultures, shoot buds	Balachandran *et al.*, 1990
Propagation	Somatic embryo generation	Hosoki and Sagawa 1977, Nirmal Babu *et al.*, 1997
Slow growth storage	Somatic embryogenesis	Nirmal Babu *et al.*, 1992

Contd..

Table 9.1–Contd..

Application	Technique	Reference
Distribution	Shoot cultures at 22ºC	Nirmal Babu *et al.,* 1992
	Shoot cultures in paraffin oil	Kacker *et al.,* 1993
	Microrhizomes	Bhat *et al.,* 1995, Nirmal Babu *et al.,* 2003
Slow growth storage	Shoot cultures	Dekkers *et al.,* 1991, Nirmal Babu *et al.,* 1997
Cryopreservation	Synseeds	Sharma *et al.,* 1994
Slow growth storage	Plantlets	Tyagi *et al.,* 1996, Nirmal Babu *et al.,* 1997, Geetha, 2002, Peter *et al.,* 2002
Slow growth storage and cryopreservation	Plantlets and shoot tips	Ravindran *et al.,* 2004
Section 1.01: *Curcuma* spp		
Section 1.02: Propagation	Shoot buds	Cervera and Madrigal 1981, Nirmal Babu *et al.,* 1997
Section 1.03: Slow growth storage	*In vitro* plantlets at 22ºC	Geetha *et al.,* 1995
Section 1.04: Distribution	Microrhizomes	Raghurajan *et al.,* 1997
Section 1.05: Synthetic seeds	Shoot buds	Gayatri *et al.,* 2005
Slow growth storage	Plantlets	Nirmal Babu *et al.,* 1997, Geetha, 2002, Peter *et al.,* 2002
Slow growth storage and cryopreservation	Plantlets and shoot tips	Ravindran *et al.,* 2004
Distribution	Microrhizome	Sunitabala *et al.,* 2001
Slow growth	Plantlets	Tyagi *et al.,* 1996
Section 1.06: *Vanilla* spp.		
Section 1.07: Disease eradication	Apical meristem	Cereveta and Madrigal 1981
Section 1.08: Propagation	Root	Philip and Nainar 1986, Minoo, 2002
Section 1.09: Slow growth storage	*In vitro* plantlets	Nirmal Babu *et al.,* 1996
Section 1.10: Distribution	Synthetic seeds	Sajina *et al.,* 1997, Minoo, 2002
Section 1.11: Cryo preservation	Synthetic seeds	Ravindran *et al.,* 2004
Section 1.12: Pollen cryo preservation	Pollen	Minoo, 2002
Slow growth storage and cryopreservation	Plantlets and shoot tips	Ravindran *et al.,* 2004
Section 1.13: Herbal spices		
Section 1.14: Slow growth storage	*In vitro* plantlets	Nirmal Babu et al.1996
Section 1.15: Capsicum		
Section 1.16: Slow growth storage	*In vitro* plantlets	Peter *et al.,* 2002, Ravindran *et al.,* 2004
Section 1.17: Cryopreservation	Seed	Peter *et al.,* 2002, Ravindran *et al.,* 2004
Section 1.18: Cryopreservation	Pollen	Alexander *et al.,* 1991
Section 1.19: Cryopreservation	Pollen	Rajasekharan and Ganeshan, 2003
Section 1.20: *Foeniculum vulgare*		
Section 1.21: Cold storage	Embryogenic suspension cells	Umetsu *et al.,* 1995

*Ashmore, 1997 and Nirmal Babu *et al.,* 1999, 2005.

These techniques standardized are presently used to conserve other *Piper* species also. The culture conditions for *in vitro* conservation of various spices are given in Table 9.2.

Table 9.2: *In vitro* Storage of Spices Germplasm–Media Constituents and Survival of Cultures*

Species	*Basal Medium*	*Carbon Source (g/l)*	*Growth*	*Survival Period (Days)*	*Survival (%)*	*Establishment in Nursery*
Elettaria cardamomum	MS	30S	Fast	200	55	85
	½ MS	20S	Normal	250	45	
	½ MS	10S+10M	Slow	360	85	
		15S+15M	Slow	360	90	
Zingiber officinale	½ MS	10S+10M	Slow	360	85	85
Curcuma longa	½ MS	10S+10M	Slow	360	90	90
C.aromatica	½ MS	10S+10M	Slow	360	85	90
C.amada	½ MS	10S+10M	Slow	360	80	85
Kaempferia galanga	½ MS	10S+10M	Slow	360	90	100
				450	85	
K.rotunda	½ MS	10S+10M	Slow	360	80	90
Alpinia purpuria	½ MS	10S+10M	Slow	240	85	90
Amomum subulatum	½ MS	10S+10M	Slow	180	90	90
Piper nigrum	½ WPM	30S	Fast	250	60	
	½ WPM	20S	Normal	360	70	
	½ WPM	15S+15M	Slow	360	85	80
Piper barberi	WPM	25S+5M	Slow	360	80	
	½ WPM	30S	Normal	360	70	85
Piper colubrinum	WPM	20S+10M	Slow	360	75	80
P.longum	WPM	20S+10M	Slow	450	70	80
Vanilla spp.	MS	10S+10M	Slow	90	100	90
Seed and herbal spices	MS	20S	Slow		95	
		15S+15M				

MS: Murashige and Skoog basal medium; S: Sucrose ; M: Mannitol. *Nirmal Babu *et al.*, 2000, 2005.

Zingeberaceous Crops

Technology for *in vitro* conservation of Zingeberaceous crops like ginger, turmeric, Kaempferia, cardamom and their related species was standardized by Geetha (2001). Other reports on *in vitro* conservation of Zingeberaceous crops are by Nirmal Babu *et al.* (2000), Peter *et al.* (2002) and Ravindran *et al.* (2004).

Tissue cultures of cardamom could be stored up to 360 days with 85 per cent suvival and up to maximum period of 420 days with 70 per cent survival in half strength MS medium with $10gl^{-1}$ each of sucrose and mannitol in screw capped culture tubes. Under these conditions, the cultures attained miniature but healthy appearance (Geetha *et al.*, 1995, Nirmal Babu *et al.*, 1999 and Geetha 2001).

Slow growth methods for short term conservation of ginger (Figure 9.3), turmeric, kasturi turmeric, mango ginger, *Kaempferia* spp.(*K.galanga and K. rotunda*) are also standardized. All these taxa could be stored upto 1 year without subculture in half strength MS medium $10 g l^{-1}$ each of sucrose and mannitol in sealed culture tubes (Geetha *et al.*, 1995, Geetha 2002). *Alpinia purpurata and Amomum subulatum* cultures could be stored up 8 and 6 months respectively in the same medium. Refining the technique is in progress to increase subculture interval for over 360 days.

Vanilla

Shoot tip cultures of V. *planifolia* and *V. aphylla* could be stored in full strength MS medium with $10\ g l^{-1}$ each of sucrose and mannitol, without subculture easily upto 13 months (Nirmal Bahu *et al.*, 1996, Minoo 2002). *In vitro* storage of another species of *Vanilla* (*V. walkeriae*) was reported by Agrawal *et al.* (1992).

Seed and Herbal Spices

Cultures of seed and herbal spices like anise, oregano, lavender, mint, spearmint, thyme, savory, chive and ocimum could be established in minimal media (*i.e.*, half strength MS medium with $20\ g l^{-1}$ sucrose and full strength MS medium with $15\ g l^{-1}$ each of sucrose and mannitol. (Nimal Babu *et al.*, 1996, 1999). These cultures can be maintained only for 90 days after which transfer to fresh medium is required.

Microrhizomes

In vitro induction of microrhizomes in ginger (Figure 9.5) was reported by many workers (Bhat *et al.*, 1994, Sharma and Singh 1995, Nirmal Babu *et al.*, 1997, 2003, Sunitibala *et al.*, 2001; Shirgurkar *et al.*, 2001, Peter *et al.*, 2002 and Ravindran *et al.*, 2004). The microrhizome derived plants have more tillers but the plant hight is smaller. They gave fresh rhizome yield ranging from 100- 800 g per plant with an estimated yield of 10 kg per 3 m^2 bed. *In vitro* formed rhizomes are genetically more stable compared to micropropagated plants and form an important source of disease-free planting material ideally suited for germplasm exchange, transportation and conservation (Nirmal Babu *et al.*, 2003). *In vitro* microrhizome induction was also reported in turmeric (Raghu rajan 1997, Nirmal Babu *et al.*, 1997, 2003, Ravindran *et al.*, 2003; Sunitibala *et al.*, 2001, Shirgurkar *et al.*, 2001 and Peter *et al.*, 2002). In turmeric, micropropagation by *in vitro* microrhizomes (Figure 9.6) is an ideal method for production of disease free planting material and also for conservation and exchange of germplasm (Raghurajan 1997, Nayak 2000 and Sunitibala *et al.*, 2001). Since minimal level of growth regulators are used and number of subculture cycles are reduced in microrhizome production, the pathway may be better suited for production of genetically stable planting materials. Microrhizomes can be produced *in vitro*, independent of seasonal fluctuations. Nayak (2000) reported *in vitro* microrhizome production in four turmeric varieties (Ranga, Rashmi, Roma and Surama). Micro rhizome production depended on size of multiple shoot used. Microrhizomes produced varied in size (0.1-2.0 g) and could be planted directly in field. Microrhizomes regenerated plantlets are smaller in size compared to conventuional method but gave reasonably big rhizomes (Nirmal Babu, 2003). Geetha *et al.* (2005) reported microrhizome induction in *Kaempferia galanga*.

Encapsulation of Somatic Embryos and Shoot Apices and Development of Synthetic Seeds

Under a broader definition, synthetic seeds include encapsulated buds, bulbs or any form of meristem that can develop into a plant. Encapsulation protects both embryos and micro-plantlets. It provides favourable conditions for handling, storage and mechanical seeding. It also provides a suitable environment to support acclimatization of autotrophic micro-plantlets to field conditions.

Encapsulation is a technique used in the production of 'synthetic seeds' by coating somatic embryos/shoot tips in alginate beads. Several gels like agar, alginate, polyco2133 (Bordon Co), Carboxy methylcellulose, Carrageenan, gelrite (Kelco.Co.), Guargum, sodium pectate and tragacanth gum were tested for synthetic seed production. Alginate hydrogel is frequently selected as a matrix for synthetic seed because of its moderate viscosity and low spinnability of solution, low toxicity for somatic embryos and quick gellation, low cost and biocompatibility characteristics.

Artificial or 'synthetic seeds' can be an ideal system for low cost plant movement, propagation, conservation and exchange of germplasm. Encapsulation of somatic embryos/shoot tips were successfully achieved as in *Piper* species, ginger, turmeric, cardamom, vanilla, camphor cinnamon and a few herbal spices by encapsulationg *in vitro* developed small shoot buds, somatic embryos and other propagules in 3- 5 per cent calcium alginate. These Synseeds could be stored upto 9 months in sterile water or in MS medium with over 75–80 per cent viability and could be cultured on multiplication medium to develop plantlets. (Sajina *et al.*, 1997, Minoo 2002, Nirmal Babu *et al.*, 2005). The other reports on synthetic seeds are available in ginger (Sharma *et al.*, 1994) turmeric (Gayatri *et al.*, 2005) and vanilla (Sajina *et al.*, 1997). *In vitro* regenerated shoot buds, protocorms and regenerating calli of vanilla were successfully encapsulated in 3- 4 per cent sodium alginate (Figure 9.7) and stored in sterile water for 10 months, at 22±2°C (Minoo, 2002). Synthetic seeds were also developed in camphor and cinnamon by encapsulating the somatic embryos and shoot buds in sodium alginate (Sajina *et al.*, 1997). Reports on synthetic seeds are available in celery (Redenbaugh *et al.*, 1986; Pratap, 1992). Sajina *et al.* (1997) reported development of synthetic seeds in lavender and fennel by encapsulating somatic embryos and shoot buds in sodium alginate.

Jaydip *et al.* (2000, 2001) reported alginate encapsulation of axillary buds of *Ocimum americanum* L. (hoary basil), *O.basilicum* L. (sweet basil), *O.gratissimum* L. (shrubby basil), and *O. sanctum* L. (sacred basil).

Germplasm Exchange

Germplasm exchange using *in vitro* culture is practiced to certain extent in potato, other tuber crops, orchids and anthuriums. To reduce possibility of introduction of new pathogens and pests along new plant introductions, it is imperative that we use *in vitro* technology for plant introduction wherever possible, especially in spices. Utilization of microrhizomes in ginger and turmeric can be a positive development in this regard.

Cryopreservation

For long-term conservation of problem species, cryopreservation is the only method currently available. Visible progress is made in the development of new cryopreservation techniques and protocols are established for over 100 different plant species.

Cryopreservation is an attractive option for long-term storage. Liquid nitrogen (–196°C) is routinely used for cryogenic storage, since it is relatively cheap and safe, requires little maintenance and is widely available. Below –120°C, the rate of chemical or biophysical reactions is too slow to cause biological deterioration (Kartha 1985). Only in long term, there might be a small risk of ionising radiation causing genetic changes in materials stored at cryogenic temperatures.

An array of plant materials could be considered for cryopreservation as dictated by actual needs *vis-a-vis* preservation. These include meristems, cell, callus and protoplast cultures, somatic and zygotic embryos, anthers, pollen or microspores and whole seeds (Kartha, 1985; Withers, 1985).

Plant germplasm stored in liquid nitrogen (–196°C) does not undergo cellular divisions. In addition, metabolic and other physical processes are stopped at this temperature. As such, plants can be stored for very long periods and both the problem of genetic instability and the risk of loosing accessions due to contamination or human error during subculturing are overcome. Most cryopreservation endeavours deal with recalcitrant seeds, *in vitro* tissues from vegetatively propagated crops, species with a particular gene combination (elite genotypes) and dedifferentiated plant cell cultures. Care must be taken to avoid ice crystallisation during freezing process, which otherwise would cause physical damage to tissues. Existing cryogenic strategies rely on air-drying, freeze dehydration, osmotic dehydration, addition of penetrating cryoprotective substances and adaptive metabolism cryopreservation methods are developed for more than 80 different plant species in various forms like cell suspensions, calluses, apices, somatic and zygotic embryos (Engelmann, 1997, 2000, Engelmann *et al.*, 1994, 1995). Their routine utilisation is still restricted almost exclusively to conservation of cell lines in research laboratories.

For small volumes, long-term storage is practicable through storage of cultures in cryopreservation at ultra-low temperature, usually by using liquid nitrogen (–196°C). At this temperature all cellular divisions and metabolic processes are virtually halted and consequently, plant materials can be indefinitely stored without alteration or modification.

Long Term Storage of Spices Germplasm by Cryopreservation

Cryopreservation of black pepper and cardamom seeds in liquid nitrogen (LN_2) was reported by Choudhary and Chandel, (1994; 1995) and Choudhury and Malik (2004). Pepper seeds are recalcitrant and the seed viability decreases with reduction in moisture content. Seeds desiccated to 12 per cent and 6 per cent moisture contents were successfully cryopreserved in liquid nitrogen at –196°C, with a survival rate of 45 per cent and 10.5 per cent respectively (Chaudhury and Chandel 1994). Choudhary and Chandel (1995) attempted cryo-conservation of cardamom seed. They tried to conserve seeds at ultra- low temperature by suspending seeds in cryovials in vapour phase of liquid nitrogen (–150°C) by slow freezing and also by direct immersion in liquid nitrogen (–196°C) by fast freezing. The result showed that seeds possessing 7.7–14.3 per cent moisture content could be successfully cryopreserved with 80 per cent germination when tested after one-year storage in vapour phase of liquid nitrogen (–150°C).

Plants could be successfully regenerated from cryopreserved seeds of capsicum (Figure 9.8) anise and technology for cryopreservation of black pepper, cardamom, ginger, turmeric and vanilla germplasm–using vitrification, encapsulation and encapsulation and vitrification methods–are available (Minoo 2002, Peter *et al.*, 2002 and Ravindran *et al.*, 2004).

Cryopreservation of encapsulated shoot buds of endangered *Piper barberi* (Figure 9.9) was reported by Peter *et al.* (2002) and Ravindran *et al.* (2004). Encapsulated shoot tips of *Piper barberi* were cryopreserved with 60 per cent success using vitrification technique. In encapsulation vitrification the encapsulated shoot tips were precultured on MS medium, supplemented with 0.3 M, 0.5 M and 0.7 M sucrose (pH 5.8) for three days followed by dehydration with PVS2 solution (100 per cent) at 0°C for 3 hours. After dehydration, the beads (10 encapsulated shoot tips in 0.8 ml PVS2 solution per 1.5 ml cryotube) were frozen rapidly by direct immersion into liquid nitrogen (–196 °C) and kept for one hour (Peter *et al.*, 2002 and Ravindran *et al.*, 2004).

Shoot tips (1.0–2.0mm) from *in vitro* grown plantlets of cardamom were subjected to progressive increase of sucrose concentrations (0.1, 0.3, 0.5, 0.7, 0.9, and 1.0) for two days each under the same cultural conditions as parent plantlets. These shoot tips were transferred to 1.8ml cryotube containing ice cold PVS2 solution (30 per cent (v/v) glycerol + 15 per cent (v/v) ethylene glycol + 15 per cent (v/v) DMSO in culture medium with 0.4 M sucrose, pH (5.8) at 0°C for 3 hours. After 3 hours equilibration at 0°C, shoot tips were directly immersed into liquid nitrogen for 1 hour. Vials were thawed in 40°C water for 1 minute. Cryoprotectant was removed and shoot tips were washed 2-3 times in 1.2M sucrose solution. About 70 per cent shoot tips were recovered on MS medium supplemented with BAP and NAA. But encapsulation vitrification method gave only 60 per cent success (Peter *et al.*, 2002 and Ravindran *et al.*, 2004).

Cryopreservation of ginger and turmeric shoot tips was successfully done with 80 per cent recovery using vitification method. Rate of recovery was only 40 per cent when encapsulated shoot tips were dehydrated in progressive increase of sucrose concentration together with 4–8 hrs. of desiccation (Peter *et al.*, 2002 and Ravindran *et al.*, 2004).

Encapsulated *in vitro* grown shoot tips of vanilla could be cryo preserved with 70 per cent success when pretreated with progressive increase of sucrose concentration (0.1M-1.0M) for one day each and dehydrated for 8 hrs (Peter *et al.*, 2002 and Ravindran *et al.*, 2004).

Successful cryopreservation of seeds, meristems and somatic or zygotic embryos were reported in Allium spp (Niwata, 1995, Hyung *et al.*, 2003, Haeng *et al.*, 2003, 2004.2005, Jung *et al.*, 2005 and Gayle *et al.*, 2004). Priliminary success was reported in cryo preservation of mint (Leigh and Remi 2003) and capsicum (Peter *et al.*, 2002).

Most of the reports are confined to a few genotypes and hence techniques standardized need to be extended to more genotypes before adopting them for routine conservation. Reports on cryoconservation of spices like *Ocimum*, *Lavendula* and *Salvia* are available from NBPGR, New Delhi.

Pollen Storage

Pollen storage has considerable value supplementing the germplasm conservation strategy by facilitating hybridisation between plants with different time of flowering and to transport pollen across the globe for various crop improvement programmes in addition to developing haploid or homozygous lines. No significant work was done in India, except a few initial reports.

Technique of pollen storage is comparable with that of seed storage, since pollen can be dried (less than 5 per cent moisture content on a dry weight basis) and stored below 0°C. There are a few reports on survival and fertilizing capacity of cryopreserved pollen more than five years old. Pollen might represent an interesting alternative for long-term conservation of problematic species (IPGRI, 1996). Pollen has a relatively short life compared to seeds (although this varies significantly among species) and viability testing can be time-consuming and uneconomical. Other disadvantages of pollen storage are small amount produced by many species, lack of transmission of organelle genomes via pollen, loss of sex-linked genes in dioecious species and general inability to regenerate into plants. Pollen, therefore, is used to a limited extent in germplasm conservation (Hoekstra,1995). An advantage is that pests and diseases are rarely transferred by pollen (excepting a few virus diseases). This allows safe movement and exchange of germplasm as pollen.

Alexander *et al.* (1991) and Rajasekharan and Ganeshan. (2003) reported Freeze preservation of capsicum pollen (*Capsicum annuum*) in liquid nitrogen (–196°C) for 42 months. Minoo (2002) reported cryopreservation of vanilla pollen for conservation (Figure 9.10) of haploid genome as well as assisted pollination between species that flower at different seasons and successful fertilisation using cryopreserved pollen (Minoo, 2002).

DNA Storage

Concurrent with advancements in gene cloning and gene transfer, there has been development of technology for removal and analysis of DNA. DNAs from nucleus, mitochondria and chloroplast are now routinely extracted and immobilized onto nitrocellulose sheets where DNA can be probed with numerous cloned genes. In addition, rapid development of polymerase chain reaction (PCR) now means that one can routinely amplify specific oligonucleotides or genes from entire mixture of genomic DNA. These advances, coupled with prospect of loss of significant plant genetic resources throughout world, have led to establishment of DNA bank for storage of genomic DNA.

Conserved DNA will have numerous uses viz, molecular phylogenetics and systematics of extant and extinct taxa, production of previously characterized secondary compounds in transgenic cell cultures, production of transgenic plants using genes from gene families, *in vitro* expression and study of enzyme structure and function and genomic probes for research laboratories.

Vast resources of dried specimens in the world's herbaria hold considerable DNA that would be suitable for PCR. It seems likely that integrity of DNA would decrease with age of specimens. Because there are many types of herbarium storage environments, preservation and collections, there is need for systematic investigations on effect of modes of preparation, collection and storage on integrity of DNA in world's major holdings.

Advantage of storing DNA is that it is efficient and simple and overcomes many physical limitations and constraints which characterize other forms of storage (Adams 1988, 1990, 1997, Adams and Adams 1991, Adams *et al.*, 1994). Disadvantage lies in problems with subsequent gene isolation, cloning and transfer but, most importantly, it does not allow regeneration of live organisms. DNA banking is yet to catch up in spices. DNA samples of over 600 genotypes of spices is stored in DNA bank of IISR, Calicut.

Callus and Cell Culture Systems

Quatrano (1968) and Nag and Street (1973) reported first successful experiments on cryopreservation of plant cells. Since then, a large number of cell suspension and calli cultures were successfully cryopreserved (Engelmann *et al.*, 1994). In general, callus cultures are more difficult to cryopreserve than cell suspensions, because of relative volume of callus, its slow growth rate and cellular heterogeneity (Withers 1987). One successful cryopreservation procedure that is applicable to all different cell suspensions or calli cultures has not been developed yet. Research focuses on optimizing factors on which successful cryopreservation of plant organs cells suspensions and calli depend, such as: (i) starting material, (ii) pretreatment, (iii) cryopreservation procedure, and (iv) post-thaw treatment.

Plant cells cultured *in vitro* produce wide range of primary and secondary metabolites of economic value. Production of phytochemicals from plant cell cultures was presently used for pharmaceutical products. Production of flavour components and secondary metabolites *in vitro* using immobilised cells is an ideal system for spices. Production of saffron and capsaicin was reported using such system (Ravishankar *et al.*, 1993; Johnson *et al.*, 1996; Venkataraman *et al.*, 1989). Johnson *et al.* (1996)

Figure 9.1: ***In vitro*** **Multiplication of Endangered–*Piper barberi***

Figure 9.2: ***In vitro*** **Multiplication of Endangered–*Vanilla aphylla***

Figure 9.3: ***In vitro*** **Conservation of Ginger on Slow Growth Medium**

Figure 9.4: ***In vitro*** **Gene Bank at IISR**

Figure 9.5: Microryzomes of Ginger for Production of Disease Free Planting Materials

Figure 9.6: Micro Rhizomes of Turmeric

Figure 9.7: 'Syn Seeds' of Vanilla for Disease Free Germplasm Conservaion and Exchange

Figure 9.8: Germinating Cryopreserved Seeds of Capsicum

Figure 9.9: Recovery and Growth of Cryopreserved 'Syn Seeds' of Endangered *Piper barberi*

Figure 9.10: Growth of Cryopreserved Pollen of Vanilla Indicating Viability

reported biotransformation of ferulic acid vanillamine to capsicin and vanillin in immobilised cell cultures of *Capsicum frutescens*. Reports on *in vitro* synthesis of crocin, picrocrocin and safranel from saffron stigma (Himeno and Sano, 1995) and colour components from cells derived from pistils (Hori *et al.*, 1988) are available for further scaling up. Callus and cell cultures were established in nutmeg, clove, camphor, ginger, lavender, mint, thyme and celery. Cell immobilization techniques are standardized in ginger, sage, anise and lavender (Ilahi and Jabeen, 1992; Ravindran *et al*, 2004; Sajina *et al.*, 1997).

Studies on conservation of cell lines are yet to become popular in spices. Suspensions of embryogenic cell lines of fennel, conserved at 4 °C for up to 12 weeks produced normal plants upon transfer to normal laboratory conditions (Umetsu *et al.*, 1995).

Genetic Stability

An important prerequisite for any conservation technique is that the regenerants produced from conserved material should be true-to-type. There are ample evidences to indicate that under certain culture conditions, the materials undergo genetic changes (somaclonal variations) and as a consequence lose their integrity and uniformity. This would be highly undesirable in spice varieties where the purpose is not only to conserve a genotype but also retain its specific quality traits. Thus testing for genetic stability of *in vitro* conserved materials is of utmost importance. Besides morphology, cytology and isozyme profiling, sophisticated biochemical and DNA-based techniques have enabled more critical analysis of the genetic stability of *in vitro* materials.

Estimation of Genetic Fidelity of Micropropagated Spice Crops Using Molecular Markers

Morphological characters coupled with RAPD profiles using 24 operon primers indicated genetic fidelity among randomly selected micropropagated plants of black pepper 'Subhakara' and 'Aimpiriyan', indicating that micropropagation protocol can be used for commercial cloning of black pepper (Nirmal Babu *et al.*, 2003). Genetic uniformity of micropropagated *Piper longum* using RAPD profiling was reported by Ajith (1997) and Parani *et al.* (1997) for conservation.

Peter *et al.* (2002) and Ravindran *et al.* (2004) reported that conserved materials of all species conserved by them showed normal rate of multiplication when transferred to multiplication medium after storage. The normal sized plantlets when transferred to soil established with over 80 per cent success. They developed into normal plants without any deformities and were morphologically similar to mother plants. RAPD profiling of these conserved plants also showed their genetic uniformity.

Thus in vitro conservation can form an ideal supplement for field gene banks in conserving active germplasm in spices and form a source for multiplication and exchange of disease free materials of various genotypes as and when required. A schematic diagram of in vitro conservation strategy for conserving spices germplasm is given in Chart 9.1. Conservation of genetic resources in in vitro gene banks is now an established convention and two gene banks for conservation of spices germplasm, function at IISR, Calicut and National Bureau of Plant Genetic Resources, New Delhi to develop *in vitro* conservation strategies for short and long-term conservation of these crops. About 1000 accessions of spices germplasm are currently kept in in vitro repository of IISR (Figure 9.4) and NBPGR (Tables 9.3 and 9.4).

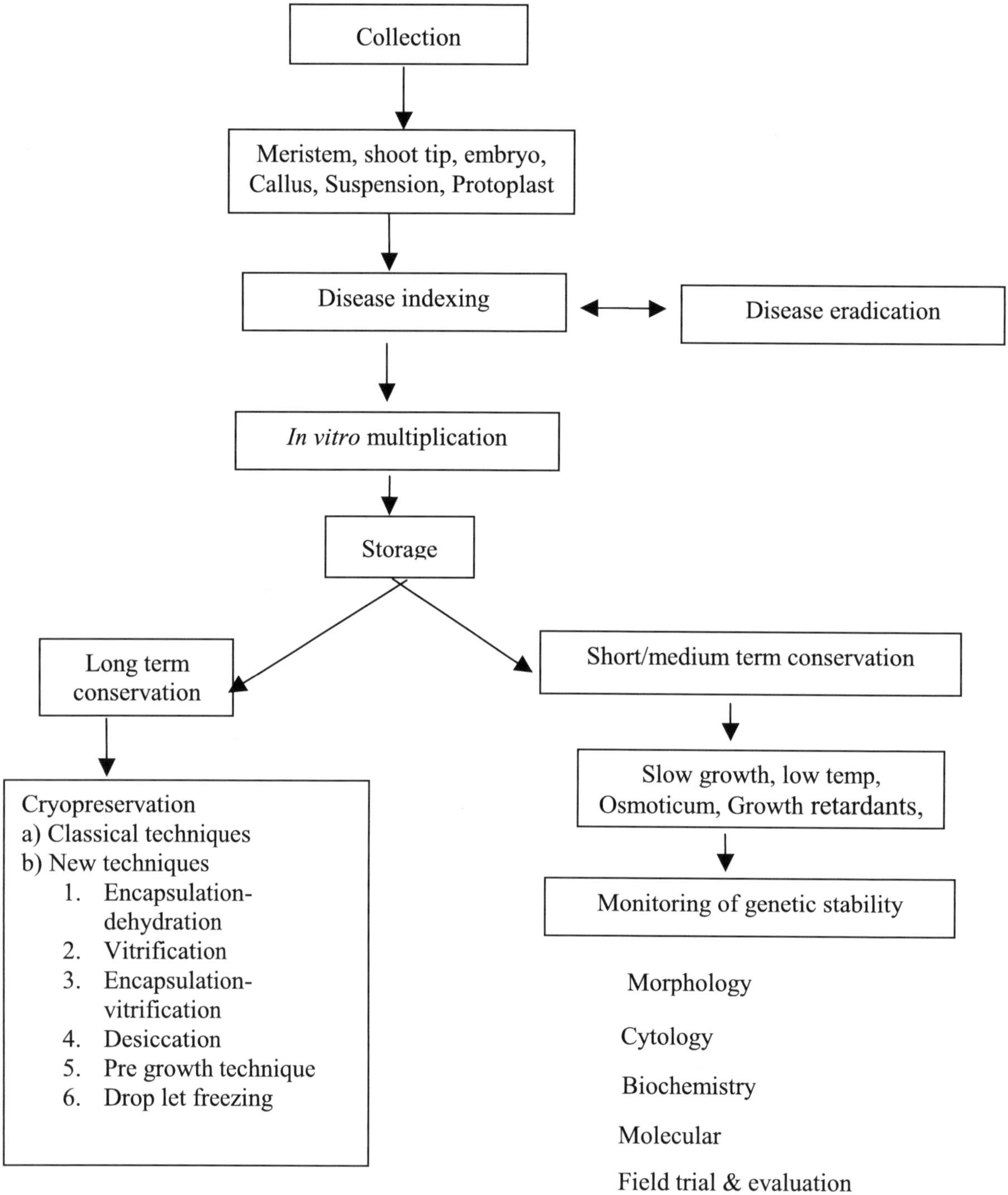

Chart 9.1: Schematic Representation Conservation of Spices Germplasm

Table 9.3: Present Status of *In vitro* Gene Bank at IISR, Calicut

Species	*No. of Accessions*
Piper nigrum and related species	50
Elettaria cardamomum	30
Zingiber officinale	70
Curcuma longa	90
Kaempferia, Alpinia, Amomum and *Hedychium* spp	5
Vanilla spp. and accessions	200
Murraya koenigi, Cinnamomum verum, C. camphora	3
Capsicum annuum	3
Acorus calamus	1
Total	**452**

Table 9.4: Important Spices Conserved in *In vitro* Repository at IISR, Calicut

Plant Species	*Important Characters*
***Piper* spp.**	
Piper nigrum	
Panniyoor	High yield
Subhkara	High yield and quality
Sreekara	High yield and quality
Pournami	Tolerant to root knot nematode
Panchami	High oleoresin
Kaluvally	Tolerant to drought
IISR Thevam	Good yield, quality medium
Girimunda	High yielding hybrid, suited for high altitude areas
Malabar Excel	High yielding hybrid, suited for high altitude areas, High oleoresine, piperine and essential oil
P. colubrinum	Resistant to *Phytophthora*
P. barberi	Endangered species
P. schmidtii	Endangered species
P. silentvalleyensis	Endangered species
P. longum (Viswam)	Released variety
Section 1.01: Cardamom	
NKE lines	Natural katte escapes
APG 150	Dwarf, multibranched, vazhukka
PV1 x CL 37	Hybrid
CCS 1	Released variety
RR 1	Rhizome rot tolerant

Contd..

Table 9.3–Cond..

Plant Species	*Important Characters*
Section 1.02: Ginger	
RRE	Rhizome rot escape
Varada	Released variety
Gurubathany	Bold rhizome
Bhainse	Extra bold rhizome
Jamaica	Low fibre
Baharica	High yield
Tajfingiva	Extra bold rhizome
Himachal Pradesh colln	High volatile oil
Jiuggjam	Bold rhizome, high yield
Maran	Tetraploid
Mahima	Resistant to root knot nematode, Bold rhizome
Rejatha	High quality, Bold rhizome
Section 1.03: Turmeric	
Suguna	High yield, early maturing, high curcumin
Sudarsana	High yield, high curcumin content, tolerant to rhizome rot
Pratibha	Seedling progeny with high yield and quality
Prabha	High yield and quality
Roma	High curcumin content
Vanilla	
Vanilla aphylla	Endangered species
Vanilla walkeriae	Endangered species

References

Adams, R.P., 1988. The preservation of genomic DNA: DNA Bank Net. *Amer. J. Bot.*, 75: 156.

Adams, R.P., 1990. The preservation of Chihuahuan plant genomes through *in vitro* biotechnology: DNA Bank-Net, a genetic insurance policy. In: *Third Symposium on Resources of the Chihuahuan Desert Region*, (Eds.) Powell, A.M., Hollander, R.R., Barlow, J.C., McGillivray, W.B. and Schmidly, D.J. Printech Press, Lubbock, TX, pp. 1–9.

Adams, R.P., 1997.Conservation of DNA: DNA banking. In: *Biotechnology and Plant Genetic Resources: Conservation and Use*, (Eds.) Callow, J.A. Ford-Loyd, B.V. and Newbury, H.J. Biotechnology in Agriculture Series, No. 19. CAB International, pp. 163–174.

Adams, R.P. and Adams, J.E., 1991. *Conservation of Plant Genes: DNA Banking and In Vitro Biotechnology*. Academic Press, New York.

Adams, R.P., Miller, J.S., Golenberg, E.M. and Adams, J.E., 1994. *Conservation of Plant Genes 11: Utilization of Ancient and Modern DNA*. Missouri Botanical Garden Press. St. Louis, MI, 276 pp.

Agrawal, D.C., Morwal, G.C. and Mascarenhas, A.F., 1992. *In vitro* propagation and slow growth storage of shoot cultures of *Vanilla walkeriae* Wight: An endangered orchid. *Lindleyana*, 7(2): 95–99.

Ajith, A., 1997. Micropropagation and genetic fidelity studies in *Piper longum* L. In: *Biotechnology of Spices, Medicinal and Aromatic Plants,* (Eds.) Edison, S., Ramana, K.V., Sasikumar, B., Nirmal Babu K. and Santhosh J. Eapen. Indian Society for Spices, Calicut, India, p. 94–97.

Alexander, M.P., Ganeshan, S. and Rajasekharan, P.E., 1991. Freeze preservation of capsicum pollen (*Capsicum annuum*) in liquid nitrogen (–196°C) for 42 months: Effect on viability and fertility. *Plant Cell Incompatibility Newsletter*, 23: 1–4.

Ashmore, S.E., 1997. Status report on the development and application of *in vitro* techniques for conservation and use of plant genetic resources. International Plant Genetic Resources Institute, Rome, Italy. 67 p.

Balachandran, S.M., Bhat, S.R. and Chandel, K.P.S., 1990. *In vitro* clonal multiplication of turmeric (*Curcuma longa*) and ginger (*Zingiber officinale* Rosc.). *Plant Cell Reports*, 3: 521–524.

Bertaccini, A., Botti, S., Tabanelli, D., Dradi, G., Fogher, C., Previati, A., da Re, F., Nicola, S., Nowak, J. and Vavrina, C.S., 2004. Micropropagation and establishment of mite-borne virus-free garlic *Allium sativum*), *Acta Horticulturae*, 631: 201–206.

Bhat, S.R., Chandel, K.S.P. and Kacker, A., 1994. *In vitro* induction of rhizome in ginger *Zingiber officinale* Rosc. *Indian Journal of Experimental Biology*, 32(5): 340–344.

Broome, O.C. and Zimmerman, R.N., 1978. *In-vitro* propagation of black pepper. *Hort. Sci.*, 43: 151–153.

Cereveta, E. and Madrigal, R., 1981. *In vitro* propagation (*Vanilla planifolia* A.). *Environmental and Experimental Botany*, 21: 441.

Chand, P.K., Mandal, J. and Pattnaik, S., 2000. Alginate encapsulation of axillary buds of *Ocimum americanum* L. (hoary basil), *O.basilicum* L. (sweet basil), *O. gratissimum* L. (shrubby basil), and *O. sanctum* L. (sacred basil). *In vitro Cellular and Developmental Biology* (Plant), 36(4): 287–292.

Chandel, K.P.S. and Pandey, R., 1991. Plant genetic resources conservation: Recent approaches. In: *Plant Genetic Resources Conservation and Management: Concepts and Approaches*, (Eds.) Paroda, R.S. and Arora, A.K. IBPGR, Regonal Office for South and Southeast Asia (ROSSEA), New Delhi, pp. 248–272.

Chandrappa, H.M., Shadakshari, Y.G., Sudharshan, M.R. and Raju, B. ,1997. Preliminary yield trial of tissue cultured cardamom selections. In: *Biotechnology of Spices, Medicinal and Aromatic Plants,* (Eds.) Edison, S., Ramana, K.V., Sasikumar, B., Nirmal Babu K. and Santhosh J. Eapen. Indian Society for Spices, Calicut, India, p. 102–105.

Choudhary, R. and Chandel, K.P.S., 1995. Studies on germination and cryopreservation of cardamom (*Elletaria cardamomum* Maton.) seeds. *Seed Science and Biotechnology*, 23(1): 235–240.

Chaudhury, R. and Chandel, K.P.S., 1994. Germination studies and cryopreservation of seeds of black pepper (*Piper nigrum* L.): A recalcitrant species. *Cryoletters*, 15: 145–150.

Choudhury, R. and Malik, S.K., 2004. Genetic conservation of plantation crops and spices using cryopreservation. *Ind. J. Biotechnology*, 3: 348–358.

Conci, V.C. and Nome, S.F., 1991. Virus free garlic (*Allium sativum* L.) plants obtained by thermotherapy and meristem tip culture. *J. Phytopathol.*, 132: 186–192.

Dekkers, A.J., Rao, A.N. and Ghosh, C.J., 1991. *In vitro* storage of multiple shoot culture of gingers at ambient temperatures of 24-29°C. *Scientia Horticulturae*, 47(1-2): 157-167.

Du, Y.Q., Zhu, J.Z. and Shen, W.P., 2004. Study on virus elimination technology of garlic in Jiading by shoot tip culture. *Acta Agriculturae Shanghai*, 20(1): 9–12.

El-Gizawy, A.M. and Ford-Lioyd, B.V., 1987. An *in vitro* method for the conservation and storage of garlic (*Allium sativum*) germplasm. *Plant Cell Tissue Organ Culture*, 9: 147–150.

Engelmann, F., 1997. *In vitro* conservation methods. In: *Biotechnology and Plant Genetic Resources: Conservation and Use,* (Eds.) Callow, J.A., Ford-Loyd, B.V. and Newbury, H.J. Biotechnology in Agriculture Series, No. 19. CAB International, UK, pp. 119–161.

Engelmann, F., 2000. Importance of cryopreservation for the conservation of plant genetic resources. In: *Cryopreservation of Tropical Plant Germplasm*, (Eds.) Engelmann, F. and Takagi, H. Current Research Progress and Application. Japan International Research Centre for Agricultural Sciences, Japan/International Plant Genetic Resources Institute, Rome, Italy, pp. 8–20.

Engelmann, F., Benson, E.E., Chabrillange, N., Gonzalez-Arnao, M.T., Mari, S., Michaux-Ferriere, N., Paulet, Glazmann, J.C. and Charrier, A., 1994. Cryopreservation of several tropical plant species using encapsulation/dehydration of apices. In. *Proc. VIIIth IAPTC Meeting*, Firenze, Italy.

Engelmann, F., Dambier, D. and Ollitrault, P., 1994. Cryopreservation of cell suspensions and embryogenic calluses of citrus using a simplified freezing process. *CryoLetters*, 15: 53–58.

Engelmann, F., Dumet, D., Chabrillange, N., Abdelnour-Esquivel, A., Assy-Bah, B., Dereuddre, J. and Duval, Y., 1995. Cryopreservation of zygotic and somatic embryos from recalcitrant and intermediate-seed species. *Plant Genet. Resources Newsletter*, 103: 27–31.

Frankel, O.H., 1975. Genetic resources survey as a basis for exploration. In: *Crop Genetic Resources for Today and Tomorrow*, (Eds.) O.H. Frankel and J.H. Hawkes. Cambridge University Press, Cambridge. pp. 99.

Gayatri, M.C., Roopadarshini, V., Kavyashree, R. and Kumar, C.S., 2005. Encapsulation and regeneration of aseptic shoot buds of turmeric (*Curcuma longa* L.). *Plant Cell Biotechnology and Molecular Biology*, 6(3/4): 89–94.

Gayle, M., Volk, Nicholas Maness and Kate Rotindo, 2004. Cryopreservation of Garlic (*Allium sativum* L) using Plant Vitrification Solution 2. *CryoLetters*, 25: 219–226.

Geetha, S.P., 2002. *In vitro* technology for genetic conservation of some genera of Zingiberaceae. *Ph.D. Thesis*, Calicut University.

Geetha, S.P., Manjula, C. and Sajina, A., 1995. *In vitro* conservation of genetic resources of spices. In *Proceedings Seventh Kerala Science Congress* January, 27-29, Palakkad, Kerala, pp. 12-16.

Geetha, S.P., Minoo, D., Sajina, A., Samsudeen, K., Rema, J., Nirmal Babu, K. and Peter, K.V., 1996. *In vitro* storage: Prospects for conservation of germplasm of spices. In: *Abst. National Symposium on Horticultural Biotechnology*, Indian Institute of Horticulture Research, Bangalore, p.2.

Haeng-Hoon Kim, Eun-Gi Cho, Hyung-Jin Baek, Chang-Yung Kim, ER Joachim Keller and Florent Engelmann, 2004. Cryopreservation of garlic shoot tips by vitrification: Effects of dehydration, Rewarming, unloading and regrowth conditions *CryoLetters*, 25: 59–70.

Haeng-Hoon Kim, Jung-Bong Kim, Hyung-Jin Baek, Eun-Gi Cho, Young-Am Chaand Florent Engelmann, 2004. Evolution of DMSO Concentration in Garlic Shoot tips during a Vitrification Procedure. *CryoLetters*, 25: 91–100.

Haeng-Hoon Kim, Ju-Won Yoon, Jung-Bong Kim, Florent Engelmann and Eun-Gi Cho, 2005. Thermal analysis of Garlic Shoot tips during a Vitrification Procedure. *CryoLetters*, 26(1): 33–44.

Himeno, H. and Sano, K., 1995. Synthesis of crocin, picrocrocin and safranal by saffron stigma like structures proliferated *in vitro*. *Agricultural Biology and Chemistry*, 51 (9): 2395–2400.

Hori, H., Enomoto, K. and Nakaya, H., 1988. Induction of callus from pistils of *Crocus sativus* L. and production of colour components in the callus. *Plant Tissue Culture Letters*, 5: 72 –77.

Hosoki, T. and Sagawa, Y., 1977. Clonal propagation of ginger (*Zingiber officinale* Rosc.) through tissue culture. *Horticulture Science*, 12: 451-452.

Huang, J.H., 1995. *In vitro* propagation and preservation of ginger germplasm resources. *Scientia Agricultura Sinica*, 28(2): 24–30.

Hoekstra, F.A., 1995. Collecting pollen for genetic resources conservation. In: *Collecting Plant Genetic Diversity: Technical Guidelines*, (Eds.) Guarino, L., Rao, V.R. and Reid, R. CAB International, Wallingford, UK, pp. 527–550.

Hyung-Jin Baek, Haeng-Hoon Kim, Eun-Gi Cho, Young-Am Chae and Florent Engelmann, 2003. Importance of explant size and origin and of preconditioning treatments for cryopreservation of garlic shoot apices by vitrification. *CryoLetters*, 24: 381–388.

IBPGR, 1986. Design, planning and operation of *in vitro* genebanks, IBPGR, Rome.

Ilahi, I. and Jabeen, M., 1992. Tissue culture studies for micropropagation and extraction of essential oils from *Zingiber officinale* Rosc. *Pakistan Journal of Botany*, 24(1): 54–59.

IPGRI, 1996 Pragamme activities, germplsm maintenance and use. In: *Annual Report*, IPGRI. Rome. pp. 56–65.

Iwananga, M., 1994. Role of International organisations in global genetic resource management. In: *Proc. 27th International Symposium on Tropical Agriculture Research*, Japan International Research Centre for Tropical Agricultural Sciences, Ministry of Agriculture, Forestry and Fisheries, Tsukuba, Japan, August, 25–26, 1993, p.1–6

Jaydip, M., Sitakanta, P. and Chand, P.K., 2000. Alginate encapsulation of axillary buds of *Ocimum americanum* L. (hoary basil), *O. basilicum* L. (sweet basil), *O. gratissimum* L. (shrubby basil), and *O. sanctum* L. (sacred basil). *In Vitro Cellular and Developmental Biology Plant*, 36(4): 287–292.

Jaydip, M., Sitakanta, P. and Chand, P.K., 2001. Growth response of the alginate-encapsulated nodal segments of *in vitro*-cultured basils. *J. Herbs Spices and Medicinal Plants*, 8(4): 65–74.

Johnson, T.S., Ravishanker, G.A and Venkataraman, L.V., 1996. Biotransformation of ferulic acid vanillamine to capsaicin and vanillin in immobilised cell cultures of *Capsicum frutescens Plant Cell, Tissue and Organ Culture*, 44(2): 117–123.

Jung-Bong Kim, Haeng-Hoon Kim, Hyung-Jin Baek, Eun-Gi Cho, Yong-Hwan Kim and Florent Engelmann, 2005. Changes in sucrose and glycerol content in garlic shoot tips during freezing using pvs3 solution. *CryoLetters*, 26(2): 103–112.

Kacker, A., Bhat, S.R., Chandel, K.P.S. and Malik, S.K., 1993. Plant regeneration *via* somatic embryogenesis in ginger. *Plant Cell Tissue and Organ Culture*, 32(3): 289–292.

Kartha, K.K., 1985. In: *Cryopreservation of Plant Cells and Organs*, (Ed.) Kartha, K.K. CRC Press, Boca Raton, Florida, pp. 243–267.

Keller, J., 1991. *In vitro* conservation of haploid and diploid germplasm in *Allium cepa* L. *Acta Hortic.*, 289: 231–232.

Kuruvilla, K.M., Madhusoodanan, K.J., Sudharshan, M.R., Natarajan, P. and Thomas, J., 2005. Performance evaluation of tissue culture Vs open pollinated seedling of cardamom. *Proceedings of ICAR National Symposium on Biotechnological interventions for improvement of Horticultural Crops: Issues and Strategies*, Kerala Agricultural University, Trissur, Kerala, India, p. 81–83.

Leigh, E. Towill and Remi Bonnart, 2003. Cracking in a vitrification solution during cooling or warming does not effect growth of cryopreserved mint shoot tips. *CryoLetters*, 24: 341–346.

Lukose, R., Saji, K.V., Venugopal, M.N. and Korikanthimath, V.S., 1993. Comparative field performance of micropropagated plants of cardamom (*Elettaria cardamomum*). *Indian Journal of Agricultural Sciences*, 63(7): 417–418.

Madhusoodanan, K.J., Kuruvilla, K.M., Vadiraj, B.A., Radhakrishnan, V.V. and Thomas, J., 2005. On farm evaluation of tissue culture vanilla plants vis-à-vis vegetative cuttings. *Proceedings of ICAR National Symposium on Biotechnological Interventions for Improvement of Horticultural Crops: Issues and Strategies*, Kerala Agricultural University, Trissur, Kerala, India, p. 89–90.

Mc Cown, B.H. and Amos, R., 1979. Initial trials of commercial micropropagation with birch. *Proc. Inter. Plant Pro. Soc.*, 29: 387–393.

Minoo, D., 2002. Seedling and somaclonal variation and their characterization in *Vanilla*. *Ph.D. Thesis*, Calicut University, Kerala, India.

Murashige, T. and Skoog, F., 1962. A revised medium for rapid growth and bioassays with tobacco tissue cultures. *Physiol Plant*, 15: 473–493.

Nag, K.K. and Street, H.E., 1975. Freeze preservation of cultured plant cells, 11. The freezing and thawing phases. *Physiol. Plantarum*, 340: 254–260.

Nayak, S., 2000. *In vitro* microrhizome production in four cultivars of turmeric (*Curcuma longa* L.) as regulated by different factors. In: *Proc. Centennial Conference on Spices and Aromatic Plants: Challenges and Opportunities in the New Century*, 20–23 September.

Nirmal Babu, K., Geetha, S.P., Manjula, C., Sajina, A., Minoo, D., Samsudeen, K., Ravindran, P.N. and Peter, K.V., 1996. Biotechnology: Its role in conservation of genetic resources of spices. In: *Biotechnology for Development*, (Eds.) Das, M.R. and Satish, M. State Committee on Science, Technology and Environment, Kerala, p. 198–212.

Nirmal Babu, K., Geetha, S.P., Manjula, C., Ravindran, P.N. and Peter, K.V., 1994. Medium term conservation of cardamom germplasm: An *in vitro* approach. In: *Abstract Second Asia-Pacific Conference on Agricultural Biotechnology*, Madras, India, p. 57.

Nirmal Babu, K., Geetha, S.P., Minoo, D., Ravindran, P.N. and Peter, K.V., 1999. *In vitro* conservation of cardamom (*Elettaria cardamomum* Maton.) germplasm. *Plant Genetic Resources Newsletter*, 119: 41–45.

Nirmal Babu, K., Geetha, S.P., Minoo, D., Ravindran, P.N. and Peter, K.V., 1999. *In vitro* conservation of germplasm. In: *Biotechnology and its Application in Horticulture*, (Ed.) S.P. Ghosh. Narosa Publishing House, New Delhi, pp. 106–129.

Nirmal Babu, K., Minoo, D., Geetha, S.P., Ravindran, P.N. and Peter, K.V., 2005. Advances in Biotechnology of Spices and Herbs. *Ind. J. Bot.Res.*, 1(2): 155–214.

Nirmal Babu, K., Minoo, D., Geetha, S.P., Samsudeen, K., Rema, J., Ravindran, P.N. and Peter, K.V., 1998. Plant biotechnology: It's role in improvement of spices. *Indian J. Agril. Sciences*, 68(8 Special Issue.): 533–547.

Nirmal Babu, K., Ravindran, P.N. and Peter, K.V., 1997. *Protocols for Micropropagation of Spices and Aromatic Crops*. Indian Institute of Spices Research, Calicut, India, p. 35.

Nirmal Babu, K., Ravindran, P.N. and Peter, K.V., 2000. Biotechnology of spices. In: *Biotechnology of Horticulture and Plantation Crops*, (Eds.) Chadha, K.L., Ravindran, P.N. and Leela, S. Malhotra Publishing House, New Delhi, pp. 487–527.

Nirmal Babu, K., Ravindran, P.N. and Sasikumar, B., 2003. Field evaluation of tissue cultured plants of spices and assessment of their genetic stability using molecular markers. Final Report submitted to Department of Biotechnology, Government of India, pp. 94.

Nirmal Babu, K., Rema, J., Sree Ranjini, D.P., Samsudeen, K. and Ravindran, P.N., 1996. Micropropagation of an endangered species of Piper, *P. barberi* Gamble and its conservation. *Journal of Plant Genetic Resources*, 9(1): 179–182.

Nirmal Babu, K., Geetha, S.P., Minoo, D., Rajalakshmi, K., Girija, P., Ravindran, P.N. and Peter, K.V., 1999. *In vitro* approaches for conservation and exchange of spices germplasm. In: *Biodiversity, Conservation and Utilization of Spices, Medicinal and Aromatic Plants*, (Eds.) Sasikumar, B., Krishnamoorthy, B., Rema, J., Ravindran, P.N. and Peter, K.V. Indian Institute of Spices Research, Calicut, pp. 45–53.

Niwata, E., 1995. Cryopreservation of apical meristems of garlic (*Allium sativum* L.) and high subsequent plant regeneration. *CryoLetters*, 16: 102–107.

Novak, F.J. and Havel, L., 1981. Shoot production from *in vitro* cultured flower heads of *Allium porrum* L. *Biol. Plant*, 23: 266–269.

Parani, M., Anand, A. and Parida, A., 1997. Application of RAPD finger printing in selection of micropropagated plants of *Piper longum* for conservation. *Current Science*, 73(1): 81–83.

Peter, K.V., Ravindran, P.N., Nirmal Babu, K., Sasikumar, B., Minoo, D., Geetha, S.P. and Rajalakshmi, K., 2002. *Establishing In vitro Conservatory of Spices Germplasm*. ICAR Project report. Indian Institute of Spices Research, Calicut, Kerala, India, pp. 131.

Philip, V.J, Joseph, D., Triggs, G.S, and Dickinson, N M., 1992. Micropropagation of black pepper (*Piper nigrum* L.) through shoot tip cultures. *Plant Cell Report*, 12: 41-44.

Philip, V.J. and Nainar, S.A.Z., 1986. Clonal propagation of *Vanilla planifolia* (Salisb) Ames. using tissue culture. *Journal of Plant Physiology*, 122: 211-215.

Pratap, K.P., 1992. Artificial Seeds. *Vatika*, 1: 27–30.

Quatrano, R.S., 1968. Freeze preservation of cultured flax cells using DMSO. *Plant*, 43: 2057–2061.

Raghurajan, V., 1997. Micropropagation of turmeric (*Curcuma longa* L.) by *in vitro* microrhizomes. In: *Biotechnology of Spices, Medicinal and Aromatic Crops*, (Eds.) Edison, S., Ramana, K.V., Sasikumar, B., Nirmal Babu, K. and Santhosh, J.E. Indian Society for Spices, p. 25–28.

Rajasekharan, P.E. and Ganeshan, S., 2003. Pollen cryopreservation in Capsicum species: A feasibility study. *Capsicum and Eggplant Newsletter*, 22: 87–90.

Ravindran, P.N., Nirmal Babu, K., Saji, K.V., Geetha, S.P., Praveen, K. and Yamuna, G., 2004. Conservation of Spices genetic resources in *in vitro* gene banks. ICAR Project report. Indian Institute of Spices Research, Calicut, Kerala, India pp. 81.

Ravishankar, G.A., Sudhakar, J.T. and Venkataraman, L.V., 1993. Biotechnological approach of *in vitro* production of capsaicin. In: *Proceedings of the National Seminar on Post Harvest Technology of Spices*, Trivandrum, p. 75–82.

Redenbaugh, K., , D.P., James, W.N., Mary, E.K., Peter, R.V. and Keith, A.W., 1986. Somatic seeds: Encapsulation of asexual plant embryos. *Biotechnology*, 4: 9–83.

Rekha Chaudhury and Malik, S.K., 2004. Genetic conservation of plantation crops and spices using cryopreservation. *Indian Journal of Biotechnology*, 3(3): 348–358.

Sajina, A., Minoo, D., Geetha, S.P., Samsudeen, K., Rema, J., Nirmal Babu, K., Ravindran, P.N. and Peter, K.V., 1997. Production of synthetic seeds in a few spice crops. In: *Biotechnology of Spices, Medicinal and Aromatic Plants*, (Eds.) Edison, S., Ramana, K.V., Sasikumar, B., Nirmal Babu, K. and Santhosh, J.E. Indian Society for Spices, Calicut, India, p. 65–69.

Sharma, T.R. and Singh, B.M., 1995. *In vitro* micro rhizome production in *Zingiber officinale* Rosc. *Plant Cell Reports*, 15(3/4): 274–277.

Sharma, T.R., Singh, B.M. and Chauhan, R.S., 1994. Production of encapsulated buds of *Zingiber officinale* Rosc. *Plant Cell Reports*, 13: 300–302.

Shirgurkar, M.V., John, C.K. and Nadgauda, R.S., 2001. Factors affecting *in vitro* microrhizome production in turmeric. *Plant Cell Tissue and Organ Culture*, 64(1): 5–11.

Rao, S.N.K., Narayanaswamy, S., Chacko, E.K. and Doraiswamy, R., 1982. Regeneration of plantlets from callus of *Elettaria cardamomum* Maton. *Proceedings of Indian Academy of Science (Plant Science)*, 91: 37-41.

Sunitibala, H., Damayanti, M. and Sharma, G.J., 2001. *In vitro* propagation and rhizome formation in *Curcuma longa* Linn. *Cytobios*, 105(409): 71–82.

Towill, L.E. and Walters, C., 2000. Cryopreservation of pollen. In: *Cryopreservation of Tropical Plant Germplasm: Current Research Progress and Application*, (Eds.) Engelmann, F. and Takagi, H. Japan International Research Centre for Agricultural Sciences, Tsukuba, Japan, pp. 115–129.

Tyagi, R.K., Bhat, S.R. and Chandel, K.P.S., 1996. *In vitro* conservation strategies for spices crop germplasm-Zingiber, Curcuma and *Piper* species. In: *Proc. 12th Symp. Plantation Crops*, (PLACROSYM-XII) 27–29 November, Kottayam, India, p. 77–82.

Umetsu, H., Wake, H., Saitoh, M., Yamaguchi, H. and Shimomura, K., 1995. Characteristics of cold preserved embryogenic suspension cells in fennel *Foeniculum vulgare* Miller. *Journal of Plant Physiology*, 146(3): 337–342.

Venkataraman, L.V., Ravishanker, G.A., Sarma, K.S. and Rajasekaran, T., 1989. *In vitro* metabolite production from saffron and capsicum by plant tissue and cell cultures. In: *Tissue Culture and Biotechnology of Medicinal and Aromatic plants*, (Eds.) Kukreja *et al.* CIMAP, Lucknow, India, p. 147–151.

Withers, L.A., 1980. *Tissue Culture Storage for Genetic Conservation*. IBPGR Technical Report. International Board for Plant Genetic Resources, Rome, p. 91.

Withers, L.A., 1985. Cryopreservation of cultured cells and meristems. In: *Cell Culture and Somatic Cell Genetics of Plants, Vol. 2: Cell Growth, Nutrition, Cyto-differentiation and Cryopreservation*, (Ed.) I.K. Vasil. Academic Press, Orlando, Florida, pp. 253–316.

Withers, L.A., 1987. Long-term preservation of plant cells, tissues and organs. *Oxford Surveys of Plant Mol. and Cell Biology*, 4: 221–272.

Withers, L.A., 1991. Biotechnology and plant genetic resources Conservation. In: *Plant Genetic Resources Conservation and Management: Concepts and Approaches*, (Eds.) R.S. Paroda and R.K. Arora. IBPGR, New Delhi, pp. 273–297.

Withers, L.A. and Williams, J.T., 1986. *In vitro Conservation*. International Board of Plant Genetic Resources: Research Highlights 1984–85, IBPGR, Rome, 21 p.

Biodiversity in Horticultural Crops Vol. 1 *Pages* **234–244**
Editors: **Professor K.V. Peter & Dr. Z. Abraham**
Published by: **DAYA PUBLISHING HOUSE**

Chapter 10

Black Pepper

V.A. Parthasarathy, K.V. Saji and K. Johnson George
Indian Institute of Spices Research, P O Marykunnu – 673 012, Calicut

India is one of the 12 centres of origin of cultivated plants figured with two hotspots–the Western Ghats and the Eastern Himalayas, in an identification of 18 biodiversity hotspots carried out in the eighties (Myers 1988). Recently Myers *et al.* (2000) updated the list to 25 hotspots in their revised classification and the two hotspots extended to India are the Western Ghats/Sri Lanka and the Indo-Burma region (covering the Eastern Himalayas).

Phylogenic Relationship of Piperaceae

The genus name *Piper* was derived probably from the Greek name for black pepper, Peperi (Ravindran, 2000). The family name Piperaceae was first used by L.C. Rich in Hamboldt, Bopland and Kunth's *Nova Genera et Species Plantarum* in 1815 (Yunker, 1958). Among the major authorities in the Piperaceae, F. A. W. Miquel was the first one to monograph the family. Miquel (1843) in his *Systema Pipraeaurum* sub divided the family into two tribes *viz.* Piperaceae and Peperomeae with 15 genera and 304 species in the tribe Piperaceae and 5 genera and 209 species in Peperomeae respectively. De Candolle monographed the family Piperaceae in 1869. His work consists of about 1000 species distributed almost equally in *Piper* and *Peperomia* in the *Prodromous.* De Candolle's 50 years work on the revision of *Piperaceae* family was published posthumously in 1923, where he considered only three genera *viz. Piper, Peperomia* and *Verhuellia.* American botanist Trelease continued de Candolle's work after his death for over a period of 30 years along with his colleague Yunker and eventually published a two volume work on Piperaceae of Northern South America (Trelease and Yunker, 1950).

The genus *Piper* shows extreme reduction in floral characters. Flowers are naked, ovary 1 celled and 1- ovuled, stigma sessile and ovule orthotropous. Saururaceae and Chloranthaceae also show such reduced reproductive characters. Rendle (1925) and Hutchinson (1926) considered these three as distinct families. But Bentham and Hooker (1880) merged Piperaceae and Saururaceae into a single family and retained Chloranthaceae as a separate family. Rousseau (1927) suggested to amalgamate these three families into a single family.

Piper Linn., *Peperomia* Ruiz and Pav. and *Lepianthus* Raf (*Pothomorphe* Miq) are the three important genera included in Piperaceae. Datta and Dasgupta (1977) suggested separation of *Peperomia* into a separate family. The orthotropy and basilar position of ovule in Piperaceae show its close affinity to Urticaceae, Loranthaceae, Polygonaceae and even monocotyledonous family Araceae (Rendle 1925). In certain anatomical features, Piperaceae members are related to Ranunculaceae (Datta and Dasgupta 1977). Rendle (1925) considered *Piper* as one of the most primitive dicots. Bessey (1915) and Hutchinson (1926) considered the family as an independent and terminal off shoot of Ranalian ancestry (Lawrence 1951). Met Calfe and Chalk (1950) considered wood of *Piper* and *Peperomia* more specialized than *Chloranthus*. Contrary to this, Rousseau (1927) assumed that the genus *Piper* is the basic group from which *Chloranthus* is derived by minor modification and *Peperomia* by major modification indicating that *Chloranthus* is more specialized.

Phytogeographical distribution shows that *Piper* and *Peperomia* originated probably in tropical America at a comparatively high altitude and dry condition (Ravindran, 2000). Recent studies suggested that the taxa representing major geographic areas could be potentially from three monophyletic groups: Asia, the South Pacific and the Neotropics (Jaramillo and Manos, 2001). *Piper* is more primitive than *Peperomia* (Bentham and Hooker, 1880). According to Datta and Dasguta (1977), evolution of *Piper* and *Peperomia* was related to adaptation with two different conditions. *Piper* adapted gradually with damper and lower lands with elaboration of vascular arrangements. *Peperomia* evolved in drier and higher lands, which required reduction of bundle surface volume ratio, a xerophytic adaptation.

Biodiversity and Distribution of Black Pepper

Piper nigrum belongs to family Piperaceae. The *Piper* species are mostly woody perennial climbers, herbs or shrubs and distributed pantropically. The International Plant Name Index (www.ipni.org) has recorded 6704 *Piper* species in their website. Maximum diversity of *Piper* species occurs in American tropics (700 spp.) followed by Southern Asia (300 spp.), where the economically important species *Piper nigrum* L. (black pepper) and *P. betle* L. (betel vine) originated. (Jaramillo and Manos, 2001). Ridley (1924) documented 75 *Piper* species from Peninsular Malaya region. De Candole (1910) enumerated 133 species of *Piper* occurring in Philippines. Later Quisumbing (1930) conducted systematic study of Philippine Piperaceae and he documented only 87 species from Philippines.

Distribution of *Piper* species ranges from sea level to high ranges of Andes to and the sub Himalayas. (Royle 1839). Trans–Gangetic region and the South Deccan are the two independent centres of origin of the genus *Piper* in India (Hooker, 1886). However, studies conducted by Rahiman (1987) stated that there are three major centers for *Piper* in India. The centers postulated by Rahiman are (1) the sub-Himalayan and North East Indian centre extends from Siwalik range near Pakistan to Mismi hills in the Arunachal Pradesh, through Kumaun, Nepal, Sikkim and Bhutan. The second one *viz.* the Western Ghat centre extends from Vada near Mumbai to Mahendragiri near Kanyakumari through Khandala ghats, Mahabaleswar, Goa, Khanpur, Bababudani hills, Anamalai hills and Cardamom hills. The third centre is the Eastern Ghat extending from Chittoor area in Andhra Pradesh to Puri area of Orrisa. *Piper* species are also distributed in Andaman and Nicobar Islands.

Wild Plant Diversity

The *Piper* species originated from the Southern Asia are of more importance since the economically important species like *P. nigrum* (Black pepper), *P. betle* (Betel vine) *P. longum* (Long pepper), *P. chaba*

(Java long pepper) and *P. cubeba* (Cubeb) originated from this region. The most important species of the genus *P. nigrum* originated from wet humid regions of Western Ghats of South India. The sub–mountainous tracts of Western Ghats are center of origin of black pepper–*Piper nigrum* Linn. About 111 species are of Indian origin (Index Kewensis 1906-1970). Even though ecosystem diversity does not contribute much to biodiversity of black pepper, species diversity and varietal diversity are considerable (Hooker, 1886, Gamble, 1925; Rahiman, 1981; Ibrahim *et. al.*, 1985; Sasikumar *et. al.*, 1999).

Table 10.1: The Species Diversity of *Piper* Occurring in South India

Species	*Remarks*
P. argyrophyllum Miq.	A slender perennial climber occurring at medium elevations. Leaf five ribbed and minutely hairy along the veins
P. attenuatum Buch-Ham.	The common wild pepper occurring in low and medium elevations. Leaves seven ribbed from the base, the outer pair reaching only half to two third of the leaf.
P. barberi Gamble	An endangered species listed in the Red Data Book
P. betle Linn.	The common betel vine.
P. galeatum (Miq) CDC.	A stout woody climber occurs at medium elevations. Leaves thick and berries are bold.
P. hymenophyllum Miq.	Occurs at medium to high elevations. Leaves pubescent.
P. hapnium Ham.	Very rare endangered species occurs at low elevations, tendency to climb over the trees and spikes are erect and larger than long pepper
P. hookeri Miq.	A variant of *P.hymenophyllum*
P. longum Linn.	The common long pepper, distributed all over in India at low elevations. Creeping habit, spikes are erect, cylindrical and fused laterally
P. mullesua Ham.	A slender climber found at high elevations. Spikes erect, globose and fused laterally
P. nigrum Linn.	Occurs at low elevations. Dioecious under natural wild conditions while cultivated forms are monoecious.
P. pseudonigrum Velayudhan	A robust climber with broad leaves with slightly pedicellate flowers, bold berries.
P. schmidtii Hook f	A vigorous climber occurs at high elevations above 1500 m. amsl.
P. silentvalleyensis Ravindran and Asokan	A new species reported from IISR, Calicut. Spikes erect and minute fruits. Only bisexual species reported from Western Ghats.
P. sugandhi Ravindran, Babu and Naik	A stout woody climber resembles *P.nigrum* but having stipitate flowers and cupular bract
P. trichostachyon CDC.	A medium elevation species resembling *P.galeatum*
P. wightii Miq.	Occur at high elevations only. Bracts are peltate in nature

The first major study on Indian *Piper* was done by Hooker (1886) in his *Flora of British India* where he reported 45 species under six sections. He also gave a list of undeterminable species. Rama Rao (1914) in his 'Flowering Plants of Travancore' listed 14 species of Piper from Western Ghats. They are *P. galeatum, P. trichostanchyon, P. longum, P. betel, P. hapnium, P. brachystachyum, P. hookeri, P. nigrum, P. sylvestre, P. hymenophyllum, P. argyrophyllum, P. wightii, P. subpeltatum* and *P. longicaule*. Fisher (1921) described six species from Anamalai hills. Fyson (1932) in his Flora of Nilgiris and Pulney hill-tops reported four species of *Piper* which include *P. schmidtii* and *P. wightii*. The most authoritative floristic study of Western Ghats was that of J.S Gamble (1925), who in his 'Flora of the Presidency of Madras' included 14 species of *Piper* along with keys for identification of each species. Species reported by him are *P. galeatum, P. trichostachyon, P. longum, P. hapnium, P. brachystachyum* (syn. *P. mullesua*), *P. hookeri, P. barberi, P. nigrum, P. hymenophyllum, P. attenuatum, P. argyrophyllum, P. schmidtii, P. wightii and P. betle*. The other floristic enumerations of *Piper* species were those of Santapau (1960), Rao and Razi (1981) Rahiman (1981) and Ravindran (1991).

Recently Ravindran *et al.* (1987) Velayudhan and Amal Raj (1992) and Nirmal Babu *et al.* (1993) reported four new taxa from Kerala. They are *P. silentvalleyensis, P. nigrum* var. *hirtellosum* and *P. pseudonigum* from Silent Valley biosphere reserve and *P. sugandhi* and *P. sugandhi var. brevipilis* from Sugandhagiri cardamom plantations, Wynad, Kerala. Gajurel *et al.* (2000) reported a new record of *P. acutistigmum* C.DC in India which is a native of Myanmar. Gajurel *et al.* (2001a and 2001 b) reported two new species *viz. P. haridasanii*, a rare species and *P. arunchalensis*–a new species endemic to Arunachal Pradesh, which is morphologically distinct from other related taxa but very similar to *P. mullesua* except for the hirsute nature. South Indian species are economically the most important, as they are closely related to cultivated black pepper.

Domesticated Diversity

Cultivar diversity (Tables 10.2–10.4) is one of the principal components of diversity in black pepper. The cultivars are evolved directly from the wild *P.nigrum*. Natural selection and conscious selection by human endeavor for various traits created diversity in cultivars. Most of the vernacular names of black pepper varieties indicate a specific feature of vine such as colour or appearance of vine, leaf shape, spike features or place from which vine originated initially. More than 100 cultivars are known today. About 60-65 of them are prevalent in cultivation. Many of cultivars are known by different names at different places. Some of the local varieties are confined to only in small pockets and are cultivars of less importance. Cultivar diversity is maximum in Kerala followed by Karnataka. Cultivar diversity occurring in Kerala and Karnataka are listed below.

Economically Important *Piper* spp.

Includes important cultivated species like *P. nigrum*, the black pepper, *P. betle*, the betel vine, *P. longum* the long pepper etc. The dried berries, powdered forms and oils and oleoresin extracted from berries are used all over the world. *Piper betle*. L. is another economically important species, widely cultivated in India for its leaves for pan industry. Essential oil from leaves is used in respiratory catarrh and also as an antiseptic (Ravindran, 2000). Plant is considered by the tribal as useful in treating madness, strangulation of intestine and venereal sore (Jain and Tarafder, 1970).

Long pepper (*Piper longum* L.) is used in Ayurvedic and Unani systems of medicine. Long pepper forms an important Ayurvedic preparation such as *Trikadu* (dry ginger long pepper-black pepper) and panchakolam. Both fruits and dried roots are used for medical preparations. The long pepper of commerce appears to be derived from four *Piper* species. Dried spikes of *P. peepuloides* are also sold as long pepper. Spikes of *P. hapnium* and globoid spikes of *P. mulleusa* are also sold as long pepper in the Indian Market. (Sasikumar, 2003). Indonesian or Java long pepper (*P. chaba*) also known as Gaja thippali or Bengla thippali is another variety of long pepper. The species is cultivated in Assam and West Bengal and also in Indonesia/Malaysia. True Indian long pepper is derived from *P. longum*, grown wild in Assam, West Bengal, Nepal, North Eastern Region, Bihar, Uttar Pradesh, Kerala, Tamil Nadu and Andhra Pradesh. It is also cultivated in Kerala, Andhra Pradesh and Maharashtra in a small scale. "Viswam" is an improved variety of *Piper longum* from Kerala Agricultural University, Thrissur, Kerala.

Table 10.2: Black Pepper Diversity Occurring in Kerala

Sl.No	Name of the Cultivar	Remarks
A. Cultivar diversity occurring in Central and Southern Kerala		
1.	Aimpiryan	High yielding, performance excellent in higher elevations, good in quality. But late maturing, vigorous vines.
2.	Arakkulammunda	Moderate and regular bearer, medium in quality, well adapted
3.	Balankotta	Cultivar with large droopy leaves, moderate and irregular bearing.
4.	Chengannurkodi	Moderate yielder from South Kerala, medium in quality.
5.	Cheppakulamundi	Moderate yielder from Central Kerala, medium in quality
6.	Jeerakamundi	Cultivar with small leaves and short spikes, alternate bearing nature, small berries.
7.	Karimunda	Most popular cultivar suitable for most of the black pepper growing areas, high yielder and medium in quality, shade tolerant.
8.	Kottanadan	A high yielding cultivar from South Kerala, drought tolerant type.
9.	Kuthiravally	A cultivar with long spikes, high yield and good quality.
10.	Kuttianikodi	A moderate yielder from Central Kerala with relatively long spikes and good spiking intensity.
11.	Malamundi	A moderate yielder, medium in quality
12.	Narayakodi	Popular in South Kerala, moderate yielder with medium quality. Not easily affected by foot rot.
13.	Neelamundi	A good yielder from Central Kerala medium in quality, tolerant to *Phytophthora* infection
14.	Nedumchola	A cultivar with small leaves and short spikes, moderate yielder.
15.	Neyyattinkaramundi	A cultivar from Central Kerala, medium in quality and yield.
16.	Perumkodi	A cultivar from Central Kerala, moderate in yield and quality.
17.	Poonjaranmunda	A cultivar originally from Central Kerala, sporadically found in gardens of North Kerala. Moderately good in yield and quality.
18.	Thommankodi	A cultivar from Central Kerala, moderately good in yield and quality.
19.	Thulamundi	A Central Kerala cultivar, medium in yield and quality.
20.	Vattamundi	A moderate yielder from Central Kerala.
21.	Vellanamban	Relatively moderate yielder and medium in quality characterized by light green colour of young shoot tip.
B. Cultivar diversity occurring in Northern Kerala		
22.	Cheriyakaniakadan	Popular in North Kerala, moderate and early bearing variety.
23.	Kalluvally	A promising North Kerala cultivar, good yielder, medium in quality with high dry recovery, drought tolerant.
24.	Kottan	A cultivar found in North Kerala, moderate in yield and medium in quality
25.	Manjamundi	A moderate yielder from North Kerala, medium in quality.
26.	Perambramunda	A cultivar from North Kerala, moderate yielder with medium quality.
27.	Vadakkan	A cultivar from North Kerala, medium in quality and yield with relatively large berries.
28.	Valliyakaniyakadan	A cultivar with larger leaves, medium in yield and quality.

Table 11.3: Black Pepper Diversity Occurring in Karnataka

Sl.No	Name of the Cultivar	Remarks
1.	Bilimallegesara	Moderate yielder with light green spikes
2.	Kurimalai	A cultivar from Karnataka, moderate yielder with medium quality.
3.	Karimaratta	A moderate yielder with uniform bearing
4.	Karimalligesara	Moderate yielder with dark green spikes
5.	Malligesara	A common cultivar from Karnataka, relatively good in yield.
6.	Uddagara	A popular cultivar of Karnataka, good in yield and medium in quality.

Table 11.4: Less Important Black Pepper Cultivars of Kerala and Karnataka

Kerala		Karnataka
1. Annarvarayan	24. Karuthakaniakadan	1. Arasinagunda
2. Aralumuriyan	25. Karuthapirimunda	2. Doddige
3. Aranavalan	26. Kothamangalamkodi	3. Doddalae
4. Arimulaku	27. Kumbhakkodi	4. Giddaghere
5. Arikotta	28. Kurialmundi	5. Ghanthuvali
6. Arivally	29. Kuzhivelikodi	6. Kudhirugunda
7. Cheruvally	30. Kottiyanikody	7. Kumpta Local
8. Chettanvally	31. Maramodiyan	8. Nastigunda
9. Chankupazhuppan	32. Marankodi	9. Palulauta
10. Chumalakody	33. Mundi	10. Shimoga Local
11. Cholamalkoti	34. Natesankoti	11. Sullia Local
12. Irumaniyan	35. Orumaniyan	12. South Kanara Local
13. Kallubalankotta	36. Padappan	13. Sagar Local
14. Kanjirakkodan	37. Oerumkarimunda	14. Vokkalgunje
15. Kaplanganmundi	38. Pirimundi	15. Vokkalu
16. Karimkodi	39. Thekkan	
17. Karimkotta	40. Thottamundi	
18. Kumbanadan	41. Uthirankota	
19. Kottaram	42. Vellamundi	
20. Malabar	43. Veluthakaniyakadan	
21. Malanadan	44. Velutha pirimunda	
22. Karivilanchi	45. Wayanadan	
23. Karivally	46. Yohannankodi	

Cubeb or tailed pepper (*Piper cubeba* L) cultivated in Indonesia and Malaysia has high therapeutic value. Fruits and oil extract of *P. cubeba* are used in treatment of asthma. Kawa pepper- *P. methysticum* G. Frost- a native of South Pacific Islands is used to make a non-alcoholic but narcotic drink, Kawa–a ceremonial drink among Polynesians since ancient period (Ravindran 2000). Roots of *P. sylvaticum*

Roxb are used widely in indigenous medicine as an effective antidote against snake poison. (Banerji and Dhara, 1974). In Peninsular Malaysia, decoction made from leaves of *Piper sarmentosum* Roxb. is used for curing pains in bones and applied to foreheads of children suffering from head aches. *Piper umbellatum* (stomach ache, cough) and *P. poryphyrophyllum* (leprosy, stomach ache and skin diseases) are the other two *Piper* species used in Peninsular Malaysia (Tawan *et al.*, 2002). *Piper kadsura* (Choisy) Ohwi, a native of Taiwan is used for treatment of asthma and arthritis in Taiwan (Linn and Lu, 1996). Recent studies indicated that this plant has inhibitory properties on Hepatitis B (Ray-Ling *et al.*, 2001)

Besides these economically important species, there are a few *Piper* species having ornamental value because of attractive foliage. *P. auritum, P. borneense, P. decurrens, P. magnificum, P. metallicum, P. ornatum, P. porphyrophyllum, P. rubronodosum* and *P. rubrovenosum* are valued for ornamental purpose (Mark Greiffith, 1992).

Conservation

Even though, early movement of settlers across length and breadth of Kerala helped spread of landraces (cultivars) to new areas, advent of improved high yielding black pepper varieties is becoming a threat to many of the old cultivars. Depletion of forest area, rampant destruction of forest trees and undergrowth, and changes in agro-ecological conditions all have led to sharp decline in populations of wild pepper and their related taxa. Species of *Piper* that are the most affected by deforestation are *P. barberi, P. hapnium, P. silentvalleyensis, P. wightii and P. schmidtii.* Many of the taxa are now confined to only a few specific locations and may soon be extinct, if not collected and conserved. Concerted efforts were made by Indian Institute of Spices Research, Calicut, All India Coordinated Research Project on Spices, National Bureau of Plant Genetic Resources, Regional station, Trissur and Kerala Agricultural University to collect and conserve black pepper and its wild relatives.

In-situ Conservation

In-situ conservation of biodiversity is safer and more desirable than *ex-situ* conservation in case of wild relatives. Here species conserved is subjected to co-evolve with environmental changes.

It is important that sanctuaries of wild relatives of crop species and other economically important species are established in the Western Ghat and North Eastern region, considered to be centers of origin of *Piper* species in India.

In India, the first gene sanctuary was established in Garo hills in Meghalaya for *Citrus* (Bhag Singh, 1981). Velayudhan *et al.* (1999) suggested establishment of gene sanctuary for *Piper* species at Silent Valley since diversity is more here. Based on surveys conducted and distribution studies conducted at IISR (Utpala *et al*–Un published) the Western Ghat region spreading from Waynad to Palaghat District and Kollam, Tirunelvelli and Thiruvananthapuram district are hotspots for *Piper* in South India. It will be appropriate to establish *Piper* gene sanctuaries in these places. Endangered species like *P. barberi* is found only in the Agasthymala hills and Thirunalvelli area. Wild species like *P. schmidtii* and *P. wightii* are distributed only in Nilgiris, Kodaikanal and Munnar (above 1500 m msl). For conserving these high elevation species, a gene sanctuary has to be established in Nilgiri forests under protected conditions.

Ex-situ Conservation

Field Genebank

Systematic surveys of all pepper growing areas, forests of Western Ghats and parts of North Eastern regions were conducted to collect available variability of cultivated forms and wild relatives and these accessions are being maintained at black pepper germplasm conservatory at IISR, Calicut as field genebank and in the nursery. Presently, IISR is in a unique position of having the largest germplasm collection of black pepper in the world consisting of 2257 accessions (Table 11.5). Besides these accessions, over 1200 hybrids and 120 OP- lines are also maintained.

Table 11.5: Black Pepper Germplasm Conserved at IISR, Calicut

Wild accessions	1060
Black pepper (cultivated)	1188
Exotic collections	9
Total	**2257**

Conservation of germplasm is done by adopting following strategies:

1. In the nurseries, each accession is trailed on bamboo splits in serial order and are under continuous multiplication
2. In the clonal repository, 10 rooted cuttings of each line are maintained
3. In the field genebank, the accessions are planted for evaluation
4. In the *in-vitro* genebank, important accessions are conserved using tissue culture tools.

This four pronged conservation strategy is used because of threat of disease and pests (eg. *Phytophthora* foot rot and nematodes).

Besides the germplasm collections of IISR, Calicut a few accessions of black pepper germplasm are also maintained under All India Coordinated Research Project on Spices at various places (Table 11.6).

Table 11.6: Black Pepper Germplasm Maintained at AICRPS Centers

Sl.No	*Name of the centre*	*Germplasm holding*
1.	Pepper Res. Station, Panniyur, Kerala	167(147 + 20)
2.	Pepper Res. Pln. Sirsi	110
3.	Chinthapalli	58
4.	Dapoli	66
5.	Yercaud	106
6.	Pundibari	9

Source: Project Coordinator Cell, IISR, Calicut

In-vitro Conservation

Germplasm conserved *in-situ and ex-situ* by means of field genebank are susceptible to many threats like natural calamities and disease epidemics. For safe guarding the germplasm as an alternative source, *in- vitro* conservation of important *Piper* species are attempted at IISR Calicut. The slow growth technologies for in-vitro conservation have been standardized and at present the *in-vitro* genebank is holding 35 accessions including exotic collections.

Besides IISR Calicut, the NBPGR New Delhi is also conserving *Piper* species (7 Nos) (Ann. Report-2003) in their national conservatory.

The current status of species biodiversity assessment and conservation indicates that vast areas of the North East are yet to be fully explored to get an understanding of the variability available and the emergency to collect and conserve them. National Bureau of Plant Genetic Resources (NBPGR) Regional Station, Madakkathara, Thrissur has to expedite this activity and document the variability available. Concurrently conserve the resources collected or transfer them to the respective commodity research institutes for further evaluation and use in crop improvement. These explorations also will bring to light the species still underexploited that may find way in local cuisine or medicine. Besides, introduction of exotic germplasm with desirable features has to be envisaged for improvement of this spice and to meet global demand.

References

Bentham, G. and Hooker, J. D. 1880. *Genera plantarum*. 3 (1): L. Rewe and Co. London.

Bessey, C. 1915. The phylogenetic taxonomy of flowering plants. *Ann. Mo. Bot. Gard.* 2: pp.109-164.

Bhagh Singh. 1981. Establishment of first gene sanctuary in India for *Citrus* in Garo Hills, Concept Publishing company.

Data, P. C. and Dasgupta, L. 1997. Root anatomy and distribution of commom *Piper* and *Peperomia* species *Geobios* 4: 143-146.

De Candole, C. 1910. A revision of Philippine Piperaceae. *Philippine J. Science.5 Bot.* 405-463.

Fyson, P.F. 1932. The Flora of South Indian Hill Stations- Ootacamund, Coonur, Kotagiri, Kodaikanal, Yercaud and Country around. Vol. I. Madras. Pp. 198-200.

Gajurel, P. R., Rethy, P. and Kumar, Y. 2000. *Piper acutistigmum* C.DC. (Piperaceae) a new record for India. *Rheedea*, 10 (2): 139-141.

Gajurel, P. R., Rethy, P. and Kumar, Y. 2001a. *Piper haridasanii*- a new species of *Piper* (Piperaceae) from Arunachal Pradesh, North Eastern India. *J. Econ. Taxon. Bot.* 25 (2): 293-296.

Gajurel, P. R., Rethy, P. and Kumar, Y. 2001 b. A new species of *Piper* (Piperaceae) from Arunachal Pradesh, North Eastern India. Botanical Journal of the Linnean Society. 137 (4). pp 417-419.

Gamble, J. S. 1925. Flora of the Presidency of Madras. Vol. 2, Botanical Survey of India, Calcutta (reprint 1961).

Hooker, J.D. (1886) *The Flora of British India*. Vol V, L. Reeve and Co. London (Rep.): 78-95.

Hutchinson, J. 1926. The families of flowering plants-I. Dicotyeodons. London, pp. 328.

Ibrahim, K. K., Sukumara Pillai, V. and Sasikumaran, S. 1985. Variability, heritability and genetic advance for certain quantitative characters in black pepper. *Agric. Res. J. Kerala* 23: 45-48.

Index Kewensis. 1895. Vol. 2, The Clarenden Press, Oxford, pp. 534-542. Supplements 1-14 (1906-1970).

Jaramillo, M. A. and Manos, P. S. 2001. Phylogeny and patterns of floral diversity in the genus *Piper* (Piperaceae). *American J. botany*, 88(4): 706-716.

Lawrence, G. H. M. 1951. Taxonomy of vascular plants (Indian ed. 1967) Oxford and IBH Publishing Co., New Delhi. pp. 444-445.

Linn, T. T and Lu, S. Y. 1996. Piperaceae. (In): Flora of Taiwan. 2nd ed. Editorial committee of the flora of Taiwan. Vol. II. 624-631.

Linnaeus, C. 1753. *Species Plantarum* Vol.1: 28-30.

Mark Greiffith. 1992. Index to Garden Platns. Mc Millon. P. 900.

MetCalfe, C.R. and Chalk, L. 1950. *Anatomy of the Dicotyledons*, Clarendon Press, Oxford.

Miquel, F.A.W. 1843. Systema Piperacearum. Rotterdam, The Netherlands.

Myers, N. 1988. Threatened biotas: hotspots' in tropical forests. *Environmentalist*, 8: 187-208.

Myers, N., Mittermeier, R. A., Mittermeier, C. G., da Fonseca, G. A. B. and Kents, J. 2000. Biodiversity hotspots for conservation priorities. *Nature*, 403: 853-858.

Nirmal Babu, K., Naik, V.G., and Ravindran, P.N. 1993. Two new taxa of *Piper*, Piperaceae from Kerala, India with a note on their origin and interrelationship. J. Spices Aromatic crops 2: 26-33.

Quisumbing. 1930. Philippine Piperaceae. *Philippine J. Science.* 43 (1): 1-246.

Rahiman, B. A. 1981. Biosystematics studies in varieties and species of *Piper* occurring in Karnataka region. *Ph.D. Thesis*, University of Mysore, Mysore.

Rahiman, B. A. 1987. Geographical distribution of the genus *Piper* Linn. in India. *Curr. Sci.*, 56 (13): 666-668.

Rahiman, B. A. and Nair, M. K. 1983. Morphology, cytology and chemical constituents of *Piper* species from India. *J. Plantation Crops*. 11(2): 72-90.

Rama Rao, M. 1914. Flowering Plants of Travancore, Govt. Press. Thiruvanathapuram. Pp. 336-338 (reprint).

Rao, R. R. and Razi, B. A. 1981. A synoptic flora of Mysore district., New Delhi.

Ravindran, P.N. 2000. Other economically important species of *Piper*. (In) Black pepper *Piper nigrum* (Ed. P. N. Ravindran). Harwood academic publishers pp 497-509.

Ravindran, P. N. 1991. Studies on black pepper and some of its wild relatives. Ph.D thesis, Univ. Calicut.

Ravindran, P. N., Nair, M.K. and Nair, R.A. 1987. New taxa of *Piper* (Piperaceae) from Silent valley Forest, Kerala, J. Eco. Tax. Bot., 10: 167-169.

Ravindran, P. N., Nirmal Babu, K., Sasikumar, B. and Krishnamuthy, K. S. 2000. Botany and Crop improvement of black pepper. In. Ravindran, P. N. *Black Pepper*, Harwood academic publishers, Singapore. pp 23-142.

Ray-Ling Huang, Chieh-Fu Chen, Hui-Yi Feng, Lie-Cheven Lin and Cheng-Jen Chou. 2001. Anti hepatitis virus of seven compounds isolated from *Piper kadsura* (Choisy) Ohwi. J. China Med. 12 (3): 179-190.

Rendle, A. B. 1925. The classification of flowering plants. Vol. II. *Dicotyledons*. University press, Cambridge. pp. 88-92.

Ridley, H. N. 1924. The flora of Malaya Peninsula. Vol. 3. pp. 25-51.

Rocha, S.F.R. and Lin Chau Ming. 1999. *Piper hispidinervum*: A sustainable source of safrole. Pp-479-481. (In) J. Janick (ed) Perspectives of new crops and uses. ASHS Press, Alexandria, VA.

Rousseau, D. 1927. Contribution a 'l' anatomie compare'e des Piperaces. Acad. Roy. Belgique cl. Sci. mem. 9: 1-45.

Royle, J. F. 1839. Illustration of botany and other branches of the Natural History of the Himalayan mounts and of the flora of Cashmere. Vol. 1 (Repr. 1970) Today and Tomorrow, printers and publishers, Delhi. 331-333.

Santapau, H. 1960. The flora of Khandala on the Western Ghats of India. Rec. Bot. Surv. Ind., 16(1): 256-257.

Sasikumar, B. 2003. Long pepper for forever young. Spice India. Aug. 2003, p-2-3.

Sasikumar, B., Krishnamoorthy, B., Saji, K. V., Johnson George, K., Peter, K. V. and Ravindran, P. N. 1999. Spices diversity and conservation of plants that yield major spices in India. *Plant Genetic Resource Newsletter* No. 118. pp: 19-26.

Tawan, C. S., Ipor, I. B., Fashihuddin, B. A. and Sani, H. 2002. A brief account on the wild *Piper* (Piperaceae) of the Crocker Range, Sabah. ASEAN Review of Biodiversity and Environmental Conservation (ARBEC), July- September.

Trelease, W. and Yuncker, T.G. 1950 *The Piperaceae of Northern South America.* Vol.1 Univ. Illinois, USA.

Velayudhan, K.C. and Amalraj, V.A. 1992 *Piper pseudonigrum*- a new species from Western Ghats. *J. Eco. Tax. Bot.*, 16: 247-250.

Velayudhan, K.C., Amalraj, V. A, Abraham, Z., Joseph John, K., Abdul Nizar, M and Asha. K.I. 1999. Wild Crop genetic resources of silent valley with special reference to *in-situ* conservation of *Piper* species. Silent Valley–Whispers of reason. Kerala forest department, p. 217-223.

Yuncker, T.G. 1958. The Piperaceae: A family profile. *Brittonia*, 10: 1-17.

Biodiversity in Horticultural Crops Vol. 1 *Pages* **245–255**
Editors: **Professor K.V. Peter & Dr. Z. Abraham**
Published by: **DAYA PUBLISHING HOUSE**

Chapter 11

Ginger and Turmeric

B. Sasikumar
Indian Institute of Spices Research, P O Marikunnu – 673 012, Calicut

Biodiversity of ginger and turmeric is concentrated in South Asia, South East Asia, Pacific Islands, Central and South Americas, and African regions. Cultivated diversity is maximum in South and South East Asian countries.

India is known as 'Land of Spices'. Since time immemorial a variety of spices are grown in this country. India is a world leader in production and export of ginger and turmeric. In fact, 'Alleppey turmeric', 'Rajapuri turmeric', 'Cochin ginger' 'Wyanadan ginger' etc. are synonymous with superior quality in these spices.

Ginger: Origin, Producing Countries and Systematics

Ginger (*Z. officinale* Rosc.) originated in South East Asia and is under cultivation since ancient times in India and China. Incidentally, centre of diversity of ginger is also in this region. Though a putative wild type of *Zingiber officinale* is collected from Western Ghats of India, its occurrence in wild habitat is yet to be confirmed. (Sasikumar *et al.*, 1995).

At present ginger is cultivated in India, China, Jamaica, Taiwan, Sierra Leone, Nigeria, Fiji, Mauritius, Indonesia, Brazil, Costa Rica, Ghana, Japan, Malaysia, Bangladesh, Philippines, Sri Lanka, Soloman Islands, Thailand, Trinidad and Tobago, Uganda, Hawaii, Guatemala and Australia.

Ginger (*Zingiber officinale* Rosc.) belongs to family Zingiberaceae and natural order Scitamineae.

Baker (1894) recognized 4 sections in Zingiberaceae and *Zingiber officinale* is included in Section II *Lampuzium* Haron. Holttum (1950) included *Zingiber* in tribe *Hedycheae* along with other genera like *Curcuma, Hedychium* and *Kaempferia.* Burtt and Smith (1983), disagreed with Holttum and proposed *Zingiber* as a separate tribe.

Species Diversity

Though *Z. officinale* Rosc. is the most important member of genus, economically there are other species of medicinal and ornamental value (Table 11.1) (Sabu and Dave Skinner, 2005).

Table 11.1: Economically Important Species Diversity in *Zingiber*

Species	*Uses*	*Remark*
Zingiber officinale Rosc.	Spice and condiment, medicinal use	Cultivated in India, Africa, Carribans and South East Asian countries
Zingiber mioga Rosc.	Vegetable. Shoot and flowers are edible	Japanese or 'myoga ginger'
Z. montanum (Keonig) Link ex Diet (Syn. *Z. cassumunar* Roxb.)	Used in traditional medicine	Found in India, Sri Lanka and South East Asian countries
Z. zerumbet (L.) Smith	Medicinal/ornamental use	'Shampoo ginger', common in tropical Asia and USA
Z. amaricanus Bl.	Medicinal use	Grown widely in USA occurs in the teak forests of South East Assia.
Z. aromaticum Val.	Ornamental/medicinal/flavouring	'Puyang ginger' of Indonesia, native of tropical Asia.
Z. argenteum J. Mood and I. Theilade	Ornamental	Endemic to Malaysia
Z. bradleyanum Craib	Ornamental	Cultivated in USA
Z. chrysanthum Rosc.	Ornamental	-
Z. citriodorum J. Mood and I. Theilade	Ornamental	Common in Thailand and USA
Z. clarkii King ex Benth	Ornamental	Native of Sikkim, India, lateral spikes.
Z. collinsii J. Mood and I Theilade	Ornamental	Common in USA and Europe
Z. corallinum Hance	Chinese medicine/ornamental	-
Z. eborium J. Mood and I. Theilade	Ornamental	'White ginger' or 'ivory ginger' of Indonesia, now common in the West
Z. griffithii Baker	Ornamental	Malaysian species
Z. gramineum Noronha	Ornamental	'Palm ginger' cultivated in USA
Z. junceum Gagnepain	Ornamental	'Yellow Delight' common in USA
Z. lambi J. Mood and I Theilade	Ornamental	Malaysian species
Z. longipedunculatum Ridley	Ornamental	Cultivated in Australia
Z. malaysianum C. K. Lin	Ornamental	'Midnight beauty', common in USA
Z. neglectum Valet	Ornamental	Common in USA
Z. niveum J. Mood and I. Theilade	Ornamental	Common in USA
Z. ottensii Valet	Medicinal/ornamental	Common in South East Asia
Z. pachysiphon B. L. Burtt and R. M. Sem	Ornamental	Cultivated in Australia
Z. rubens Roxb.	Medicinal/ornamental	Indo-Malaysian species introduced to USA
Z. spectabile Griff.	Ornamental/medicinal	Malaysian species common in USA
Z. vinosum J. Mood &.I. Theilade	Ornamental	Malaysian species introduced to USA

Baker (1894) described a total of 24 species from Indo-Malayan region. Gamble (1925) recorded seven species from South India such as *Z. officinale* Rosc., *Zingiber roseum* Rosc., Z. *nimmonii* Dalz., Z. *wightianum* Thwaites, Z. zerumbet (Lin.) Smith and Z. *neesanum* (Grah.) Ramamoorthy Syn. Z. *pupurem* Rosc. Kumar (1999) reported *Zingiber clarkei* is endemic to Sikkim Himalayas. Other economically important (medicinal) *Zingiber* species found in Sikkim are Z. *capitatum* var. *elatum* and Z. *chrysanthum* etc. (Kumar, 1999).

Cultivar Diversity

Good variability in cultivated ginger is in India and China. In both these countries, most of land races are known after their place of domestication and hence all of them may not be distinct genetically (Tables 11.2 and 11.3). Kerala, Orissa, Meghalaya, Arunachal Pradesh, Karnataka, West Bengal and Andhra Pradesh are the leading Indian states in ginger production totaling to 70000 ha. in area. In China, ginger is grown in about 13450 ha. In India good variability for yield and quality of ginger is encountered in Kerala and North East India (Table 11.4). Though no sexual reproduction is reported in ginger, geographical spread accompanied by genetic differentiation into locally adapted population augumented by mutation could be the main factor responsible for diversity in this clonally propagated crop. Early movement of settlers across length and breadth of Kerala and story of shifting cultivation in North Eastern India are well documented sociological events. (Sasikumar *et al.*1999).

Table 11.2: Cultivar/Varietal Diversity in Ginger in India

Sl.No.	*Cultivar/Variety*	*Sl.No.*	*Cultivar/Variety*
1.	Anamika	26.	Rajagarh Local
2.	Assam	27.	Rejatha*
3.	Arippa	28.	Rio-de-Janeiro+
4.	Bajpai	29.	Sargi guda
5.	Burdwan	30.	Saw Thing laidum
6.	China+	31.	Singhihara
7.	Ernad-chernad	32.	Seirra Leone+
8.	Ernad Manjeri	33.	Suprabha*
9.	Ellakallan	34.	Suruchi*
10.	Edappalayam	35.	Suravi*
11.	Himachal	36.	Saying makhim (Pink ginger)
11.	Himgiri*	37.	Taiwan+
13.	Jorhat	38.	Taffingiva+
14.	Juggigan'	39.	Thang-chang
15.	Jamaica+	40.	Thingpui
16.	Karakkal	41.	Thing laidum
17.	Kuruppampady	42.	Thodupuzha
18.	Kunduli Local	43.	Tura
19.	Maran	44.	Uttarpradesh
20.	Mahim	45.	Valluvanad
21.	Mahima*	46.	Varada*
22	Mananthody	47.	Vengara
23	Nadia	48.	Wyand Local
24	Narasapattam	49.	Wynad Kunnmangalam
25	Poona	50.	Zahirabad

+: Exotic; *: Improved variety

Table 11.3: Cultivar Diversity of Ginger in China

Category	*Variety/Cultivar*
Sparse seedling/big fleshy rhizome	Gandzhou, Shandong Laiwu
Dense seedling/fleshy type	Guangzhow, Zhejiang
Edible/Medicinal	Fujian red bud, Hunan yellow heart, chicken claw ginger, Xingguo ginger
Edible/processed	Guangzhou (fleshy type), Fuzhou giner (Purple ginger), Tongling (White ginger), Fujian bamboo ginger, Zunyi (big white ginger), Leifeng ginger.
Ornamental ginger	Laishe ginger
	Flower ginger
	Tea ginger
	Strong ginger
	Hengchum ginger
	Hekou ginger
Other types	Zaoyang (Hubei province)
	Zunji Big (Giuzhou)
	Chenggu Yellow (Shaxi)
	Yulin round Fleshy (Guangxi)
	Bamboo root ginger (Sichuan)
	Mianyang (Sichuan)
	Xuanchang (Ahuii)
	Yuxi yellow (Yunnan)
	Laiwu Slice ginger (Shandong)
	Yellow claw (Zhejiang)
	Taiwan fleshy (Taiwan)

Source: Ravindran *et al.*, 2005.

In West Africa two main types of ginger are grown (Tindall,1968).Graham (1936) reported that 5 kinds of ginger are recognized in Jamaica *viz.* St. Mary, Red Eye, Blue Turmeric, Bull Band China Blue. Ridley (1912) reported 3 forms of ginger from Malaysia *viz.*'halyia betle'(true ginger) 'halyia udang' (red ginger) and 'halyia bara' (yellow ginger). A red variety of ginger, *Z. officinale* var. *rubra* (pink ginger) and a'withered skin variety' are also reported from Malaysia. Rosales (1938) reported two cultivars from Philippines *viz.* 'Native' and Hawaiin'. 'Taffingiwa' is the common ginger in Nigeria, next important being 'Yasun Bari' (black ginger). (Ravindran *et al.*, 2005). Ginger is cultivated in about 14500 ha. in Nigeria.

In Japan, ginger cultivars are grouped into three types based on their stature (1) dwarf statured, early maturing, profuse tillering types with small rhizomes *e.g*:'Kintoki; (2) medium statured plants with moderate tillers and medium size rhizome *e.g* 'Sanshu' and (3) tall plants with big rhizome and less tillers *e.g* 'Oshoga' (Ravindran *et al.*, 2005). Sanshu (4x Sanshu) and Buderim Gold are tetraploid ginger varieties grown in Japan (Adaniya and Shirai 2001) and Queensland, respectively. Ginger is grown in about 9900 ha. in Indonesia.

Table 11.4: Ginger Cultivars Identified for Various Uses

End use	*Cultivar*
High yield (fresh)	Rio-de-Janeiro, Suprabha, Varada, Mahima, Rejatha, Juggigan, Himachal, Himgiri, Maran, Wynad Local, Karakkal, Thingpui, Nadia, Suruchi.
Bold rhizome	Mahima, Rejatha, Varada, Gorubathani, Bhaise, Suravi
High dry recovery	Varada, Mahima, Zahirabad, Jorhat Local, Kuruppampadi, Ernad, Chernad, Mowshom, Rejaha, Maran.
Low fibre	Varada, China, Nadia, Poona, Zahirabad, Mahima.
High oleoresin	Wynad Kunnamangalam, Ernad Chernad, Nadan (Pulpally), Nadan, Himachal, China, Rio-de-Janerio, Maran.
High gingerol & shogaol	Wyand, Baharica, Rio-de-Janeiro, Kunduli, Kunnamangalam, Ambalavayalan, Ernad Chernad, Swathing Pui, Maran, Jorhat Local.
High essential oil	Pulpally, Sabarimala, Valluvanad, Rio-de-Janerio, Nadan (Pulpally), Thodupuzha, Himgiri, Ellakallan, Mananthody
High zingiberene	Baharica, Amaravathy (6) gingerol
Salted ginger	Rejatha and Varada
Ginger candy	Varada

Nepal is another country rich in cultivar diversity of ginger. Ginger is grown in about 6082 ha. in Nepal under different agro-ecological situations. Most of Nepalese ginger cultivars are bold types, known after their locality of cultivation like 'Tehratum', 'Salyan', 'Naval Parasy', 'Palpa', 'Bhaktapur, 'Bhojpur' and 'Illam' (Sasikumar, 2000). Though about 50 ginger cultivars are known in Nepal, some of them may not be genetically distinct as these cultivars are recognized after their place of cultivation, mainly.

Propagation

Ginger is propagated through rhizome saved from previous harvest.

Turmeric

Genus *Curcuma* is gaining global importance as a potential source of new drug(s) to combat a variety of ailments as the genus contains molecules credited with many therapeutical, insect repellant and aromatic properties.

Origin and Producing Countries

Curcuma is mainly of South Asian origin and centre of diversity of *Curcuma* also falls with in this region. Turmeric though grown in India, Pakistan, Malaysia, Myanmar, Vietnam, Thailand, Philippines, Japan, China, Korea, Sri Lanka, Nepal, South Pacific Islands, East and West Africa, Malagasi, Carribean Islands and Central America, India is the major producer and exporter of this spice. Turmeric is grown in a bout 1.5 lakh ha. in India. Andhra Pradesh, Orissa, Tamil Nadu, Meghalaya, West Bengal and Maharashtra are the major turmeric producing states in India.

Species Diversity of *Curcuma* and Systematics

Curcuma species widely differ in their habitat. Habitat ranges from sea level (sandy costal) to high altitude as above 200 m in Western Ghats and Himalayas. In India, *Curcuma* species diversity is concentrated in North- Eastern India, South India and Andaman and Nicobar Islands. Other countries rich in *Curcuma* species diversity are China, Thailand, Malaysia, Philippines and Vietnam. Economically important *Curcuma* species are given in Table 11.5.

Table 11.5: Economically Important *Curcuma* spp.

Species	*Uses*
C. *longa* L. syn C. *domestica* Val.	Spice, medicine, dye, religious local delicacies, insect repellent, aroma therapy and perfume.
C. *amada* Roxb, C.*mangga* Val.& Zijp.	Spice, medicine, pickles and salads
C.*zedoaria* Roxb.	Folk medicine, arrow root industry
C.*ochrorrhiza* Val and Van Zijp	Malayan traditional medicine
C *pierreana* Gagnep.	Vietnamese traditional medicine
C. *aromatica* Salsb.	Medicine, toiletry articles, insect repellent
C. *kwangsiensis* S. G. Lec and C.F.Liang syn. C. *chuanyujin*, C. *phaeocaulis* Val.	Chinese traditional medicine
C. *caesia* Roxb.	Spice and medicine
C. *comosa* Roxb.	Traditional medicine of Thailand
C. *angustifolia* Roxb, C. *zedoaria* Roxb. C. *caulina* F. Grah, C.*psuedomontana* F. *Grah.* C. *montana* Roxb, C. *rubescens* Roxb. C. *leucorrhiza*, C. *xanthorrhiza* Roxb C. *decipiens* Dalz. C.*malabarica* Vel *et al.* C. *raktakanta* Mangaly and Sabu, *C. haritha*	Arrow root industry
C. aeruginosa Roxb.	
C. alismatifolia Gagnep, *C. thorelii*, C. *roscoeana* Wall.	Ornamental (Cut flower)

Source: Sasikumar (2005).

Though about 100 *Curcuma* species are proposed globally, it is now recognized that some of the species may be synonyms such as *C. zedoaria* syn. *C. xanthorrhiza* (India), *C. albicoma* syn *C. sichunanensis; C. chuanyujin* syn. *C. kwangsiensis, C. wenyujin* syn. *C. aromatica*(China) etc. (Sasikumar 2005). Velayudhan *et al.* (1999) recorded a total of 117 species (Table 11.6), obviously including synonymous entities as separate species.

Genus *Curcuma* belonging to Hedycheae contains about 80 species out of which 41 are distributed in India including atleast 10 spices endemic to India.

Velayudhan *et al.* (1996) proposed a new conspectus of genus in India. These authors found that Indian species can be accommodated in two subgenera of Valeton (1918) *viz. Eucurcuma* and *Paracrucuma. Eucurcuma* contains 3 main sections namely *tuberosa, nontuberosa* and *stolonifera* based on presence or absence of tubers and stolons, each section again encompassing one or more subsections based on floral, vegetative and rhizome characters. Under sub genus *Paracurcuma*, only the anther spurs lacking species *viz. C. aurantiaca* and *C. ecalcarata* are included. Velayudhan *et al.* (1999) also proposed existence of 21 distinct morphotypes in *Curcuma longa*. These 21 morphotypes are grouped into six taxonomic varieties namely *C.longa* var. *typica, C. longa* var. *atypical, C. longa* var. *camphora, C. lona* var *spiralifolia, C. longa* var. *musacifolia* and *C.longa* var *platifolia*. Most of the Indian morphotypes belong to C. *longa* var. *atypia*.

Table 11.6: Species Diversity of *Curcuma*

Sl.No.	*Species*	*Origin*	*Sl.No.*	*Species*	*Origin*
1.	*Curcuma aeruginosa* Roxb.	Myanmar	39.	*C. purpurescens* Blume.	Java
2.	*C. albiflora* Thaw.	Sri Lanka	40.	*C. reclinata* Roxb.	India
3.	*C. amada* Roxb.	India	41.	*C.roscoeana* Wall.	Myanmar
4.	*C. amarissima* Rosc.	India	42.	*C. rotunda* Linn.	
5.	*C. americana* Lam.		43.	*C. rubescens* Roxb.	India
6.	*C. angustiflolia* Dalz and Gibs.	India	44.	*C. rubricaulis* Link	
7.	*C. angustifolia* Roxb.	India	45.	*C. strobilifera Will.*	Myanmar
8.	*C. aromatica* Salisb.	India	46.	*C. strobinia* Wall.	India
9.	*C. attenuata* Wall.	Myanmar	47.	*C. sumatrana* Miq.	Sumatra
10.	*C. australasiaca* Hook. f.	Australia	48.	*C. viridiflora* Roxb.	Sumatra
11.	*C. caesia* Roxb.	India	49.	*C. xanthorrhiza* D. Dietr.	Malaya
12.	*C. careyana* Wall.	India	50.	*C. xanthorrhiza* Roxb.	
13.	*C. caulina* F. Grah.	India	51.	*C. zedoaria* Rosc.	India
14.	*C. comosa* Roxb.	Myanmar	52.	*C. zedoria* Roxb.	
15.	*C. cordata* Wall.		53.	*C. zerumpet* Roxb.	
16.	*C. cordifolia* Roxb.		54.	*C. bakertiana* Hunsl.	New Guinea
17.	*C. decipiens* Dalz.	India	55.	*C. coccinia* Wall.	
18.	*C. alata* Roxb.	Myanmar	56.	*C. sylvestris* Redl	Malayasia
19.	*C. erubescens* Wall		57.	*C. urceolata* Rich.	
20.	*C. ferrugenia* Roxb.	India	58.	*C. ceratotheca* Schum.	Celebes
21.	*C. glaucophylla* Wall.		59.	*C. sessilis* Gag.	Burma
22.	*C.grandiflora* Wall.		60.	*C. sparganifolia* Gagnep.	Combodia
23.	*C. latifolia* Rosc.	India	61.	*C. stenochyla* Ganep.	Cochin China
24.	*C. longa* Linn.		62.	*C. cochinchinensis*	Indochina
25.	*C. longiflora* Salisb.	India	63.	*C. harmnadi* Gagnep.	Cochin China
26.	*C. montana* Roxb.	India	64.	*C. lanceolata* Ridl.	
27.	*C. musacea* Wall.	India	65.	C. *leopoldii* Hot.	
28.	*C. neilgherrensis* Wight	India	66.	*C. pierreana* Gagnep.	Indochina
29.	*C. officianalis* Salisb.		67.	*C. singularis* Gagnep	Indochina
30.	*C. oligantha* Trim.	Sri Lanka	68.	*C. thorelii*	Indochina
31.	*C. ornata* Wall.	Myanmar	69.	*C. trichosantha* Gagnep.	Indochina
32.	*C. pallida* Lour.	China	70.	*C. heyneana* Val. and Zijp.	Java
33.	*C. parviflora* Wall.	Myanmar	71.	*C. loerzingii* Val.	Sumatra
34.	*C. petiolaris* Hort.		72.	*C. longi-spica*, Val.	N. Guinea
35.	*C. petiolata* Roxb.	Myanmar	73.	*C. mangga* Val. and Zijp.	Java
36.	*C. spicata* Wall.	Myanmar	74.	*C. ochrorhiza*, Val. and Van Zijip.	Java
37.	*C. porphyrotannica* Zijp.	Timor	75.	*C. phaeocaulis*	Java
38.	*C. pseudomontana* F. Grah		76.	*C. soloensis* Val.	Java

Table 12.6–Contd...

Sl.No.	Species	Origin	Sl.No.	Species	Origin
77.	*C. sylvatica* Val.		97.	*C. domestica* Val.	
78.	*C. aurantiaca* Van *Zijp*	Java	98.	*C. speciosa* Link	
79.	*C. elomata* Craib.	Siam	99.	*C. viridiflora* Roxb.	Java
80.	*C. latiflora* Val.	New Guinea	100.	*C. colorata* Val.	Java
81.	*C. sulcata* Haines	India	101.	*C. euchroma* Val.	Java
82.	*C. elata* Roxb.	Myanmar	102.	*C. ochrorrhiza* Val.	Java
83.	*C. inodora* Blatter	India	103.	*C. brog* Val.	Java
84.	*C. petiolata* Roxb.	Myanmar	104.	*C. agrostis* Sare	
85.	*C. purpurea* Blatter	India	105.	*C. ranadei* Prain	
86.	*C. zanthorrhiza* Roxb.	Amboinea	106.	*C. haritha* Mangaly and Sabu	India
87.	*C. zedoaria* (Bergius) Roscoe		107.	*C. raktakanta* Mangaly and Sabu	India
88.	*C. burtii* K. Lasru and R.M Smith	Thailand	108.	*C. kudagensis* Vel. *et.al*	India
89.	*C. kwangsiensis* S.G. Lee and C.F. Liang	China	109.	*C. thalakaveriensis* Vel. *et.al*	India
			110.	*C. malabarica* Vel *et.al*	India
90.	*C. curcumorpha* Rao and Verma	India	111.	*C. karnatakensis* Amal. *et.al*	India
91.	*C. longiflora* (Wall.). Rao and Verma	India	111.	*C. niodona*	
92.	*C. ecalcarata* Sivar and Indu	India	113.	*C. nilambuernsis* Amal *et.al*	India
93.	*C. sichuanensis* Gagnep.	China	114.	*C. cannanorensis* Ansari and Nair	India
94.	*C. alismatifolia* Gagnep.	Thailand	115.	*C. vamana* Mangaly and Sabu	India
95.	*C. gracilima* Gagnep.		116.	*C. lutea* Ansari and Nair	India
96.	*C. kunstleri*		117.	*C. peethapushpa* Sandh and Sivar	India

Source: Velayudhan *et al.*, 1999.

Taxonomic classification and systematics of genus may not be fool proof as there are many problems like lack of type specimens and illustration of old species, and absence of important floral parts in herbariums, fleshy and perishable aerial portions etc. and recent molecular taxonomic studies of 16 *Curcuma* species also pin point this fact (Shyamkumar and Sasikumar, 2005).

Cultivar Diversity

Cultivar diversity of cultivated turmeric is the highest in India. In India South India, Orissa and North Eastern states are rich in cultivar diversity. A good number of varieties also exist in other producing countries like Bangladesh, Pakistan, Vietnam, Nepal, Myanmar, South Pacific Islands, Malagasi etc. Cultivar diversity of turmeric in India is given in Table 11.7.

Turmeric cultivars of India are classified as short duration varieties (G.L Puram, Kasturi, Jabedi, Dughi, Dindigram, Udayagiri Amalapuram); medium duration varieties (Kesari, Kothapeta, Amruthapani, Rajapuri, Panamalur, Gorakpur) or long duration varieties (Armoor, Duggirala, Mydukur Tekurpet, Sugandham, Avanigadda, Ethamkula). Short duration varieties are characterized by thick, long rhizome, low curcumin and early maturity (180-200 days). These varieties yield about 8-20 tonnes (fresh) rhizome/ha. Medium duration varieties mature in about 230-240 days, and possess medium long, with close internodes, thick rhizome. These varieties yield about 14-35 tonnes (fresh)/ ha. Long duration varieties are characterized by long, stout, smooth and hard rhizomes. These cultivars take about 9 months for maturity and yield about 15-37 tonnes/ha. (fresh rhizome).

Table 11.7: Cultivar/Varietal Diversity in Turmeric

Sl.No.	*Cultivar/variety*	*Sl.No.*	*Cultivar/Variety*
1	Armur	35	Lokhande
2	Alleppey	36	Lekadong
3	Alleppey Supreme*	37	Mundage
4	Avanigadda	38	Megha turmeric*
5	Amruthapani	39	Mydukkur
6	Amalapuram	40	Nandyal
7	Balaga	41	Pattani
8	Bilaspur	42	Panamalur
9	BSR-1*	43	Pakistan
10	BSR-2*	44	Perumnadan
11	Bullapura	45	Prabha*
12	C-A-72 Udayagiri	46	Prathibha*
13	C-A-12	47	Rajapuri
14	CLL- 324	48	Ranga*
15	CLL- 328	49	Rasmi
16	Chinnanadan	50	Rajendra Sonia*
17	Chaya paspu	51	Roma*
18	Co-1*	52	Sobha*,
19	Deshi	53	Sona*
20	Duggirala	54	Shimla
21	Dunrigam	55	Sugantham*
22	Dughi	56	Suguna*
23	Erode Local	57	Suvarna*
24	Ethamukkala	58	Sudarsana*
25	Gorakpur	59	Suranjana*
26	Guntur	60	Suroma*
27	GL Puram	61	Thekkurpetta
28	Jabedi (G-67)	62	TSundar
29	Kanti*	63	Thalachira
30	Kasturi Tanaka	64	Varna*
31	Kasturi	65	Vonimitta
32	Katpadi Local	66	Waynadan
33	Kedaram*	67	Yelachage
34	Kothapetta		

*: Improved variety.

Turmeric varieties/cultivars such as Alleppey, Lekadong Prabha, Prathibha, Kedaram, Alleppey Supreme, Wyand Local, Mananthody, Aieng, Edappalayam, Roma and Sugandham are rich in curcumin (above 5 per cent) Turmeric varieties also differ in their curcuminoid levels too. Variety x location interaction is very high for curcumin content.

Propagation

Turmeric is propagated, like ginger, vegetatively by means of rhizomes saved from previous harvest. Unlike in ginger, viable sexual reproduction and true turmeric seedling production are there in this spice (Sasikumar *et al.*, 1996) though cultivated *C. longa* is triploid (2n=63). Major contribution of cultivar diversity in turmeric too must be due to geographical spread of varieties accompanied by genetic differentiation into locally adapted populations, caused by mutation plus selection.

References

Adaniya, S. and Shirai, D., 2001. *In vitro* induction of tetraploid ginger (*Z. officinale*) and pollen fertility and germinability. *Sci. Hort.*, 83: 277–287.

Burtt, B.L. and Smith, R.M., 1983. Zingibeaceac. In: *A Revised Handbook to the Flora of Ceylon, Vol. IV*, (Ed.) M.D Dasanayake. Amerind Pub., New Delhi., pp. 488–532.

Baker, J.G., 1894. Scitamineae. In: *J.D Hooker's Flora of British India*, 6: 198–264.

Graham, J.A., 1936. Methods of ginger cultivation in Jamaica. *J. Jamaican Agric. Soc.*, 40: 231–232.

Holttum, R.E., 1950. Zingiberaceac of the Malayan Peninsula. *Gard. Bull. Singapure*, 13:1–249.

Gamble, J.S., 1925. *Flora of the Presidency of Madras*, Vol.II. Botanical Survey of India, Kolkata.

Kumar, S., 1999. A note on conservation of economically important Zingiberaceae of Sikkim, Himalaya. In: *Biodiversity, Conservation and Utilization of Medicinal and Aromatic Plants*, (Eds) Sasikumar, B. *et al* Indian Institute of Spices Research, Calicut, India, pp. 201–207.

Ravindran, P. N., Nirmal Babu, K. and Shiva, K.N., 2005. Botany and Crop Improvement of ginger. In: *Ginger: The Genus Ginger*, (Eds) Ravindran, P. N. and Nirmal Babu, K. CRC Press, Boca Raton, pp. 15–86.

Ridley, H.N., 1912. *Spices*. Mc Millan and Co. Ltd., London.

Rosales, P.B., 1938. Agronomic study of the native and Hawaiian ginger. *Philipp. Agric.*, 26: 807–822.

Sabu, M. and Dave Skinner, 2005. Other economically important *Zingiber* species. In: *Ginger: The Genus Ginger*, (Eds) Ravindran, P. N. and Nirmal Babu, K. CRC Press, Boca Raton, pp. 533–7545.

Sasikumar, B., 2005. Genetic resources of *Curcuma:* diversity, characterization and utilization. *Plant Genetic Resources: Conservation and Utilization*, p. 230–251.

Sasikumar, B., 2000. Ginger in Nepal. *Indian Spices*, 36(2 and 3): 7.

Sasikumar, B., Johnson K. George and John Zachariah, T., 1995. A note on a ginger type collected from Western Ghat forests. *J. Spices and Aromatic Crops*, 4: 160–161.

Sasikumar, B., Johnson K. George, John Zachariah, T. Ratnambal, M.J., Nirmal Babu, K. and Ravindran, P.N., 1996. IISR Prabha and IISR Prathibha: Two high yielding and high quality turmeric (*C. longa*) varieties. *J. Spices and Aromatic Crops*, 5: 34–40.

Syamkumar, S. and Sasikumar, B., 2005. Molecular marker based genetic diversity analysis of *Curcuma* species from India. *Euphytica* (Communicated).

Sasikumar, B., Krishnamoorthy, B., Saji, K.V., Gearge, J.K., Peter, K.V. and Ravindran, P. N., 1999. Spices diversity and conservation of plants that yield major spices in India. *Plant Genetic Resources Newsl.*, 118: 19–26.

Tindall, H.D., 1968. Commercially grown vegetables. In: *Commercial Vegetable Growing*. Oxford University Press, London.

Valeton, T.H., 1918. New notes on Zingiberaceae of Java and Malaya. *Bull. Jard. Butenzer Ser.*, II, 27: 1–81.

Velayudhan, K.C., Amalraj, V.A. and Muralidharan, V.K., 1996. The concepectus of the genus *Curcuma* in India. *J. Econ. Tax. Bot.*, 20: 345–382.

Velayudhan, K.C., Muralidharan, V.K., Amalaraj, V.J., Gautam, P.L., Mandal, S. and Dinesh Kumar, 1999. *Curcurma* genetic resources. Scientific Monograph No. 4, National Bureau of Plant Genetic Resources, New Delhi, p. 149.

Biodiversity in Horticultural Crops Vol. 1
Editors: **Professor K.V. Peter & Dr. Z. Abraham**
Published by: **DAYA PUBLISHING HOUSE**

Pages **256–277**

Chapter 12

Tree Spices

B. Krishnamoorthy, J. Rema and P.A. Mathew
Indian Institute of Spices Research, P O Marikunnu – 673 012 Calicut

Biological diversity of Indian subcontinent is one of the richest in the world owing to its vast geographical area, varied topography and climate. Because of its richness in overall species diversity, India is recognized as one of the 12 mega diversity regions of the world. Among 1.7 million species globally described and recorded in scientific literature, India has about 1,26,200 species (Khoshoo, 1995). India ranks tenth in the world both in respect of richness of flowering plants (17,000 spp.) and mammals (372 spp.) besides being the center of diversity and origin of about 167 important cultivated plant species and domesticated animals. As far as agro-biodiversity is concerned, India ranks seventh among centers of diversity and origin of crop plants (Khoshoo, 1995). Of the 25 hotspots (Myers *et al.*, 2000) recognized in the world, Eastern Himalayas and Western Ghats are in India. Western Ghats, which is one of the nine biogeographic regions of India, possesses various types of tropical forests, ranging from wet evergreen to dry deciduous. Nearly 63 per cent of tree species of low and medium elevation evergreen forests of Western Ghats are endemic. This high level of diversity and endemism in the Western Ghats have conferred on them hot spot status. Three endemic centres in Kerala namely Agasthyamalai, Anamalai high ranges and Silent valley are identified.

Seventeen tree spices are grown in India, the commercially important ones being nutmeg, cinnamon, cassia, allspice, cambogia, kokam, tamarind, pomegranate and curry leaf. Recent global developments and the loss in genetic diversity necessitated development of programmes on conservation of biodiversity in tree spices and their sustainable use in agriculture as spice and medicine. Variability available in these crops is limited and also many species are becoming endangered and hence, conservation of genetic variability in these crops is very important.

Collection and conservation of biodiversity in tree spices is one of the major mandates of Indian Institute of Spices Research (IISR), Calicut, Kerala, India. Conservation of tree spices is also being carried out by a few State Agricultural Universities in India. The germplasm conservatories at IISR,

Calicut consist of 1180 accessions of tree spices which include all the major tree spices and their related taxa.

Cinnamon

The true cinnamon, *Cinnamomum verum* Bercht. and Presl. (Syn. *C. zeylanicum* Blume), (Lauraceae) commonly known as Sri Lankan cinnamon, is an evergreen tree reaching a height of 6-15 m, with bisexual flowers. Cinnamon of commerce is the dried inner bark of *C. verum*. Besides cinnamon, bark oil, bark oleoresin and leaf oil are products sought after by food, pharmaceutical and perfume industry for varied uses.

Species Diversity

Family Lauraceae consists of 11 genera Cryptocarya, Apollonias, Beilschmiedia, Cinnamomum, Alseodaphne, Machilus, Phoebe, Actinodaphne, Litsea, Neolitsea and Cassytha (Gamble 1925). The genus *Cinnamomum* Schaeffer (Lauraceae) comprises about 250 species (Willis 1973) of trees and shrubs of tropics and subtropics. It is distributed in South-East Asia, China and Australia (Mabberly 1987), growing mainly in tropical rain forests at varying altitudes. In India, it is represented by 26 species (Hooker 1886). Kostermans (1983) reported 12 species from South India. Many of the species of *Cinnamomum* have medicinal and spice value and are of great demand commercially. *C. verum* Bercht. and Presl (true or Ceylon cinnamon), *C. cassia* Presl. (Chinese cinnamon, *Cassia lignea*), *C. burmannii* Blume (Indonesian cassia), *C. loureirii* Nees (Vietnamese cassia), *C. tamala* (Buch-Ham.) Nees and Eberm. (Indian cassia, Indian bay leaf or tejpat) and *C. camphora* (camphor tree) are a few of the economically important species of the genus.

Diversity in *Cinnamomum* is mainly species diversity. As most of the species occur in the wild, semi-domesticated gene pools of *Cinnamomum* do not occur.

Sri Lanka is the primary center of origin of *Cinnamomum* and several species of related taxa occur in Sri Lanka. Cinnamon originated in central hills of Sri Lanka in places like Kandy, Matale, Belihull Oya, Haputale and the Sinharaja forest range (Ranatunga *et al.*, 2004). Wijesekera *et al.* (1975) reported several species endemic to Sri Lanka. Western Ghats and the adjoining areas are considered secondary center of origin of *C. verum*. The Cinnamomum species distributed in two centers of diversity in India are distinctly different. The south Indian species and the north Indian species are distinctly different and are listed in Tables 12.1 and 12.2. A few species in both the places. Cinnamon was first introduced into India from Sri Lanka by the British during 18th century and is cultivated in Kerala, Karnataka and Tamil Nadu and is more prevalent in the hilly regions of Western Ghats. The oldest cinnamon plantation is the Anjarakandy estate (110 ha. approximately) in Cannanore District of Kerala, which was planted by British.

False cinnamon is obtained from *C. cassia* (L) Brecht. and Presl. (syn. *C. aromaticum* Nees). It originated and is cultivated in China. Its bark is used for flavoring food and beverages and also used in pharmaceutical preparations and perfumery. The other sources of cassia are *C. loureirri* Nees., *C. burmannii* Blume and *C. tamala Nees*. *C. cassia* (Chinese cassia) is indigenous to China-Vietnam region and *C. burmannii* (Indonesian cassia) is native to Sumatara and Java regions of Indonesia. *C. tamala* (Indian cassia) is found in the forests of North India and Nepal. The earlier publications indicated that Vietnam cassia is *Cinnamomum loureirrii*. Now it is confirmed that the Vietnam cassia is nothing but *C. cassia*. Kostermans (1964) lists seven varieties of C. *burmannii namely* var. *angustifolium* Messner; var. *chinense* (Bl.) Meissner, var. *kiamis* (Nees) Meissner, var. *lanceolatum* Miquel, var. *microphyllum* Miquel, var. *sumatrense* Teijsm and Binnenad and var. *suleavene* Miquel.

Table 12.1: Distribution of *Cinnamomum* Species in South India

Cinnamomum Species	*Place of Distribution*
C. chemungianum Mohan and Herry	Western Ghats
C. filipedicellatum Kosterm.	Western Ghats, Anamalais, Nilgiris
C. goaense Kosterm	North Western Ghats
C. gracile Hook. f.	Western Ghats
C. heyneanum Nees	Western Ghats
C. keralaense Kosterm.	Western Ghats
C. macrocarpum Hook. f.	Nilgiris, Anamalais
C. malabatrum (Burm.f.)Bl.	Western Ghats
C. nicolsonianum Manilal and Shylaja	Western Ghats
C. perrottettii Meissn.	Niligiris, Western Ghats
C. riparium Gamble	Niligiris, Western Ghats
C. sulphuratum Nees	Western Ghats
C. travancoricum Gamble	Western Ghats
C. verum Brest. and Presl.	Sri Lanka, Western and Eastern Ghats
C. walaiwarense Kostern	Western Ghats
C. wightii Meissn.	Niligiris

Table 12.2: Cinnamomum Species Available in North India

Species	*Distribution*	*Major Constituent*
C. assamicum	North East India	–
C. bejolghota (Buch-Ham)	Eastern Himalayas, Myanmar	1, 8- cineole, Linalool
C. camphora (L.) Bercht and Presl.	Sub-Himalayan region	Camphor
***C. cassia* (L.) Brecht and Presl.**	Mizoram	Cinnamaldehyde
C. cecidodaphne Meissner	Eastern Himalayas	Cineole, Methyl cinnamate
C. glanduliferum Nees	Himalayas, Khasi Hills	Cineole, Terpineol, Camphor
C. impressinervium Meissner.	North East India	Eugenol
C. iners	Assam	–
C. parthenoxylon	Shillong, Meghalaya	–
C. pauciflorum Nees	North East India	Cinnamaldehyde
C. sulphuratum Nees	North Cachar Hills, Assam, Western Ghats	Linalool, Citral, Cinnamaldehyde
C. tamala Nees	North East India, Assam	
C. verum	North East India	Cinnamaldehyde

C. tamala, occurring mostly in tropical and subtropical Himalayas and extending to North-East India up to an altitude of 2000 m, is the main source of spice tejpat. It also grows in Nepal, Bangladesh and Myanmar. It is a moderate sized evergreen tree which is the source of *tejpat*, *tejpat* oil and Indian cassia bark. *Tejpat* is the dried leaf of *C. tamala* and *tejpat* oil is the oil obtained by distillation of leaves.

The dried bark of stem is the Indian cassia bark. *C. tamala* is listed as a threatened species as plant population declines day by day due to over exploitation and habitat destruction in India (Kumar *et al.*, 1997).

C. tamala is native to India and originated in the Himalayas. It is distributed in the Indian sub continent, Indo-China region, Bangladesh and Nepal. In tropical and sub-tropical Himalayas, it is distributed upto an altitude of 900-2500 m (Ahmed *et al.*, 2000). It occurs in North Western, Eastern and Sikkim Himalayas. It is also found in Meghalaya (Khasi Hills and Jaintia Hills), Assam (North Cachar Hills); Jammu and Kashmir (Basantgarh and Rajouri); Himachal Pradesh (Drang Forest in Dauladhar ranges, Hamirpur, Shimla, Kangra, Chamba, Mandi, Solan, Nahan, Palampur) and Uttaranchal (Jaunsar, Tehri Garwal and Kumaon). Commercial cultivation of *C. tamala* is very limited and is reported in certain parts of country. *C. tamala* plantations occur in Khasi and Jaintia Hills, Garo Hills, Mikir Hills, Manipur and Arunachal Pradesh and in limited areas in Nainital (U.P) and Kangra (Himachal Pradesh) (Bradu and Sobti 1988).

C. camphora (camphor tree) is an evergreen aromatic medium to large sized tree growing naturally in China, Japan, Taiwan and in the adjoining regions of South East Asia. In India, it is grown in a few plantations. It is used commercially for production of camphor used in perfume industry. Hirota and Hiroi (1967) and Wan-Yang *et al.* (1989) recognized five chemical races among Chinese camphor trees. Khein *et al.* (1998) recognized eight chemovarieties among camphor trees in Vietnam. Various chemotypes occur in camphor yielding camphor, linalool, safrole and cineole as major constituents (Wan-Yang *et al.*, 1989). The Formosan camphor tree (*C. camphora ssp. Formosana*) has at least seven distinct chemovarieties (Tetenyi, 1970) and they are chvar *borneol;* chvar *camphor;* chvar *safrole;* chvar *sesquiterpene alcohol;* chvar *cineol;* chvar *linalool* and chvar *sesquiterpene*. The chvar *linalool* (designated by later taxonomists as *C. camphora* var *linaloolifera* (Fujita, 1967) consists of two further chemoforms: chforma 86 per cent linalool and chforma 71 per cent linalool. In *C. camphora,* several varieties are reported namely, var. *cuneata* Blume ex Meissner; var *glaucescens* A Braun, var *linaloolifera* Fujita, var. nominalis Hayata, var. *normale* Hayata ex Ito, var. *parviflora* Miquel, var. *procera* Blume ex Meissner, var. *quintuplinervia* Miquel, var. *rotundata* Meissner, var. *rotundifolia* Makino ex Honda, var. *typical* Petyaev. The other important *Cinnamomum* species are given in Table 12.3.

Diversity Analysis

In a cluster analysis of South Indian species, *C. verum* was more closely related to *C. malabatrum* (Shylaja 1984). Cluster analysis of the taxa using centroid linkage led to grouping of *C. verum* with some collections of *C. malabatrum* while *C. cassia* and *C. camphora* formed unique groups. Principal component analysis was done to identify characters responsible for species divergence. *C. camphora* formed a separate group from other species based on bud type, stomatal type, phyllotaxy and epidermal thickness. *C. perrottettii, C. riparium* and *C. macrocarpum* formed another group based on hairiness, hair size, and hair frequency. *C. cassia* diverged from rest of species by nature and length of inflorescence Species such as *C. camphora, C. cassia, C. riparium, C. macrocarpum* and *C. perrottettii* remained independent entities with respect to principal components studied, thereby indicating their taxonomic independence. *C. verum* and *C. malabatrum* show considerable closeness, indicating their taxonomic affinity. Phenotypic analysis of a large number of *Cinnamomum* spp. was carried out by Bakker *et al.* (1992) indicating ten clusters. According to this indication *C. burmannii* and *C. verum* are in Cluster 1, together with many other species. Within this cluster *C. burmannii* occupies an independent sub-cluster position, while *C. verum* is in another sub-cluster together with 21 other species (including *C. tamala, C. culitlawan, C. keralaense, C. sintok,*). *C. cassia* formed a separate cluster.

Table 12.3: Other Important *Cinnamomum* Species

Species	*Distribution*	*Major Constituent*
C. angustifolium	Madagascar	Phellandrene, P-cymene
C. appelianum	China	Cineole
C. bodineere	China	Citral
C. capparu-coronde	Sri Lanka	1,8- cineole, linalool, eugenol
C. cordatum	Malaysia	Linalool (leaf)
C. culitlawan	Indonesia	–
C. deschampsii	Malaysia	–
C. glaucescens.	Nepal	1,8 cineole,
C. japonicum	Japan	p- cymene, 1,8 cineole
C. kanahirai.	China, Taiwan	Terpinen–4-ol
C. massoie	Papua, New Guinea	
C. mercadoi	Philippines	
C. mindanaense	Philippines	
C. oliverii	Australia	
C. osmophloem	Tiwan	Trans Cinnamaldehyde
C. parthenoxylon	Vietnam, China, Indonesia	1,8, cineole type and linalool type
C. platlphyllum	China	Trans–Methyl isoeugenol
C. porrectum	China	Citral
C. septenrionale	China	Trans–Methyl isoeugenol
C. tenuipilis	China	Linalool
C. wilsonii	China	Cinnamicacetate

Source: Shylaja *et al.*, 2004.

Chemotaxonomy

Chemotaxonomical studies on *C. verum* and a few of its related taxa occurring in Western Ghats of Kerala, were carried out by Shylaja (1984) and Ravindran *et al.* (1992). They analyzed flavonoids, terpenoids and steroids and found much variation among species. *C. verum, C. camphora, C. cassia* and *C. riparium* are chemically very distinct among themselves and from other species. Much intraspecific chemical variability was noticed in *C. malabatrum*. They also found that Sri Lankan and Indian accessions of *C. verum* were chemically identical. Intraspecific chemical variability exists in many species of *Cinnamomum*. Fujita (1967) attempted a sub-specific classification of many species based on volatile oil composition. Such intraspecific variability might have evolved as a result of interbreeding, segregation, chance mutations, and isolation mechanisms. Chemotypes of a few *Cinnamomum* species are given in Table 12.4.

Based on available data regarding chemical composition of *C. tamala* leaf oils of both wild and cultivated sources, it was found that various chemotypes exist. Various chemotypes are eugenol type (Gulati 1979, Sood *et al.*, 1979, Hussain *et al.*, 1988), cinnamaldehyde type (Bradu and Sobti 1988) cinnamaldehyde-linalool type (Nigam and Ahmed 1990, and linalool type (Nath *et al.*, 1994). The above analysis indicates that either different chemotypes exist or species used for chemical analysis was taxonomically misclassified. Population of *C. tamala* growing in North East India was classified into four distinct variants based on leaf morphology and leaves of many related species are being sold in market as *tejpat* (Baruah and Nath 2000). Nath *et al.* (1999) analyzed four variants of *C. tamala* and reported chemical composition of leaf oil.

Table 12.4: Chemotypes of *Cinnamomum* species

Species	*Chemotypes*
C. bejolghota	Linalool type, Linalool–terpineol type, Linalool–phellandrane type
C. bodinieri	Citral type, Camphor type,
C. cecidodaphane	Methyl eugenol type, Safrole type
C. culitlawan	Safrole type, Eugenol type,
C. glanduliferum	Safrole type, Methyl eugenol type, Methyl eugenol + Safrole type
C. sulphuratum	Linalool type, Citral type, Cinnamaldehyde type, Methyl cinnamate type
C. tamala	Eugenol type, Cinnamic aldehyde type, Linalool type, Cinnamic aldehyde + Linalool type
C. verum	Cinnmic aldehyde type, Cinnamic aldehyde + eugenol type, Cinnamic aldehyde + Safrole type
C. wilsonii	Citral type, Cinnamicacetate type, Cinnamic aldehyde + Safrole type
Cinnamomum camphora	Camphor type, Linalool type, Cineole type, Safrole type, Borneol type,

Source: Ravindran *et al.*, 2004.

Cultivar Diversity

In India, high coefficient of variation for dry and fresh bark yield/plant, bark oleoresin, leaf oil, leaf size index and recovery of bark(per cent) were observed in cinnamon. Association analysis for nine characters in cinnamon revealed significant correlation of fresh weight of bark and leaf oil with dry bark yield. Bark oil was negatively correlated with leaf oil (Krishnamoorthy *et al.*, 1992). Correlation and path analysis studies conducted by Joy *et al.* (1998) indicated that the economic yield characters namely fresh leaf yield, leaf oil yield and eugenol yield were highly correlated among themselves. Plant height and canopy spread were highly and positively correlated with yield component (Joy *et al.*, 1998). In Sri Lanka, 8 different types of cinnamon, based on pungency of bark and petiole, texture of bark and structure of leaves, are cultivated and they are known in Sri Lanka by their local names (Ranatunga *et al.*, 2004).

Quality

Over 160 cinnamon trees of Indian origin and 66 trees of Sri Lankan origin were analyzed for their essential oil composition and oleoresin content. Indian trees yielded bark essential oil in the range of traces to 4.28 per cent and Sri Lankan trees in the range of traces to 3.85 per cent. Indian collections had 21 elite trees yielding more than 2 per cent essential oil from bark while in Sri Lankan collection such trees were five in number. Gas liquid chromatographic examination of bark oil sample from elite trees showed cinnamaldehyde to be the major content in both Indian and Sri Lankan trees.The percentage was much higher in Sri Lankan trees (70-84) as against that in Indian trees (55-71). The second major compound was eugenol in Sri Lankan trees constituting 5.0 -7.5 per cent of oil whereas in Indian trees, it was cinnamylacetate, which accounted for 4-26 per cent of oil. Eugenol was only the third or forth-major constituent in Indian bark oils varying generally from 2.5-5.4 per cent in different lines with exception of top yielder from Indian collection with 11.6 per cent eugenol. Acetone oleoresin yield varied from 5.8-19.0 per cent. Compositional analysis of leaf oil samples showed Indian trees to contain more eugenol (83-93 per cent) in their leaves compared to Ceylon collection (66-85 per cent) (Babu and Krishnamoorthy, 2002).

A study on seedling progenies of cinnamon revealed a clear variation in performance of progenies (Krishnamoorthy *et al.*, 1991a). Four different flush colours were noted among cinnamon collections, *viz.*, pure purple, purple dominated with green, green dominated with purple and pure green. Colour of young leaf flushes of cinnamon was related to its quality. Krishnamoorthy *et al.* (1988) reported that ratio of plants having young leaf flushes with purple and purple pigmented ones to green ones was 1:1. Plants with purple leaf flushes recorded about 29 per cent more bark oil compared to green flushes, showing that flush colour may be taken as a criterion for selection of quality seedlings at nursery stage.

At present there are 7 varieties of cinnamon released from various institutes in India. They are IISR Navashree, IISR Nithyashree (Indian Institute of Spices Research Calicut.), YCD-1(Horticultural Research Station, Yercaud, Salem District), PPI (C) 1 (Horticultural Research Station, Pechipari, Kanykumari District) Konkan Tej (Dr. B.S. Konkan Krishi Vidyapeeth, Dapoli, Maharashtra), RRL (B) C-6 (Regional Research Laboratory, Bhubaneswar, Orissa) Sugandhini (Aromatic and Medicinal Plants Research Station, Odakkali, Ernakulam District, Kerala).

Conservatory of tree spices at IISR consists of seedling progenies of 26 Chinese cassia lines collected from Sri Kundara Estate, Valparai, Tamil Nadu. Quality analysis of these accessions revealed a high co-efficient of variability for bark oil and leaf oil (28.1 per cent and 30.1 per cent respectively) but pungency and flavour of accessions were more or less uniform as indicated by cinnamaldehyde content. Bark oil ranged from 1.2 to 4.9 per cent, leaf oil 0.4 to 1.6 per cent and bark oleoresin 6 to 10.5 per cent. Leaf and bark oils possess cinnnmaldehyde as major constituent and a few accessions C1, D1 and D3 were identified as superior, based on their overall chemical and flavour profile and regeneration capacity. (Krishnamoorthy *et al.*, 1999).

High Elevation Species

C. wightii, *C. perottettii* and *C. macrocarpum* are seen in Nilgiris at an elevation above 2500 m.

Rare and Endangered Species

Owing to severe deforestation, there is every possibility of a few species becoming extinct in near future. The wild population of *Cinnamomum* is a real threat due to indiscriminate bark extraction from them. *C. nicolsonianum* Manilal and Shylaja and *C. heynianum* are endangered. *C. nicolsonianum* is a very rare and endangered species having large leaves and very small axillary panicles that are a few flowered which occur in low elevations in Western Ghats. The tree is extremely rare and possibly in the verge of extinction (Manilal and Shylaja 1986). *C. sulphuratum*, listed as red is found in abundance in Kemmangundi (Western Ghats) (Aravind *et al.*, 2005).

Conservation

Conservatory of Cinnamon at IISR, Calicut includes *C. verum*, *C. cassia*, *C. camphora* (an economically important tree yielding camphor oil), *C. citriodorum*, lemon grass oil smelling *Cinnamomum* spp. collected from Munnar (Idukki District) *C. malabatrum*, *C. perrottetti*, *C. wightii*, *C. macrocarpum*, *C. sulphuratum*, *C. riparium and C. tamala*. (Table 12.5.). The germplasm of cinnamon is also assembled at Konkan Krishi

Vidyapeeth, Dapoli, Maharastra, Tamil Nadu Agricultural University (Horticultural Station Yercadu), Regional Research Laboratory, Bhubaneswar and Kerala Agricultural University (Odakkali).

Table 12.5: Genetic Resources of Tree Spices Conserved at Indian Institute of Spices Research, Calicut

Crop	*Areas Surveyed*	*Important Species*
Cinnamon and Cassia	Calicut, Wyanad, Ernakulam, Kottayam (Kerala); Kanyakumari, Annamalai, Kolli and Shevroy Hills (Tamil Nadu); Kodagu (Karnataka); Ratnagiri (Maharashtra); Meghalaya (North East India) Andaman and Nicobar Islands	*Cinnamomum camphora* *C. cassia* *C. citriodorum* *C. filipedicellatum* *C. glaucescens* *C. macrocarpum* *C. malabatrum* *C. perrottetti* *C. riparium* *C. sulphuratum* *C. tamala* *C. verum* *C. wightii*
Nutmeg	Thenmala, Cheekilodu, Balussery, Mannuthy, Sugandhagiri, Munnar (Kerala); Nagercoil, Kolli and Shevroy Hills (Tamil Nadu); Ratnagiri (Maharashtra) Andaman and Nicobar Island	*Myristica amygdalina* *M. andamanica* *M. attenuate* *M. beddomeii* *M. fatua var. magnifica** *M. fragrans* *M. malabarica* *M. prainii* *Gymnocranthera canarica* *Knema andamanica*
Clove	Ambanadu Estate, Quilon, Wyanad, (Kerala); Kolli and Shevroy Hills, Nagercoil (Tamil Nadu); Ratnagiri (Maharashtra)	*Syzygium aromaticum* *S. caryophyllatum* *S. cuminii* *S. fruiticosum* *S. heynianum* *S. jambolana* *S. jambos* *S. lanceolatum* *S. zeylanicum* *Eugenia uniflora*
Allspice	Ambalavayal, Mannuthy, Kanyakumari (Kerala)	*Pimenta dioica*
Garcinia	Kasaragod, Chembanode, Kozhikode, Kumarakam, Kannur, Silent Valley, Thaliparamba, Thrissur (Kerala); Moodabidri, Puttur, Pudhuvethu, Sediapur, Kidu, Mercara (Karnataka) Vengurla (Maharashtra) Andaman and Nicobar Islands	*Garcinia cowa* *G. gummi-gutta* *G. heynianum* *G. hombroniana* *G. imberti* *G. indica* *G. tinctoria* *G. mangostana* *G. morella* *G. malabarica*

Garcinia

Garcinia belongs to family Clusiaceae and according to old botanical classification, *Garcinia* is placed within family Guttiferae which includes about 1300 species. Guttiferae is further divided into 42 genera and five sub-families namely, Kielmeroideae, Hypericoideae, Calopylloideae, Moronbeioideae and Clusioideae. Of these, subfamily Clusioideae consists of two tribes *viz.* Clusieae and Garcinieae and Garcinieae in turn has two genera namely, *Garcinia* and *Mammea* (Muhammed *et al.*, 1994).

Trees in this genus can be either dioecious or polygamous. In dioecious species, reproductive organs are unisexual. In polygamous species, male, female and hermaphrodite flowers are found in same plant. Male flowers in *Garcinia* are noted for their distinctive pistilodes.

Species Diversity

Genus *Garcinia* (Cluciaceae) includes more than 200 species of trees and shrubs, distributed in tropics of the world chiefly in Asia and Africa. About 35 species exist in India including exotic ones (Rodrigues, 2001) many of which are economically important. Out of 35 species reported to exist in India, 7 are endemic to Western Ghats, 6 in Adaman and Nicobar Islands and 4 in North East India (Rodrigues, 2001). Some of these are detailed in Table 12.6. The genus *Garcinia* includes *Garcinia atroviridi, G. dulcis, G. echinocarpa, G. gummi-gutta, G. hombroniana, G. indica, G. lanceaefolia, G. livingstonei, G. mangostana, G. microstigma, G. morella, G. paniculata, G. pedunculata, etc.* (Roberts 1984). These species can be useful as sources of gamboges, dyes, hydroxyl citric acid (HCA), as good rootstock or for breeding purpose. Among the commercially important species, *G.gummi-gutta* and *G. indica* are cultivated mainly for their spice, which is obtained from dried rind. *G. mangostana* and *G. morella* are also cultivated to a certain extent for their fruit and gamboges respectively. Little attention was given to document, and conserve other economically important species of this genus in the era of Intellectual Property Rights and Biodiversity Conservation.

Garcinia indica Choisy (*Kokum*) is an evergreen graceful tree growing with conical canopy attaining 10 meters in height. Main shoot is orthotropic whereas branches are plagiotropic exhibiting branch dimorphism. Three sex forms are seen in Kokum namely female, male and hermaphrodite types though it is generally dioecious. The dried rind of its fruit is the traditional 'kokum' or 'binda' of commerce used for garnishing curries. It is a popular tree spice having tremendous potential and also has many medicinal properties. In South India, it is used instead of tamarind in curries. Juice of fruit is used as a mordant and expressed oil of seed is the kokam oil of natives, extensively used to adulterate ghee. Seeds of fruits yield valuable edible fat known in commerce as kokam butter. The recent understanding on effect of hydroxyl citric acid (HCA) in preventing obesity in humans has boosted its economic value. It is also the richest source of red pigments in plant kingdom. The tree grows extensively in the Konkan region of Maharashtra, Goa, coastal areas of Karnataka and Kerala, evergreen forests of Assam, Khasi, Jantia hills, West Bengal and Gujarat. Kokum' or *Garcinia indica* Choicy is a spice with great medicinal values occurring in western coast in a semi wild state. It is a native of Western Ghats of Kerala (India) and Malaysia. It grows in evergreen forests of Western Ghats in South India and its habitat extends from Konkan southward to Travancore and into the Shola forest of Nilgiris where it reaches an altitude of up to 2000 m above mean sea level. In Kerala, it is very popular in the Central Travancore areas. Kerala seems to be one of the centers of origin of cambodge (*G. gummi-gutta*) where maximum diversity is seen. It is fairly common and abundant in the forests of western Sri Lanka from sea level to 600 m and in Malaysia. It is widely distributed in evergreen forests of Western Ghats from South Kanara and Mysore to South Kerala up to low lying reclaimed lands bordering backwaters. The kokum is very common in the region of west coast of India from S. Kanara towards north. In Kerala the tree is quite rare except as few trees in state agriculture farms. The tree also exists in Mysore, Coorg and Wayanad.

Table 12.6: Important *Garcinia* Species Available in India

Species	*Distribution*
Garcinia andamanica King	Andaman Islands
G. anomala Planc.	Khasi Hills
G. atroviridis Griff.	North Eastern districts of Assam
G. cornea L.	Bengal
G. cowa Roxb.	Eastern parts of India, Assam, Bihar Bengal, Orissa, Andaman
G. dulcis (Roxb.) Kurz	Introduced into India from Malaysia
G. echinocarpa Thw.	Tirunelveli forests
G. gummi-gutta (L). Robsmall [*G. cambogia* (Gaetn) Desr.]	Western Ghats, Maharashtra, Goa, Karnataka,Kerala, Shola forests of Nilgiris.
G. hanburyii Hook.	South India
G. hombroniana Pierre	Nicobar Island
G. imbertti Bourd	S. India
G. kydia Roxb.	Andman Islands
G. lanceaefolia Roxb.	Assam, Khasi Hills
G. livingstonei T.Anders.	Introduced to India form East Africa
G. malabarica Talbot	S. India
G. mangostana L.	Introduced to S. India
G. microstigma Kurz	Andaman Islands
G. morella Desr.	Assam, Khasi Hills, Western Ghats
G. paniculata Roxb.	Foot hills of Himalayas, Assam, Khasi Hills
G. pedunculata Roxb.	Assam, Manipur
G. speciosa Wall.	Andaman islands
G. spitcata Hook. (*G. ovalifolia* Hook. f.)	Western Ghats form Konkan Southwards
G. stipulate T. Anders.	Eastern Himalayas
G. succifolia Kurz	S. India
G. travancorica Beddome	Western Ghats
G. wightii T. Anders.	S. Indian Forests
G. xanthochymus Hook. (*G. tinctoria* Wight *G. pictorius* Roxb.)	Eastern Himalayas, Western Ghats, Andaman islands

G. gummi-gutta is found commonly in evergreen forests of Western Ghats, from Konkan South to Travancore and in the shoal forests of Nilgiris up to an altitude of 200 m. In Western Ghats of Kerala, 'kudampuli' grows at an altitude of 1300 to 2000 m above msl. The best suited regions for its growth are those having high humidity and an altitude of 400-100 m. It is a small or medium sized tree with round crown and horizontal or drooping branches, leaves dark green and shining, elliptic obovate, 5-

7 cm long and 2.5-7.5 cm broad, fruits yellow or red when ripe, with 6-8 grooves ; about 6-8 seeds, surrounded by a succulent aril (CSIR 1956).

Cultivar Diversity

Kokum

Good variability in vegetative and reproductive characters is reported in kokum. This variability is of great use in the improvement of crop in coming years. Most of the collection of variability is done by Dr. Balasaheb Sawant Konkan Krishi Vidyapeeth, Dapoli, Maharshtra, followed by ICAR research complex, Goa and University of Agricultural Sciences (USA), Dharwad. Pujari and Joshi (2001) reported two types of branching pattern *i.e.*, horizontal and drooping. The latter type is not useful because it interferes with harvesting of fruit. Since grafts are to be planted, selections with orthotropic growth pattern need to be collected. Male, female and hermaphrodite sex forms are recognized. From Ratnagiri, Karnik *et al.* (2001) recorded 11 flower types and hermaphrodite type was actually located for first time from here. Generally fruits mature by late May or June and rain interferes with harvesting and processing. Preference is for early maturing types. Shinde *et al.* (2001) reported occurrence of early maturing types (February- March). Excellent variability for various fruit characters are reported. Colour varies from bright red, dark purple, white or even lemon yellow. However only red forms have commercial value. Gradations in red colour are seen. Another character is big size or boldness. Generally, fruit weight is only 34.0g. Fruits weighing 85 g are collected (Pujari and Joshi, 2001). Rind is quite thin (5mm) whereas thick skinned types (> 50mm) are preferred and are available. Gawanker *et al.* (2001) reported a promising thick rind type MLDK-5, which is under evaluation, out of five selections made at Ratnagiri. Though fruits are acidic, sweet forms are also reported, which may be as good as mangosteen. Fruit flies are the most important pests and selection needs to be made for resistant types. Based on earlier studies, one selection has been released as Konkan amruta by Dr. Balasaheb Sawant Konkan Krishi Vidyapeeth, Dapoli, Maharashtra.

Garcinia

Trees differ in size, shape and bearing habit. A few trees bearing fruits through out the year were observed. Trees vary with regard to flowering habit also. Dioecious and hermaphrodite types are seen. Lot of variation exists with regard to number of fruits, fruit size, shape, seeds per fruit and weight of fruit and seed. Seedless types are also observed rarely. The hydroxycitric acid content in the cultivars also varies significantly.

Conservation

Currently collection of biodiversity in kokum is actively being persued by Dr. Balasaheb Sawant Konkan Krishi Vidyapeeth, Dapoli, Maharashtra in the Konkan region. Several clones are already under evaluation. Variability in South Canara types is yet to be collected. Variability in Goa is being collected by ICAR Research complex for Goa at Old Goa. A few collections also were made by UAS, Dharwad. The Indian Institute of Spices Research, Calicut, (Table 12.5), National Bureau of Plant Genetic Resources (NBPGR) Regional Station at Trichur and Regional Agricultural Research Station, Kumarakom(KAU) are engaged in collection of biodiversity of Garcinia especially Kokum and Cambogia. Variability in other species in Andaman and Nicobar Islands and in North Eastern States of India need to be collected for conservation. Correct identities of many species also need to be confirmed.

Rare and Endangered Species

Garcinia mestonii Bailey, found in Australia is a rare species (Hammer and Khoshbakht 2005) and *G. indica* listed in red list in Charmandi (Western Ghats) is at present found in abundance there (Aravind *et al.*, 2005).

Nutmeg

Nutmeg (*Myristica fragrans* Houtt.) is an important tree spice, yielding two spices, namely, nutmeg (dried seed) and mace (dried aril surrounding seed). It is an evergreen, conical tree reaching a height of 10 meters, belonging to family Myristicaceae.

Species Diversity

The primitive family Myristicaceae, has about 18 genera and 300 species. Members of the family are pantropical, being associated with rainforests of Asia, Africa, Madagascar, South America and Polynesia. India has four genera namely *Horsfieldia, Gymnacranthera, Knema* and *Myristica* and altogether 15 species.The Table 12.7 indicates wild and related spices of Myristicaceae occurring in India. Members occur in evergreen forests of Andaman and Nicobar Islands, Meghalaya and Western Ghats. *Myristica* with 120 species is the largest of genus and New Guinea has the largest number of species. *Knema* spp. are found in Southeast Asia and Malaysia (Ahmedullah and Nayar, 1986).

Table 12.7: Members of *Myristicaceae* occurring in India

Members of Myristicaceae	*Distribution*
Myristica amygdalina Wall.	Nagercoil
M. andamanica Hook. f.	Andaman Islands
M. attenuate Wall.	Western Ghats
M. beddomeii King	Western Ghats
M. dactyloides	Western Ghats
M. gibbosa Hook. f. and Thoms.	Khasia Mountains
M. glabra Blume	Silhet (Assam), Tinnevelly (Tamil Nadu), Andaman Islands
M. glaucescens Hook.f. and Thoms.	Tinnevelly, Andaman islands
M. irya Gaertn. Fruct.	Tinnevelly, South Andaman islands
M. kingii hook. F.	Sikkim Himalaya
M. longifolia Wall.	Sikkim, Himalaya, Assam, and the Khasia Hills
M. magnifica Hood. f.	Western Ghats, parts of Tinnevelly, and
M. malabarica Lamk.	S. Karnataka
M. prainii King	Andaman Islands.
Gymnacranthera canarica Warb	Karnataka, Kerala.
Knema attenuate Warb.	Western Ghats
K. andamanica spp. *nicobarica*	
K. andamanica spp. *andamanica*	Nicobar, Andaman Islands

Myristica is the most primitive genus of the family (Sinclair 1958). Warming (1890) and Talbot (1902) are of the opinion that Myristicaceae is closely related to Lauraceae. There are evidences from morphological and anatomical studies that Myristicaceae is closely related to Annonaceae and Canellaceae (Wilson and Maculans 1967). At present, Myristicaceae is a member of Magnoliales or its taxonomical equivalents (Cronquist 1981; Dahlgren 1983). Taxonomists consider Myristicaceae as an archaic one and group it along with Magnolias, the most primitive of flowering plants.

Nutmeg is indigenous to Moluccas islands in Indonesia. Semi-domesticated gene pools of *Myristica* species do not exist as most of these species occur in wild. *M. fragrans* is typically dioecious, with male and female flowers on different trees. Occasionally, male trees carrying a few female flowers are observed. Hermaphrodite trees having bisexual flowers are rarely noticed. Nutmeg is cultivated in Ernakulam, Trichur, Kottayam and Trivandrum districts of Kerala and also in a few areas in Tamil Nadu, Karnataka, Maharashtra, Goa, Andhra Pradesh, North East India and Andaman and Nicobar islands. The oldest nutmeg population in Kerala are in Kalady and Pala and are more than 150 years old. Nutmeg is usually grown in river banks as it grows luxuriously in silts deposited by rivers. It is wind pollinated. Inflorescence is a branched raceme in male and simple cyme in female. Male inflorescence has more flowers (up to 10) while female has less (up to 3).

Ecosystem Diversity

Myristica swamps are dominated by members of Myristicaceae. Krishnamoorthy (1960) reported *Myristica* swamp, for the first time, as a special type of habitat from Travancore. These swamps were found in valleys of Shendurney, Kulathupuzha and Anchal forest ranges in Southern Western Ghats. *Myristica* swamps are also reported in Uttara Kannada District of central Western Ghats in Karnataka. These swamps are isolated and situated in localities from near sea level to about 450 m altitude. The northernmost swamp known is associated with a sacred grove in the Satari taluk of Goa. *Myristica* swamps are also reported from New Guinea (Corner, 1976).

Western Ghats have three genera and five species of Myristicaceae; all of them are trees associated with evergreen to semi-evergreen forests. Of these *Gymnacranthera canarica* and *Myristica fatua* var. *magnifica* are exclusive to the swamps. *M. malabarica* is occasional in the swamps and more frequent in the evergreen forests. *M. malabarica*, often produces stilt roots and flying buttresses, even though it is seldom associated with swamps, indicating its possible origin in swamps. Myristicaceae was the most dominant family of the swamps forming 32 per cent of total number of trees. Within the Myristicaceae, *G.* canarica accounted for 78 per cent, followed by *M. fatua* var. *magnifica* (19 per cent), *K. attenuata* (2 per cent) and *M. malabarica* (1 per cent). Since two species namely. M. *fatua* and *G. canarica* seem to be the most characteristic of *Myristica* swamps, distribution of these species can be an indicator of distribution of *Myristica* swamps.

Cultivar Diversity

In a study to identify sex segregation in nutmeg, it was observed that out of 90 progenies, 40 were males, 45 females and 5 bisexuals. A preliminary analysis of genetic variability in 28 trees (14 years old) indicated variability only for fruits/tree. Correlation analysis revealed a significant negative correlation of fruits/tree and mace weight. Seed weight had a positive significant correlation with mace weight (Krishnamoorthy *et al.*, 1991b).

In a more systematic study on genetic variability of nutmeg, progenies from 16 mother trees of different localities (five progenies for each mother tree), lack of adequate genetic variability was evident for many of the important attributes. This particular study revealed that even though morphological variation exists for leaf shape and canopy shape in nutmeg populations, exploitable genetic variation for crop improvement in nutmeg is lacking (Krishnamoorthy *et al.*, 1996).

Progeny evaluation aims in analyzing breeding behaviour of certain elite trees for sex ratio and prepotency so that such promising trees are selected for seeds/scion collection. Significant differences were observed among populations for plant height, number of main shoots, number of years for flowering, fruit weight and ratio of mace weight to seed weight. Phenotypic coefficient of variation is more than the genotypic coefficient of variation, indicating role of environment in the expression of these characters.

Additive genetic factors are attributed for variations due to comparatively low estimates of heritability and genetic advance of these traits. Mace weight to seed weight ratio has a very high heritability and very good genetic advance. Selection for this trait may be very effective. Non-significant variation was observed for canopy size, number of erect shoots, number of fruits and girth. This may be due to narrow genetic pool introduced into India, from which present day population evolved. There is an urgent need to introduce genetic variability.

Variability of growth, flowering, and fruit set of 39 seedling progenies of nutmeg was studied at Ratnagiri, Maharashtra. Considerable variation among genotypes with respect to growth parameters was noticed. High magnitude of phenotypic coefficient of variation and genotypic coefficient of variation indicated good amount of variation among genotypes for these characters. High estimates of heritability and genetic advance for fruit set suggested that it is under control of additive gene action (Haldankar *et. al.* 2004).

Promising Accessions

A few nutmeg accessions (A9-4, A9-20, A9-22, A9-25, A9-79, A9-86, A4-12, A4-22, A4-52, A11-29, A11-70) were identified as promising at IISR, Calicut. Five accessions namely N-72, N-29, N-70, N-49 and N-74 were identified as promising from College of Agriculture, Dapoli (Haldankar *et al.*, 2004). A super clonal selection A9-4 was released as IISR Viswashree from Indian Institute of Spices Research, Calicut. A hermaphrodite variety Kokan Suganda was released from Konkan Krishi Vigyan Peeth, Dapoli.

Quality Lines

Sixty-five nutmeg germpalsm accessions conserved at Indian Institute of Spices Research were evaluated for chemical composition of nutmeg and mace and high variability was observed among accessions. Essential oil content ranged from 3.9 to 16.5 per cent in nutmeg and 6 to 26.1 per cent in mace. Myristicin content ranged from 1.1 to 45.6 per cent in nutmeg oil and 0.21 to 36.6 per cent in mace oil; the elemicin content ranged from 1.0 to 29.7 per cent in nutmeg oil and 1.0 to 30.2 per cent in mace oil. Safrole content ranged from 0.1 to 22.1 per cent in essential oil and 0.2 to 21.8 per cent in mace oil. (Maya *et al.*, 2004).

Accessions with high oil yield in nut and mace, high butter content, high oleoresin in nut and mace, high myristicin and elemicin, low myristin and elimicin were identified from germpalsm available at Indian Institute of Spices Research and the details are given in Table 12.8.

Conservation of Germplasm

Germplasm collections of nutmeg at IISR include 484 accessions. Important species conserved at IISR include *M. fragrans*, *M. fatua* var. *magnifica*, *M. malabarica*, *M. beddomeii*, *M. andamanica*, *M. attenuata*, *M. prainii*, *M. amygdalina*, *Gymnacranthera canarica* and *Knema andamanica* (Table 12.5). Unique types are a tree bearing 1-4 seeds/fruit, an endangered species, M. *fatua* var *magnifica* and a few elite lines. Another bold accession A9-69 and a 'bald type' are of importance. The germplasm is conserved in field repository at Indian Institute of Spices Reseach Calicut. Germpalsm of nutmeg is also conserved at NBPGR Regional Station, Thrissur, Kerala Agricultural College, Tamil Nadu Agricultural College, College of Agriculture, Konkan Krishi Vigyan peeth, Dapoli.

Table 12.8: High Yielding Accession of Nutmeg

Category	Accession	Remarks
Essential oil in nutmeg	A9-18	16.5%
Essential oil in mace	A9-18	26.1%
Butter	A11-12	44%
Oleoresin in nutmeg	A9-30 A9-116	5.3% (acetone extract)
23 per cent (ethanol extract)		
Oleoresin in mace	A4-22	32.2% (acetone extract)
High myristicin and elemicin in nutmeg oil	A4-17, A4-20, A9-4-12, A9-4-13, A9-4-15	
High myristicin and elemicin in mace oil.	A4-17, A9-4-1, A9-4-3, A9-4-8, A9-4-11	
High myristicin in both nutmeg and mace oils.	A4-17	
Low myristicin and elemicin levels in nutmeg oil	A4-22, A9-69, A9-71, A9-95 and A9-102	
Low myristicin and elemicin levels in mace oil	A9-1, A9-44, A9-71 and A9-95	
Low myristicin, low elemicin and low safrole coupled with high sabinene in both nutmeg and mace oils	A9-71 and A9-95	

Rare/Threatened Species

Myristica swamps are endangered habitats. The species *M. magnifica* var *fatua, M. malabarica* and G. canarica are rare/threatened and need to be conserved. *M. malabarica* is an endangered crop plant (Hammer and Khoshbakht 2005). *M dactyloides* included as red listed species from Charmadi and Kammangundi (Western Ghats) is now found widely distributed (Aravind *et al.,* 2005).

Conclusion

Conservation of *Myristica* swamps with their ancient and unique biota, and associated ecological value, is of paramount importance as they are priceless possessions for evolutionary biologist. These swamps are virtually live museum of ancient life of great interest to biologists. The forest departments of the respective states, with help of biologists and conservationists should protect these fragmented swamps for prosperity. The surviving *Myristica* swamps are declared as specially protected areas. A major threat to the Travancore *Myristica* swamps was their conversion into rice fields. It is important to prevent conversion of swamps into paddy fields and betelnut gardens to protect these swamps. Since *Myristica* swamps are the most endangered and fascinating habitats/ecosystems, they have good potential for developing ecotourism. If myristica swamps can be associated with sacred groves, traditional conservation practices can be adopted.

Clove

Clove of commerce is the dried, aromatic, fully grown but unopened flower buds of *Syzygium aromaticum* (L.) Merr. and Perry (Syn. *Eugenia caryophyllus* (Sprengel) Bullock and Harrison) (Myrtaceae).

Clove is indigenous to Moluccas where it occurs as a second storey forest tree on the lower mountain slopes. The Dutch Government restricted cultivation of clove exclusively to Moluccas but by 19th century beginning, it was smuggled to Mauritius and later on established in the islands of Zanzibar and Pemba. These islands, now part of Tanzania, have become the world's largest producer of clove followed by Indonesia. Under natural conditions, both aromatic and non-aromatic trees occur and the selective culture of aromatic trees led to present clove population in the world all over (Verheiji and Snijders, 1989). In India, clove cultivation is limited to parts of Kerala and Kanyakumari, Tirunelveli and the Nilgiris of Tamil Nadu. The clove plantation within India originated from a few seedlings introduced from Mauritius.

Clove belongs to genus *Syzygium* Gaertn. (Myrtaceae) with about 500 species. Distinction between *Syzygium* and its closely related *Eugenia*, is a matter of controversy. Henderson (1949) maintained only the single genus *Eugenia* explaining that the above distinction is transgressed. Schmid (1972) seems to provide convincing evidence including floral anatomy for separation of *Eugenia* and *Syzygium*. The former name is now restricted to New World Taxa while all Asiatic species are included in *Syzygium*. The correct name of clove as accepted today is *Syzygium aromaticum* (L.) Merr and Perry.). Genus *Syzygium* Gaertn. has spread to many tropical countries. Many species of *Syzygium* occur in Indian sub-continent, but they are all very distant species resembling clove only on taxonomic grounds. A detailed listing of these species is not attempted here. Important species of *Syzygium* occurring in Western Ghat forests are *S. aromaticum, S. cuminii, S. fruticosum, S. jambos, S. zeylanicum, S. travancoricum* and *S. jambolana.*

Flowers of clove are hermaphrodite, borne at terminals in trichotomonous panicles, with 3-20 flowers/panicle. Floral biology of cloves favours self pollination. Self pollination is reported to be a more probable mechanism for pollination in clove, as maximum pollen viability and stigma receptivity are attained simultaneously. Flowers are frequently visited by ants, thrips and bees, suggesting possibility of transferring pollen from anther to stigma of same flower or cross pollination.

Genetic base of clove in India is narrow because the original number of introductions is a few. Clove tree appears to be of uniform type and no distinct varieties/cultivars are recognized. It is also known that the tree population in Zanzibar and Madagascar are also rather uniform, but in Indonesia, 3 types are distinguished. These types differ in tree habit, leaf size, clove size and colour. In India, differences are observed in tree shape, bearing habits, cropping season, yield, colour of petiole, shape and size of clove.

Seed and seedling characters of progenies of 14 elite clove trees revealed no appreciable variation for 100 fruit weight, 100 seed weight, fruit breadth, fruit length, seed breadth, seed length, germination (per cent), root and shoot length and number of leaves of seedlings at 40 days after sowing, one and two years old plant height among the progenies of elite lines at seedling stage (Krishnamoorthy and Rema, 1994). Morphological variability in seedling characters of progenies of indigenous accessions was very narrow.

Characteristic odour and flavour properties of clove bud, leaves and stem are determined by composition of their aromatic steam volatile oils. Odour and flavour characteristic of these three types of oils differ significantly owing to relative abundance of minor and trace components present. Major component of the volatile oil is eugenol. In buds, the aromatic steam volatile oil can be as high as 17 per cent. The oil recovery in stem ranges from 4 to 7 per cent and leaf is 0.5 to 3 per cent.

Yield in clove is determined by the age, flower buds/inflorescence, number of inflorescence etc. The flowers/inflorescence vary from 3 to 20 and inflorescences/shoot from 1-7 in different accessions.

At present, crop improvement programme in clove is limited to selection of mother trees based on their regular and heavy bearing nature by surveying clove growing tract, besides selection in germplasm. Recently with introduction of dwarf gene source in conservatory, it is proposed to develop a dwarf, bushy and early bearing tree with bold cloves. This would favour easy harvest and other cultural operations. High oil yielding accessions would be beneficial for distillation of oil. Breeding trees with high leaf oil and bud oil would also be advantageous, but is not easy to achieve these aims due to lack of wide variability in population. The only way is to introduce further variability from Zanzibar and Indonesia and then initiate a cross breeding programme. Clonal propagation of selected mother trees, avoiding a juvenile phase may result in a breakthrough in productivity.

Conservation

Surveys were conducted in major clove growing areas of Tamil Nadu and Kerala and a total of 408 accessions are conserved in germplasm repository of clove at Indian Institute of Spices Research, Calicut. Special efforts were made to identify and collect diverse types from Ashamboo hills, the southern most hills of India, where clove was first introduced. The distinctly diverse collection of clove includes three morphological variations *viz.*, dwarf and bushy types, accession with exceptionally bold and thick clove buds (king clove) and narrow leaved variants (Krishnamoorthy and Rema 1994), whereas a reddish purple petioled Zanzibar type from Zanzibar island and one from Sri Lanka are the only introductions at present. Very rarely clove tree with very small buds called *lilliput* clove is also observed. A dwarf gene source in clove has important breeding value in developing dwarf genotypes with compact canopy size. Some of the impotant species of *Syzygium* conserved at IISR are *S. aromaticum, S. heynianum, S. jambos, S. jambolana, S. fruiticosum, S. cuminii, S. caryophyllatum, S. lanceolatum* and *S. zeylanicum* (Table 12.5).

Syzygium paniculatum Gaertner is listed among the vulnerable crop plant list (Hammer and Khoshbakht 2005).

Allspice

Allspice, *Pimenta dioica* (L.) Merr. (syn: *P. officinalis* Lindl., *Myrtus pimenta* L., *M. dioica* L. and *Eugenia pimenta* DC (Merrill 1947) is a polygamodioecious evergreen tree, the dried unripe fruits of which provide the culinary spice, pimento, of commerce. Leaves are aromatic and are used in distillation. Berries are used for preparation of essential oil and oleoresin. It belongs to family *Myrtaceae* and is known in English as allspice or pimento. Family *Myrtaceae* consists of about 3000 woody species, most of which are grown in tropics. The genus *Pimenta* Lindl. consists of about 18 species of aromatic shrubs and trees native to tropical America (Willis 1973). The genus is closely related to *Myrtus* L. and *Eugenia* L. The commercially important *Pimenta* spp. are *Pimenta dioica* (L.) Merr. providing the spice pimento (allspice) and *P. racemosa* (Mill) Moore, bay or bay rum tree providing oil of bay.

Allspice is a small, functionally dioecious evergreen tree, 7–10 m tall, slender trunk profusely branched at their extremities. Bark, smooth and shiny, pale silvery brown, shedding in strips of 25–75 cm long at intervals. Leaves, borne in clusters at ends of branches, simple, opposite, entire, thinly, coriaceous, punctate with pellucid glands, aromatic when crushed. Flowers structurally hermaphrodite, but functionally male or female, white, aromatic, 8–10 mm diameter.

Tree is indigenous to West Indies. Jamaica is the major producer and exporter of the spice. Trees are also found in Central America (Mexico, Honduras, Guatemala, Costa Rica and Cuba) and in neighbouring Caribbean islands, although its original home is in dispute.

Allspice was introduced into West Indian Islands (Grenada, Barbados, Trinidad and Puerto Rico) from its place of origin. Attempts to introduce it into countries in tropical regions namely, India, Sri Lanka, Fiji, Malaysia, Singapore and Indonesia (Java, Sumatra) have not succeeded fully. In India, a few trees are available in Maharashtra, Tamil Nadu, Karnataka and Kerala. In India, the plant is grown in gardens, especially in Bengal, Bihar and Orissa. There are a few allspice trees available in Nagarcoil, Kallar, Burliar (Tamil Nadu) and Horticultrual Research Station, Ambalavayal.

The IISR, Calicut has a germpalsm collection of 180 trees. Variants are rarely reported in allspice. Two seedling variant types with dwarf/semi-dwarf habit and short internodes and bushy nature possessing a large number of branches are being conserved in the field germplasm repository of Indian Institute of Spices Research, Calicut, Kerala, India (Krishnamoorthy *et al.*, 1997). The leaves were smaller (about 1/3 the size) when compared to normal leaves. The variants were multiplied clonally through approach grafting and all the clones exhibited the parental character (Mathew *et al.*, 1999). This dwarf/semi-dwarf plant type in allspice with large number of branches offers great potential in utilizing them in crop improvement programmes.

Conclusion

Our biotic resources are under increasing pressure due to over exploitation of plant and animal resources, changing land use, habitat loss and fragmentation, growing pollution of soil, water and air and the green house effect. The effective forest cover in India is limited to less than 11 per cent of its geographic area, whereas the forest cover required for ecological stability is 33 per cent. According to the State Forests Report, 1997 of the Ministry of Environment and Forests, the forest cover has gone down from 638,879 sq. km in 1995 to 633,397 sq. km. The enormous genetic diversity is being lost mainly due to genetic erosion, genetic vulnerability and genetic wipe out. In addition, increasing population, development pressures on land resources, deforestation changes in land use pattern and natural disasters contribute to the genetic wipe out of promising diversity and loss of genetic resources. Now with increasing importance of plant genetic resources especially in the context of intellectual Property Rights, Plant Breeders' Rights and Biodiversity Convention, there is an urgent need to conserve biodiversity. Identification and documentation of rare, threatened and endemic species is important in conservation of biodiversity. These species usually have specific ecological niches, edaphic gradients, and vulnerable habitats. *In situ* conservation is the best way to conserve plants, including varied germplasm of different species. The hot spots have to be safe guarded to save species under risk or that have found a place in red data book. Since *Myristica* swamps are highly endangered habitats, it would be better to term them as "burning spots" within a so-called "hot spot."

There is no record of systematic conservation of tree spices in their place of origin. As *ex-situ* conservation, IISR Calicut established germplasm repository of tree spices. The repository contains a wide range of cultivated, wild and related species of nutmeg, cinnamon, garcinia, allspice and clove collected from within India and a few from outside the country. An alternative to save genetic resources is to conserve them under controlled condition. Cryopreservation of appropriate tissue is one such options. Germplasm repository has to be maintained and preserved to be available for continued improvement of economically important spice crops. Since the endemic species are found in restricted areas, often requiring special niche for survival, many of them, especially those with very restricted distribution, are extinction-prone, particularly if their habitats are disturbed. At the global level, Hotspots are areas of high conservation priority, because if unique species are lost they can never be replaced.

References

Ahamedullah, M. and Nayar, M.P. 1986. *Endemic Plants of the Indian Region*. Botanical Survey of India, Calcutta, p. 261.

Aravind, N.A., Manjunath, J., Rao, D., Geneshaiah, K.N., Shaanker, R.U. and Vanaraj, G., 2005. *Current Science*, 8(2): 258–265.

Babu, L., Mathew, P.A. and Krishnamoorthy, B., 2002. Biodiversity in cinnamon germplasm: A qualitative evaluation. *National Seminar on Strategies for Increasing Production and Export of Spices*, Calicut, 24–26 October, p. 4.

Bakker, M.E., Gerritsen, A.F. and Van Der Schaaf, P.J., 1992. Leaf anatomy of *Cinnamomum* Schaeffer (Lauraceae) with special reference to oil and mucilage cells. *Blumea*, 37: 1–30.

Baruah, A., Nath, S.C. and Boissya, C.L., 2000. Systematics and diversities of *Cinnamomum* species used as Tejpat spice in North-east India. *J. Econ. Taxon. Bot.*, 24(2): 361–374.

Bradu, B.L. and Sobti, S.N., 1988. *Cinnamomum tamala* in N.W. Himalayas; evaluation of various chemical types for perfumery value. *Indian Perfumer*, 32(4): 334–340.

Corner, F.J.H., 1976. *The Seeds of Dicotyledons*. Cambridge University Press, Cambridge.

Cronquist, A., 1981. *An Integrated System of Classification of Flowering Plants*. Columbia University Press, New York.

CSIR, 1956. *The Wealth of India: Raw Materials*, New Delhi, India, 4: 101–103.

Dahlgren, R., 1983. General aspects of angiosperm evolution and macro systematics. *Nord. J. Bot.*, 3: 119–149.

Fujita, Y., 1967. Classification and phylogeny of the genus *Cinnamomum* in relation to the constituent essential oils. *Bot. Mag.* Tokyo, 80: 261–271.

Gamble, J. S., 1925. *Flora of Madras Presidency*, Vol. II. (1967 reprint). Botanical Survey of India, Calcutta.

Gawanker, M.S., Shingre, D.V. and Jamadagni, B.M., 2001. Variability for morphological properties of fruits in Kokum. Abstracts, National Seminar on Kokum, May 12–13, Regional Fruit Research Station Vengurla, Maharashtra, p. 2.

Gulati, B.C., 1979. *Proc. Symp. Production and Utilization of Forest Products*, RRL, Jammu, 1: 130–139.

Haldankar, P.M., Joshi, G.D., Patil, B.P. and Haldavnekar, P.C., 2004. Variability for growth, flowering and fruit set in seedling progenies of nutmeg (*Myristica fragrans* Houtt.). *Journal of Spices Aromatic Crops*, 13(1): 28–33.

Hammer, K. and Khoshbakht, K., 2005. Towards a red list for crop plant species. *Genetic Resources and Crop Evolution*, 52: 249–265.

Henderson, M.R., 1949. The genus *Eugenia* (Myrtaceae) in Malaya. *Garden's Bull.*, 12: 1–293.

Hirota, N. and Hiroi, M., 1967. Later studies on the camphor tree, on the leaf oil of each practical form and its utilization. *Perf. Essential Oil Record*, 58(6): 364–367.

Hooker, J.D., 1886. *The Flora of British India*, Vol. V. Bishan Singh Mahendrapal Singh, Dehradun (reprint, 1973).

Hussain, A., Virmani, O.P., Sharma, A., Kumar, A. and Misra, L.N., 1988. *Major Essential Oil Bearing Plants of India*. Central Institute of Medicinal Plants, Lucknow, pp. 213–214.

Joy, P.P., Thomas, J., Mathew, S. and Ibrahimm, K.K., 1998. Growth leaf oil yield and quality investigations in Cinnamon (*Cinnamomum verum*). *J. Med. Aromatic Plants*, 28: 401–406.

Karnik, A.R., Raorane, G.P. and Bhagwat, N.R., 2001. Flower types biology and sex forms in kokum (*Garcinia indica* chiosy) (Abstract). *National Seminar on Kokum*, May 12–13, Regional Fruit Research Station Vengurla, Maharashtra, p. 3.

Khein, P.V., Chaien, H.T., Duang, N.X. and Leclercq, P.A., 1998. Chemical segregation of pogeny of camphor trees with high camphor and linalool content. *J. Essent. Oil Res.*, 10: 607–612.

Khoshoo, T.N., 1995. Census of India's biodiversity: Tasks ahead. *Curr. Sci.*

Kostermans, A.J.G.H., 1964. *Bibliographia Lauracearum*, Bogor Bot. Gardens, Dhakarta.

Kostermans, A.J.G.H., 1983. South Indian species of *Cinnamomum schaeffer* (Lauraceae). *Bulletin of the Botanical Survey of India*, 25(14): 90–133.

Krishnamoorthy, B. and Rema, J., 1994. Characterization of seedling progenies of elite lines of clove. *Indian Cocoa Arecanut and Spices J.*, 18(3): 82–84.

Krishnamoorthy, B. and Rema, J., 1994. Three promising morphological variants in clove (*Syzygium aromaticum*) (L.) Merr. and Perry) from Tamil Nadu, India. *Journal of Spices and Aromaticum Crops*, 3(2): 168–169.

Krishnamoorthy, B., Gopalam, A. and Jose Abraham, 1988. quality parameters of *cinnamom* (*Cinnamomum verum*) in relation to flush colour. *Indian Cocoa Arecanut and Spices*, 12(2): 38.

Krishnamoorthy, B., Rema, J. and Sasikumar, B., 1991a. Progeny analysis in cinnamon. *Indian Cocoa, Arecanut and Spices J.*, 14(3): 124–125.

Krishnamoorthy, B., Sasikumar, B., and Rema, J., 1996. Genetic variability and segregation of sex in nutmeg (*Myristica fragrans* Houtt.). *Journal of Plantation Crops*, 24 (Suppl.): 468–472.

Krishnamoorthy, B., Sasikumar, B., Rema, J. and George, J.K., 1997. Genetic resources of treespices and their conservation in India. *Plant Genetic Resources Newsletter*, 111: 53–58.

Krishnamoorthy, B., Sasikumar, B., Rema, J. and Gopalam, A., 1992. Variability and association studies in cinnamom (*Cinnamomum verum*) *J. Spices and Aromatic Crop*, 1(2): 148–150.

Krishnamoorthy, B., Sasikumar, B., Rema, J., Sayed, A.A.M. and Jose Abraham, 1991b. Variability and association in nutmeg. *Indian Cocoa, Arecanut and Spices J.*, 14(3): 121–122.

Krishnamoorthy, K., 1960. In: *Tropical Moist Evergreen Forest Symposium*. Forest Research Institute Dehra Dun.

Krishnamoorthy, B., John Zachariah, T., Rema, J., and Mathew, P. A., 1999. Evaluation of *sected Chinese cassia (Cinnamomum cassia) accessions for chemical quality*. *Journal of Spices and Aromatic Crops*, 8(2): 193–195.

Kumar, S., Singh, J., Shah, N.C. and Ranjan, V., 1997. *Indian Medicinal and Aromatic Plants Facing Genetic Erosion*. CIMAP, Lucknow, p. 46, 220.

Mabberly, D.J., 1987. *The Plant Book*. The Cambridge University Press, Cambridge.

Manilal, K.S. and Shylaja, M.1986. A new species of Cinnamomum Schaeffer (Lauraceae) from Malabar. *Bull. Bot. Surv.* India, 28: 111–113.

Manilal, K.S., 1988. *Flora of Silent Valley Tropical Rain Forests of India*. Mathrubumi Press, Calicut.

Mathew, P.A., Krishnamoorthy, B. and Rema, J., 1999. Seedling variants in allspice (*Pimenta dioica* L. Merr.). *Journal of Spices and Aromatic Crops*, 8(1): 93–94.

Maya, K.M., John Zachariah, T. and Krishnamoorthy, B., 2004. Chemical composition of essential oil of nutmeg (*Myristica fragrans* Houtt.) accessions. *J. Spices Aromatic Crops*, 13(2): 135–139.

Merrill, E.D., 1947. The technical name of allspice. *Contr. Gray Herb*, 165: 30–38.

Muhammed, M., Rosen, R., McCarty, M., Conte, A., Patil, D. and Butrym, E., 1994. *Citrin: A Revolutionary herbal Approach to Weight Management*. New Editions Publishing, Burlingame, CA.

Myers, N., Mittermeler, R.A., Mittermeler, C.G., Fonseca da, G.A.B. and Kent, J., 2000. *Nature*, 403: 853–858.

Nath, S.C., Baruah, A. and Hazarika, A.K., 1999. Essential oils of leaves of *Cinnamomum* Schaeffer members. *Indian Perfumer*, 43(4): 182–190.

Nath, S.C., Hazarika, A.K. and Singh, R.S., 1994. Essential oil of leaves of *Cinnamomum tamala* Nees. and Eberm. from North East India. *J. Spices and Aromatic Crops*, 3(1): 33–35.

Nayar, M.P., 1996. *Hot Spots of Endemic Plants of India, Nepal and Bhutan*. The Director, TBGRI, Trivandrum, p. 252.

Nigam, M.C. and Ahmed, A., 1990. The essential oil of *Cinnamomum tamala II PAFAI*, 12(3): 24–35.

Pujari, K.H. and Joshi, G.D., 2001. Physicochemical characterization of spreading type of kokum. In: (Abstracts) *National Seminar on Kokum*, May 12–13, Regional Fruit Research Station Vengurla, Maharashtra, p. 5.

Ranatunga, J., Senanayake, U.M. and Wijesekera, R.O.B., 2004. Cultivation and management of Cinnamon. In: *Cinnamon and Cassia The genus Cinnamomum*, (Eds.) Ravindran, P.N., Babu, K.N. and Shylaja, M. CRC Press, Washington, D.C.

Ravindran, N.R., Manilal, K.S. and Ravindran, P.N., 1992. Chemotaxonomy of *Cinnamomum* Schaeffer occurring in Western Ghats. *J. Indian Bot. Soc.*, 71: 37–41.

Ravindran, P.N., Babu, K.N. and Shylaja, M., 2004. *Cinnamon and Cassia The genus Cinnamomum* CRc Press, Washington D. C.

Roberts, E., 1984. Vegetable *Materia Medica of India and Ceylon*. Bishen Singh Mahendra Pal Singh, Sehra Dun, India.

Rodrigues, B.F., 2001. Some medicinal uses of *Garcinia indica* (Thouars) Choisy (Abstracts). *National Seminar on Kokum*, May 12–13, Regional Fruit Research Station Vengurla, Maharashtra, p. 21.

Schmid, R., 1972. A resolution of the *Eugenia–Syzygium* controversy (Myrtaceae). *American J. Bot.*, 59: 423–36.

Shinde, A.K., Godse, S.K., Dalvi, M.B., Bhole, S.R. and Patil, B.P., 2001. Survey of kokum seedling for elite type from Sindhudurg District. (Abstract) *National Seminar on Kokum*, May 12–13, Regional Fruit Research Station Vengurla, Maharashtra P.5.

Shylaja, M., 1984. *Studies on Indian Cinnamomum. Ph.D. Thesis*, University of Calicut.

Shylaja, M., Ravindran, P.N. and Babu, K.N., 2004. Other useful spices of cinnamomum. In: *Cinnamon and Cassia the genus Cinnamomum*, (Eds.) Ravindran, P.N., Babu, K.N. and Shylaja, M. CRC Press, Washington, D.C.

Sinclair, J., 1958. *Flora Malesianae precursors XL II, the genus Myristica in Malesia and outside Malesia*, pp. 1–540.

Sood, R.P., Padha, C.D., Talwar, Y.P., Jamwal, R.K. and Rao, P.R., 1979. Essential oil from the leaves of *Cinnamomum tamala* Nees et Eberm. Growing in Himachal Pradesh. *Indian Perfumer*, 23: 75–789.

Talbot, W.A., 1902. *The Tree, Shrubs and Woody Climbers of Bombay Presidency*, 2nd edn. Govt. Central Press, Bombay, 385 pp.

Tetenyi, P., 1970. *Infraspecefic Chemical Taxa of Medicinal Plant*, Chem. Pub. Co., New York.

Verheiji, E.W.M. and Snijders, C.H.A., 1989. *Syzygium aromaticum* (L.) Merril and Perry. In: *Plant Resources of South East Asia*, (Eds.) Westphal, E. and Jansen, P.C.M. PUDOC, Wangeniangen.

Wan-Yang, S., Wei, H., Guang-yu, W., De-xuan, G., Guang-yuan, L. and Yin-gou, L., 1989. Study on chemical constituents of the essential oil and classification of types form *Cinnamomum camphora*. *Acta Bot. Sinica*, 31: 209–214.

Warming, E., 1890. Cited by Garratt, 1933. Bearing of wood anatomy on the relationships of the Myristicaceae. *Tropical Woods*, 36: 20–44.

Weiss, E.A., 1997. *Essentail Oil Crops*. CAB International, UK.

Wijesekera, R.O.B., Ponnuchamy, S. and Jayewardene, A.L., 1975. *Cinnamon*. Ceylon Institute of Scientific and Industrial Research, Colombo, Sri Lanka.

Willis, J.C., 1973. *A Dictionary of the Flowering Plants and Fems*, 8th Edn., Revised by H.K. Airy Shaw; (Cambridge, U.K., Cambridge Univ. Press).

Wilson, T.K. and Maculens, L., 1967. The morphology of the Myristicaceae.I. Flowers of *Myristica fragrans* and *M. malabarica*. *Amer. J. Bot.*, 54: 214–220.

Biodiversity in Horticultural Crops Vol. 1 *Pages 278–287*
Editors: **Professor K.V. Peter & Dr. Z. Abraham**
Published by: **DAYA PUBLISHING HOUSE**

Chapter 13

Cardamoms

J. Thomas, K.J. Madhusoodanan and V.V. Radhakrishnan
Indian Cardamom Research Institute, Spices Board of India,
Myladumpara, Idukki, Kerala

Cardamom of commerce (also known as small cardamom) is the true cardamom, the 'queen of spices', and is obtained from the plant botanically known as *Elettaria cardamomum* Maton, a member of the family *Zingiberaceae*. In addition, there are many other plants belonging to the genera *Amomum* and *Aframomum*, both belonging to the ginger family, producing aromatic seeds. Among them the most important and the one that is being grown commercially is *Amomum subulatum* Roxb, the Nepal cardamom or large cardamom. The others are clubbed together as false cardamoms.

Small cardamom is a native of the moist evergreen forests of Western Ghats of Southern India. Cardamom is also grown in Sri Lanka, Papua New Guinea, Tanzania, Vietnam and in Guatemala. It is grown on a plantation scale in Guatemala, which incidentally is now the largest producer of cardamom. Area under cardamom in India is around 73,237 ha, spread over three southern states of Kerala, Karnataka and Tamil Nadu at an altitude of 900 to 1200 above MSL and the tract extends from Thirunelveli (Tamil Nadu) in the south up to Sirsi (Karnataka) covering over 1000km through cardamom hills of Kerala. Area under the crop in Guatemala is about 40,000ha, Sri Lanka 300ha and other producing countries less than 300ha. Total annual world production is on the increase and is presently around 32,000 tonnes of which India contributes 12000 to 14000 tonnes, Guatemala 16000 to 18000 tonnes and remaining by others. Major consumers are Saudi Arabia, India, Japan and Scandinavian countries.

Cardamom is popular as a spice and food additive due to its delicious flavour. Its volatile oil contributes to the fragrance. Its pharmacological properties give a place in various systems of medicine. The biological activities of its constitutes find it a place in confectionary, seasoned and prepared foods.

Large cardamom is indigenous to moist deciduous and evergreen forests of sub-Himalayan tract. It is an important cash crop of Sikkim from where it spread to other North-Eastern states of India and neighbouring countries like Nepal and Bhutan. India is the second largest producer of large cardamom and the crop presently occupies an area of 30,039 ha with an annual production of 6,154 tonnes (Anonymous, 1994).

Genetic Diversity of Cardamom

Based on nature of panicles, three varieties of small cardamom are recognized. The var. *Malabar* is characterized by prostrate panicle and var. *Mysore* possesses erect panicle. The third var. *Vazhukka* is a natural hybrid between the two and its panicle is semi erect or flexuous. In India, a number of other cultivars of cardamom are also recognized. In general, they can be considered as ecotypes of var. *Mysore, Malabar* or *Vazhukka*. Most common among them are *Manjurabad, Bijapur, Kannielam, Makara elam, Thara* and *Nadan*.

Genetic Resources

The out breeding nature and lack of compatibility barriers contributed towards the rich diversity of the crop under cultivated and wild conditions. Seed propagation also adds to the same. Explorations for germplasm collection were initiated way back in 1950 and thereafter, many collections were made by as many as six research organizations in India. During 1994, a descriptor for cardamom was brought out by the International Plant Genetic Resources Institute (IPGRI), Rome, Italy. Among the cardamom collections, genotypes having marker characters include terminal panicle, narrow leaves, pink coloured tillers, compound panicles, elongated pedicel etc. Asexuality, cleistogamy and female sterility are a few of the variations observed among the collections.

Explorations hence resulted in the collection of a large number of variants in cardamom and these are being conserved under *ex situ* conditions for characterization and utilisation. The field gene bank of the Indian Cardamom Research Institute at Myladumpara in Idukki district of Kerala and its Regional Stations at Saklespur and Thadiyankudisai have around 800 cardamom accessions and is recognized as the National Centre for Genetic Resources of cardamom. The gene bank has collections from plantations and natural habitats. The centres under All India Co-ordinated Research Project on Spices also maintain cardamom collections from respective localities.

Characterization

Characterization involves morphological, agronomical as well as chemical characters. Many variations in morphological and chemical characters and in yield were recorded in these collections (Madhusoodanan *et al.*, 1994).

Utilisation

Cardamom diversity was exploited in crop improvement programmes employing various techniques particularly selection and hybridisation. These techniques are primarily targeted to evolve superior varieties/selections having high yield, good quality capsules and tolerant to abiotic and biotic stresses. A few high yielding varieties were released (ICRI-1, ICRI-2, ICRI-3, ICRI-4, Pv-1, Pv-2, Mudigere P-1, CCS-1, RR-1 etc) and a few lines are under evaluation for release (MCC-12, MCC-21, MCC-73, MCC-85, MCC-346 etc.). Efforts are made to incorporate desirable traits through hybridisation programme. MHC-10, MHC-13, MHC-18, MHC-23, MHC-26 and MHC-28 are a few elite hybrids evolved. Performance of these clones/hybrids is presently under investigation at different agro climatic zones.

In recent years, a few improved ' land races' in cardamom were identified by farmers. 'Njallani Green Gold' is the most popular one adapted for large-scale cultivation in the cardamom Hill Reserve of Idukki district. Besides, landraces like Vali Green Bold, Vander cardamom, Palakkudi, Panikulangara and PNS Vaigai are also getting popularised.

Large Cardamom

The genus *Amomum* comprises of a large number of species occurring in India, Malaysia, Indonesia and in other Pacific ocean islands. In India, there are two independent centers of distribution for the genus and they are North-Eastern India and the Western Ghats. About eighteen species are reported from the North-Eastern region and from Sikkim Himalaya, six species are reported, out of which *Amomum subulatum* is economically the most important species. Its cultivation is mainly confined to an elevation ranging from 900-2000 m above MSL at a temperature varying between 10 and 30°C and rainfall 2000-3000mm per annum. A large number of cultivars are identified and the most popular among them are *Ramsey, Sawney, Serumna, Varlangey* and *Golsey*. Out-breeding nature of crop coupled with geographical isolation of the hills contributed to enormous crop diversity.

Exploration for germplasm in large cardamom has a long history and earlier it was reported that germplasm accessions were collected and conserved by the ICAR Research Complex at Gangtok, Sikkim. Consequent to the establishment of the ICRI Regional Station at Gangtok, efforts are focused on conducting explorations and collection of genetic resources of large cardamom from throughout Sikkim, Darjeeling District of West Bengal and adjoining areas.

Diversity available in large cardamom exploited so far resulted in identification of a few breeding stocks and a few elite clones. Variants include plants with narrow leaves, dwarf type, terminal spike bearing types etc. Selection and hybridization are breeding strategies mainly employed in large cardamom. ICRI-Sikkim-1 and ICRI-Sikkim-2 are the two released varieties of large cardamom. In addition to these many high yielding clones having a yield potential of above 1000 kg/ha are in the pipeline for release and it includes SCC-72, SCC-75 and SCC-81.

False Cardamoms

The false cardamoms are important locally as spices and condiments. They are also used in various ailments. Studies related to cytology, biosystematics and chemistry are attempted and detailed studies are totally lacking in this group.

False Cardamoms Belonging to the Genus *Elettaria*

E. ensal (Gaertn.) Abheywickrme

This is a Sri Lankan wild cardamom. It is morphologically similar to true cardamom, but plants are more robust, bearings erect panicles and much elongated fruits (3-5cm). The taxonomic entity of this species is still under debate. According to Burtt (1980) and Burtt and Smith (1983) it is under *Elettaria cardamomum* only and not treated as a separate species. Abheywickreme (1959) considered it as a separate species due to its very robust nature and distinct variation in fruit size. Fruits contain more seeds and its flavour differs from true cardamom. Fruits are used by local people for flavouring and also in tribal medicine preparations.

E. longituba (Ridl.) Holtt.

This is a Malaysian wild cardamom and is endemic to Malaysian region. It is a perennial and its flowers appear singly at longer intervals, only a few flowers occur per cincinnus, flowering stolons (panicle) is 3-4 m long, anther dehiscing by pores without a hairy flap, fruit is long, globose or pyriform, smooth and slightly ribbed.

***E. rubida* R.M. Sm.**

It occurs in Indonesia region. Inflorescence red with orange flowers, anther cristate, leaves obovate with attenuate base (Sakai and Nagamasu, 2000).

***E. stolonifera* (K. Schum.) Sakai and Nagamasu**

It is available in Indonesian region. Plant is 0.8 to 1.5 m tall. Inflorescence 25 to 60cm long, trailing on ground, flowers white, anthers 3mm long, anthers dehiscing in upper half only, anther crest 3–lobed and fruit unknown (Sakai and Nagamasu, 2000).

***E. kapitensis* Sakai and Nagamasu**

It is abundant in Indonesian-region. Plant is about 0.7m tall, inflorescence 40 to 60cm long, flowers white, calyx 17mm long, corolla tube and calyx free above the ovary, anther dehiscing throughout their length, anther crest about 2mm long and fruit unknown (Sakai and Nagamasu, 2000).

***E. surculosa* (K. Schum.) B.L. Burt and R.M. Smith**

It occurs in Indonesian region. Plant is about 1.2 to 2m tall, inflorescence up to 2m long, creeping on ground, flowers white, anther dehisces by small pores, pore covered with a flap with long hairs, anther crest up to 2mm long, fruit about 40x16mm, ellipsoid, sparsely pubescent and reddish brown (Sakai and Nagamasu, 2000).

***E. linearicrista* Sakai and Nagamasu**

It occurs in Indonesian-region. Plant is about 0.6 to 1.5m tall, leaves narrowly oblong, inflorescence about 40cm long, axis densely pubescent, flowers white, anther dehiscing by longitudinally elongated pores at upper middle of the thecae and fruit unknown (Sakai and Nagamasu, 2000).

***E. longipilosa* Sakai and Nagamasu**

It is grown in Indonesian region. Plant is 1 to 1.3m tall, leaves densely pubescent on lower surface, inflorescence about 50cm long, creeping just below ground, flowers white, anther thecae dehiscing by a small pore just above mid-point, pore covered with a flap with very long hairs, anther crest deeply three lobed and fruit unknown (Sakai and Nagamasu, 2000).

***E. brachycalyx* Sakai and Nagamasu**

It spreads in Indonesian-region. Plant is about 1 m tall, inflorescence about 60cm long, flowers white, calyx about 6mm long, fissured for two thirds of its length, anther thecae dehiscing by pores covered with a flap of long hairs, anther crest about 1mm long, 3-lobed and fruit unknown (Sakai and Nagamasu, 2000).

False Cardamoms Belonging to the Genus *Aframomum*

***A. corrorima* (Braun) (Syn. *A. melegueta*)**

It is also known as Korarima cardamom. This species grows in wild and cultivated sporadically in Ethiopia, Nigeria and nearby regions and this species is endemic to that area. It is a perennial aromatic herb having strong, fibrous, subterranean scaly rhizomes and with a leafy stem of about 1-2 m high. Rhizomes are subterete, brown–reddish brown, smooth, glabrous, scales brown, thin and deciduous. Leafy stem unbranched, formed by leaf sheaths.

Plants grow naturally at 1700-2000 m altitude in Ethiopia, and in other tropical West African countries. They produce flowers from January to September and fruits mature in 2-3 months after flowering. Usually bees are the pollinators and propagated both by seeds and rhizome parts. Harvesting is done when fruits turn red.

It was in demand as a spice and as a stimulating carminative. It is used for a variety of ailments by local people. Fruit pulp around the seed is eaten especially before maturity and is chewed as stimulant. Seeds are used as a spice in food to flavour all types of dishes in tropical African countries, and also used in flavouring tea and coffee, and for special kinds of breads. In Ghana, seeds are used in anaema preparation. In Logos and Sierra Leone, seeds are treated a remedy for inflamed conditions of tonsil, and in fevers. The whole plant decoction is taken as a febrifuge, the roots have a cardamom like taste, and a decoction is given for constipation. Roots are regarded as a vermifuge for tapeworm. The juice of young leaves acts as a styptic. This was used in UK and USA in the preparation of medicines for cattle and also in the preparation of the spiced wine (called hippocras), flavoured with this cardamom, cinnamon and ginger. The seed has anti-ulcer, gastro protective, anti-microbial and anti-feedent activities (Ravindran *et al.,* 2002). Seeds of this species are used for peptic ulcer in the Bangangate region of Western Cameroon (Noumi and Dibakto, 2000).

A. danielli **K. Schum**

It is a perennial herb in West Africa, growing to about 75cm and having leaves of about 25-30cm. Flowers are red in colour, many in an inflorescence, lip pink, yellow at base and margin white. Fruits (capsule) are red and smooth; seeds smooth, shining olive brown with a turpentine like taste. Pulp is agreeably acidic, eaten by people to refresh and relieve thirst in fever. The seed extract has anti-microbial and anti-inflammatory activities.

A. granum–paradisi **K. Schum (black *Amomum*)**

The immature fruit is dark in colour, hence known as black *Amomum*. Occurs in Ghana, Gambia, Sierra Leone and South Nigeria. The fruit capsule is ten or more grooved and succulent. Seeds are sub-globose or slightly angular having smooth and shining surface. Seeds are aromatic and possess camphoraceous taste and flavour. Whole plant is used medicinally. Leaves are cooked with food for imparting flavour.

A. latifolium **K. Schum**

Indigenous to tropical African country of Sierra-Leone, seeds are like grape stones, smooth, shining and weakly aromatic. Ripe fruit is acidic and is eaten by natives; given to refresh and relieve thirst in fever and is good against fatigue. The decoction of whole plant is mixed with leaves of *Morinda lucida* (Rubiaceae) and used by the tribe Susus as a daily bath to restore strength after fever.

A. sanguineum **K. Schum**

It is a native of West Africa. Seeds of this plant are used in cosmetics, to flavour tea and rice and in traditional medicines to prevent throat infections and to treat stomachache, dysentery and snake bites.

A. angustifolium **Schum**

It is also known as Madagascar cardamom. Occurs in East Africa and islands of the Indian Ocean. Seeds are used as condiment (Arora, 1985).

A. hanburyi Schum

Occurs in Cameroon. Seeds are used as condiment (Arora, 1985).

False Cardamoms Belonging to the Genus *Amomum*

A. aromaticum Roxb.

It is also called as Bengal cardamom. It is a perennial herb and grows to about 1m height. It is indigenous to northern West Bengal, Assam, Meghalaya, Sikkim, North Bihar and northern West Bengal, extending to Nagaland and Utter Pradesh. Cultivated in northeastern regions at foothills of Himalayas. Leaves oblong, lanceolate, glabrous, 15-30 cm x 5-10 cm, flowers pale yellow, in small globose spikes, corolla-tube white tinged with brown, capsules rugose, 25cm long, trigonous, oblong or globose, with 7-13 narrow membranous longitudinal wings, seeds many in each cell. Seeds are used as spice and are medicinal, used in place of *A. subulatum*, the large cardamom (Ravindran *et al.*, 2002). Rhizomes are used as condiment and flowering shoot in curries (Arora, 1985).

A. dealbatum Roxb.

It is a herbaceous perennial with large-lanceolate leaves and globose short, peduncled spike, distributed in Khasi hills and eastern Himalayas. Seeds are feebly aromatic in taste and odour. Seeds are used as substitute for large cardamom.

A. longiligulare Wu

This spice occurs in Vietnam, and is used locally for medicinal purposes and for flavouring. Locals used this species against dyspepsia, vomiting, diarrhoea and dysentery.

A. tsao-ko Crevost Lemarim

It is otherwise called as Vietnam cardamom. A herb, 2-3 m high, leaves oblong-lanceolate, lamina 60-70 x 18-20 cm, glabrous, spike globose, peduncle short, outer bract 2.5 cm long, ovate, corolla tube pale brown 2.5cm long, lip pale yellow. Anther crest large, petalloid, lobes rounded. Capsules 2.5 cm long, oblong, trigonous and three valves with numerous seeds in each cell. Cultivated in South China, Vietnam especially at the forest regions of the Hoang Lien Son Mountain.

Seeds of this species are used as spice in Vietnam and China in place of true cardamom. It is also used as a local remedy for many ailments including dyspepsia, diarrhoea and indigestion. Seven anti-oxidative compounds were reported from its seeds.

A. krervanh Pierre. Krervanh

Indigenous to Thailand. It is used in traditional medicine in Thailand. People of Cambodia and Indo-China used fruits as condiments and to flavour curries, sausage and cordials (Arora, 1985). An anti-malarial compound is isolated from its seeds, used against *Plasmodium falciparum*, which causes malaria in India (Kamchonwongpaisan *et al.*, 1995).

A. xanthiodis Wall

It is also known as Tavoy cardamom. Distributed in the forests of Indo-Burma, Malayan region. It is a perennial herb, leafy stem reaching 1.5-2 m; spikes globose, a few flowered and shortly peduncled; fruits (capsule) echinate, oblong, trigonous, pale brown.

The fruits are used against diarrhoea, dysentery, cramps, hiccups, nausea, splenitis and stomachache, and is an anodyne, antitussive, apenient, carminative and sedative. Seed hastens the solution of coins, fish-bones or other foreign substances accidentally swallowed. Seeds are effective in relieving painful urination, and for regulating frequency of motions. For this, powdered seed mixed with butter is administered. The Chinese and Burmese use this as an important condiment (Ravindran *et al.*, 2002). It is also called as wild Bastard Siamese cardamom (Arora, 1985).

A. compactum Soland (Syn. *A. kepulaga* Sprague and Burkill)

Round cardamom, cultivated in Malaysia and Java. Fruits are used as spice/condiment (Arora, 1985). It is also known as Siam cardamom and is reported to be anti-toxic, anti-emetic, carminative and stomachic. In China, it is a folk remedy for ague, cachexia, cancer, catarrh, cold, cough, cramps, dyspepsia, gout, heartburn, hepatises, nausea, ophthalmic, rheumatism and vomiting often used in combination with other plant products and rarely alone. The plant is used in folk remedies for indurations of liver and uterus and for cancer. Fruits are eaten together with cloves to alleviate addiction to cigarettes and fruits are also eaten to counter feeling of wanting to vomit (Ong and Nordiana, 1999).

A. constalum Benth.

Endemic to eastern Himalayan Forests. Perennial herb with stout leafy stem 1.5–2m; spikes 5-7.5cm, oblong shortly peduncled fruit ovoid, strongly ribbed, smooth, seeds many, aromatic. In Chinese medicine, seeds are employed for ailments of stomach and for asthma, pulmonary afflictions and general debility. Tribals use seeds for relieving stomachache.

A. pavieanum Pierre and Gagnep

It is endemic to Southeast Asia. An infusion of fruits of this plant is used in the treatment of diarrhoea and general debility following certain dysenteries. Rhizome is used as seasoning agent in eastern and southern parts of Thailand.

A. acre Valeton

An Indonesian species mainly found in Sulawesi region. Fruits and inner part of petioles are used for pickling, often pickled in vinegar and subsequently used as a spice and a flavorant.

A. ochreum Ridley

Occurs in Peninsular Malaysia. Seeds of this species are used locally as a substitute for cardamom. The plant is 3-4 m tall, produces inflorescences (spikes) from base, flowers having obovate, 3-lobed labellum, orange-yellow with red veins and spots, anther appendage transversely oblong, faintly red-spotted (de Guzaman and Siemonsma, 1999).

A. testaceum Ridley

Occurs in Thailand, Vietnam, Peninsular Malaysia and Borneo; occasionally cultivated. Seeds are aromatic and locally used like true cardamom. Rhizomatous herb with leafy shoots about 3 m tall, inflorescence is an oblong spike emerging from the rhizome, flowers have obovate lebellum, white with a broad dull yellow patch towards the apex and paler yellow median band flanked by purple lines, stamens with 8 mm long filament and a 3-lobed anther appendage (de Guzman and Siemonsma, 1999). Fruits slightly ribbed and pinkish, slightly hairy. Seeds are brown in colour, aril thin and aromatic.

A. xanthophlebium Baker (syn. *A. stenoglossum* Baker)

Occurs in peninsular Malaysia and Borneo. Flowers are used for flavouring curries. Rhizomatous herb, leafy shoots up to 4-5 m tall, inflorescences are ellipsoidal spikes arising from rhizome, flowers have obovate labellum, edges crinkled, white suffused with red stripes and spots and yellow stripes towards the apex, anther connective 3-lobed, filament 1 cm long, pink. Fruit obovoid, smooth oppressed silky hairy (de Guzman and Siemonsma, 1999). Flowers of this species are used by local people for flavouring various dishes. Seeds are used by tribals in local medicinal preparations.

***A. globosum* Lour**

It is a round Chinese cardamom. Cultivated in China. Seeds are used as condiments (Arora, 1985).

***A. maximum* Roxb**

It is known as Java cardamom. Cultivated in Malaysia and Java. The people in Java used this spice as local condiment (Arora, 1985).

Conservation of Cardamom Diversity

Conservation refers to the management of natural resources including maintenance, sustainable utilization, restoration and enhancement of natural environment. In cardamom, emphasis is given to *ex situ* conservation. It runs to problems due to large extend of crop area, frequent occurrence of natural calamities like drought, wind and excess rainfall and loss of plants due to pests and diseases. To overcome this, *in vitro* conservation techniques are practiced and studies on this line were taken up in cardamom (Nirmal Babu *et al.*, 2001).

In situ conservation refers to conservation of crop genetic resources in particular ecosystem, protected in its natural habitat without external interference. This would be possible with the establishment of gene sanctuaries, national parks, biosphere reserve etc. In cardamom *in situ* conservation is not yet practiced. Its feasibility in the *Cardamom Hill Reserve* needs to be studied, as the crop is an integral part of forest ecosystem.

Gene Erosion

In recent years, depletion of cardamom germplasm is increasing and this is due to (a) exponential rate of natural habitat degradation in the Cardamom Hill Reserve (CHR) (b) mono-culturing of varieties/ land races for large scale cultivation (c) frequent occurrence of environmental stress (drought), (d) incidence of pests and diseases and (e) lack of proper conservation practices. Among these, habitat degradation is the most deleterious one and is mainly due to deforestation. In spite of the ban on green felling, degradation of forests continues due to anthropogenic reasons. Large-scale cultivation of improved clones replacing existing ones will also impose threat to crop diversity. Cardamom is a shade loving plant, requires 50-55 per cent sunlight for its growth and development. In recent years cardamom hills in Idukki experienced severe variation in weather conditions and drought becomes very severe, which lead to loss of valuable genes. Besides, intensive cultivation in the CHR pauses threat to long-term sustainability of crop.

Ecosystem Degradation

Shift of forests to agro-forests through cardamom cultivation, affects the tree composition of tropical evergreen forest. Richness of species like Akil (*Dysoxylum binectriferum*), Peenari (*Nathapodytes nimmoniana*), Chandana vembu (*Cedrela toona*), Karimaram (*Diospyros candolleana*), Korangatti (*Attrocarpus fraxinifolnius*) has narrowed down in the CHR due indiscriminate felling and are replaced by species like Murukku (*Erythrina lithosperma*), Sliver oak (*Grevillea robusta*), Karuna (*Vernonia monosis*), Jack (*Artocarpus heterophyllus*), Unnam (*Grewia tiliaefolia*), Silk Cotton (*Bambusa malabaricum*) etc. This trend also affected diversity richness of fauna of CHR. Colonies of honey bees species *viz., Apis cerana indica* and *A. dorsata* were plenty in the past. But now their population has come down with the resultant loss of source of nectar and pollen due to chemical pollution from agriculture. Further, incidence of prolonged drought, enhancement of ambient temperature, soil temperature, reduction in humidity, soil moisture, inherent fertility status, soil flora and fauna and fluctuation even in the monsoon pattern were also induced as a result of intensive cultivation practices. A perusal of the metrological data revealed reduction of total number of rainy days and total rainfall to the tune of 35 per cent and 30 per cent respectively in 2002 compared to previous years. This also affects diversity richness of crop. Ecosystem restoration is the only strategy to conserve biodiversity in Cardamom Hill Reserve and in the Sub-Himalayan tract.

Strategies for Conservation of Diversity

Conservation strategies must be based on sound scientific knowledge of habitat, needs of people and intensity of threats to diversity richness. For long-term sustainability of cardamom and forest wealth in the ecosystem, laws should strictly be adhered to not to undertake any activities leading to natural habitat degradation. Declaration of protected cardamom reserve areas alone will not meet the challenges in the conservation due to proximity of plantations. Diversity conservation should be possible only with active support and participation of local people including tribal communities. Participatory forest management approach would be practiced in cardamom hills.

With introduction of improved clones/landraces, the shade regulation management quite often leads to extremes resulting in natural habitat degradation. To realize the potential yield of these improved types, improved agrotechniques including shade management are to be carried out. Shade management is to be done judiciously as per norms stipulated in the Cardamom Reserve and drought management strategies such as adoption of contour planting, maintenance of water conservation pits and water harvesting systems. Further, integrated pest management, organic cultivation of cardamom and adoption of resistant lines (biotic and abiotic stress) undoubtedly improve the soil conditions and biotic wealth of the reserve forests.

Considering limitations in the *ex situ* conservation programme and the advantages in preserving ecological niche required for crops like black pepper and cardamom, the feasibility of *in situ* conservation in the protected areas needs to be ascertained. Cardamoms are not only flavouring, but also more functional as medicinal or food preservative in the Indian tradition. Thus conserving them is mandatory to uphold the rich tradition/culture of the society. Since India has a rich tradition of conservation, cardamom ecosystem needs to be conserved for sustainable utilization of its richness and its diversity and it would be possible with the co-operation of the inhabitants including tribals.

References

Anonymous, 2004. *Spices Statistics*–2004, Spices Board, Cochin.

Abheywickreme, 1959. A provisional checklist of the flowering plants of Ceylon. *Ceylon J. Sci.* (*Biol. Sci.*). 2: 119-240.

Arora, R.K., 1985. Genetic resources of less known cultivated food plants. NBPGR Sci. Monogr. No. 9. National Bureau of Plant Genetic Resources, New Delhi, India, pp. 102–103.

Burtt, B.L., 1980. Cardamoms and other Zingiberaceae in *Hortus Malabaricus*. In: *Botany and History of Hortus Malabaricus*, (Ed.) Manilal, K.S. Oxford and IBH, New Delhi, pp. 139–148.

Burtt, B.L. and Smith, R.M., 1983. Zingiberaceae. In: *A Revised Handbook of the Flora of Ceylon, Vol. IV,* (Ed.) Dassanayeke, M.D. Amerind Publications, New Delhi.

de Guzman, C.C. and Siemonsma, J.S. (Eds.) 1999. Plant Resource of South-East Asia, No. 13, In: *Spices.* Backhuys Publications, Leiden.

Holttum, R.E., 1950. The Zingiberaceae of the Malay Peninsula. *The Gard. Bull., Singapore,* 13: 1-249.

Kamchonwongpaisan, S., Nilanonta, C., Tarnchompoo, B., Thebtaranonth, C., Thebtaranonth, Y., Yuthavong, Y., Kongsaeree, P. and Clardy, J., 1995. An antimalarial peroxide from *Amomum krervanh* Pierre. *Tetrahedron Lett.,* 36: 1821–1824.

Korikanthimath, V.S., 2001. Cardamom (small). In: *Handbook of Herbs and Spices,* (Ed.) Peter, K.V. Woodhead-Publishing House, U.K., p. 123–132.

Madhusoodanan, K.J., Kuruvilla, K.M. and Priyadarshan, P.M., 1994. Genetic resources of cardamom. *Advances in Horticulture,* 9(1): 121–130.

Madhusoodanan, K.L. and Rao, S.R., 2001. Cardamom (large). In: *Handbook of Herbs and Spices,* (Ed.) Peter, K.V. Woodhead Publishing House, UK, p. 134–140.

Nirmal Babu, K., Geetha, S.P., Minoo, D., Anu, A., Vimala, J., Dayamathi, K.P.M., Krishnamoorthy, B., Ravindran, P.N and Peter, K.V., 2001. Tissue and cell culture research on spices. *Indian Spices.* 38 (3): 24–27.

Noumi, E. and Dibakto, T.W., 2000. Medicinal plants used for peptic ulcer in the Bangangte region, Western Cameroon. *Fitoterapia.* 71: 406–412.

Nybe, E.V., Mini Raj, N. and Peter, K.V., 2006. *Spices.* New India Publishing Agency New Delhi, p. 43–66, 66–67.

Ong, H.C. and Nordiana, M., 1999. Malay ethno-medico botany in Machang Kelantan, Malaysia. *Fitoterapia.* 70: 502–513.

Ravindran, P.N., Shylaja, M. and Nirmal Babu, K., 2002. False cardamoms. In: Cardamom. The genus *Elettaria.* Ravindran, P.N. and Madhusoodanan, K.J 3(eds.). Taylor and Francis, London. pp. 330–340.

Sakai, S. and Nagamasu, H., 2000. Systematic studies of Bornean Zingiberaceae: II. *Elettaria* of Sarawak. *Edinb. J. Bot.* 57: 227–243.

Sudharshan, M.R. and Korikanthimath, V.S., 2005. Cardamom. In: *Handbook of Industrial Crops,* (Eds.) Chopra, V.L. and Peter, K.V. Haworth Press, USA, pp. 43–74.

Biodiversity in Horticultural Crops Vol. 1 *Pages 288–308*
Editors: **Professor K.V. Peter & Dr. Z. Abraham**
Published by: **DAYA PUBLISHING HOUSE**

Chapter 14

Large Cardamom (*Amomum subulatum* Roxb.)

***M.R. Sudharshan*[1] *and U. Gupta*[2]**
Indian Cardamom Research Institute, Spices Board,
[1]Regional Station, Sakleshpur – 573 134
[2]Regional Station, Tadang ,Sikkim

Amomum subulatum Roxb., belongs to order Zingiberales and family Zingiberaceae. It is known as Boro alach, Buro elachi and Morung elachi in Bengali, Xiang dou kou in Chinese, sort kardemomme in Danish, zwarte kardemom in Dutch, Bengal cardamom, black cardamom, fragrant cardamom, greater cardamom, greater Indian cardamom, Nepal cardamom, Indian cardamom, winged Bengal cardamom in English, mustakardemumma in Finish, Nepal-cardamom, schwarzer cardamom in German, kali elaichi in Hindi, cardamom nero in Italian and cardamom nero in Spanish (Michel Porcher 2001).

Cardamoms are the dried seed capsules of a small group of species of plants belonging to the family *Zingiberaceae* which contain seeds possessing a pleasant characteristic aroma and flavour. These are broadly grouped into two categories:

1. Small cardamom–popularly known as *Chota Elaichi* (*Elettaria cardamomum* Maton) or the true cardamom. It is also known as "Queen of Spices'.
2. Large Cardamom–*bada elaichi* (*Aframomum* and *Amomum* species).

Amomum subulatum Roxb. is the greater Indian or Nepal cardamom also called large cardamom, the most important cash crop of Sikkim. From here it spread to the North Eastern States, part of West Bengal and the neighboring countries, Nepal and Bhutan. It is used extensively as a spice in South Asian countries, and as a substitute for true cardamom in the Middle East regions. It is also important in tribal and indigenous medicine. Sikkim has the largest area of 30,039 ha under large cardamom cultivation with a production of 6154 metric tonne (Anonymous 2004).

Order *Zingiberales* (formerly known as *Scitamineae*) to which the family *Zingiberaceae* belongs originated as wild plants in tropical evergreen forests. *Zingiberaceae,* the largest family of this order, is found throughout tropics.

North-Eastern India has the largest concentration of Zingiberaceous flora (Next to Malaysia) especially in number of genera. The family is represented here by 19 genera and 88 species out of 22 genera and about 170 species reported from India. (Ved Prakash and Mehrotra, 1996).

The family Zingiberaceae provided important aromatic spices that are widely used. The important spices of this family are *Zingiber* (ginger), *Curcuma* (turmeric), *Alpinia* (galangal), *Kaemferia*, all representing rhizomatous spices and *Elettaria* (small cardamom), *Amomum* and *Aframomum* (large cardamoms) representing seed spices.

The International Standards Organization(ISO) has officially recognized nine species as cardamoms under three main groups (Pruthi 1977):

Group I: *Elettaria cardamomum*-True cardamom

Group II: 4 species of *Aframomum*

(a) *Aframomum augustifolium* (Sonn) K. Schum.–Madagascar cardamom
(b) *Aframomum hanburyi* K. Schum.–Cameroon cardamom
(c) *Aframomum korarima* (Pereira) Engler–Korarima cardamom
(d) *Aframomum melegueta* (Roscoel) K.Schum–Grains of paradise or Guinea grains.

Group III: 4 species of *Amomum*

(a) *Amomum aromaticum* Roxb.–Bengal cardamom
(b) *Amomum kepulaga* Spraeque et–Round cardamom Burkill. Syn. *A. cardamom* Roxburgh- or Chester cardamom or Siam cardamom.
(c) *Amomum krervanh* Pierre et Gagnipain–Cambodian cardamom
(d) *Amomum subulatum* Roxburgh–Greater Indian cardamom, Nepal cardamom or large cardamom.

The *Amomum* species are known in the North East India and South East Asian countries, while the *Aframomum* species are known in the African regions of Sierra Leone, guinea Coast, Madagascar and Tanzania. The fruits of the *Amomum* and *Aframomum* are much larger in size in comparison with *Elettaria cardamomum* and it is easy to distinguish them but the seed morphology and anatomy are similar in all the three genera (Madhusoodanan *et al.*, 2001).

All the cultivars of large cardamom cultivated commercially belong to *Amomum subulatum* Roxburgh species. Earlier explorers like Baker J.G.(1890) listed eight species of Amomums from Easter tropical Himalayan region. viz *A. aromaticum* Roxb., *A. corynostachium* Wall.,*A. costatum* Benth., *A. dealbatum* Roxb., *A.kingii* Baker, *A. pauciflorum* Baker, *A.linguiforme* Benth. and *A.subulatum* Roxburgh. However, later workers (Ved Prakash and Mehrotra 1996) listed only six species of Amomums, out of which the occurrence of *A.fulviceps* in North East India is to be confirmed. Kumar (2001), has recorded only five species of Amomums from the state of Sikkim. *A. aromaticum* Roxb., *A. corynostachium* Wall., *A. dealbatum* Roxb., *A.kingii* Baker, and *A.subulatum* Roxburgh. It is stated that *A. kingii* has been reported only form a single collection and further collections are required for critical study.

Amomum subulatum is a perennial herb consisting of subterranean rhizomes; widely creeping and several leafy aerial shoots/tillers arise from rhizomes. Leafy stem elongate. Number of such rhizomatous leafy shoots varies from 15 to 140 in a single plant (a clump). Height of leafy shoot ranges from 1.7 to 2.6 m depending on cultivar and possess 9-13 leaves in each tiller. Leaves are distichous, simple, oblong- lanceolate, glabrous beneath with a prominent mid rib, 30-80 x 8-10 cm, subsessile. Spike globose shortly peduncled, outer bracts obtuse with a horny cusp *ca* 3cm long, ovate, red to brown; inner short obtuse, lip obovate-cuneate emarginated, anther crest small truncate, capsule echinate. Flowers are bracteate, bisexual, zygomorphic, epigynous, and cuspinated. Calyx *ca* 4 cm long; corolla tube *ca* 3 cm long, white, lobes 2-3 cm long, lip oblong, cuneate; *ca* 3cm long, narrow at tip often bifid, yellow with deep yellow veins; anther *ca* 1cm long with crest *ca* 4 mm, entire, wavy with inflexed sides; filament *ca* 5mm long with crest *ca* 4 mm., entire wavy with inflexed sides; filament *ca* 5 mm long, capsules ellipsoid, flattened, *ca* 2.5cm long, longitudinally striate with crenulate wings, red brown. Seeds *ca* 2.5 mm broad, roundish, black. Each spike contains about 10-15 fruits (capsules) and rarely up to 20-25 capsules, depending on cultivars. Flowering season begins early at lower altitude with peak flowering during March-April, whereas it starts at higher altitudes in May with peak during June-July. Harvesting begins during August-September at lower altitudes and in October-December at higher altitudes (Gupta and John 1987).

Large cardamom is grown in cold humid conditions under shade of trees at an altitude between 800–1500 m above mean sea level (MSL) with an average precipitation of 3000–3500 mm, spread over about 200 days and the temperature ranging between 6° C in December-January to 30° C in June-July (Singh, 1988). Frost, hailstorm and snowfall are the major deleterious factors affecting large cardamom. Crop grows well in moist, but well drained loose soil. Depth of soil varies from a few centimeters to several meters depending upon the topography and soil formation. Soil is acidic and rich in organic matter (Mukherji, 1968).

Cultivars

There are mainly three cultivars of large cardamom, *viz. Ramsey, Sawney* and *Golsey* (Rao *et al.*, 1993).

Table 14.1: Features of Three Different Cultivars of Large Cardamom

Character	*Cultivars*		
	Ramsey	*Golsey*	*Sawney*
Altitude	High	Low to middle	Middle
Extent of cultivation	60%	30%	7%
Habit	Tall, vigorous, wide clump growth	Less vigorous, leaves upright, clumps medium	Tall, vigorous, leaves bend downwards
Stem colour	Maroon with dense foliage	Green to maroon	Pinkish
Flowers	Yellowish and small; corolla tip with pink tinge at base	Yellowish orange	Yellowish with pink tinge at base of corolla
Capsules	Smaller 16 to 30 seeds	Bold & round 40-50 speeds	Medium 30-40 seeds
Essential oil	1 to 1.8%	2.3 to 5%	1.8 to 2.5%
Shade requirement	High shade	Less shade	Moderate to high
Disease reactions	Susceptible to Chirkey and Foorkey at lower altitudes	Tolerant to Chirkey and Foorkey	Susceptible to viral diseases

Varlangey (*Barlangey*) and *Bebo* are also main cultivar types in addition to the above three by Gyatso *et al.* (1980). Other variants of above cultivars *Ramnag, Ramla, Madhusey, Mongney* are also seen in cultivation in small areas in Sikkim. Another cultivar *Seremna or Lephrakey* (a Galsey type) is also getting importance and is spreading to more areas in lower altitudes (Upadhyaya and Ghosh, 1983).

Ramsey (*Ramshai*)

The name *Ramsey* was derived from two Bhutia words–'*Ram*' meaning colour and 'sey/shai' for yellow. This cultivar is well suited for higher altitudes, even above 1500 m. on steep slopes. Plants are tall approximately 2.5m, grown up clumps of 8-10 years age group possess 60-140 tillers. Leaves are thin and long. The tillers colour is green to maroon. Second half of May is the peak flowering season. Capsules are small, the average being 2.27 cm in length with 2.5 cm diameter, with 30-35 capsules in a spike, each containing 16-30 seeds. Harvest is during October-November. Biennial peak bearing is noticed.

This cultivar is more susceptible to viral diseases like *foorkey* and *chirke* especially if planted at lower altitudes. It occupies a major area under large cardamom in Sikkim and Darjeeling district of West Bengal. Two strains of this cultivar *viz.,Kopringe* and *Garadey* from Darjeeling having stripes on leaf sheath, are reported to be tolerant to *chirke* virus (Karibasappa *et al.*, 1987).

Sawney

It is a Nepali word which means that the variety is harvested in the month of *Sawan* in Nepali, corresponds to August. Plants of this variety are also tall but the leaves are wider and shorter compared to that of Ramsey. This cultivar is widely adaptable, especially suited for mid and high altitudes *i.e.* around 1300-1500 m. It is robust in nature and consists of 60-90 tillers in each clump. Colour of tillers is similar to *Ramsey*. Each productive tiller on an average produces two spikes. Average length and diameter of a spike is 6 and 11 cm. Flowers are longer (6.23 mm) and yellow in colour with pink veins. Second half of May is the peak flowering time (Rao *et al.*, 1993).

Capsules are bigger and bold and number of seeds in each capsule are more (35) than in *Ramsey*. Harvest begins in August-October and may extend up to November in high-altitude areas. This cultivar is susceptible to both *chirke* and *foorkey* viral diseases. Cultivars such as *Red Sawney* and *Green Sawney* derived their names from capsule colour. *Mongney*, a strain found in south and west districts of Sikkim, is a non-robust type with its small round capsules resembling mostly that of *Ramsey*.

Golsey/Golshai/ (*Dzongu Golsey*)

The name has derived from Hindi and Bhutia words; '*Gol*' means round in Hindi and '*sey*' means yellow in Bhutia. This cultivar is suitable to low altitude areas below 1300 m above msl especially in Dzongu area in North Sikkim. Plants are not robust like other cultivars, and consist of 20-50 straight tillers with erect leaves. Alternate, prominent veins are extended to the edges of leaves (Biswas *et al.*, 1986). Unlike *Ramsey* and *Sawney*, tillers are green in colour. Each productive tiller on an average produces two spikes. Flowers are bright yellow. On an average, each spike is 5.3 cm long with 9.5 cm diameter and contains an average of seven capsules. Capsules are big and bold, 2.46 cm in length and 3.92 cm in diameter and contain about 60-62 seeds. This cultivar becomes ready for harvest in August-September.

Golsey is tolerant to *chirke* and less susceptible to *foorkey* and susceptible to leaf streak diseases. The cultivar is known for its consistent performance, though not a heavy yielder. Many local cultivars are known in different locations such as Ramnag from north Sikkim. Ram meaning 'colour' and nag for black, which refers to its dark pink capsules. *Seto-Golsey* is from west district of Sikkim with robust leafy stems/tillers and green capsules. *Madhusey* with elliptic and pink coloured capsules is having robust leafy stem and has sweet seeds compared to other cultivars (Rao *et al.*,1993).

Ramla

The plants are tall and vigorous like *Ramsey* and have capsule characters like *Dzongu Golsey*; the colour of tiller is pink like *Ramsey* and Sawney. Cultivation is restricted to a few mid-altitude plantations in North Sikkim. The capsules are dark pink with 25-38 seeds/capsule. *Ramla* appears to be a natural hybrid between *Dzongu Golsey* and *Ramsey*.(Varadarasan and Biswas,2002). They are susceptible to *foorkey* but are moderately tolerant to *chirke* disease.

Varlangey/Barlangey

This cultivar grows in low, medium and high altitude areas in South Regu (East Sikkim) and at high altitudes at Gotak (Kalimpong sub-division in Darjeeling district of West Bengal). Its yield performance is exceptionally high at higher altitude areas *i.e.*1500 m and above. It is a robust type and total tillers may range from 60 to 150. Colour of tillers is like that in Ramsey *i.e.* maroonish-green to maroon towards collar zone; girth of tillers is more than that of Ramsey. Each productive tiller on an average produces almost three spikes with an average of 20 capsules/spike. Size of capsules is bigger and bold with 50-65 seeds. Harvest begins in last week of October. This cultivar is also susceptible to foorkey and chirke diseases.

Bebo

This cultivar is grown in Basar area of Arunachal Pradesh. The plant has unique features of rhizome and tillering. The rhizome rises above the ground level with roots penetrating deep into the soil and young tillers are covered under thick leafy sheath. It is supposed to be tolerant to *foorkey* disease. The spikes have relatively long peduncle(10-15 cm) and the capsules are bold, red or brown or light brown; seeds contain low level of essential oil (2 per cent v/w) (Dubey and Singh, 1999).

Seremna (*Sharmney or Lephrakey*)

This cultivar is grown in a small pocket at Hee-Gaon in West Sikkim at low altitude and is known for its high yield potential. Plant features are almost similar to Dzongu Golsey but the leaves are mostly dropping, hence named as Sharmney. Total tillers range from 30 to 49 and is not robust in nature. On an average 2-3 spikes emerge from each productive tiller with an average of 10.5 capsules/spike, each having 65-70 seeds (Vardarasan and Biswas, 2002).

Diversity in the Cultivated Varieties

Morphological attributes of eleven accessions of large cardamom collected from Sikkim and Dargjeeling district of West Bengal were studied. Three accessions were Sawney, five were Ramsey one was Golsey and two barlangey cultivars. It was found that the accessions varied in the number of mature tillers, number of productive spikes, number of capsules per spike number of seeds per capsule and ratio of mature tillers to productive spikes. Based on the study, a vast potential for improvement of large cardamom was indicated (Rao *et al.*, 1993). The capsule morphology of nine cultivars of large cardamom was studied and significant differences in number of capsules, size, colour, shape and number of seeds were recorded. The exact shape and colour of fresh capsules were recorded as showin in Table 14.2 (Gupta 1986).

High genotypic coefficients of variation for mature tillers/clump, panicles/clump and capsule to panicles ratio were observed when five-year-old plants of Sawney, Pink-Golsey, Ramsey, Ramnag and Madhusey varieties were studied under uniform conditions (Karibasappa *et al.*, 1987). All the varieties showed high heritability coupled with high genetic advance for mature tillers, panicles/clump, panicle weight and capsule yield.

Plate 14.1: Large Cardamom Growing Regions in Sikkim State

Plate 14.2: Large Cardamom Habitat

Plate 14.3: Plant Habit

Plate 14.4: A High Yielding Clone ICRI-Sikkim-1

Plate 14.5: The Spike

Plate 14.6: Yielding Plant

Plate 14.7: Ripe Capsules

Plate 14.8: Large Cardamom Capsules

Plate 14.9: Chinese Cardamom

Plate 14.10: *Amomum dealbatum*

Plate 14.11: *Amomum kingii*

Table 14.2

1.	Sawney Tall	Roundish obovate, red, ridges prominent
2.	Gosey Tall	Roundish obovate, dull pink, top dark red, ridges prominent.
3.	Dwarf Bush*	Obovate, light pink, ridges medium prominent.
4.	White Ramna	Roundish, dull light yellow, ridges less prominent.
5.	Sawney small (Syn. Madhusey)	Roundish obovate, deep red, ridges prominent.
6.	White Golsey (Syn.Sethuram)	Roundish obovate, very light yellow,ridges less prominent.
7.	Ramsey	Oblong to ovate, light red, ridges prominent.
8.	Golsey Dwarf	Roundish obovate, dull pink ridges prominent.
9.	Red Sawney (Syn.Ramnag)	Oblong to obovate, dark red, ridges medium prominent.

* Name coined by the authors for a high yielding selection.

Exploration for germplasm in large cardamom has a long history and earlier it was reported that germplasm accessions were collected and conserved in the Indian council of Agricultural Research Complex, Gangtok. These germplasm accessions were screened for biological relationships using enzyme studies (peroxidase) through electrophoretic method (Karibasappa 1986).

Consequent to establishment of Indian Cardamom Research Institute, Spices Board, regional station, Gangtok, efforts have been made to conduct explorations and collection of genetic resources of large cardamom throughout Sikkim, Darjeeling district of West Bengal and adjoining areas. Considering the importance of genetic resources in large cardamom, National Bureau of Plant Genetic Resources (NBPGR) New Delhi, under the National Agricultural Technology Project (NATP) on plant biodiversity has supported the collection of large cardamom accessions. The germplasm collections showed high variations (>20 per cent CV) for yield and major yield contributing traits like productive tillers, spikes per clump and capsules per spike. Apart from this, accessions with narrow leaves, terminal inflorescence, and dwarf type etc. were encountered.

Fifty accessions were subjected to variability studies employing principal component analysis using yield and yield contributing traits. This has resulted in the identification of three main components which together contribute around 70 per cent variation. The diversity available in large cardamom exploited so far has resulted in the identification of a few breeding stocks and a few elite clones like ICRI Sikkim-1 and ICRI Sikkim-2 (Gupta and Gopakumar, 2002).

In recent years fast depletion of genetic resources is happening due to variation in the habitat as a result of frequent land slides, torrential rains, flood and deforestation which are very common in Sikkim Himalaya. Besides valuable materials are being lost due to human activities leading to urbanization as well as pest and disease outbreak. All these factors cause threat to diversity of the crop. The breeding nature of the crop and geographical isolation prevailing in the hills of Himalaya might have contributed to the variability in large cardamom. The richness of bio-diversity in North-East India supports the view of origin of some these important spices in the region. The presence of several wild relatives and the tremendous variability within the cultivated species support the view of large cardamoms' origin in Sikkim (Subba 1984, Rao *et al.*, 1993).

In India, the area under large cardamom cultivation during 70's was 11,204 ha with a production of 2400-2500 tonnes (Gyatso *et al.*, 1980), it increased to 25,635 ha with a production of 4000 tonnes in 80's. The cultivation has further expanded to an area of 30,039 ha with an annual production of 6154 tonnes (2004). The steady expansion of area coupled with cultivation of recognized high yielding varieties are likely to adversely affect the native diverse species and variability with in the species.

Cultivation of cardamom (*Amomum villosum* Lour.) is reportedly causing loss of biodiversity in Yunnan (China) and in Vietnam. It is feared that in Yunnan extensive cardamom cultivation might cause the disappearance of nationally protected species of plants from nature reserves (Wang Baorong *et al.*, 1997).

Nature of ecological impact of cardamom cultivation (*Amomum aromaticum* Roxb.) in the forest of the Hoang Lien Mountains in three villages in three communes in Southern Van Ban District, lao Cai Province, Vietnam was studied. Forest catchments in the area represented significant areas of sub-montane forest, which was identified as a priority vegetation type in nature reserve. Though the area under cardamom cultivation currently is small the rate of change in area of cultivation having increased over 200 per cent over the last three years in all three villages. These changing factors along with uncertainty of the duration of the change in the landscape indicate cardamom cultivation may have significant implications for some species in the long-term (Sebastian Buckingham *et al.*, 2003).

Cytology

Chromosome number is 2n=48 (Vankatasubbu 1946 and Sharma and Bhattacharya 1959). Variability was also reported with 2n=26, 34, 42 and 44 in *Amomum subulatum* (Sharma and Bhattacharya 1959).

Plant Propagation

Propagation of large cardamom is done through seeds, rhizomes (sucker multiplication) and tissue culture techniques. Cultivars suitable for specific areas, altitudes, agro-climatic conditions and mother plant/clump of known performance are selected for collection of seed, rhizome and vegetative bud.

Raising Seedlings

Healthy plantations, free from viral diseases in particular, are selected for seed capsules. Gardens with productivity of 1000 kg/ha or more during the past 3 years are considered.

Spikes are harvested at maturity and seed capsules are collected from the lowest two circles in the spike. After de-husking, seeds are washed well with water to remove mucilage covering of seeds, mixed with wood ash and dried under shade.

Dried seeds are treated with 25 per cent nitric acid for 10 min. for early and higher percentage of germination (Gupta, 1989). The acid-treated seeds are washed thoroughly in running water to remove the acid residue and are surface dried under shade. Seeds are sown immediately after acid treatment.

There are two seasons for sowing seeds: pre-winter (September-October and early November) and post-winter (late February-March). September sowing results in quicker and better germination (within 25-30 days), 30-40 per cent for acid-treated seeds (Gupta,1989).

Seedbed is prepared in a well-drained area. The soil is cut to a depth of 30 cm and exposed to sun for a week. Bed of 15-25 cm height, 90 cm width and convenient length is prepared, incorporating well-decomposed compost/cattle manure. Seeds (acid treated) at the rate of 100 gm for a bed size of 1x 3 m are sown in furrows along the width and they are covered with a thin layer of soil. The space between the furrows is maintained at 10 cm. After sowing, the beds are covered with thick mulch (with paddy straw or dry grass) and watered regularly to keep the bed moist.

The seedbeds are examined for germination between 25-30 days after sowing. Once the germination starts overhead shading of convenient height with Bamboo poles and agro-shade net/ bamboo mat (50 per cent shade, black agro-shade net is found ideal) is provided. The mulch from the bed is removed and cut into small pieces and are spread over in between the seed rows. The beds are watered regularly to keep the bed moist and weeding is attended whenever necessary.

Once seedlings in the primary seed bed reach 3-4 leaf stage (in February/March if seeds are sown in September/October or April/May if the seeds are sown in February/March) they are transplanted either to polybags or into secondary nursery beds in February/March or April/May respectively (Gupta,1989).

Topsoil in virgin land/forest area rich in leaf mold is collected and mixed with well-decomposed cattle manure to get good potting mixture. A porting mixture of 5:1, topsoil: cattle manure is prepared and filled in polybags of 24 x 24cm size with perforations at the base for drainage. Polybags are arranged under overhead shaded shed. Primary seedlings are transplanted in polybag (one seedling per polybag) during February/March or April/May.

These seedlings are planted in the main plantation in July/August. The polybags are removed and seedlings with the soil ball intact are planted. Sometimes the seedlings are maintained in polybag till April next year and planted in May-June in the main field (Anonymous, 1998).

Seedlings from primary nursery are sometimes transplanted in beds. Beds of the size. And nature similar to that of primary beds is prepared: and seedlings at 3-4 leaf stages are transplanted in March/

April/May, maintaining a spacing of 15 cm between the seedlings. A layer of well-decomposed cattle manure is applied and incorporated in the soil. Watering is given at regular intervals to keep the soil moist. The entire secondary nursery is maintained under overhead shade (preferably with black agro-shade net). Seedlings are maintained for 10-12 months. Expected growth of seedlings is about 45-60 cm heights with 5-10 tillers each. These seedlings are transplanted in June-July in the main field (Anonymous, 1998).

High yielding and disease free planting materials are selected for propagation through rhizome. Trenches of 60 cm width, 60 cm depth and convenient length are made across the slopes. Trenches are filled with topsoil, leaf mold and decomposed leaf litter. Rhizomes with one mature tiller and two young shoots or vegetative buds are planted at a spacing of three feet in the trenches during June-July. Thick mulching with dry leaf/grass is applied at the base of the rhizome and watering if required, is done regularly to keep the soil moist. Each clump is split into units of two to three tillers and are used for planting in the main field during June/July or used for further multiplication (Anonymous, 1998).

Large cardamom can be multiplied on a large scale through micropropagation. Protocols for micropropagation were developed at Indian Institute of Spices Research (Sajina *et al.*, 1997; Nirmal Babu *et al.*, 1997).

Nirmal Babu *et al.* (1977) and Sajina *et al.* (1997) accomplished both multiple shoots and rooting in the same medium, in MS basal + BAP (1 mg/1) and IBA (0.5 mg/1) with 3 per cent sucrose and gelled with 0.7 per cent agar at pH 5.8 and 12 h photoperiod at a light intensity of 2500 lux. This combination produced 8-12 shoots/culture and roots/shoots.

Plantation Management

Large cardamom is grown in forest loamy soils having soil depth a few inches to several feet. Colour of soil ranges from dark yellowish brown to very dark grayish brown. Texture is sandy, sandy loam, silty loam or clay. In general, soil is acidic having pH ranging from 5 to 5.5 or more, and with 1 per cent or more organic carbon (Bhutia *et al.*, 1985; Biswas *et al.*, 1986). As the terrain is gentle to deep slope, chances of water logging is less, however, water-logged conditions are not suitable and adequate drainage is quite essential for better stand of the crop (Singh *et al.*, 1998).

In general Large cardamom is cultivated on hill slopes, and often in terraced lands (earlier under paddy cultivation), after raising adequate shade trees. Large cardamom is a shade-loving crop. It grows well under dense shade (60-70 per cent of full day light interception) to light shade (about 30 per cent full day light interception) condition (Singh *et al.*, 1989). The day light intensity required for optimum growth of large cardamom ranges from 5000 to 20,000 lux. In areas having insufficient shade, planting shade tree saplings of different species is done in June-July.

The most common shade trees are *Utis* (*Alnus nepalensis*), 600-2000 m above mean sea level (amsl): *Chilaune* (*Schima wallichi*, 550–1515 m amsl): *Panisaj* (*Terminalia myriocarpa*, 400–1000 m amsl): *Pipli* (*Exbucklandia populnea* 900-2000 m amsl) *Malato* (*Macaranga denticulata*, 670–1515 m amsl): *Asarey* (*Cole brookianum* 850-2000 m amsl): *Gogun* (*Saurauvia nepalensis*, 1400–2000 m amsl): *Karane* (*Symplocos ramosissima* Wall 1500-2400 m amsl): *Bilaune* (*Maesa chesia* 670–1515 m amsl) etc. The Large cardamom based agro-forestry system under the influence of Alnus was more productive with faster rate of nutrient cycling (Sharma and Ambasht, 1988).

Planting

For planting of large cardamom, pits are opened at spacing suitable for the variety/cultivar. In case of robust variety/cultivar such as Sawney, Varlangey, Ramsey etc., spacing followed is 150 x 150 cm while a spacing of 120 x 120 cm is used for Golsey (Dzongu). Pits are opened in April-May. The size of pits usually is 30x30x30 cm.

The ideal season for planting is June-August depending on rains. While planting, too much soil is avoided on the base of seedling, which may otherwise cover the collar region that leads to rotting. Staking is very essential for better anchorage during the initial stage of establishment.

Mulching is done immediately after planting as well as in October-December in the existing plantation. This practice helps to cover the exposed roots, to conserve soil moisture in the ensuing dry months and helps in recycling of nutrients.

Plant Nutrition

Application of NPK fertilizer in three splits, once in April-May after the first summer showers and second split in June and third in September-October before monsoon ceases, increases yield. A fertilizer schedule of 18.4, 6 and 18.6 g NPK/clump produced more number of tillers and spikes/ clump in Sawney. However Golsey did not respond to fertilizer application. Soil application of NPK at the rate of 20:30:40 kg/ha (in two splits: April-May and September) was also found to increase growth and yield characters (Anonymous, 1998). Soil application of Potassium chloride shows significant effect on growth and yield characters.

Large cardamom cultivation is a natural farming with least external inputs and interference with the soil-ecosystem. Organic recycling of nutrients is very essential for sustained productivity.

Weeding

About 51 species of weeds have been recorded in large cardamom plantations (Anonymous, 1984a). Depending on the intensity of weed growth, two to three rounds of weeding are required in a year. First weeding is done in February-April, before application of first split of fertilizer and just before flowering. Second and third round of weeding is done before harvesting in August-September-October along with removal of dried leaves, unproductive tillers etc. The weeded materials are used for mulching.

Irrigation

Yield performance is better in plantations where irrigation is given during dry winter and summer months. Watering during November-March is found essential to maintain a sustainable good yield in the plantation. Water is tapped through pipes and is provided to plants through surface channel/ hose in different directions.

Crop Improvement

Large cardamom is essentially cross-pollinated crop and is entomophilous. The flower morphology is adapted for such a mode of pollination. Each spike consists of about 40-50 flowers, which open in the acropetal sequence over a period of about 15-25 days. Flower opening starts at early morning hours, *i.e.* 3-4 a.m., anthers dehisce almost simultaneously whereas stigma receptivity starts an hour later and last for 24 h. The stigma was found to be receptive even after 36 h from the time of flower opening during rain-free days (Gupta and John,1987; Karibasappa, 1992).

Bumble bees (*Bombus* sp.) are the main pollinators, through a variety of honeybees and other insects do pollinate the flowers. Bumble bees are effective pollinators due to its compatible size with the flowers and is having brush-like hairy structures on its dorsal thorax which helps in carrying pollen mass and depositing it on stigmatic surface while entering the flowers. The highest foraging activity of bumble bees is seen during 6-7a.m., but becomes slow or dull in rainy conditions.

Pests

More than 22 insect pests are associated with large cardamom, and only a few of them cause substantial damage to the crop (Bhowmik, 1962; Pangtey and Azad Thakur, 1986).

Leaf Caterpillar

Leaf eating caterpillar (*Artona chorista* Jordon, Lepidoptera: Zygaenidae), is a major pest of Large cardamom in Sikkim and West Bengal (Yadava *et al.*, 1992: Singh and Varadarasan,1998). Its outbreak was recorded in 1978 in Sikkim where about 800 ha of large cardamom plantations were severely defoliated (Subba, 1980). The leaf caterpillar is first recorded as *Clelea plumbiola* Hampson on large cardamom by Bhowmik (1962).

Artona chorista occurs sporadically in epidemic form in Sikkim and West Bengal every year. Usually the incidence of the pest is observed from June to July and October-March in the field. Severe damage was recorded in Lower Dzongu, Phodong, Ramthung Basti (north Sikkim), Soreng, Hee, Chako(west Sikkim), Kewizing (south Sikkim): Assamlinzey, Dalapchand and Rongli (east Sikkim) and Gotak (Darjeeling Dist. of West Bengal).

The larvae are gregarious in nature and feed underneath the cardamom leaf: the infested leaf can easily be identified from a distance and these can be collected along with larvae and destroyed in June-July and October-December.

Even though insecticides like Quinalphos and Endosulfan (0.05 per cent) are effective, growers are advised to avoid use of such chemicals which kill the natural enemies also. This crop is not exposed to any chemical insecticides (Subba, 1979, 1980, 1984). Therefore naturally-occurring parasitoids keep the pest population under check (Varadarasan *et al.*, 2002).

Hairy Caterpillars

Hairy caterpillars are a group of defoliators infesting large cardamom. Severe infestation by hairy caterpillars affects the crop yield. Among the hairy caterpillars, *Eupterote sp.* are the predominant ones found in cardamom plantations. The Major species are: *Eupterote fabia* Cramer and *Eupterote sp.* (Lepidoptera:Eupterotidae).

Aphids

Aphids cause more damage as a vector rather than as a pest. The aphids are associated with the transmission of viral diseases (Foorkey and Chirke) of large cardamom. The aphid population is recorded high during summer months at lower altitudes. The major species are:

1. *Pentalonia nigronervosa f.caladii* Goot (Hemiptera: Aphididae).
2. *Micromyzus kalimpongensis* (Hemiptera:Aphididae).
3. *Rophalosiphum maidis* Fitch (Hemiptera:Aphididae).
4. *R. padi* Lin. (Hemiptera:Aphididae). (Varadarasan and Biswas,2002).

Shoot Fly

Shoot fly, *Merochlorops dimorphus* Cherian (Diptera: Chloropidae), recorded as a major pest of Large cardamom is damaging young shoots. Low to moderate damage by shoot fly is recorded in large cardamom plantation in Sikkim and East Bengal. In the main field, more damage is recorded at higher altitudes than in the lower. Kumar and Chatterjee(1993) reported another species of shoot fly, *Bradysia* sp. (Diptera:Sciaridae), damaging large cardamom. For managing this pest, infested young shoots should be removed at ground level and destroyed.

Stem Borer

Glypheterix sp.(Lepidoptera: Glyphiperidae) is a minor pest and is specific to large cardamom. Stem borer is recorded from March to November at Gotak (West Bengal), Pangthang, Khasay, Gamdong (east Sikkim), Kabi and Mangan (north Sikkim). The larva feeds on the central portion of the shoot and as a result, terminal leaf of the plant gets dried up and this symptom is known as 'dead heart'. It is a minor pest and can be controlled by removing infested shoots along with caterpillars.

Diseases

The crop is susceptible to a number of diseases, which are mainly viral and fungal in origin. Large cardamom productivity is affected seriously by viral diseases. However fungal diseases are not major constraints. There are two viral diseases on large cardamom causing severe damage to the plantations. 'Chirke' is serious as far as rate of spread is concerned. 'Foorkey' is serious as far as yield loss is concerned. Among fungal diseases, flower rot, clump rot, leaf streak and wilt are known to cause damage to the plant and ultimately reduce the crop yield.

Chirke

This virus disease is characterized by mosaic with pale streak on leaves. The streaks turn pale brown resulting in drying and withering of leaves and plants. The flowering in diseased plants is extensively reduced and only one to five flowers develop in one inflorescence, as against 16-20 in an inflorescence of healthy plants (Raychaudhary and Ganguly,1965) and by the end of third year of crop, the loss is around 85 per cent. The disease is readily transmitted by mechanical sap inoculation and in field it is spread by aphid.

Foorkey

This disease is characterized by dwarf tillers with small, slightly curled pale green leaves. The virus induces remarkable reduction in size of leafy shoots and leaves of the infected plants and also stimulates proliferation of large number of stunted shoots arising from the rhizome. The spikes/ inflorescences are transformed into leafy vegetative parts and fruit formation is altogether suppressed. The diseased plants remain unproductive and gradually degenerate. Foorkey symptom appears both on seedlings and grown up plants (Varma and Capoor, 1964).

Unlike Chirke, Foorkey virus is not transmitted through sap but by the aphids *Pentalonia nigronervosa* Cog. and *Micromyzus kalimpongensis* Basu (Basu and Ganguly, 1968; Sharma *et al.*, 1972). The primary spread of disease from one area to another is through infected rhizomes and further spread within the plantation by aphids. Infected rhizomes can be killed by injecting Agroxcone-40. The management strategies include; regular rouging of diseased plants. The diseased plants are uprooted and destroyed as and when they are traced. Uprooted plants are taken to an isolated place, chopped into small pieces buried in deep pits for their quick decomposition.

Leaf Streak Disease

It is a fungal disease caused by *Pestalotiopsis royenae* (D.Sacc) Steyaert., and is a serious disease among foliar diseases and is prevalent round the year. The disease symptom is the formation of numerous translucent streaks on young leaves along the veins. The infection starts from emerging folded leaves; infected leaves eventually dry up causing loss of green part and reduce the yielding capacity of the plant. Dzongu Golsey is found more susceptible to leaf streak

Three rounds of 0.2 per cent spray of copper oxychloride at 15 days interval, two schedules in a year *i.e.* February-March and September-October can control this disease.

Harvesting and Postharvest Technology

First crop comes to harvest about 2-3 years after planting of sucker or seedling. However, stabilized yields are obtained only from the 4th year up to 10-12 years. Harvesting season starts in August/September in low altitudes and continues up to December at high altitudes. Usually harvesting is done in one round and hence the harvested produce often contains capsules of varying maturity. Harvesting is done when the seeds of top capsules in the spike attain dark gray colour. A special type of knife, locally known as *Elaichi chhuri* is used for harvesting. The stalk of the spike is cut very close to the leafy shoot. After harvest, individual capsules are separated manually. Capsules after harvest are cured to reduce moisture level to 10-12 per cent. The traditional curing is called *Bhatti* curing system (direct heat drying). Large cardamom is also cured by flue pipe curing system (indirect heat drying).

Bhatti System

Drying of cardamom is generally done in kilns locally called Bhatti. A Bhatti consists of a platform made of bamboo mats/wire mesh, laid over a four-walled structure made of stone pieces with a V shape opening in the front for feeding fire wood. Capsules are spread over the platform and are dried by direct heat generated from the firewood. About 70-80 kg of firewood is required for curing 100 kg of green cardamom in this traditional kiln (Sundriyal *et al.*, 1994). Both green and dry woods are used and, as a result, huge volume of smoke is generated that passes through cardamom (Singh *et al.*, 1978). Depending on the thickness of the cardamom spread, it takes 60-72 h for curing (John and Mathew, 1979). Colour of cardamom cured under this system is dark brown or black. If smoke percolates through cardamom, it loses original colour, gets smoky smell and fetches low price (Karibasappa *et al.*, 1987).

Portable curing chambers, Flue pipe curing system and gasifier system of curing are developed to overcome the draw backs of Bhatti system (Joseph Johny *et al.*, 1996).

Chemical Composition

Large cardamom seeds (dried) contain: moisture (8.5 per cent), protein (6 per cent), volatile oil (2.8 per cent), crude fiber (22 per cent), starch (43.2 per cent), ether extract (5.3 per cent), alcohol extract (7 per cent) and (4 per cent). They are found to contain (for 100 g seeds), calcium 666.6 mg, magnesium 412.5 mg, phosphorous 61 mg, fluoride 14.4 ppm. Seeds contain the glycosides perunidin 3,5-diglucoside, leucocyanidin 3-0-β-D-glucopyranoside and subulin, an aurone glucoside. Cardamomin–a chalcone, alpinetin–a flavanone, are also reported (Shankaracharya et al.,1990).

The powdered seeds on steam distillation yield 1-3.5 per cent of a dark brown, mobile essential oil. A study has shown that the volatile oil content was 2.44 per cent in Sawney, 2.42 per cent in Pink Golsey, 2.25 per cent in Ramnag and 1.66 per cent in Ramsey (Shankaracharya *et al.*, 1990). The oil has the following properties: sp.gr (20 C): 0.9142, ref.index (26 C):1.46, optical rotation: -18 3; acid value 2.9; saponification value 14.53; and saponification value after acetylation 40.2.

Large cardamom oil is characterized by flat cineol odour, harsh aroma and inferior flavour as against the warm spicy, aromatic odour of cardamom (Table 14.3). The Large cardamom oil is almost devoid of α-terpinyl acetate, and is very rich in 1,8-cineole (Balakrishnan *et al.*, 1984; Pruthi 1993).

Table 14.3: Composition of Volatile Oil of Large Cardamom in Comparison with the Oil of Small Cardamom

Constituents	*Large Cardamom (Range) (per cent)*	*Small Cardamom (Range) (per cent)*
α–terpinene	0.5–11.13	0.37–2.5
α–pinene	2.0–3.11	1.10–13.00
β–pinene	2.4–3.67	0.2–4.9
Sabinene	0.2–9.10	2.5–4.9
Camphene	0.44	0.02–0.13
υ–terpinene	0.2–16.2	0.04–11.2
Limonene	6.38–10.3	0.12–2.1
ρ–cymene	0.20–0.30	0.40–0.70
1,8–cineole	63.3–75.27	23.4–51.30
Linalool	0.41	2.1–4.5
Geraniol	0.12	0.25–0.38
α–terpineol	4.9–7.2	0.86–1.90
Terpinen–4 ol	1.42–2.0	0.14–15.3
Nerolidol	0.12–1.0	0.23–1.60
Nerlacetate	0.14	0.02–0.09
α–terpinyl acetate	5.10	34.60–52.5
α–bisabolene	1.3–3.6	0.07–0.83
β–terpineol	0.8	0.70–2.10

Properties and Uses

Large cardamom is used as an ingredient as well as a flavouring agent with masala and curry powders; in flavouring sweet dishes, cakes and pastries; as a masticatory and for medicinal purposes. The seeds are used for chewing along with betel quid (betel leaf, arecanut lime, with or without tobacco). In gulf countries, large cardamom is used as a cheaper substitute for spicing tea in place of small cardamom. In the Indian systems of medicine; Ayurveda and Unani, it is used as preventive as well as curative for throat troubles, congestion of lungs, inflammation of eyelids, digestive disorders and even in the treatment of pulmonary tuberculosis (Kirtikar and Basu,1952). The seeds are fragrant adjuncts for other stimulants, bitters and purgatives. The seeds have a sharp and good taste and are tonic to heart and liver. The pericarp is reported to be good for alleviating headache and heals stomatites (Anonymous,1950). Decoction of the seeds is used as a gargle in afflictions of teeth and gums. With melon seeds they are used as a diuretic in cases of gravel of the kidneys. They promote elimination of bile and are useful in congestion of liver. They are also used in the treatment of gonorrhea. In large doses with quinine, they are used in neuralgia. The seed oil is applied to the eyes to allay inflammation.

The direct uses of large cardamom are in pickles, pulao, meat and vegetable dishes. It is also used in industrial sector for flavouring toothpastes, sweets, soft drinks, toffees, flavoured milk and alcoholic drinks. The ripe fruits are eaten raw by people of Sikkim and Darjeeling and are considered a delicacy (Gyatso *et al.*, 1980).

In general *Aframomum* is used as a spice, *Elettaria* both as a spice and as medicine (and a masticatory), and *Amomum* used as an ingredient in several traditional medicines in China (traditional Chinese medicine -TCM), India, Korea and Vietnam.

References

Anonymous, 1950. *Amomum. Garden Bull.*, 13: 192–214.

Anonymous, 1984. *Cardamom Statistics* 1981–82. Pub. Cardamom Board, Cochin–18.

Anonymous, 1984a. *Large Cardamom: Package of Practices*. Pub. Cardamom Board, Gangtok, Sikkim.

Anonymous, 1998. *Annual Report 1997–1998*, Spices Board, Sugandha Bhavan, Cochin.

Anonymous, 2004. *Spices Statistics, 5th Edition*, Spices Board India, Cochin.

Baker, J.G., 1890–92. *Scitaminae*. In: J.D. Hooker's *The Flora of British India*, London, 6: 233–243.

Balakrishnan, K.V., George, K.M., Machulla, T., Narayana Pillai, O.G., Chandran, C.V. and Verghese, J., 1984. Studies in Cardamom–1. Focus on oil of *Amomum subulatum* Roxb. *Indian Spices*, 21(1): 9–12.

Basu, A.N. and Ganguly, B., 1968. A note on the transmission of 'Foorkey disease' of large cardamom by the aphid. *Micromyzus kalimpongensis* Basu. *Indian Phytopathology*, 21: 127.

Bhowmik, T.P., 1962. Insect pests of large cardamom and their control in West Bengal. *Indian J. Ent.*, 24: 283–286.

Bhutia, D.T., Gupta, R.K. and Biswas, A.K., 1985. Fertility status of the soils of Sikkim. *Sikkim Science Society Newsletter*, 5.

Biswas, A.K., Gupta, R.K. and Butia, D.T., 1986. Characteristics of different plant parts of large cardamom. *Cardamom*, 19(3): 7–10.

Dubey, A.K. and Singh, K.A., 1999. Large cardamom: A spice crop of India. *Indian Farming*, 49(2): 17–18.

Gupta, P.N., 1986. Studies on Capsule morphology of large cardamom cultivars (*Amomum subulatum*, Roxb.). *J. of Plantation Crops*, 16(Sppl.): 371–75.

Gupta, U., 1989. Studies on seed germination of seeds of large cardamom. *Spice India*, 2(3): 14.

Gupta, U. and John, T.D., 1987. Floral biology of Large Cardamom. *Cardamom*, 20(5): 8,15.

Gupta, U. and Gopakumar, B., 2002. ICRI Sikkim–1 and ICRI Sikkim–2 two high yielding large cardamom selections. Pub: ICRI, Spices Board, Tadong, Sikkim.

Gupta, U., Kuruvilla, K.M., Madhusoodanan, K.J. and Thomas, J., 2003. Biodiversity of large cardamom in Sikkim Himalaya. In: *National Seminar on New Perspectives in Spices, Medicinal and Aromatic Plants*, ICAR Research Complex Goa, 27–29 November.

Gyatso, K., Tshering, P. and Basnet, B.S., 1980. Large Cardamom of Sikkim. *Krishi Samachar*, 2(4): 90–95.

John, J.M. and Mathew, P.G., 1979. Large cardamom in India. *Cardamom*, 3(10): 13–20.

Joseph Johny, Raghavan, B., Nanjundaiah, G. and Shakaranarayana, M., 1996. Curing of large cardamom (*Amomum subulatum* Roxb.) Fabrication of a new dryer and a comparative study of its performance with existing dryers. *J. Spices and Aromatic Crops*, 5(2): 105–110.

Karibasappa, G.S., 1986. Annual Report. *ICAR Research Complex for North East Hill Region*. Sikkim Centre.

Karibasappa, G.S., 1992. Phenology of flowering in large cardamom at Gangtok. *Indian J. Agric. Sci.*, 59, 205–209.

Karibasappa, G.S., Dhiman, K.R. and Rai, R.N., 1989. Half sib progeny analysis for variability and association among the capsule characters and path studies on oleoresin and its cineole content in large cardamom. *Indian J. Agric. Sci.*, 53: 621–625.

Karibasappa, G.S., Dhiman, K.R., Biswas, A.K., and Rai, R.N., 1987. Variability association among quantitative characters and path analysis in large cardamom. *Indian J. of Agric. Sciences*, 57(12): 884–888.

Kirtikar, R.P. and Basu, B.D., 1952. *Indian Medicinal Plants*, Vol. 4. Balu, L.M., Pub., Allahabad.

Kumar, A. and Chaterjee, S.V., 1993. Record of new pest *Bradysia* sp. and its predator, *Phaonia simulans* on large cardamom. *J. Appl. Zool. Res.*, 3(1): 103–104.

Kumar, S., 2001. *Zingiberaceae of Sikkim.* Deep Publications, pp. 11–17.

Madhusoodanan, K.J. and Y.S. Rao, 2001. Cardamom (Large). In: *Handbook of Herbs and Spices*, (Ed.) K.V. Peter. Woodhead Publishing Ltd., Cambridge, England.

Michel, H. Porcher, 2001. Sorting *Amomum* names: Down loaded from michel@mullara.met. unimelb.edu.au.

Mukherji, M.K., 1968. Large Cardamom Cultivation in Darjeeling District of West Bengal. *Cardamom News*, 2(11): 1–8.

Nirmal Babu, K., Ravindran, P.N. and Peter, K.V., 1997. *Protocols for Micropropagation of Spices and Aromatic Crops*, Indian Institute of Spices Research, Calicut.

Pangrey, V.S. and Azad Thakur, N.S., 1986. Insect pests of large cardamom. *Indian Farming*, 35(12): 17–19.

Pruthi, J.S., 1977. Cardamom: Greater cardamom and lesser cardamom. *Spices and Condiments.* National Book Trust of India, New Delhi, pp. 53–63.

Pruthi, J.S., 1993. *Major Spices of India: Crop management and Post Harvest Techonolgy*, ICAR Publications, New Delhi, p. 114–179.

Rao, Y.S., Gupta, U., Anand Kumar and Naidu, R., 1993. A note on Large Cardamom (*Amomum subulatum* Roxb.) Germplasm collection. *J. Spices and Aromatic Crops*, 2(142): 77–80.

Rao, Y.S., Anand Kumar, Sujatha Charerjee, Naidu, R. and George, C.K., 1993a. Large cardamom (*Amomum subulatum* Roxb.): A review. *J. Spices and Aromatic Crops*, 2(1&2): 1–15.

Rao, Y.S., Gupta, U., Anand Kumar and Naidu, R., 1990. Phenotypic variability in large cardamom. In: *Proc. Nat. Symp. on New Trends on Crop improvement of Perennial spices.* Rubber Research Institute of India, Kottayam.

Raychaudhary, S.P. and Ganguly, B., 1965. Further studies on 'chirke' disease of large cardamom (*Amomum subulatum* Roxb.). *Indian Phytopathology*, 18: 373–377.

Sabastian, Buckingham and T'u Minh Trep, 2003. A rapid characterization of Cardamom *Amomum aromaticum* (Roxb.) cultivation in three villages in Van Ban District, Lao Cai Province. Downloaded from Internet.

Sajina, A., Mini, P.M., John C.Z., Nirmal Babu, K., Ravindran, P.N. and Peter, K.V., 1997. Micropropagation of large cardamom (*Amomum subulatum* Roxb.), *J. Spices and Aromatic Crops*, 6: 145–148.

Sankaracharya, N.B., Rhagavan, B., Abraham, K.O. and Shankaranarayana, M.L., 1990. Large cardamom–Chemistry, technology and uses. *Spice India*, 3(8): 17–25.

Sharma, D.C., Raychaudhuri, S.P. and Capoor, S.P., 1972. Countdown on the cardamom virus. *Farm and Parliament*, 7(12): 13–15.

Sharma, E. and Ambasht, R.S., 1988. Nitrogen accretion and its energetics in Himalayan alder plantations, *Functional Ecology*, p. 229–235.

Sharma, A.K. and Bhattacharya, N.K., 1959. Cytology of several members of Zingiberaceae and a study of the inconsistency of their chromosome complement. *Cellule*, 59: 299–345.

Singh, K.A., Rai, R.N. and Bhutia, D.T., 1989. Large cardamom (*Amomum subulatum* Roxb.) plantation: an age old agrogorestry system in eastern Himalayas. *Agroforestry System*, 9: 241–257.

Singh, B.P., 1988. Large cardamom in Sikkim. *Spices News Letter*, 22(10): 21.

Singh, G.B., 1978. Large cardamom. *Cardamom*, 10(5): 3–13.

Singh, G.B., Gupta, P.N. and Pant, H.G., 1978. Large cardamom, a foreign exchange earner from Sikkim. *Indian Farming*, 3: 7–8.

Singh, J. and Varadarasan, S., 1998. Biology and management of leaf caterpillar, *Artona chorista* Jordon (Lepidoptera: Zygaenidae), a major pest of large cardamom (*Amomum subulatum* Roxb.). In: *Recent Advances in Plantation Crops Research.* Allied Publishers Ltd., New Delhi.

Subba, J.R., 1980. Field problems of large cardamom cultivation in Sikkim. *Cardamom*, 12: 3–5.

Subba, J.R., 1984. *Agriculture in the Hills of Sikkim*. Sikkim Science Society, Gangtok, Sikkim. pp. 286.

Sundriyal, R.C., Rai, S.C., Sharma, E. and Rai, Y.K., 1994. Hill agro forestry, systems in South Sikkim, India. *Agroforestry Systems*, 26: 215–235.

Upadhyaya, C. and Gosh, S.P., 1983. Wild cardamom of Arunachal Pradesh. *Indian Hort.*, 27(4): 25–27.

Varadarasan, S. and Biswas, A.K., 2002. Large cardamom (*Amomum subulatum* Roxb.) in *cardamom: The Genus Elettaria*, (Eds.) P.N. Ravindran and K.J. Madhusoodanan. Taylor and Francis, London.

Varma, P.M. and Capoor, S.P., 1964. 'Foorkey disease of large cardamom. *Inidan J. Agric. Sci.*, 34: 56–62.

Ved Prakash and Mehrotra, B.N., 1996. V. Diversity and conservation: Zingiberaceae of N.E.India: Diversity and Taxomomic status. *Proc. 2nd Symp. Family Zingiberaceae*, pp. 262–273. South China Institute of Botany.

Venkatasubbu, K.R., 1946. A preliminary survey of chromosome numbers in Scitamineae of Bentham and Hooker. *Proc. Ind. Acad. Sci.*, 23B: 281–300.

Wang, Baorong, Su Wenhua, Yan Haizhong and Huang Jianguo, 1997. Impact of planting *Amomum* on key protected plants in Mengyang Nature Reserve of Xishuangbanna and its countermeasures. *Chinese Journal of Applied Ecology*, 8(Suppl): 75–81.

Yadava, S., Kumar, A. and Naidu, R., 1992. Biology of *Artona chorista* Jordon (Lepidoptera: Aygaenidae) on large cardamom. *Proc. Bioecol. And Control Ins. Pests*, pp. 73–78.

Biodiversity in Horticultural Crops Vol. 1 *Pages* ***309–319***
Editors: **Professor K.V. Peter & Dr. Z. Abraham**
Published by: **DAYA PUBLISHING HOUSE**

Chapter 15

Kokum, Malabar Tamarind and Mysore Gamboge

Z. Abraham and R. Senthilkumar
National Bureau of Plant Genetic Resources
Regional Station, Vellanikkara, KAU P.O., Thrissur, Kerala – 680 656

Genus *Garcinia* L. (Family: Clusiaceae) gained commercial prominence recently due to anti-obesity properties attributed to the extract from fruits. *Garcinia* is named after Laurent Garcin, an 18th century French botanist who travelled widely throughout India. Genus *Garcinia* is distributed in tropics of Old World, especially, Asia, with approximately 200 species. In India, 38 indigenous and wild, and 3 exotic species occur. Malaysia and Africa with large number of endemic species are the two main centres of development of genus *Garcinia*. A few of the species of family Clusiaceae possess medicinal properties, whereas most of the plants are known for their oil glands or secretary canals or cavities, which contain yellow or brightly coloured resins (Muhammed *et al.*1994). Trees in genus *Garcinia* can be either dioecious or polygamous. In dioecious species, reproductive organs are unisexual. In polygamous species, male, female and hermaphrodite flowers are found in same plant. Male flowers in *Garcinia* have their distinctive pistillodes (Raven *et al.*, 1986). Flowers of *Garcinia* may be solitary, fascicled and umbelled or panicled. In male flowers, stamens exist either free or joined to form a ring or lobular mass that surrounds a rudimentary ovary. Anther filaments are short and thick; though sometimes two-lobed; anthers are straight/horse-shoe shaped with annular dehiscence. In female flowers, staminodes are free or joined together. The ovary consists of 2 to 12 cells with solitary ovules positioned at inner angle of each cell. The berry encapsulated by a tough rind, sits on top of calyx. Most *Garcinia* berries contain several large seeds suspended in a pulpy, sweet or sweet-sour interior (CSIR 1956, Roberts 1984). In India, several species of *Garcinia* occur naturally in South-Western Ghats, North Eastern States and Andaman and Nicobar Islands. Among these, *Kokum*, Malabar tamarind and Mysore gamboge are important.

Kokum or *Kokam*

Kokum is a potential under-exploited minor fruit and oilseed crop, currently gaining much commercial and medicinal importance. Botanically *kokam or kokum* is known as *Garcinia indica* (Dupetit-Thouars) Choisy. It is also known as *brindon* in Goa, *bhirind* or *ansul* in Marathi and Konkani, *murugal* in Kannada and *punampuli* in Malayalam (Sullivan *et al.*, 1974). The tree is commonly known as *kokum* butter tree, mangosteen oil tree or *brindonia* tallow tree due to fat content in seeds. This species being endemic is oriental in origin, found in South Western Ghats in tropical rain forests in Ratnagiri, Konkan and Kodagu of Karnataka and Kasaragod and Wyanad of Kerala. The crop prefers a warm and moderately humid tropical climate with a total rainfall range of 2500-5000 mm and grows under a mean annual temperature of 20-30°C, 60-80 per cent humidity and upto an altitude of 800 meters above mean sea level. In Karnataka and Goa, it is common throughout ghat forests and found in home gardens and forests throughout south-western coastal region upto Kasaragod in Kerala. The *kokum* tree reaches a height of 10 to 17 metres and bears mostly dark purple to black or rarely white or creamish yellow fruit. Rind of *kokam* is sticky with curled edges when dry. It is smaller and thinner than most of species of genus *Garcinia*. It flowers during November-February and fruits ripen in April-May (CSIR 1956). Flowers, which can be axillary or terminal, exist in solitary form or as spreading fascicles. Male flowers are characterized by numerous stamens and two celled anthers with exceedingly short filaments. Female flowers are either sessile or on short pedicels, bundled two or three together. Ovary is 4-8 celled with sessile stigma. Fruit is spherical but un-furrowed, 2.5 to 3.0 cm in diameter and encases 5 to 8 seeds (Muhammed *et al.*, 1994, Subash Chandra 1996). Freshly harvested fruits are reddish green and turn to full red-purple colour in a day or two. Flesh of fruit is juicy and has a sweetish acid taste. Normal shelf-life of fresh fruit is about five days. Common method practiced for preservation is sun drying. For this, fresh fruits are cut into halves and fleshy portion with seed is removed. Fruit rind or skin is then repeatedly soaked in juice of pulp during sun drying. Product obtained after sun drying is referred to as *amsul* or unsalted *kokum* in commerce. Salted *kokum* (*agar*) is also marketed; wherein common salt is used during soaking and drying of rind. *Lonavala kokum* and *Pakali kokum* are trade names (Sampathu and Krishnamurthy 1982) or geographical indications associated with this product. Fruit has an agreeable flavour and a sweetish acid taste valued for its nutritional and medicinal properties. *Kokum* has been traditionally used as an acidulant. It is used in Konkan region, chiefly in the form of *kokum* as a garnish, to give an acid flavour to curries and also for preparing cooling syrups (CSIR 1956). For traditional fish curry of Konkan coast and Goa, *kokum* rind is a usual ingredient. Dried rind, strained in water, when sweetened with jaggery is a must for marriage feasts and functions in Uttar Kannad District of Karnataka. It promotes digestion. Wine red syrup of commerce, extracted from fresh rind of ripe fruit with sugar, is stored in households of this region for making cool drinks in summer (CSIR 1956). Sweet and soft pulpy cover of seeds is eaten or made into curries. Fruit is also pickled (Subhash Chandra 1996). Dried rind is used as an excellent substitute for tamarind and lime, in preparation of fish and non-vegetarian curries for imparting a special flavour and taste. Fruit of *G.indica* is anthelmintic and cardiotonic and useful for treatment of piles, dysentery, tumours, pains and heart complaints. Syrup from fruit juice is given in bilious affections. Root is astringent (Krishnamurthy *et al.*, 1981, Sampathu and Krishnamurthy 1982). Seeds yield a valuable edible fat, *kokum* butter. It is extracted mostly as a cottage industry by crushing kernels, boiling pulp in water and skimming off fat from top or churning crushed pulp with water. Oil is also obtained by solvent extraction method. Yield of fatty oil is about 25 per cent. *Kokum* butter sold in market consists of egg shaped lumps or cakes of light grey or yellowish colour with a greasy texture and a bland oily taste. It is used mainly as an edible fat and sometimes as an adulterant of ghee. Refined and deodorized fat is white in colour and compares favourably with high class hydrogenated fats (Nadkarni 1954,

CSIR 1956). Solid fat obtained from seeds is processed and refined to produce a creamy white butter. It is also referred to as Goa butter and can be found in many cosmetic products such as creams, lotions and foundations with a softness rating of 9. *Kokum* butter is often used as a substitute for cocoa butter due to its uniform triglyceride composition. It melts when it comes into contact with skin. Often underrated, *kokum* butter is one of the most hard and stable vegetable butters, firmer than mango butter, but not quite as firm as cocoa butter. *Kokum* butter is more quickly absorbed into skin without a greasy feeling. *Kokum* butter is composed of beneficial compounds that help to regenerate skin cells. It is commonly used in skin healing lotions, creams and body butters, as well as soaps, cosmetics and toiletries. *Kokum* butter is nutritive, demulcent, astringent and emollient. It is suitable for ointments, suppositories and other pharmaceutical purposes. It is used for local application to ulcerations and fissures of lips, hands and nipples. Cake left after extraction of oil is used as manure (CSIR 1956). *Kokum* butter is used as a specific remedy against diarrhoea and dysentery. It is now used in cosmetics and medicines known as *vrikshamla* in Ayurveda. Various parts of tree like root, bark, fruit and seed oil are used for treating piles and abdominal disorders (Subash Chandra 1996). *Kokum* rind contains 2 to 3 per cent anthocyanin pigments. It is a promising source of natural colourant for acid foods. Processing conditions are standardized at Central Food Technology Research Institute, Mysore for extraction and purification of pigments on commercial scale. Preliminary studies showed that syanidin-3-sambubioside and cyanidin-3-glucoside are present as major pigments in the ratio of 4:1. Food applications of *kokum* colour are in processed fruit products, alcoholic and non-alcoholic beverages, preservatives, and instant foods (Krishnamurthy *et al.*, 1982). A new fat soluble pigment, garcinol was isolated from fruit rind. Identity of this pigment was established by chemical and spectral studies (Krishnamurthy *et al.*, 1981). Acid in dry *kokum* rind was identified as hydroxycitric acid. Pharmaceutically active ingredient, (-) hydroxy citric acid (HCA) that accelerates fat burning and inhibites fatty acid synthesis, occurs in fruits (Akhtar Hussain *et al.*, 1992). Antony *et al.* (1998) performed quantitative analysis of HCA and HCAL [(-) hydroxy citric acid lactone] in *G.indica* and estimated presence of 19 to 26 per cent of total acids in fruits. Conventional way of propagation is by seeds. As crop is cross pollinated, seedling progeny shows heterogeneity and enormous variability is seen. In certain pockets of Konkan, soft wood grafting and approach grafting are resorted to, which favour early yield and high density planting (Subhash Chandra 1996). Tissue culture method is also being attempted for micropropagation (Rao *et al.*, 2000).

Composition of *G. indica* fat (*kokum* butter) is shown in Table 15.1 (Jameison *et al.*, 1943).

Table 15.1

Characters	*Value*	*Character*	*Value*
Melting point	39.5–40.0º C	Sap. Equiv.	299.5
Iodine value	37.4	Unsapen matter per cent	1.4
Free fatty acids (per cent as oleic)	7.2		
Components of fatty acids (per cent by weight)			
Palmitic	2.5		Stearic 56.4
Arachidic	-	Olein	39.4
Linolein	1.7		
Components of glycerides (per cent by mol)			
Tritearin	1.5	Oleodistearin	68
Oleopalmitostearin	8	Palmitodiolein	20
Triolein	2		

Composition of fresh *kokum* rind is given in Table 15.2 (Sampathu and Krishnamurthy 1982).

Table 15.2

Character	*Value*	*Character*	*Value*
Moisture (%)	80.00	Protein (N × 6.25) %	1.92
Crude fibre (%)	14.28	Total ash (%)	2.57
Tannine (%)	2.85	Pectin (%)	5.71
Starch (%)	1.00	Crude fat (%)	10.00
Acid (as hydroxy citric acid) (%)	22.80	Pigment (%)	2.40
Ascorbic acid (%)	0.06	Carbohydrates by difference (%)	35.00

(Values are expressed on moisture free basis)

Area under cultivation of *kokum* is 1200 ha in South Maharashtra and a total production of 10200 tonnes with ripe fruit yield of 8.5 tonnes/ha was reported. Chromosome number of *kokum* is 2n=48 (Thombre 1964) and 2n=54 (Krishnasamy and Raman 1949). Mostly, populations consist of only female trees in which flowers contain staminodes instead of stamens, without any male tree around. One of the possibilities of less number of male trees is gradual removal of the trees being non-productive. A few female trees observed were more than 100 years old and still fruiting profusely. When propagated through seeds, male and female trees and very rarely trees with both male and female flowers are obtained. Absence of differentiated embryo, endosperm or embryonic axis indicated that so called "seed" is not a true "seed". Germination characteristics of "seed" showed clear cut presence of polarity. Regeneration of multiple seedlings from whole "seed" and "seed" pieces further indicated apomictic nature of "seed". Based on "seed" characteristics and field examination of natural and raised populations, facultative agamospermy in *G. indica* is reported (Malik *et al.*, 2005). *Kokum* "seed" fat is exported to Netherlands, Italy, Japan, Singapore, United Kingdom and Malaysia. Total export of *kokum* during 2001-2002 as dried pericarp was 2.56 tonnes valued Rs.216 million. Data on world area, production and productivity are not available as it is native and endemic to India, but reported introduced to France, Bourbon and Mauritius as a crop. Available lone variety 'Konkan Amruta' needs to be tested in multi-locations to assess stability in yield and to determine projected yield over years. It was released by Konkan Krishi Vidyapeeth in 1997 as an early maturing variety with yield of 138 kg/ tree and an average fresh fruit weight of 35 g.

At present, National Bureau of Plant Genetic Resources, Regional Station, Thrissur established in its field genebank, 33 accessions of *kokum*, collected from Dakshin Kannad (3 accessions), Uttar Kannad (18), Shimoga (5) and Chikmagalur (2) districts of Karnataka. Kerala collections are from Kasaragod (1 accession) and Kannur (1) districts. Among these collections, a promising genotype (IC136687-3) was registered (INGR 04063) for its total number of fruits and fresh fruit weight/tree/ year. Of the accessions, which flowered during 1998, five trees in 3 fruiting accessions were characterized for 16 quantitative traits. Among these, two genotypes were promising in subsequent bearing on the basis of yield and weight of 100 dry kernels (IC136687-2) or fruits/tree and yield of dry rind (IC136685-1).

Malabar Tamarind

Malabar tamarind is an indigenous, tropical under-exploited, semi-domesticated spice and oilseed crop. It is botanically known as *Garcinia cambogia* (Gaertn.) Desr. and is found wild in evergreen or semi-evergreen forests of South Western Ghats from South Maharashtra extending southwards to Kerala and Tamil Nadu. It is referred in Malayalam vernacular as *'Kodampuli'* or *'Kudampuli'*. It is found wild in evergreen forests upto an altitude of 1300 to 2000 m above mean sea level (MSL). The best suited regions for its growth are those having high humidity and an altitude of 400-1000 m (CSIR 1956). It is a small or medium sized tree with round crown and horizontal or drooping branches, leaves dark green and shining, elliptic obovate, 5-7 cm long and 2.5-7.5 cm broad, fruits ovoid or spherical in shape, 5 cm in diameter, yellow or red when ripe, with 6-10 grooves and about 6-10 seeds, surrounded by a succulent aril (CSIR 1956). Dioecious in nature, productive bisexual or female trees outnumber males (referred to as *'varangu'* in Malayalam vernacular) which are usually not retained once they exhibit sexuality. Every year, from February to April, trees put forth flowers. Flowers are pale yellow in colour, 10 to 40 flowers are seen as bunches in male plants. These flowers fall off. In female plants, three or four flowers are seen in a bunch and the female or bisexual flowers are bigger than male flowers. After flowering, it takes three weeks for fruit set. It takes 120 to 135 days for fruits to ripen. Ripening usually coincides with South West Monsoon (June-July). Ripe fruits are orange yellow and are ellipsoid or spherical in shape. Fruits vary in size weighing 50 to 180 g. Fruits abscise on ripening and are collected after they fall to ground. Seeds are enclosed in a pulpy translucent aril which is often sweet and edible, but sometimes acrid. On pricking, fruits exude a yellow resinous liquid which hardens slowly into a brown gummy mass (Nadkarni 1954, Raven *et al.*, 1986, Muhammed *et al.*, 1994). It can be propagated by seeds. Plants propagated through seeds do not generally exhibit parental characteristics. Seedlings segregate into productive bisexual or female and unproductive male trees. Sex is revealed only after flowering for which seedlings take 7-8 years. Unproductive trees are usually removed after they exhibit their sex. To overcome this drawback, softwood grafting is usually practiced. Grafts flower in 3 to 4 years of planting, resulting in dwarf bushy shrubs favouring high density planting. Grafting can be taken up anytime of the year, if mist chamber facility is available. If not, June to October is suitable (Spices Board 2001). One year old seedlings are used for soft wood grafting (Muthulakshmi and Sarah 1999). Planting of grafts can be taken up at onset of first monsoon showers. Recognising commercial importance, tissue culture work was initiated recently and success is encouraging. Multiple shoot regeneration from axillary buds is observed (Rao *et al.*, 2000). Dried fruit rinds are hard and dark brown. It is rich in acids (about 30 per cent), which comprises of tartaric acid (10.6 per cent), reducing sugars (as glucose) 15.0 per cent, and phosphoric acid (as calcium triphosphate) 1.52 per cent. Of the total acids present, nearly 90 per cent are non-volatile. Principle acid in fruit rind is (-) hydroxycitric acid based on chemical and spectroscopic studies and not citric acid (Lewis and Neelakandan 1965). By titration and measurement of conductivity, the acid was proved to be an α-hydroxy tribasic acid. Hydroxycitric acid (1,2-dihydroxypropane-1,2,3-tricarboxylic acid) has two asymmetric centres. Being a γ-hydroxy acid, it cyclisizes readily to corresponding lactose (Lewis and Neelakandan 1965). Lowenstein (1971) found that (-) hydroxy citric acid (HCA) strongly inhibited fatty acid synthesis *in vivo* or in living systems. Sullivan *et al.* (1972) reported that fatty acid and cholesterol synthesis were blocked significantly in test animals, which were given (-) HCA before. In another study, Sullivan *et al.* (1974) noted that rats fed on (-) HCA eat less than did control animals. Sullivan and Triscari (1977) reported that (-) HCA lowered body fat level with no loss of body protein or lean mass in test animals that had been experimentally made obese. HCA inhibits lipogenesis, lowers production of cholesterol and fatty acids, increases production of glycogen in liver, suppresses appetite, and increases body's production of heat by activating the process of thermogenesis (Lowenstein 1971). It is a potential dietary supplement for weight loss and appetite control. Several products of (-) HCA like citrin, citrimax, garcinia spray and garcinia puff are available commercially in market (Muhammed *et al.*, 1994). They are gaining popularity as drugs for weight loss (Dallas and Michael 1990). Dried rind is also used for polishing gold and silver, and as a substitute for acetic and formic acids in coagulation of rubber latex. Wood (3700-4625 kg/m^3) is gray and close grained. It is not durable, but heart wood of old trees is distinctly hard and durable. Wood is used for posts, match boxes and splints (Rama Rao 1914, Gamble 1922, Lewis 1934, CSIR 1956). Constituents like cambogin and camboginol were isolated from latex of tree and their structure was confirmed by UV, IR and NMR studies (Rao and Sakariah 1988). Common method followed is that, collected fruits are first split open length-wise. After removing seeds, the rinds are sun/oven/smoke dried or alternatively smoke and sun dried. To ward off fungal infestation, rind is coated with coconut oil and salt. Among different drying methods, rind from smoke drying was soft and flexible, shiny black in colour and with good shape retaining capacity and consumer appeal (Muthulakshmi *et al.*, 2001). Fruits are edible but too acidic to be eaten raw. They are valued for their dried rind, used in Kerala, as a condiment for flavouring curries in place of tamarind or lime. Dried rind is used in fish curries for imparting unique delicate flavour (Muthulakshmi and Sarah 1999). In Sri Lanka, fruits are picked under-ripe, thick pericarp cut into sections, dried in sun and preserved for future use. Dried material is used with salt in curing of fish. A decoction of fruit rind is given for rheumatism and bowel complaints. It is also employed in veterinary medicine as a rinse against disease of mouth in cattle. Resin possesses purgative properties (Rama Rao 1914, Chandrasena 1935, Kirtikar and Basu 1984). Extracts activate digestion and are used in treatment of worms, parasites, tumours and dysentery. Seeds yield 31 per cent of an edible fat, resembling *kokum* butter. Fat has a granular structure with following constants: melting point –29.5°C, acid value –5.0; saponin value –0.2; unsaponified matter –1.0 per cent and titre –51.2°. Fat is rich in oleic acid (CSIR 1956).

NBPGR Regional Station, Thrissur has so far assembled 126 accessions of Malabar tamarind, collected from Dakshin Kannad (2 accessions), Uttar Kannad (25), Kodagu (10), Chickmagalur (8), Shimoga (1) and Belgaum (8) districts of Karnataka; Thrissur (17 accessions), Kottayam (3), Kannur (8), Alappuzha (11), Ernakulam (9), Malappuram (3), Kozhikode (3), Kollam (6), Pathanamthitta (5) and Thiruvananthapuram (1), Kerala and South Goa (4 accessions) district of Goa. Twenty-four bearing trees belonging to 22 accessions were characterised for 13 quantitative fruit, seed and stalk characters. Seven trees, IC244086-1, IC244115, IC244077-1, IC244100-2, IC244111-1, IC244110 and IC244106-2 were superior in fresh fruit weight, fresh rind weight and fresh rind thickness above 90 g, 60 g and 7 mm, respectively. Seedlessness or soft seeded nature was recorded in two accessions. An analysis of data on yield of 14 trees belonging to 12 accessions for 5 successive years (2001-2005) showed that IC244100-2 and IC244111-1 are superior in yield over years and IC136681 is superior for fruits/tree/year as shown in Table 15.3.

IC244100-2 is promising for prolonged bearing and dual season bearing, namely, season (May-July) and off-season (October-November). IC244100-3 is exclusively off-season and can be recommended for regions where harvesting coincides with south-west monsoon rains in Kerala and Karnataka causing problems in processing of harvested fruits to obtain dried rind. IC136677-1, IC136679-1, IC136681, IC244077-1, IC244081-2 and IC244083-1 bear fruits exclusively during season and are short duration bearing and can be recommended to those farmers having sufficient drying and processing facilities.

Dried rind is the commercial part that can be sold readily in local markets of Kerala for its condiment value. A study was undertaken during *kharif* 2003 on dry recovery of rind in 24 trees belonging to 19 accessions. This showed that dry recovery of rind, which is a promising trait, was more than mean of 8 per cent in IC244106-2, IC244097-3, IC244096-2, IC136677-1, IC244086-2, IC244115, IC244101-3, IC244075-1 and IC244100-1, when fresh fruit weight as a whole was taken into consideration for calculating dry recovery (per cent). Similarly, time taken for drying isolated fresh rind of fruits to obtain dried rind was less than mean of 9.1 per cent in IC244084-1, IC244101-3, IC244110, IC244075-1 and IC244100-1. Superior accessions identified for a combination of two characters with value as condiment on the basis of fresh rind weight >60 g and thickness >7.3 mm are as shown in Table 15.4.

Table 15.3

Accessions	*2001*			*2002*			*2003*			*2004*			*2005*		
	Frt No.	*FFY*	*SFW*	*Frt No.*	*FFY*	*SFW*	*Frt No.*	*FFY*	*SFW*	*Frt No.*	*FFY*	*SFW*	*Frt No.*	*FFY*	*SFW*
IC136677-1	152	11.7	76.8	72	3.1	43.3	536	36.7	68.5	4	0.3	75.0	139	7.6	54.7
IC136677-2	312	19.0	60.8	2	0.2	108.5	520	23.0	44.2	104	3.2	31.2	391	22.0	56.4
IC136679-1	488	20.5	42.0	14	0.6	46.4	694	29.3	42.2	108	5.1	47.2	304	17.0	55.9
IC136681	1537	18.8	12.2	256	5.6	21.9	2354	35.5	15.0	3208	44.8	14.0	2284	44.2	19.4
IC244075-1	7	0.5	69.3	379	31.3	83.0	36	1.8	51.4	13	1.1	85.4	587	45.8	78.0
IC244077-1	502	49.1	97.8	297	22.9	77.1	2538	111.5	44.0	137	11.8	86.1	468	33.6	71.8
IC244081-2	516	41.5	80.4	283	18.9	66.8	1584	87.7	55.8	230	21.3	92.6	968	86.7	89.6
IC244086-1	27	1.3	48.6	7	0.7	100.3	512	33.1	64.8	19	1.6	81.6	82	7.0	86.0
IC244100-2	874	73.3	83.9	360	31.5	87.4	4111	214.1	52.1	386	36.7	95.0	1751	165.6	94.6
IC244100-3	159	13.7	86.0	253	28.1	111.3	1779	70.6	39.7	72	7.1	99.0	381	45.3	118.8
IC244106-2	22	1.8	82.7	125	11.3	90.9	1413	61.5	43.5	3	0.4	143.3	672	70.5	105.0
IC244110	193	18.0	93.6	397	33.1	83.5	3247	180.6	55.5	206	28.8	140.1	1099	104.7	95.3
IC244111-1	399	41.7	104.0	1048	85.4	81.4	3464	156.5	45.2	443	48.4	109.1	1994	164.6	82.6
IC244113-1	138	8.3	60.0	88	7.0	80.0	682	37.3	54.8	30	2.2	73.6	1239	85.6	69.1
Mean	380.4	22.8	71.3	255.8	20.0	77.3	1676.4	77.1	48.3	354.5	15.2	83.8	882.8	64.3	76.9
Minimum	7	0.5	12.2	2	0.2	21.9	36	1.8	15	3	0.3	14	82	7	19.4
Maximum	1537	73.3	104	1048	85.4	111.3	4111	214.1	68.5	3208	48.4	143.3	2284	165.6	118.8
SD	413.1	21.2	24.8	268.9	22.6	25.3	1286.6	65.1	12.8	833.0	17.5	36.1	704.9	52.3	25.1
PCV (%)	108.6	92.8	34.8	105.1	113.3	32.8	76.7	84.5	26.6	235.0	115.1	43.1	79.9	81.3	32.7

[Frt No.: Number of fruits; FFY: Fresh fruit yield (kg); SFW: Single fresh fruit weight (g)].

Table 15.4

Accessions	*Rind Weight (g)*	*Rind Thickness (mm)*
IC244075-1	62.5	10.9
IC244077-1	63.7	7.9
IC244086-1	65.4	8.7
IC244097-3	63.0	9.6
IC244100-2	62.9	7.4
IC244106-2	74.9	12.1
IC244110	87.2	11.8
IC244111-1	115.1	11.5
IC244115	99.2	7.6

Superior accessions identified for a combination of two characters with value as a tree-borne oilseed on the basis of seeds/fruit >5 and seed weight >1.5 g are as shown in Table 15.5.

Table 15.5

Accessions	*Seeds/Fruit*	*Seed Weight (g)*
IC244077-1	6.1	2.1
IC244100-2	5.8	1.6
IC244109-1	5.1	1.6
IC244111-1	6.4	2.4

Three accessions found superior for dual purpose namely, condiment and oilseed are as shown in Table 15.6.

Table 15.6

Accessions	*Rind Weight (g)*	*Rind Thickness (mm)*	*Seeds/Fruit*	*Seed Weight (g)*
IC244077-1	63.7	7.9	6.1	2.1
IC244100-2	62.9	7.4	5.8	1.6
IC244111-1	115.1	11.5	6.4	2.4

IC244111-1 and IC244100-2 were consistently superior and hence, registered as genetic stocks. IC244100-2 (INGR No. 04061) is early bearing (6 years after transplanting), high yield in number of fruits (1496), high fresh fruit yield (104.2 kg), single fresh fruit weight (82.6 g), high seeds/fruit (6), high seed weight (1.6 g), high dry rind thickness (3.5 mm) and high yielding during last 5 years. IC244111-1 (INGR No. 04062) is also early bearing (7 years after transplanting), high yield in number of fruits (1470), high fresh fruit yield (99.3 kg), single fresh fruit weight (84.4 g), high seeds/fruit (6) and high seed weight (2.4 g), high dry rind thickness (2.5 mm) and high yielding during last 5 years. Absence of differentiated embryo, endosperm or embryonic axis indicated that so called "seed" is not a true seed. Germination characteristics of "seed" showed clear cut presence of polarity. Regeneration of multiple seedlings from whole "seed" and "seed" pieces further indicated apomictic nature of "seed". Based on "seed" characteristics and field examination of natural and raised populations, facultative agamospermy in *G. cambogia* is indicated (Malik *et al.*, 2005).

Mysore Gamboge

Mysore gamboge [*Garcinia xanthochymus* Hk. f. ex T.And; syn. *G.tinctoria* (DC.) W.Wight] is an important under-utilized crop distributed in lower hills of Eastern Himalaya, Western Ghats and Andaman Islands (Rema and Krishnamoorthy 2000). It is dioecious and seed propagated plant comes to bear in 8-10 years (Varghese and Thomas 2000). Male trees are not seen but pseudo-bisexual trees are known. Acidic, pleasant fruits are edible, though not very palatable and are used for making sherbets, medicaments, preserves and jams. Dried fruit rind is used as a condiment in West Coastal Karnataka. Gum resin from stem bark and fruit makes a good watercolour (gamboge) used in dyeing. About 4-years-old seedlings are useful as root-stock for grafting and inarching mangosteen; sometimes grown in gardens as an ornamental tree for its dense foliage (Singh 1993). Very old trees are met with in degraded areas but young trees are not to be seen. Genetic erosion has already set in as only isolated trees are seen in degraded forest areas near villages in Karnataka. In Kerala, mostly planted trees are seen in botanical gardens. Soft wood grafting of *G. xanthochymus* was standardized on 9 months old root-stocks of same species with 90 per cent success. Grafted plants flowered within 3 years after grafting and set fruits in fourth year, while seed propagated plants did not flower even 6½ years after planting in field (Krishnamoorthy *et al.*, 2006). Mean weight of fruits of grafted plants was 47 g, mean length 4.4 cm and diameter 4.6 cm. Length of peduncle was 4.2 cm, calyx persistent and 5 in number, seeds 1 to 5 in a fruit, brown, about 2.5 cm in length and 1.5 cm in width and embedded in bright orange yellow pulp. So far we have established in field genebank, 12 accessions of Mysore gamboge. Absence of differentiated embryo, endosperm or embryonic axis indicated that so called "seed" is not a true "seed". Germination characteristics of "seed" showed clear cut presence of polarity in this species. Regeneration of multiple seedlings from whole "seed" and "seed" pieces further indicated apomictic nature of "seed". Based on "seed" characteristics and field examination of natural and raised populations, obligate agamospermy in *G. xanthochymus* is indicated (Malik *et al.*, 2005).

A large number of clones suited to different climatic conditions in Maharashtra, Goa, Karnataka, Tamil Nadu and Kerala are needed in each of these species for which local materials are to be tested locally and expanded further to multi-locations. At present, various useful products from these species are processed in cottage industries. For commercial exploitation on industrial scale, there is need to develop appropriate technology at horticultural, processing and marketing stages.

In addition to *kokam*, Malabar tamarind and Mysore gamboge, mangosteen (*G. mangostan* L.), sea-shore mangosteen (*G. hombronian* Pierre) and *G. morella* L. are also important, but only very little effort is made to collect, characterize and conserve these species. Efforts should be made to collect all economically useful species of *Garcinia* occurring in India by different crop based institutes from South-Western Ghats, Andaman and Nicobar Islands and the North-Eastern states, besides Easten Ghats region.

References

Akhtar Husain, Virmani, O.P., Popli, S.P., Misra, L.N., Gupta, M.M., Srivastava, G.N., Abraham, Z. and Singh, A.K., 1992. *Dictionary of Indian Medicinal Plants.* Central Institute of Medicinal and Aromatic Plants, Lucknow, pp. 218.

Antony, J.I.X., Josan, P.D. and Shankaranarayana, M.L., 1998. Quantitative analysis of (-) hydroxy citric acid and (–) hydroxy citric acid lactone in *Garcinia* fruits and *Garcinia* products. *J. Food Science and Technology*, 35: 399–2.

Chandrasena, J.P.C., 1935. *The Chemistry and Pharmacology of Ceylon and Indian Medicinal Plants.* Lucy Chandrasena, Colombo.

CSIR, 1956. The Wealth of India: Raw Materials, Vol. IV. CSIR, New Delhi, India, pp. 101–103.

Dallas, C. and Michael, R., 1990. The diet and health benefits of HCA. Keats Publishing Inc., New York.

Gamble, J.S., 1922. *A Manual of Indian Timbers*. Sampson, Low, Marstan and Co. Ltd., London.

Jameisn, G.S., 1943. *Vegetable Fats and Oils*, 2nd Edition. Reinhold Publishing Corp., New York.

Kirtikar, K.R. and Basu, B.D., 1984. *Indian Medicinal Plants* (Text 1). Lalit, Mohan Basu, Allahabad.

Krishnamoorthy, B., Mathew, P.A., Rema, J., Minod Divakaran and Jayarajan, K., 2006. Soft wood grafting of *Garcinia xanthochymus* (Hook.f.) [Syn.*G.tinctoria* (Wight)]. *J. Spices and Aromatic Crops* 15(1): 63–64.

Krishnamurthy, N., Lewis, Y.S. and Ravindranath, B., 1981. On the structure of garcinol, isogarcinol and camboginol. *Tetrahedron Letters*, 22(8): 793.

Krishnamurthy, N., Lewis, Y.S. and Ravindranath, B., 1982. Chemical constituents of *kokum* fruit rind. *J. Food Science and Technology*, 19: 97–110.

Krishnaswamy, N. and Raman, V.S., 1949. A note on the chromosome numbers of some economic plants of India. *Curr. Sci.*, 18(10): 376–378.

Lewis, F., 1934. *The Vegetable Products of Ceylon*. The Associated Newspapers of Ceylon Ltd., Colombo.

Lewis, Y.S. and Neelakandan, S., 1965. (–) Hydroxy citric acid: The principal acid in the fruits of *Garcinia cambogia. Phytochemistry*, 4: 619.

Lowenstein, J M., 1971. Effect of (–) hydroxy citrate on fatty acid synthesis by rat liver *in vivo. J. Biology Chemistry*, 246: 629–632.

Malik, S.K., Chaudhury, R. and Abraham, Z., 2005. Seed morphology and germination characteristics in three *Garcinia* Species. *J. Seed Science and Technology*, 33(2): 595–604.

Muhammed, M., Rosen, R., McCarty, M., Conte, A., Patil, D. and Butrym, E., 1994. *Citrin: A Revolutionary Herbal Approach to Weight Management*. New Editions Publishing, Burlingame, CA.

Muthulakshmi, P. and Sara, T.G., 1999. Kodampuli: A multipurpose fruit tree. *Indian Horticulture*, 44(2): 7–8.

Muthulakshmi, P., Sara, T.G. and Lyla Mathew, K., 2001. Quality parameters of *Kodampuli* rind under different drying conditions. *Spice India*, 14(1): 14–16.

Nadkarni, A.K., 1954. *Indian Materia Medica Vol. III*. Popular Book Depot, Bombay.

Rama Rao, M., 1914. *Flowering Plants of Travancore*. Govt. Press, Trivandrum.

Rao, R.N. and Sakariah, K.K., 1988. Lipid lowering and antiobesity effect of hydroxy citric acid. *J. Nutrition Research*, 8: 209–212.

Rao, Y.S., Mary Mathew, K., Kuruvilla, K.M., Lakshmanan, R., Gilu Lizy George, Madhusoodanan, K.J. and Potty, S.N., 2000. Practice, and promise of tree spices tissue culture. In: Programme and Abstracts. PLACROSYM XIV, 12–15 December 2000, Hyderabad, India (Abstract), pp. 40–41.

Raven, P., Evert, R. and Eichhorn, S., 1986. *Biology of Plants*, 4th Edn. Worth Publishers, New York.

Rema, J. and Krishamoorthy, B., 2000. *Garcinia* specis of economic importance: Distribution and uses. *Indian Spices*, 37(1): 20–23.

Roberts, E., 1984. *Vegetable Materia Medica of India and Ceylon*. Bishen Singh Mahendra Pal Singh, Dehra Dun, India.

Sampathu, S.R. and Krishnamurthy, N., 1982. Processing and utilization of kokum. *Indian Cocoa, Arecanut and Spices Journal*, 6(2): 12–14.

Singh, N P., 1993. Clusiaceae. In: *Flora of India, Vol. 3*, (Eds.) Sharma, B.D. and Sanjappa, M. Botanical Survey of India, Calcutta, pp. 86–151.

Spices Board, 2001. *Camboge: Package of Practices*. Cochin, Kerala, pp. 8.

Subash Chandra, M.D., 1996. Nature watch: The *kokum* tree. *Resource*, 1(1): 86–99.

Sullivan, A.C. and Triscari, J., 1977. Metabolic regulation as a control for lipid disorders. Influence of (–) hydroxy citrate on experimentally induced obesity in the rodent. *American J. Clinical Nutrition*, 30: 767–776.

Sullivan, A.C., Hamiltan, J.G., Miller, O.N. and Wheahkey, V.R., 1972. Inhibition of lipogenesis in rat liver by (–) hydroxy citrate. *Arch. Biophys.*, 150: 183–190.

Sullivan, A.C., Triscari, J., Hamiltan, J.G. and Miller, O.N., 1974. Effect of (–) hydroxy citrate upon the accumulation of lipid in the rat. *Appetite Lipids*, 9(2): 129–234.

Thombre, M.V., 1964. Studies in *Garcinia indica* Choicy. *Science and Culture*, 30(9): 453–454.

Varghese, S. and Thomas, J., 2000. *Garcinia tinctoria:* A lesser known evergreen tree of multiple uses. *Indian Horticulture*, 45: 21.

Biodiversity in Horticultural Crops Vol. 1 *Pages* ***320–329***
Editors: **Professor K.V. Peter & Dr. Z. Abraham**
Published by: **DAYA PUBLISHING HOUSE**

Chapter 16

Seed Spices

S.K. Malhotra and B.B. Vashishtha
National Research Centre on Seed Spices,
ICAR, Tabiji, Ajmer – 305 206, India

India has old history of cultivation of seed spices and enjoys position of the largest producer in the world. The seed spices were introduced to India from Mediterranean and Central Asian region, perhaps a long time ago. Out of the total 20 seed spices grown in the country, the ten are prominent. These are divided into major and minor group of seed spices. The crops covered under major seed spices are coriander, cumin, fennel and fenugreek, whereas, ajowan, dill, nigella, celery and anise constitute minor group (Table 16.1). The seed spices mentioned above are dicotyledonous and most of them belong to family Apiaceae except fenugreek and nigella, which are from family Fabaceae and Ranunculaceae, respectively. The seed spices are grown in different parts of the world covering mainly Mediterranean region, South Europe and Asia. Though, seed spices are grown in larger areas in India and are exported to many other countries, but none of them is native to India except ajowan and Indian dill (Malhotra and Vijay, 2003).

The seed spice is a group, which denotes all those annuals whose dried fruits or seeds are used as spice and are well distributed over different agro-climatic regions in India. The major belt spreads from semi-arid to arid region covering a large area in Rajasthan and Gujarat. The states, Rajasthan and Gujarat have emerged as "Seed Spices Bowl" and together contribute more than 80 per cent of the total seed spices produced in the country. The prominent states other than Rajasthan and Gujarat, where one or more of seed spices grown are Haryana, Punjab, M.P., Bihar, Uttarakhand, U.P., West Bengal, Orissa, Tamil Nadu and Karnataka.

The country is bestowed with immensely rich land races diversity in seed spice crops. Much of the country's agro-biodiversity is in the custody of farming community who followed age old farming systems. However, it is being lost from the natural habitats due to expansion of agriculture production in the frontier areas. Scientific management of these valuable resources on different dimensions of activities in India *viz.* survey and collection, characterization, evaluation, documentation and conservation have assumed prime importance today.

Table 16.1: Seed Spice Crops, Botanical Name, Family and their Centre of Origin

Seed Spices	*Botanical Name*	*Family*	*Centre of Origin*
Major group			
Coriander	*Coriandrum sativum* L.	Apiaceae	Mediterranean region
Cumin	*Cuminum cyminum* L.	Apiaceae	Mediterranean region
Fennel	*Foeniculum vulgare* Mll.	Apiaceae	South Europe and Mediterranean region
Fenugreek	*Trigonella foenum-graecum*		
	Trigonella corniculata L.	Fabaceae	South East Europe and West Asia
Minor group			
Ajowan	*Trachyspermum ammi* Sprague	Apiaceae	India and Egypt
Dill	*Anethum graveolens* L.	Apiaceae	Europe, Africa and Asia
	Anethum sowa Kurz.		India for sowa
Celery	*Apium graveolens* L.	Apiaceae	Mediterranean region
Anise	*Pimpinella anisum* L.	Apiaceae	Eastern Mediterranean region
Caraway	*Carum carvi L.*	Apiaceae	Mediterranean region
Nigella	*Nigella sativa* L.	Ranunculaceae	Eastern Mediterranean region

Survey and Collection

The first meaningful effort to collect and conserve the genetic resources of seed spices started with umbelliferous seed spices in 1957 at IARI, New Delhi (Joshi *et al.*, 1963). Most of the collections were lost due to lack of proper maintenance or follow-up. Systematic genetic resource activities for different crop species with long term objectives started after 1976 with the establishment of NBPGR in general and in the early nineties by four centres of AICRP (Spices) one each at Jobner (Rajasthan), Jagudan (Gujarat), Coimbatore (T.N.) and Lam (A.P.) in particular. Later,a few more research stations were added as AICRP centres. Survey and collection work on four major seed spices namely coriander, cumin, fennel and fenugreek have been given importance and priority. The germplasm collections maintained at different centres are listed in Table 16.2. Besides AICRP Centres,a few more SAU's, CSIR and their research centres are also maintaining collections of various seed spices. Two major weaknesses of seed spices germplasm collection are: (*i*) germplasm expeditions have been made at random and unsystematically, also the collections are made by breeders whose prime objective was to develop improved varieties. A biasdness to superior types might have occurred in such collections. (*ii*) All seed spices except fenugreek are cross pollinated. The traditional varieties therefore, exist in form of complex gene mixture.

Since establishment of NRC on Seed Spices at Ajmer by ICAR during April, 2000, survey and collection of plant genetic resources were given priority. In addition to major seed spice crops, NRC on Seed Spices maintains collections of ajowan, dill, nigella, celery and anise also. Efforts are being made to introduce caraway, sweet fennel and a few others under semi-arid agro climatic conditions. Total germplasm collection being maintained at NRC on seed spices is 1047, which comprises 866 from major group (Table 16.2) and 181 from minor group (Table 16.3). The germplasm of a crop species includes land races, primitive cultivars, released varieties, genetic stocks with known desirable attributes and wild and weedy relatives of the cultivated species. None of the seed spices is native to India, but continuous natural selection and creative human selection in a wide range of agro-climatic regions and different cultures led to enormous morphological and genetic diversity in India. Development of improved cultivars has improved production and productivity of seed spices, but has endangered the accumulated diversity among traditional cultivars of these crop species. Conservation of germplasm of seed spices is of profound importance for sustaining production.

Table 16.2: Germplasm Collection of Major Seed Spices at Different Centers

Sl.No.	AICRP Centres	Coriander		Cumin		Fennel		Fenugreek	
		I	E	I	E	I	E	I	E
1.	RAU, Jobner	649	112	313	10	185	8	325	12
2.	GAU, Jagudan	70	18	173	7	284	20	48	2
3.	TNAU, Coimbatore	205	–	–	–	–	–	262	–
4.	APAU, Guntur	230	–	–	–	–	124	–	–
5.	RPAU, Dholi	102	–	–	–	40	–	103	–
6.	IGAU, Raigarh	20	–	–	–	–	–	13	–
7.	NDAU, Kumarganj	29	–	19	–	15	–	17	–
8.	CCSHAU, Hisar	76	–	–	–	65	–	65	–
9.	NRCSS, Ajmer	255	18	126	8	238	19	200	2

I: Indigenous; E: Exotic collections.

Table 16.3: Germplasm Collection of Minor Seed Spices at Different Centres

Sl.No.	Centres	Ajwan		Dill		Nigella		Celery		Anise		Caraway	
		I	E	I	E	I	E	I	E	I	E	I	E
1.	GAU, Jagudan	75	–	24	–	–	–	–	–	–	–	–	–
2.	APAU, Guntur	46	–	–	–	–	–	–	–	–	–	–	–
3.	MPUA&T, Pratapgarh	20	–	20	–	–	–	–	–	–	–	–	–
4.	NRCSS, Ajmer	77	–	54	4	14	–	9	2	8	3	7	3

I: Indigenous; E: Exotic collections.

Germplasm Maintenance

Proper maintenance of germplasm should take care of important aspect of maintaining genetic structure of original sampled population. To achieve this objective, two criteria need to be fulfilled. Breeding system must be controlled, which involves prevention of out crossing with other entries and the second, to reduce effect of natural selection in an environment other than original one (Frankel and Soule, 1981). Mode of reproduction determines genetic structure as well as pattern of ecotypic differentiation and appropriate maintenance procedure (Singh *et al.*, 2004). In self-pollinated crops, the precise gene and genotypic composition of the population are preserved in the absence of selection, whereas cross-pollinated crops mandate controlled pollination techniques for their maintenance.

Reproduction behavior of umbelliferous seed spices is studied and elaborated (Ramanujam *et al.*, 1964, Hore, 1989 and Malhotra, 2005).

Coriander is an andromonoecious crop and is a highly cross-pollinated crop. Even though, it is cross-pollinated crop, there is no inbreeding depression observed even after three generations of selfing (Romanenko *et al.*, 1992). Upto 60 per cent cross-pollination was observed in closely spaced plants (Dimri *et al.*, 1976). Therefore, genebank accessions need to be isolated from each other for genetically identical reproduction of an accession. If sufficient field area are available, isolation can be achieved by providing 80 m distance between accessions. In most of the cases, number of accessions to be maintained is large and therefore, isolation of accessions has to be achieved by technical tools such as pergamin bags. It may reduce fruit setting considerably. Alternatively if there are only a few seeds, then isolation cabins can be used for successful reproduction of populations. Cumin and fennel are also cross-pollinated crops, bees being the major pollinator. Maintenance of these crops germplasms can be achieved by barriers, which can prevent bee movement. Fenugreek is a self-pollinated crop and hence its maintenance is easy as compared to other seed spices. The accessions can be maintained by selfing using appropriate sized butter paper bags and avoiding contamination during threshing of individual accessions.

Seeds should be harvested at maturity from plants to ensure optimum viability. Seed spices have remarkable resistance to high temperature, but for long-term storage, the drying temperature should not exceed 40° C. Therefore, seeds harvested from plants have to be dried, preferably under shade, to appropriate moisture depending on the storage condition (Table 16.5). Care should be taken to ensure that seeds are free from weeds, foreign particles, pests and diseases during processing.

Germplasm Conservation

Conservation of Plant Genetic Resources for their sustainable use is important and can be achieved by two approaches namely (i) *in situ* conservation and (ii) *ex situ* conservation.

In-situ Conservation

Seed spices are annual in nature and are propagated through seeds. Thus in-situ conservation is the economical method. Under normal storage conditions, the viability of seed spices seeds remains satisfactory for two years. Fresh seeds therefore, have to be produced every third year. All of seed spices except fenugreek are cross pollinated and pollination is brought about mainly through bees. Thus, 1 to 1.5 m long musline cloth cages are used as bee barriers. In each cage, number of plants caged are 5 in fennel, dill and ajowain, whereas, 7-10 plants in cumin and coriander. The number is certainly small to regenerate the full spectrum of genetic variability of individual accession but is enough for maintenance of large germplasm by overcoming problem of genetic drift (Malhotra, 2005).

An alternative strategy is to form gene pools based on agro-morphological traits of the accessions. A limited number of non-competitive gene pools can be formed by grouping the germplasm accessions based on important traits like maturity and habit. These gene pools can then be maintained as composite populations by random mating among accessions of each group in polycross blocks maintaining proper isolation. This approach is successfully adopted in coriander where four composites are developed based on maturity and habit of accessions (Sharma, 1994).

In situ conservation refers to conservation of genetic resources within their natural habitat. It can be of two types namely conservation in genetic reserve area and on farm conservation. Among seed spices, the vegetatively propagated species of black cumin (through underground bulbs) which grow in wild state in high altitudes of Kashmir, Himachal Pradesh and Uttarakhand, the conservation through formation of genetic reserve areas could be a viable option. The on- farm conservation refers to "continuous cultivation and management of diverse set of crop populations by farmers in a particular ecosystem where a crop has evolved" (Bhatt and Singh, 2004). This is a dynamic method of conservation followed by farmers for maintaining traditional land races and also involves evolution of cultivars in due course.

Ex situ Conservation

Ex situ conservation refers to conservation of genetic resources outside their natural habitat. The *ex situ* conservation methods include seed genebank, cryopreservation, *in vitro* storage, DNA storage, pollen storage, field genebank and botanical gardens (Tao, 1985). Seeds of seed spices like coriander, cumin, fennel and fenugreek are orthodox seeds *i.e.*, they can withstand desiccation without any significant loss in viability. The conservation of seed spices germplasm can be accomplished through seed genebank.

Conservation in the form of seeds is both the safest and cheapest method of conservation. Due to orthodox nature of seeds, seeds of seed spices can be stored under sub-zero conditions by reducing moisture content to minimum. Based on storage conditions, seed storage can be classified into short, medium and long term storage (Table 16.4). Type of storage to be adopted will also be dependent on purpose of storage.

Table 16.4: Types of Seed Storage Based on Storage Conditions (Ellis *et al.*, 1980)

Storage Type	*Type of Collection*	*Storage Parameters*		*Period of Storage (Years)*
		Temperature (°C)	*R.H. (per cent)*	
Short term	Seed before processing	15 to 20	30–40	2
Medium term	Active collection	0 to 10	20–30	5–25
Long term	Base collection	–18 to –20	No control	> 25 *

*Till the viability of above 85 per cent is maintained during quick viability test.

Under room temperature, viability of seeds of seed spices remains satisfactory for 2 years. Advantage of long-term conservation over alternative procedure of regular regeneration at an interval of 2 years (in seed spices crops) is due to avoidance/reduction of natural selection, genetic drift in small populations, natural hybridization, and destruction by parasites/climatic rigors/loss through human error (Frankel and Soule, 1981). For conservation of seeds without loosing viability and maintaining genetic integrity of accession, standards are established for genebank by IPGRI (Tao, 1985). Based on genebank standards, the implied seed standard for storage of seed spices germplasm is presented in Table 16.5.

Cryopreservation

The cryopreservation is another option available for conservation of seed spices. Cryopreservation is the storage of genetic material at ultra-low temperature, usually in liquid nitrogen at–196 ° C. Although, this technique has shown enormous potential of storage for prolonging life span of stored material for indefinite period, it is regarded as a tool mainly for conservation of recalcitrant seeds and vegetatively propagated plant species but shall also be effective for seed spices which are otherwise orthodox in nature. Technologies such as DNA storage and pollen storage are still in its infancy and could be a potential tool for conservation of crop species in the near future. Since, each method of conservation in the two basic approaches *in situ* and *ex situ* conservation–has its own objectives, advantages and disadvantages, these methods should not be viewed as alternatives or confirming each other but rather as being complimentary. Biodiversity of a crop could be conserved through a "complementary conservation strategy" involving more than one method, which can be complimentary to each other (Tao, 1985).

Table 16.5: Seed Standard for Storage of Seed Spices (Tao, 1985)

Parameters	*Preferable Standards*
Seed moisture content	3–7 per cent (Base collection) and 8–10 per cent (Active collection)
Sample size/accession	3000–4000 seeds (fenugreek) 4000–12000 seeds (all seed spices except fenugreek)
Viability monitoring interval	10 years (Base collection) and 5 years (Active collection)
Regeneration	Minimum population of 100 plants/accession

Germplasm Characterization

The systematic germplasm characterisation provides information about identity/uniqueness of accessions and basic information to help and promote its utilization in plant improvement. There is assemblage of a sizeable germplasm collection of different seed spices crops at NRC of seed spices and centres under AICRP on Seed Spices. For characterization, descriptor list of ten seed spices crops *viz.* ajowan, anise, celery, caraway, coriander, cumin. dill, fennel, fenugreek and nigella has been published (Malhotra and Vashishtha, 2006), so that informations related to different accessions may be recorded uniformly by scientists, crop curators and breeders associated with seed spice germplasm management in the country.

Germplasm Evaluation

To facilitate effective utilization of plant genetic resources, it is important that these are evaluated for productivity and its components, crop duration, resistance to biotic and abiotic stresses and quality of produce. In seed spices, quality characters need special attention on account of their export value. Germplasm materials available at different centres are evaluated and utilized for crop improvement. Research work carried out under the AICRP network on conventional breeding work through selection from the germplasm resulted in release of fifty two (Table 16.6) improved varieties in seed spices (coriander-18, cumin-5, fennel-10, fenugreek-14, Ajowan (2), Dill (2), Kalongi (1). Evaluation status of existing genetic resources of seed spices in India as per major thrust areas and further their utilization are given in Table 16.7. Existing germplasm collection of seed spices at various centres in the country possesses limited variability particularly in ajowan, fennel and cumin. There is lack of genetic resources with high volatile oil content in seeds and resistance to biotic and abiotic stresses. Promising genetic stocks with potential genes are needed to be identified and utilised in crop improvement programme. At the same time, there is strong need to validate identity of accessions at biochemical and molecular levels using RFLP, AFLP and RAPD.

Information on evaluation and characterisation of seed spices are required to be documented and catalogued crop-wise for use by breeders, geneticists, biotechnologists and other users. Documentation of germplasm is necessary for sustainable use of genetic resources of seed spices.

Table 16.6: Seed Spice Varieties Released through AICRP on Spices

Crops	*Variety Released*
Coriander (18)	G Cr 1, G Cr 2, Co1, Co2,Co3, CS-287, RCr 41, Sadhana, Swathi, Sindhu, RCr 20, RCr 435, RCr 436, RCr 684, Hisar Anand, Rajendra Swathi, Hisar Sugandh, Hisar Surbhi
Cumin (5)	G C 1, G C 2, G C 3, RZ 19 and RZ 223.
Fennel (10)	PF-35, G F1, G F 2, Co 1,RF 101, UF-125, Rajendra Sourabha, Hisar Swarup, GF11, RF 143.
Fenugreek (14)	Co1, Co2, Rajendra Kanti, Lam sel.1, RMt 1, RMt 303, RMt 143, Hisar Sonali,Guj. Methi-1, Rajendra Khushba, Hisar Suvarna, Hisar Mukta, Hisar Madhvi, RMt 305.
Ajowan (2)	NRCSS-AA-1, NRCSS-AA-2
Dill (2)	NRCSS-AD-1, NRCSS-AD-2
Kalongi (1)	NRCSS-AN-1

Table 16.7: Sources Utilized in Seed Spice Improvement

Crops	*Biotic and Abiotic Stresses (Sources Utilized)*
Coriander	Wilt (ACr -01-250), stem gall (ACr -01-250), Dual Purpose use leaf and seed) (Acr-01-256),
Cumin	Wilt resistance (AC-01-3, AC-01-167), Drought tolerance (AC-01-3), High oil content (AC-01-167)
Fennel	Ramularia Blight (Sel 01-87), High oil content (Sel 01-119), early season (Sel 01-119)
Fenugreek	Downy Mildew (AM-01-10), Root rot (AM-01-10), Early and Large podded (AM-01-35), Dual Purpose use leaf and seed) (AM-01-35)
Ajowan	Powdery Mildew (AA-01-19, AA-01-61), High oil content in seed (AA-01-19), Drought tolerance (AA-01-19, AA-01-61), Early and bold seeded (AA-01-61) (Both varieties identified for release as NRCSS-AA-1, NRCSS-AA-2 by AICRP)
Dill	Powdery Mildew (AD-01- 32), High oil content in seed (AD-01- 32), Drought tolerance (AD-01- 6) Dillapiole toxin less (AD-01-43) (Both varieties identified for release as NRCSS-AD-1, NRCSS-AD-2 by AICRP)
Nigella	High yielding (AN-01-1) variety identified for release as NRCSS-AN-1 by AICRP

Source: Malhotra, 2004.

The extent of variability available in the germplasm collection in seed spices like coriander, cumin, fennel and fenugreek for important agronomic traits are discussed by Malhotra (2005). The biotic and abiotic stresses and quality characters in seed spices are major thrust areas and for which the new germplasm is required to be collected, evaluated for identification and isolation of desired types for further utilization in the breeding programmes. The wild relatives and sub species available in different seed spices are needed to be evaluated as given in Table 16.8.

Biotic Stresses

Seed spices crops are attacked by many diseases and insect pests (Table 8). Out of different diseases, wilt in cumin and coriander; root rot in fenugreek, coriander and nigella; powdery mildew in coriander, cumin, dill, caraway and fenugreek; blight in cumin and fennel and aphid in coriander, cumin, dill, fennel and fenugreek cause severe losses.

Abiotic Stresses

Seed spice crops are grown in semi-arid to arid region with deficit irrigation facilities. Occurrence of frost during winter in a few of the seed spices production areas, saline soils and brackish underground water, leading to abiotic stresses and thereby causing damage to crops are other stresses. Some of these stresses are listed cropwise in Table 16.9.

Table 16.8: Major Diseases and Insect Pests in Seed Spices

Sl.No.	Seed Spices	Major Diseases and Insect Pests
1.	Coriander	Stem gall (*Protomyces macrosporus*), Powdery Mildew (*Erisiphe polygoni*), Wilt (*Fusarium oxysporum* f.sp. *corianderi*), Stem rot (*Rhizoctonia solani*), Root rot (*Curvularia pallescens*), Aphids and Mites
2.	Cumin	Alternaria Blight (*Alternaria burnsii*) Powdery Mildew (*Erisiphe polygoni*), Wilt (*Fusarium oxysporum f.sp. cumini*) and Aphids
3.	Fennel	Alternaria Blight (*Alternaria tenuis*), Ramularia Blight (*Ramularia foeniculi*), Gall (*Protomyces macrosporus*), Powdery Mildew (*Leveillula taurica*), Wilt (*Fusarium oxysporum f.sp. corianderi*), Bacterial soft rot (*Erwinia carotovora*), Aphids and Mites
4.	Fenugreek	Leaf Spot (*Cercospora traversiana*), Powdery Mildew (*Erisiphe polygoni, Leveillula taurica*), Downey Mildew (*Peronospora trigonellae*), Collar rot (*Rhizoctonia solani*), Root rot (*Alternaria alternata*), Aphids and Mites
5.	Ajowan	Collar and root rot (*Scleroticum rolfsii*), Powdery Mildew (*Erisiphe polygoni*) and Aphids
6.	Dill	Root rot (*Fusarium oxysporum*), Powdery Mildew (*Erisiphe polygoni*), Aphids and Cut worms
7.	Nigella	Root rot (*Fusarium oxysporum*), Cut worms
8.	Anise	Alternaria leaf Blight (*Alternaria solani*), Powdery Mildew (*Erisiphe polygoni*), Aphids, Semilooper
9.	Caraway	Alternaria leaf Blight (*Alternaria solani*), Powdery Mildew (*Erisiphe polygoni*) and Aphids,
10.	Celery	Early blight (*Cercospora apii*), late blight (*Septoria apii*), Fusarium yellow (*Fusarium apii*), Stem rot (*Rhizoctonia solani*) and leaf minor

Table 16.9: Abiotic Stresses in Seed Spices

Sl.No.	Seed Spices	Abiotic Stresses
1.	Coriander	Frost damage, drought tolerance, rainfed cultivation, salinity
2.	Cumin	Frost damage, drought tolerance, limited water conditions, salinity
3.	Fennel	Limited water conditions, salinity
4.	Fenugreek	Salinity, limited water conditions,
5.	Ajowan	Frost damage, drought tolerance, rainfed cultivation, salinity
6.	Dill	Frost damage, drought tolerance, rainfed cultivation, salinity
7.	Nigella	Limited water conditions, salinity
8.	Anise	Limited water conditions, salinity

Constraints in Utilization of Genetic Resources

The NRC on Seed Spices strives hard to collect and conserve valuable gene pool of different seed spices as a national repository for further utilization in crop improvement. Constraints in effective utilization of genetic sources are given below:

1. Inherent nature of slow germination of seed spice crops, lack of complete information on pollination behaviour, high diseases and insect-pests pressure *viz.*, wilts, blights, rots,

powdery mildews, aphids and stored grain pests, quality degradation due to microbial load during storage, lack of information on postharvest techniques particularly storage methods for seed spice crops are major challenges posing effective utilization of seed spices germplasm.

2. Most of the available germplasm exists in the form of traditional/local varieties and have been subjected to natural selection for local adaptation. The germplasm exits in the form of complex gene mixtures. Proper sampling and regeneration are essential to recover and maintain full range of genetic variability for further utilization in seed spices crop improvement programme.
3. Present germplasm collection possesses a limited variability. Strong and stable resistant sources are not available against biotic and abiotic stresses. Value added germplasm need to be identified from collections in respect of spices oil and oleoresins, suitability to rainfed conditions, tolerance to frost, resistance to diseases and insect-pests. Dual purpose lines of coriander and fenugreek *i.e.* both for leaves and seeds are to be isolated.There is need for collaborative exploration in Mediterranean regions (Morocco, Algeria, Libya, Egypt, Tunisia) and adjoining parts of Mediterranean sea, Iran, Iraq, Russia and Europe.
4. Collections made are random, without proper documentation both at field level and the maintenance site. Collections are made by breeders whose prime objective was to develop improved varieties. A biasdness for superior types might have occurred in such collections.
5. Serious attention was not paid regarding survey and germplasm collection of ajowan, dill, celery, nigella, aniseed and caraway. Plant biodiversity in these crops need to be collected, conserved and utilized, as they are under-exploited spices crops.
6. Plant biodiversity of seed spices require to be evaluated for different morpho-agronomic attributes, resistance against biotic and abiotic stresses, chemical and biochemical traits and promising stocks with potential and rare gene combinations isolated.

Future Perspective to Enhance Use of Genetic Resources

1. Most of the varieties released so far are developed through selection. To solve specific problems of biotic and abiotic stress resistance, modern techniques using biotechnology involving molecular assisted plant breeding techniques are to be used.
2. To effectively counter any attempt of biopiracy of valuable germplasm of seed spices from India, the first and foremost step needed is stock taking of our assets and their accurate documentation. There is need to develop adequate research base for characterisation of accessions at biochemical and molecular level employing techniques like isozyme profiling, RFLP,AFLP and RAPD. Molecular characterization of important accessions of seed spices crops is required to validate the identity.
3. Conservation of seed spices germplasm at National Gene Bank especially with reference to National Active Germplasm Site, draws our attention as a high priority action.
4. Seed spices are still considered as under-utilised crops and serious attention has not been given regarding systematic germplasm collection, evaluation and conservation. Prevailing diversity of major and minor seed spices in remote and tribal pockets need to be revisited. There is scope for collection of valuable land races of such seed spice crops.
5. The varied agro-climatic conditions offer good scope for promotion of non-traditional seed spices (anise, caraway, celery, parsley and black caraway) to make India a broadened supply base in the world market. Systematic germplasm collection, evaluation and conservation of such spices have better future perspectives for crop improvement.

References

Bhatt, K.C. and Singh, R.K., 2004. On-farm *in situ* management of crop genetic resources. In: *Plant Genetic Resource Management*, (Eds.) B.S. Dhillon, R.K. Tyagi, Arjun Lal and S. Saxena. Narosa Publishing House, New Delhi, pp. 297–317.

Dimri, B.P., Khan, M.N.A. and Narayana, M.R., 1976. Some promising selections of Bulgarian coriander (*Coriandrum sativum* Linn.) for seed and essential oil with a note on cultivation and distillation of oil. *Indian Perfumer*, 20: 13–21.

Ellis, R.H., Roberts, E.H. and Whitehead, J., 1980. A new, more economic and accurate approach to monitoring the viability of accessions during storage in seed banks. *PGR Newsl.*, 41: 3–18.

Frankel, O.H. and Soule, M.E., 1981. The conservation of plants used by man. In: *Conservation and Evolution*, (Eds.) O.H. Frankel and Michael, E. Soule. Cambridge University Press, Cambridge, U.K., pp. 225–251.

Hore, A., 1989. Improvement of minor (Umbelliferous) spices in India. *Economic Botany*, 33(3): 290–297.

Joshi, B.S., Ramanujam, S. and Saxena, M.B.L., 1963. Improvement of some oil bearing spices plants. *Bull. Reg. Res. Lab.* Jammu, 1: 94–100.

Malhotra, S.K. and Vijay, O.P., 2003. Plant Genetic Resources of Seed Spices in India. *Seed Spices Newsletter*, 3(1): 1–4.

Malhotra, S.K., 2004. Status paper on utilization of plant genetic resources of seed spices crops in India. *Concept Paper Presented in National Workshop on Utilization of plant genetic resources at NBPGR*, New Delhi held from 5–6, October.

Malhotra, S.K., 2005. Seed spices. In: *Breeding Technology of Crop Plants*, (Ed.) A.K. Sharma. Yash Publishing House, Bikaner, pp. 535–553.

Malhotra, S.K. and Vashishtha, B.B., 2006. A manual on minimal descriptors of seed spices crops (for characterisation and evaluation). National Research Centre on Seed Spices, Ajmer, Rajasthan, India, p. 1–51.

Peter, K.V. and Nirmal Babu, K., 2005. Genetic resources of seed spices. *Indian J. Pl. Genet. Resour.*, 18(1): 19.

Ramanujam, S., Joshi, B.S. and Saxena, M.B.L., 1964. Extent and randomness of cross-pollination in some umbelliferous spices of India. *Indian J. Genet.*, 24(1): 62–67.

Romanenko, L.G., Nevkrytaja, N.V. and Kuznecova, E.J.U., 1992. Self fertility in coriander (in Russian). *Sel. Semenovod.* (Moskva), p. 25–28.

Sharma, R.K., 1994. Genetic resources of seed spices. In: *Advances in Horticulture, Vol. 9*, (Eds.) Chadha, K.L. and P. Rethinam. Malhotra Publishing House, New Delhi, India, pp. 193–205.

Singh, M., Sharma, A.K., Shahana, S. and Dhillon, B.S., 2004. Plant genetic resource management in relation to mode of sexual reproduction. In: *Plant Genetic Resources: Indian Perspective*, (Eds.) R.S. Paroda, R.K. Arora and K.P.S. Chandel. National Bureau of Plant Genetic Resources, New Delhi, India, pp. 297–317.

Tao, K.L., 1985. Standards for genebanks. *PGR Newsl.*, 62: 36–41.

Biodiversity in Horticultural Crops Vol. 1 *Pages 330–339*
Editors: **Professor K.V. Peter & Dr. Z. Abraham**
Published by: **DAYA PUBLISHING HOUSE**

Chapter 17

Cashew

M. Gangadhara Nayak and M. Gopalakrishna Bhat
National Research Centre for Cashew, Puttur – 574202 D.K., Karnataka

Cashew (*Anacardium occidentale* L.), the most promising crop of neglected and barren lands, is now assuming importance as a foreign exchange earner in several developing countries of tropical regions of both western and eastern hemispheres. At the time of first Portuguese colonization, the name used by local population (Tupi natives of Brazil) for cashew was 'acaju' (nut) which turned into caju in Portuguese language and cashew in English. Most of the names for cashew in Indian languages are also derived from 'caju' (Johnson 1973). The plant belongs to genus *Anacardium*, a member of family *Anacardiaceae* to which 74 genera with nearly 400 species are attached which are growing abundantly in the tropics.

Cashew has a long history as a useful tree but only in the present century it became an important tropical horticultural crop. Small-scale local exploitation of cashew for its nuts and cashew apples is the pattern for more than 300 years in Asia and Africa. It was not until early years of 20th century that international trade in cashew kernels began with first exports from India. Although with slow beginning, cashew became an important commercial crop in recent years. In India, use of cashew apples and nuts was started by local people, and accounts from Africa are similar; making cashew wine/liquor possessing a common practice in both Asia and Africa (Johnson 1973).

Cashew, an evergreen tree, which produces both nuts and apples has become a crop of importance owing to the highly valued kernel and wide adaptability to varying agro-climatic conditions. Even today, India is the largest producer, exporter and consumer of cashew in the world. Today, cashew has become a crop of high economy and has moved from a crop of waste lands to commercial crop, even in organized sector (Singh 1998).

Nutritional Value of Cashew Kernel

The *Maconde* tribe in Mozambique calls it the devil's nut. It was offered at wedding banquets as a token of fertility. Research carried out at University of Bologna has in fact indicated presence in cashew kernel of numerous vitamins including vitamin E, considered by many to be aphrodisiac (Massari 1994). Cashew is a versatile tree nut. It is, in fact, a precious gift of nature to mankind. The cashew kernels has a unique combination of fats, proteins, carbohydrates, minerals and vitamins. Cashew contains 47 per cent fat, but 82 per cent of it contains unsaturated fatty acids. Unsaturated fat content of cashew not only eliminates the possibility of increase of cholesterol, but also balances or reduces the cholesterol level in blood. Cashew also contains 21 per cent proteins and 22 per cent carbohydrates and the right combination of amino acids, minerals and vitamins and therefore nutritionally it stands on par with milk, egg and meat, of course without any problem associated with cholesterol. As cashew has a very low content of sugar, almost as low as 1 per cent, soluble sugar, consumer of cashew is privileged to get a sweet taste without having to worry about excess calories. Cashew nut consumption does not lead to obesity. In short, it is a good appetiser, an excellent nerve tonic, a stimulant and a body builder.

Origin and Distribution

Brazil is the original home of cashew. The earliest report of cashew was from Brazil coming from French, Portuguese and Dutch observers (Johnson 1973). The French naturalist and monk, A. Thevet was the first to describe, in 1558 a wild plant extremely common in Brazil, cashew tree and its fruits. He recounted that cashew apple and its juice were consumed and that the nuts were roasted in fire and the kernels eaten. He provided the first drawing of the cashew showing the local people harvesting cashew fruits and squeezing juice from cashew apples into a large jar (Johnson 1973, Delogu and Haeuster 1994).

There are indications that local *Tupi* Indians used cashew fruits for centuries. They probably played a major role in the dispersion in their temporary migrations towards the coast of North-Eastern Brazil where a considerable interspecific variation was recorded (Ascenso 1986). Entire cashew fruit with pseudo-apple when matures usually floats on surface of water. This could account in Brazil for coastward dispersal of the species by river draining North and East. Fruit bats might have also been involved in seed movement. Within the Amazon forest, fruit bats were the most important agents of seed dispersal of tree species (Johnson 1973). From its origin in North-Eastern Brazil, cashew spread into South and Central America.

Presence of cashew in other continents is to be attributed to man's intervention (Johnson 1973). The Portuguese discovered cashew in Brazil and spread first to Mozambique (Africa) and later into India between sixteenth and seventeenth centuries (De Castro 1994). According to Agnoloni and Giuliani (1977), it arrived in Africa during second half of 16th century, first on East coast and then on the West and lastly in the islands. In Africa, although it can be guessed that cashew was introduced at an early period by the Portuguese, there are no records which provide specific dates. Dispersal of cashew in East Africa may in part be due to the elephant, whose fondness for fruits is well known (Johnson, 1973). Attracted by colour of the false fruit, they swallowed this together with the nut which was too hard to be digested. This was then expelled with their droppings, a natural manure, and trodden far enough into the ground by the animals leaving behind to root and grow, into a seedling first and then a tree. This is how cashew spread along the east coast of Africa facing the Indian Ocean (Massari 1994). The spreading of the cashew within the South American continent was gradual and spontaneous (Delogu and Haeuster 1994).

Spread of Cashew in India

Though cashew is indigenous to Brazil, India is the country that nourished this crop and made it a commodity of international trade and acclaim. It was brought to India as seed material by Portuguese travellers in 16th Century between 1563 and 1578. Initially cashew plants were seen in and around Malabar region of Kerala and Goa in West coast and later the cashew spread to remaining areas of West coast. It was first grown in the gardens of Cochin and in the Malabar coast, and Cochin served as a dispersal point for the cashew in India and perhaps south-east Asia as well (Johnson 1973). Spread of cashew was mostly by natural dispersal means through birds, animals and human carriers. Gradually, cashew was introduced to eastern India and North-Eastern States under afforestation and soil conservation programmes (Balasubramaniam and Singh 2000). Presently, cashew plants in semi-wild state and in well managed orchards are seen in Maharashtra, Goa, Karnataka, Kerala along the West coast, Tamil Nadu, Andhra Pradesh, Orissa and West Bengal on the East coast. To a limited extent, the crop is also seen growing in Chattisgarh, Gujarath, Assam, Arunachal Pradesh, Meghalaya, Tripura, Manipur, Nagaland and Andaman and Nicobar Islands (Singh 1998).

Dispersion of Cashew to South-East Asia and Australia

After India, it was introduced into South-Eastern Asia (Delogu and Haeuster 1994). Dispersal in South-East Asia was aided by monkeys. Whether cashew reached the Philippines via India is uncertain. It may have come directly from the New World on the Manila Galleons (Johnson 1973). The cashew later spread to Australia and some parts of North-American continent, such as Florida. Finally, its present diffusion can be geographically located between 31° North latitude and 31° South latitude, both as wild species and under cultivation (Delogu and Haeuster 1994). At present, cashew is cultivated in many tropical countries, mainly in coastal areas (Van Eijnatten 1991, Ascenso 1986). In 19th century, proper plantations were raised and then spread to a number of other countries in Africa, Asia and Latin America (Massari 1994). Traditionally cashew has been cultivated on commercial scale in Brazil, India, Tanzania, Mozambique, Kenya and Madagascar, while in recent years plantations are also raised in south-east Asian countries.

Related Species of Cashew

According to Baily, genus *Anacardium* contains eight tropical American species. Parente (1972) reported ten different species but Peixoto (1960) names 20 different species several of which have edible peduncles such as *A. nanum*. A very early bearing small shrub, *A. subterranium*, a small shrub with its trunk completely underground containing water resources, *A. microcarpum* a small tree from sandy savannas, *A. spruceanum* a large tree, and the largest species of the genus, *A. gigantium* which grows in the Amazon forest with good timber value. Valeriano (1972) names five different species namely *A. occidentale, A. pumilum, A. gigantium, A. rhinocarpus* and *A. spruceanum*. He also suggests the recognition of only two species namely, *A. nanam* and *A. gigantium* which can be further subdivided based on colour (yellow or red) and shape (round, pear-shaped or elongated) of the pseudo-fruit. Description provided by Peixoto (1960) separates recognition of more than two species. It may be possible that some species originated due to intercrossing between species. It appears from published accounts that *A. occidentale* is the only species that has been introduced outside the new world. Within Central and South America as many as 20 species of *Anacardium* exist (Table 17.1).

Genetic Diversity

Cashew is botanically known as *A. occidentale*. According to Bailey (1958) *Anacardium* is a small genus of eight species indigenous to South America. It appears from the published accounts that *A. occidentale* is the only species, introduced outside the New World although as many as 20 species of *Anacardium* exist within Central and South America. Mitchell and Mori (1987) recognised ten species in the genus *Anacardium*, one of which, *A. fruticosum*, is described as new (Table 17.2). The genus has a primary centre of diversity in Amazonia and a secondary centre in the Planalto of Brazil. Ascenso (1986) reported that cashew (*Anacardium occidentale*) is the only species in the genus which attained economic importance. This morphologically polymorphic species also exhibits chromosome polymorphism (Mitchell and Mori 1987). Chromosome numbers reported in literature range from 2n=24 (Khosla *et al.*, 1973, Goldblatt 1984), 2n=30 (Machado 1944), 2n=40 (Simmonds, 1954, Goldblatt 1984) to 2n=42 (Khosla *et al.*, 1973, Goldblott 1984, Darlington and Janaki Ammal 1945, Purseglove 1988). Such chromosome polymorphism is well known in many domesticated plants (Khosla *et al.*, 1973). Wide variation within the accessions due to cross pollinated nature of crop was reported for 292 germplasm accessions of seed origin which were evaluated at CPCRI Regional Station, Vittal/Shantigodu. Considerable variability was observed for some of the economic characters like flowering season, flowering duration, harvesting duration, nuts/panicle, weight/apple, weight/nut, apple to nut ratio, shelling percentage, weight/kernel, kernel count/lb, shell thickness and mean yield/plant (Table 17.3) (Bhaskara Rao and Swamy 1994). Variability in nut and apple characters such as shape, size and colour are enormous and are of much use in crop improvement programme.

Table 17.1: Species of *Anacardium**

Botanical Name	Country
Anacardium brasiliense Barb.	Brazil
A. curatellaefolium St. Hil (= *A. subcordatum* Presl.)	Brazil
A. encardium Noronha	Malaya
A. giganteum Hancock ex Engl.	Brazil
A. humile St. Hil (= *A. subterraneum* Liais)	Brazil
A. mediterraneum Vell.	Brazil
A. nanum St. Hil (= *A. humile* Engl., *A. pumilum* Walp)	Brazil
A. occidentale L.	Brazil
A. rhinocarpus DC.	Brazil
A. spruceanum Benth ex. Engl.	Brazil
A. microsepalum Loes	Amazon region
A. corymbosum Barb.	Brazil
A. excelsum Skeels (= *Rhinocarpus excelsa*)	Brazil
A. parvifolium Ducke	Amazon region
A. amilcarianum Machado	Brazil
A. kuhlmannianum Machado	Brazil
A. negrense Pires and Fro'es	Brazil
A. rondonianum Machado	Brazil
A. tenuifolium Ducke	Brazil
A. microcarpum Ducke	Amazon region

* *Source*: Index Kewensis, 1996, Royal Botanical Gardens, Kew.

Table 17.2: Species of Anacardium and Synonyms

Species	*Synonym (s)*
A. excelsum	*Rhinocarpus excelsa* *Anacardium rhinocarpus* *Anacardia rhinocarpa* *Anacardium caracoli*
A. spruceanum	*Anacardium brasiliense*
A.occidentale	*Cassuvium pomiferum* *Acajuba occidentale* *Anacardium occidentale var. americanum* *A. occidentale var. indicum* *A. mediterraneum, A. curatellifolium* *Cassuvium reniforme* *A. occidentale var. longifolium* *A. occidentale var. gardneri* *A. subcardatum, A. microcarpum* *A. rondonianum, A. amilcarianum* *A. Kuhlmannianum, A. othonianum*
A. humile	*Monodynamus humilus* *A. subterraneum, A. pumilum* *A. pumilum var. petiolata* *A. humile vàr subacutum*
A.nanum	*A. pumila*
A. corymbosum	—
A. parvifolium	*A. tenuifolium*
A. fruticosum	*A. parvifolium var. fruticosd*
A. giganteum	—
A. microsepalum	*A. negrense*

Mitchell and Mori, 1987.

Table 17.3: Variability Recorded for Some of the Characters in the Germplasm Accessions (Seedling Progenies)

Character	*Range*
Flowering season	October–January
Flowering duration	40–127 days
Harvesting duration	30–105 days
Number of fruits per panicle	1–8
Apple weight	30–150 g
Nut weight	2.4–18.0 g
Apple : nut ratio	4:1–12:1
Shelling percentage	19.0–35.0
Kernel weight	0.5–4.5 g
Kernel count/lb	100–900
Shell thickness	1.5–5.0 mm
Mean yield/plant/year (10 annual harve`sts)	0.50–11.75 kg

Source: Bhaskara Rao and Swamy, 1994.

Exploration and Collection

Eventhough no reliable record of the introductions are available, it is presumed that the initial introductions in the Malabar coast of Kerala were from only few trees and due to the hardy nature of the crop it had spread to all the coastal regions of India naturally. All these introductions were belonging to *A. occidentale*. The initial emphasis was only on the establishment of plantations of seedling origin. As cashew is primarily a cross pollinated crop, it is highly heterozygous. Considerable segregation has resulted in large variation in the populations (Bhaskara Rao and Bhat 1996). The early attempts for germplasm collections were made during the early fifties with the sanctioning of *ad-hoc* schemes in the then composite states of Madras, Travancore, Cochin and Bombay. The research stations started under these *ad-hoc* schemes in Kerala (Kottarakkara), Karnataka (Ullal), Tamil Nadu (Vridhachalam), Anhdra Pradesh (Bapatla) and Maharashtra (Vengurla) took up the programme of collection of locally available elite plants for evaluation and further selection. These were the first attempts in collection of cashew germplasm in India. Many of the other research centres which were established subsequently have collected the seeds of germplasm from these centres, namely, Bapatla, Kottarakkara, Ullal, Vridhachalam and Vengurla. While making the initial collections of germplasm, these centres had confined their survey mainly to the respective States and hence they represent the local germplasm available in the States. Since the inception of All India Coordinated Spices and Cashewnut Improvement Project (AICS and CIP) in 1971, Central Plantation Crops Research Institute (CPCRI) Regional Station, Vittal, also took up the programme of cashew germplasm collection which mainly consisted of the seedling progenies of collections which are available at Bapatla, Vridhachalam, Vengurla, Anakkayam and a few collections made locally from Karnataka (Table 16.3). Subsequent to the establishment of National Research Centre for Cashew (NRCC) at Puttur (Karnataka) in 1986, the germplasm collection through seeds has been discontinued and only the vegetatively propagated material is being collected and conserved in the National Cashew Gene Bank (NCGB). A coordinated approach was brought in the cashew gemplasm collection by organising joint survey teams consisting of scientists of NRCC and the centres of All India Coordinated Research Project on Cashew (AICRP on Cashew) of the respective States. The main criteria adopted for collection of germplasm were the superior economic characters such as high yield, bold nut, cluster bearing, apple qualities, canopy shapes and dwarfness of the plant or any other unique characters observed during the survey. Since 1986, cashew germplasm collection surveys are being undertaken by NRCC in collaboration with centres of AICRP on Cashew and other cashew centres of cashew growing States, namely, Karnataka, Kerala, Maharashtra, Goa, Tamil Nadu, Andhra Pradesh, Orissa and West Bengal. So far all together 494 cashew germplasm accessions were made and planted in cashew field genebank at NRCC, Puttur for evaluation (Nayak and Bhat, 2004). Till the establishment of NRC for Cashew, over one thousand accessions of seedling origin were collected by different Cashew Research Stations in the country. Among the seedling accessions, 23 are exotic collections of which nine were collected from Brazil, Nairobi, Mtwara, Lindi, Nacala, Mozambique, Tanganyka, Singapore and Australia and 14 from the Republic of Panama. The details of germplasm holding is given in Tables 16.4 and 16.5. The non-traditional areas such as Garo Hills (Meghalaya) and Bastar (Chhattisgarh) were also surveyed for germplasm collection. However, more comprehensive exploration is required for targeted collection of germplasm with specific traits such as resistance to TMB, CSRB, high yield, dwarf and compact, bold nut with cluster bearing, tolerance to drought, frost, salt and to problematic soils.

Table 17.4: Germplasm Holding in NCFGB, Puttur

State	*No. of Accessions*
Andhra Pradesh	102
Assam	3
Goa	45
Karnataka	120
Kerala	70
Maharashtra	45
Meghalaya	7
Orissa	21
Tamil Nadu	42
Tripura	3
West Bengal	14
Exotic	22
Total	**494**

Table 17.5: Cashew Germplasm Holding in Different Centres Till 2005

Centre	*No. of Accessions*
East Coast	
Bapatla	131
Bhubaneshwar	89
Jhargram	125
Vridhachalam	264
West Coast	
Madakkathara	148
Pilicode	65
Vengurla	265
Maidan tract/others	
Chintamani	126
Jagdalpur	61
Total	**1274**

Germplasm Evaluation

Characterization

Germplasm evaluation, in the broad sense and in the context of genetic resources, is the description of the material in a collection. It covers the whole range of activities starting from preliminary characterization to further detailed evaluation and documentation. The evaluation of germplasm is an important part of a variety development programme designed to utilize germplasm resources. As new accessions become available, it is important to identify, characterize and document the new materials so that researchers may incorporate lines with desirable characters into utilization programmes. The important steps in this regard are (i) to compile the passport data and (ii) to characterize the accessions for heritable traits (minimum descriptors). For systematic characterisation of cashew germplasm, each accession was grown in the field genebank (@ 6 grafts/accession; spacing 6m x 6m) at NRCC, Puttur, Karnataka. Agronomically recommended package of practices were adopted. Observations were recorded on 3 selected plants in each accession after 10th year of planting and after obtaining 6 annual harvests for 68 descriptors (Photo 1). For cashew germplasm, IPBGR (presently IPGRI) "Cashew Descriptors" (IBPGR 1986) were followed. So far 320 clonal accessions conserved in NCGB have been characterised and evaluated. Three catalogues of cashew germplasm have been brought out for 255 evaluated cashew germplasm accessions (Swamy *et al.*, 1997, 1998 and 2000). By adopting K-clustering algorithm using centroid distance method under Indostat Statistical package, the 255 accessions were grouped into 22 clusters based on 19 descriptors. From each group a 10 per cent sample size was randomly selected as core entries/collections.

Germplasm Cataloguing

The National Research Centre for Cashew (NRCC), Puttur, (Karnataka) which was established in 1986, was given a priority programme of establishing the National Cashew Gene Bank (NCGB) and also the mandate to collect and conserve the available cashew germplasm. So far, a total of 494 clonal accessions have been collected and conserved in NCGB. Characterisation of cashew germplasm has been taken up at NRCC using the "Cashew Descriptors" suggested by IBPGR (Presently, IPGRI), Rome, Italy. The first set of 56 accessions planted in 1986 have been characterised and published as the "Catalogue of Minimum Descriptors of Cashew Germplasm Accessions-I", (Swamy *et al.*, 1997). The second set of 97 accessions planted in 1987 and 1988 have been characterized and documented in the "Catalogue of Minimum Descriptors of Cashew Germplasm Accessions-II", (Swamy *et al.*, 1998). The third set of 102 accessions planted in 1989 and 1990 have been characterised and documented in the "Catalogue of Minimum Descriptors of Cashew Germplasm Accessions-III" (Swamy *et al.*, 2000). These are the first few efforts made in characterisation of clonal accessions of cashew in the world and so far, 255 accessions have been characterised and catalogued.

Germplasm Conservation

In situ conservation of cashew germplasm is done only in the Amazon forests of Brazil (original home of cashew). However, *ex situ* conservation is also being followed in cashew as in all other tree crops. The cashew germplasm is maintained as an active collection in the field gene bank which locks up the land for a long time. Further, field maintenance also has the inherent danger of losing the valuable collections during the natural calamities (as it often happens in the East Coast, especially at Bapatla centre and Bhubaneswar centre due to cyclones). In order to overcome this danger as well as to effectively utilize the land for other field experiments, suitable *in vitro* preservation/cryopreservation methods with the aid of biotechnology are to be developed. An effort in this direction is being made by establishing a Biotechnology Laboratory at NRC for Cashew, Puttur. In the NCGB so far 433 clonal accessions which were collected from 1986 to 2000 have been planted @6 grafts per accession by adopting a spacing of 6m x 6m and are being maintained as active collections in the field gene bank (Table 17.4). Similarly, Regional Cashew Gene Banks (RCGBs) have been established at AICRP on Cashew Centres which are maintaining a total of 1274 accessions (Table 17.5). The germplasm accessions conserved in the NCGB at NRCC, Puttur include the diverse types–high yield, bold sized nut, semi-tall plant, cashew nut shell liquid (CNSL) free, purple plant, high shelling percentage, cluster bearing, big sized apple, early maturity etc. and the related species, namely, *Anacardium pumilum, A. othonianum* and *A. microcarpum.*

Recently a germplasm conservation block has been established by planting 4 grafts per evaluated accessions at a closer spacing of 4m x 4m so as to reduce the land utilized for maintenance of large number of germplasm collections.

Germplasm Registration

Germplasm of agric-horticultural, commercial and other economic crops, which are unique and has verified attributes of scientific/commercial value are being registered by the Plant Germplasm Registration Committee of ICAR at NBPGR, New Delhi. Nine germplasm accessions from National Cashew Field Gene Bank at Puttur for unique characters *viz.*, NRC-59 (VTH 196/18) for big apple, bold nut size and high shelling percentage, NRC-111 (Mardol-4) for mid season flowering, big apple and big nut size, NRC-116 (CNSL Free) for cashewnut shell liquid free type, NRC-120 (Nairobi) for early season flowering, bold apple and nut size, NRC-121 (Purple genotype) for purple stem and leaves and high shelling percentage, NRC-140 (VTH 155 L) for semi-tall, early season (Nov-Dec) and long flowering duration (120 days), NRC 142 (VTH 578/1) for wild relative with genetic diversity, NRC-152 (VTH 713/4) for wild relative with genetic diversity and NRC-201 (Pl.No.1254) for upright and compact habit, semi-tall type were registered with NBPGR, New Delhi (Nayak and Bhat, 2004b).

Germplasm Augmentation through Introduction from Other Countries

One of the immediate priorities of the Indian cashew germplasm programme is to enhance the genetic variability in the collections, as all the plantations were raised from the limited initial introductions and genetic base available in the country is only moderate. Therefore introduction from Central America and Brazil where considerable variability including the dwarf types are reported and as these are also the original home of cashew is essential. Introduction of *A. gigantium* from Surinam which is reported to have the biggest apple (200 g) will be advantageous, especially in Goa where the cashew apple utilization contributes substantially to the economy of the state. Introduction of dwarf clones reported from Brazil (Ascenso, 1986) will be useful in developing high density orchards which will enhance productivity considerably. One of the main problems in cashew is that all the existing germplasm accessions are susceptible to cashew stem and root borer (CSRB) which kills the trees. There is a need to screen the allied species (for their suitability as root stocks) which have relatively hard wood and also posses smooth bark. It is suggested that introduction from Brazil of other species like *A. rhinocarpus* and *A. spruceanum* which are reported to possess hard wood will be useful for testing their suitability as root stocks (Bhaskara Rao and Swamy, 1994).

References

Agnoloni, M. and Giuliani, F., 1977. *Cashew cultivation*. Library of Tropical Agriculture, Ministry of Foreign Affairs. Instituto Agronomico Per L' oltremare, Florence, 168 p.

Ascenso, J.C., 1986. Potential of the cashew crop–1. *Agric. Intl.* 38(11): 324–327.

Baily, L.H., 1949. *Manual of Cultivated Plants*. MacMillan Co., New York (USA).

Baily, L.H., 1958. *The Standard Encyclopedia of Horticulture*, Vol. 1, 17th Edition. MacMillan, New York.

Balasubramaniam, P.P. and Singh, H.P., 2000. Cashew Development in India: An integrated strategical approach. In: *Cashew Industry*. Directorate of Cashew and Cocoa Development. Cochin, pp. 17–29.

Bhaskara Rao, E.V.V. and Swamy, K.R.M., 1994. Genetic Resources of Cashew. In: *Advances in Horticulure Vol. 9: Plantation and Spice Crops*, Part 1, (Eds.) K.L. Chadha and P. Rethinam. Malhotra Publishing House, New Delhi, pp. 79–97.

Bhaskara Rao, E.V.V., 1996. Emerging trends in cashew improvement. In: *Crop Productivity and Sustainability: Shaping the Future, Second International Crop Science Congress* (Eds.) V.L. Chopra, R.B. Singh and Anupam Varma. Oxford and IBH publishing Co. Pvt. Ltd., New Delhi, pp. 675–682.

Bhaskara Rao, E.V.V. and Bhat, MG., 1996. Cashew breeding: Achievements and Priorities. *Proceedings of the Seminar on Crop Breeding in Kerala*. University of Kerala, January 25, pp. 47–53.

Darlington, C.D. and Janaki Ammal, E.K., 1945. *Chromosome Atlas of Cultivated Plants*. Allen and Unwin, London, 397 p.

De Castro, P., 1994. Summary of the study. In: *The World Cashew Economy*, (Eds.) A.M. Delogu and G. Haeuster. NOMISMA, L'Inchiostroblu, Bologna, Italy, pp. 11–12.

Delogu, A.M. and Haeuster, G., 1994. *The World Cashew Economy* NOMISMA, L'Inchiostroblu, Bologna, Italy.

Goldblatt, P. (ed.)., 1984. *Index to Plant Chromosome Numbers*, 1979–1981. Missouri Bot. Garden.

Johnson, D., 1973. The botany, origin, and spread of the cashew *Anacardium occidentale* L. *J. Plant. Crops*, 1: 1–7.

Khosla, P.K., Sareen, T.S. and Mehra, P.N., 1973. Cytological studies on Himalayan *Anacardiaceae*. *Nucleus*, 16: 205–209.

Machado, 1944. Estudos novos sobre uma planta velha cajueiro (*Anacardium occidentale* L.). *Rodriguesia*, 8: 19–48.

Massari, F., 1994. Introduction. In: *The World Cashew Economy*, (Eds.) A.M. Delogu and G. Haeuster. NOMISMA, L'Inchiostroblu, Bologna, Italy, pp. 3–4.

Mitchell, J.D. and Mori, S.A., 1987. The Cashew and its relatives (Anacardium : Anacardiaceae). *Memoirs of The New York Botonical Garden*, 42: 1–76.

Nayak, M.G. and Bhat, MG., 2004a. Biodiversity in cashew its conservation and utilization in crop improvement. *Paper Presented in National Seminar on Horticulture for Sustainable Income and Environmental Protection* held at SASRD, Nagaland University, Medziphema, Dimapur between 24 and 26th February.

Nayak, M.G. and Bhat, M.G., 2004b. Registration of cashew germplasm. Compiled information on registered germplasm has been communicated for publication in *Indian Journal of Plant Genetic Resources*. NBPGR, New Delhi.

Ohler, J.G., 1979. Cashew. Kohinklijk Instuut voor de Tropeu; Amsterdam, p. 50–78.

Parente, J.I.G., 1972. Cajueiro. Aspectos economicos e agronomicos. Circulara, instituto de Pesquisa Agropecuaria do Nordeste (Brazil) No. 19.

Peixoto, A., 1960. Caju. Servvico de In formacao Agaricola, Ministerio de Agricultura Rio de Joneiro (Brazil), 61p.

Purseglove, J.W. (ed), 1988. *Tropical Crops: Dicotyledons*. English Language Book Society, Longman, London, pp. 18–32.

Simmonds, N.W., 1954. Chromosome behaviour in some tropical plants. *Heredity*, 8: 139.

Singh, H.P., 1998. Keynote address. *National Seminar on Cashew Devlopment in India: Challenges and Opportunities*. Held at Goa on 1st and 2nd March 1998.

Swamy, K.R.M., Bhaskara Rao, E.V.V. and Bhat, MG., 1997. *Catalogue of Minimum Descriptors of Cashew (Anacardium occidentale L.) Germplasm Accessions–I*. National Research Centre for Cashew, Puttur, Karnataka, (June), 41 p.

Swamy, K.R.M., Bhaskara Rao, E.V.V. and Bhat, MG., 1998. *Catalogue of Minimum Descriptors of Cashew (Anacardium occidentale L.) Germplasm Accession–II*. National Research Centre for Cashew, Puttur, Karnataka, India (October), 54 p.

Swamy, K.R.M., Bhaskara Rao, E.V.V. and Bhat, MG., 2000. *Catalogue of Minimum Descriptors of Cashew (Anacardium occidentale L.) Germplasm Accessions–III*. National Research Centre for Cashew, Puttur, Karnataka, (October), 54 p.

Valeviano, C., 1972. Cajuoiro. Boletim do Instituto Biologi CO de Bahia (Brazil) 11, 1. p., 19–58.

Van Eijnatten, C.L.M., 1991. *Anacardium occidentale* L. In: *Plant Resources of South-East Asia No. 2: Edible Fruits and Nuts* (Eds.) E.W.M. Verheij and R.E. Coronel. Pudoc–DLO, Wageningen, The Netherlands. pp. 446.

Biodiversity in Horticultural Crops Vol. 1 *Pages* ***340–364***
Editors: **Professor K.V. Peter & Dr. Z. Abraham**
Published by: **DAYA PUBLISHING HOUSE**

Chapter 18

Rubber (*Hevea brasiliensis*)

Y. Annamma Varghese and Saji T. Abraham
Rubber Research Institute of India, Kottayam – 686 009

Para rubber tree of commerce, *Hevea brasiliensis* (Willd ex. Adr.de Juss.) Muell.Arg is the most important source of natural rubber (NR) and meets around 98 per cent of global requirement. Natural rubber industry owes its existence to the early collection of the species in 1876 by Sir Henry Wickham from its native environment in Amazon rain forest in Brazil. This triggered an interest in commercial cultivation of the tree in South and South East–Asian countries and marked the beginning of research on *Hevea*. Demand for natural rubber showed a sudden increase in the twentieth century, with a tremendous boost in demand, because of rapid growth in automobile industry with its demand for tyres. The very high demand of NR in this industry was well understood by planting community and gradually rubber established itself as a highly successful plantation crop both in estates and in small holdings.

Over the decades, importance of rubber in modern life increased considerably. Natural rubber (NR), an industrial raw material of strategic importance is among the most versatile agricultural products, and finds use in about 50,000 products, the world over. In India, around 35,000 products are made out of NR. Cultivation of rubber is the chief means of livelihood for millions of people in many rubber-growing countries of Far East, who depend directly or indirectly on wages or profits received from production of plantation rubber. From an agro–ecological point of view, rubber is an eco-friendly tree species and reproduces a stable ecosystem similar to natural forest; in addition to giving high economic returns.

Asian countries dominated natural rubber sector, occupying around 93 per cent of area under rubber cultivation among all rubber growing countries, African and Latin American countries contributing 5 per cent and 2 per cent respectively. Today, Indonesia is the largest cultivator of rubber with an area of 3,26,200 ha followed by Thailand (2,083,000 ha), Malaysia (1,282,000 ha), China (600,000ha) and India (5,78,000 ha).

History of cultivation of Rubber in India dates back to 1878, when rubber was first cultivated as a forest crop in Nilambur in the State of Kerala, in South India, using planting material brought from Royal Botanic Gardens, Heneratgoda, Ceylon(present Sri Lanka) (Petch, 1914, Dean, 1987). Rapid growth in production of NR in India is very impressive considering domination of small growers in this plantation sector. Production increased from a very low level of 23,730 tonnes in 1955-56 to a high of 707,100 tonnes in 2003. In India, among different rubber growing states, the southern state of Kerala had maximum area of 4,78,402 hectares followed by Tripura (30,270 ha), Karnataka (20,460 ha) and Tamil Nadu (18,633 ha) (Rubber Board 2005). Productivity of rubber plantations in the country, increased from 353 kg/ha in 1955-56 to 1705 kg/ha in 2004-2005, the highest productivity among NR producing countries.

Origin and Domestication

Genus *Hevea* occupies whole of Amazon river basin in Brazil, also extending towards south and north covering parts of Brazil, Bolivia, Peru, Colombia, Ecuador, Venezuela, French Guyana, Surinam and Guyana. All the 10 species grow in Brazil while seven are found in Colombia, five in Venezuela and four in Peru (Wycherley, 1992). *H. brasiliensis* extends about half of range of the genus, mainly occupying the region south of the Amazon, extending to Acre, Mato Grosso and Parana in Brazil, parts of Bolivia, Peru, north of Amazon to west of Manaus as far as the extreme south of Colombia.

Domestication of *Hevea* in the East is the most spectacular story in the history of rubber where, a novel plantation industry of some 9,00,000 ha was established in a short span of 40 years, to meet new industrial demand (Baulkwill, 1989). The species is now grown mainly in tropical regions of Asia, Africa and America and in countries like Malaysia, Indonesia, India, Sri Lanka, Thailand, China, Philippines, Vietnam, Kampuchea, Burma, Bangladesh, Singapore, Nigeria, Cameroon, D.R. of Congo, Ivory Coast, Ghana, Zaire, Liberia, Brazil and Mexico. However, major share of total production comes from tropical Asia.

The famous Wickham collection of rubber formed the base for development of entire rubber plantations in the East (Lane, 1953; Wycherly, 1968; Dean, 1987). In 1876, Wickham collected 70,000 seeds of *H. brasiliensis* from near Boim on the Rio Tapajoz and from the well-drained undulating country towards the Rio Madeira that produced excellent wild rubber. About 2,700 of these seedlings which survived the journey from S. America to London, were raised at the Royal Botanic Gardens, Kew, England; 1919 of them were dispatched during 1876 to Sri Lanka; a few went to Malaysia, Singapore and Indonesia (Baulkwill, 1989; Dean, 1987).

The Royal Botanic Gardens, Heneratgoda, in Colombo, became a major source of rubber seeds for domestic use and for export (Jones and Allen, 1992). Recognizing potential of rubber, Henry Ridley was instrumental in distributing rubber seeds widely and encouraging its cultivation when he became director of Singapore Botanical Gardens in 1888.

The Rubber Tree

In addition to cultivated species *Hevea brasiliensis*, the genus *Hevea* (family Euphorbiaceae) has 10 more species *viz. H. benthamiana, H.camergoana, H.camporum, H. guianensis, H. microphylla, H. nitida, H. paludosa, H. pauciflora, H. rigidifolia,* and *H.spruceana.*

Rubber tree is a sturdy, quick growing, erect perennial, growing to a height up to 30m in plantations and having an economic life span of over 30 years. The tree, however, grows to a height of over 40m and live for over 100 years in wild. Tree has a straight trunk with light grey bark and branches developed in such a way as to form an open leafy crown. The young plants show characteristic growth pattern of alternating periods of rapid elongation and consolidated development. Leaves are arranged in groups or storeys, each one with a cluster of spirally arranged trifoliate glabrous leaves and extra floral nectaries present in the region of insertion of leaflets.

Hevea is a deciduous tree, which sheds leaves (wintering) during December–February followed by refoliation and flowering. The plant is monoecious with unisexual flowers produced in pyramid shaped panicles in axils of leaves having numerous small male flowers and a fewer female flowers of bigger size at tip of panicles and their branchlets. Ovary is tricarpellary syncarpous, which on pollination develops into a three lobed dehiscent capsule (regma) with three large mottled seeds. Pollination is by insects and fruits ripen in 5-6 months after fertilization. Seeds contain an oily endosperm.

The economic product from rubber tree, latex, a specialized cytoplasm, is contained in a laticiferous system in bark differentiated by activity of vascular cambium. *Hevea* bark has two distinct zones (1) outer hard bast–which is more protective in function with mostly discontinuous and non-functional laticifers and (2) inner soft bast- containing productive and continuous latex vessels differentiated from vascular cambium. The tree is exploited by tapping bark–a process of controlled wounding of bark to cut open latex vessels by regularly removing a thin shaving of bark from surface of tapping cut at specified intervals. On tapping trees, latex flows out mainly because of sudden release of high turgor pressure in latex vessels and elastic contraction of walls resultant of cutting vessels open. After a while, capillary forces regulate flow until it ceases as latex coagulates and plugs the vessels. Tapping should commence early morning, as late tapping will reduce latex yield due to increased transpiration leading to lower turgor pressure. Tapping is a highly skilled work where depth of tapping should be optimum to open maximum number of vessels with out injuring cambium.

Genetic Resources in *Hevea*

Sources of Germplasm

Spectrum of *Hevea* germplasm can be broadly classified into (a) wild genotypes from primary center of diversity in Brazil and (b) cultivars developed in centers of secondary diversity. Primary center of diversity not only accommodates wild genotypes of genus but also the naturally occurring hybrids, morphotypes and variants, commercial cultivars, selections and genetic variants form the genetic resources in the secondary centers. All available genotypes are conserved in a 'gene pool' and can be utilized in breeding programmes as and when required.

Commercial cultivation is very often restricted to high yielding cultivars, which may become obsolete when newly improved ones are released. As such, it is not unlikely that the older materials disappear in course of time. These are valuable genetic materials and most of the countries where *H. brasiliensis* is an introduced crop, the variants, selections and obsolete cultivars are maintained as an insurance against gene erosion.

Genetic Diversity in Original Collection

Original genetic material of *Hevea*, referred to as the 'Wickham base' consisted of a few seedlings collected from primary center of origin of crop and introduced to South East Asian countries, by Sir Henry Wickham in 1876 (Simmonds, 1989). From this original genetic base, substantial improvement in productivity was achieved over a relatively short period through systematic breeding and selection in different rubber research institutes in S.E. Asia. This indicates that although the original collection was made from a minuscule of genetic range of the species in Boim, near the Tapajos River in Brazil, there existed sufficient variability for early crop improvement programmes. *Hevea*, being a highly heterozygous cross-pollinated species, widespread cultivation of seedling trees in early years of rubber cultivation has added to genetic variability in original gene pool. This method resulted in plantations with a very wide genetic base, which could serve as a buffer and help plants tide over times of stress. Such plantations also show greater tolerance to adverse soil and climatic conditions. These seedling plantations are the source for selection of many potential primary clones of commercial significance.

Breeding and Selection in *Hevea*

Systematic crop improvement strategies and breeding programmes chalked out by rubber breeders from different South Eastern Asian rubber growing countries played a key role in exploiting genetic variability existing in original collection of *Hevea brasiliensis*. Genetic improvement of *Hevea*, in spite of being a very elaborate and time-consuming procedure has paid rich dividends in increasing yield by several folds and making available several high yielding clones, for commercial planting. With ordinary seeds as planting material in initial years of rubber research, productivity was only around 300 kg ha^{-1} yr^{-1}. Subsequent selection of high yielding mother trees and multiplication by budding resulted in early primary clones with improved yield potential. Subsequently breeding new clones through hybridization resulted in a series of high yielding hybrid clones, which further enriched diversity in this crop.

Introduction and Exchange

Original breeding pools in rubber were built up by introduction or exchange of available clones among rubber growing countries. International agencies like International Rubber Research and Development Board (IRRDB) and Association of Natural Rubber Producing Countries (ANRPC) organize clone exchange programs among member countries. Thus popular clones evolved in different countries are being introduced to member countries and evaluated under local agro–climatic conditions and promising selections recommended for large scale planting. In India so far, 127 Wickham clones evolved in Malaysia, Indonesia, Sri Lanka, Thailand, China and Ivory Coast formed the exotic component of gene pool, (Table 18.1) in addition to recent introduction of the large collection of wild germplasm.

Mother Tree Selection

Mother tree selection or ortet selection is the oldest breeding method adopted in *Hevea* in early years of rubber cultivation aimed at selection of outstanding seedling genotypes resultant of natural cross combinations. Clones developed through ortet selection are called primary clones. Screening of extensive seedling plantations of Gough Garden series (GG1, GG2 etc) and Prang Basar Isolation Garden series (PBIG) in Indonesia, Malaysia and Sri Lanka resulted in an array of primary clones of importance such as Tjir1, PR 107, GT 1, BD 10, AVROS 255, Gl 1, PB 28/59, Mil 3/2, Hil 28 of which clones like GT 1, PB 28/59 are still widely planted. In India, earlier mother tree selections include 46 clones of which RRII 1, RRII 4, RRII 5, RRII 6, RRII 33, RRII 43, RRII 44 etc. are among potential clones (Marattukalam *et al.*, 1980). Screening of available polyclonal seedling plantations helps to selection of promising 'plus trees' and development of ortet clones thereby broadening genetic base of gene pool also.

Hybridization and Selection

The most important and largest contribution to bio-diversity in rubber was by exploitation of heterosis in this cross-pollinated tree crop. Hybridization and clonal selection, the most important conventional method of breeding in *Hevea*, has contributed a large number of hybrid clones of commercial significance in rubber plantation sector in major rubber growing countries like Malaysia, India and Sri Lanka. Early primary clones were used as parents in initial hybridization programmes, which resulted in heterotic hybrids of commercial significance like RRIM 500 and 600 series, yield levels of which were much superior to that of parent clones. Only drawback in the process was that such best clones in each series were further used as parents in subsequent hybridization programmes by a process of cyclical generation-wise assortative mating (GAM) over the past a few decades.

Table 18.1: List of Clones Introduced into India

Year	*Country of Origin*	*Clones*
Before 1956	Indonesia	Tjir 1, Tjir 16, AVROS 49, AVROS 255, AVROS 352, LCP 1320, PR 107, BD 5, BD 10
	Malaysia	PB 25, PB 86, PB 186, PB 5/60, PB 5/139, PB 6/9, PB 6/50, Pil B 84, Pil D 65, Gl 1, Ch 2, Ch 4, Ch 8, Ch 26, Ch 29, Ch 30, Ch 31, Ch 32, CHM 3, S.Reko 9, Lun N
	Sri Lanka	Mil 3/2, Hil 28, Hil 55, Wagga 6278, War 4
1956	Malaysia	RRIM 501, RRIM 526, RRIM 601, RRIM 602, RRIM 603, RRIM 604, RRIM 605, RRIM 607, RRIM 608, RRIM 609, RRIM 610, RRIM 611, RRIM 612, RRIM 615, RRIM 617, RRIM 618, RRIM 620, RRIM 621, RRIM 622, RRIM 623
	South America	IAN 45-717, FX 516, F 4542
1962	Malaysia	RRIM 513, RRIM 519, RRIM 600, RRIM 628, PB 206, PB 213, PB 217, PB 5/76
1963	Malaysia	RRIM 632, RRIM 636, RRIM 701, RRIM 707, PB 5/51, PB 5/63, PB 28/59, PB 28/83
	Indonesia	GT 1
1964	Malaysia	PB 215, PB 203, PB 235, PB 240, PB 242, PB 252, PB 253
1966	Malaysia	RRIM 703, RRIM 704, RRIM 705, RRIM 706, Ch 153
1969	Liberia	Harbel 1
	South America	IAN 45-713, IAN 45-873
1972	Sri Lanka	RRIC 7, RRIC 36, RRIC 45, RRIC 52, RRIC 89, RRIC 100, RRIC 102, RRIC 104, RRIC 105, Nab 17
1979	Malaysia	PB 260, PB 310, PB 311
1984	Thailand	KRS 25, KRS 128, KRS 163
	China	SCATC 88-13, SCATC 93-114, Haiken 1
1991	Ivory Coast	IRCA 18, IRCA 109, IRCA 111, IRCA 130, IRCA 230
1993	Malaysia	RRIM 712, RRIM 722, RRIM 728
1995	Indonesia	BPM 24, PR 255, PR 261
	Sri Lanka	RRIC 110, RRIC 130

The past five decades definitely paid rich dividends by development of a series of high yielding hybrid clones by hybridization and selection process by various rubber research institutes. Rubber Research Institute of Malaysia developed clones of RRIM 500 to 1000 series, while Prang Basar Institute in private sector selected certain PB clones of commercial significance. Indonesian Research Institute for Estate Crops in Java and Sumatra (BPPM) evolved PR, AVROS, BPM, LCB, PPN and RR clones. RRIC/RRISL clones originated from Sri Lanka, KRS clones from Thailand, IRCA clones from Ivory Coast and Haiken, YRITC and SCATC clones from China.

Crop improvement programs in India, initiated in 1954 evolved a large number of hybrid progenies, which were evaluated to identify potential recombinants. Rubber Research Institute of India came out with a series of clones starting from early RRII 100 series to the most recent RRII 400 series. Early hybrid clones developed by RRII include RRII 100, 200 and 300 series. Among clones of RRII 100 series, RRII 105 is a highly successful and popular clone in terms of realized and potential yield which occupy about 85 per cent of rubber grown area. Clones RRII 116 and RRII 118 are outstanding for growth vigour, but with only medium yield. Clones RRII 203 and 208 and RRII 300 and 308 are among selections in 200 and 300 series respectively. The most recent RRII 400 series, from a set of 23 hybrid clones of cross combination RRII 105 x RRIC 100 of which nine clones were selected based on their early performance at age of four and a half years, showed promising trend in the mature phase too and recorded significant yield increase in the range of 23-46 per cent over RRII 105 during first 8 years of tapping in small scale trials (Saraswathyamma, 2003). Of nine high yielding hybrid clones, five clones have been included in current planting recommendations of Rubber Board of India *viz.* RRII 414 and RRII 430 in Category I; RRII 417 and RRII 422 in Category II and RRII 429 in Category III (Table 18.2).

Table 18.2: Performance of RRII 400 Series Selections in Small-scale Trial

Clone	*Girth* at Opening (cm)*	*Mean Annual Firth increment (cm)*	*Dry Rubber Content (DRC per cent)*	*Mean Yield Over 10 Years of Tapping (gram/tree/tap)*	*Yield** increase Over RRII 105 (per cent)*
RRII 414	57.49	17.02	30.64	74.31	38
RRII 417	54.73	10.76	34.63	70.74	32
RRII 422	52.4	8.71	34.27	65.68	22
RRII 429	62.77	14.33	28.81	76.51	42
RRII 430	53.42	8.44	35.26	64.17	20
RRII 105	49.04	12.41	33.46	53.77	–

Source: Saraswathyamma, 2004

*Circumference of trees measured at 125 cm height from bud union.

**Yield expressed as grams of dry rubber per tree per tap.

Crop Improvement in Non-traditional Rubber Growing Areas

Non-traditional rubber growing regions in India are spread over states of Tripura, Assam, Meghalaya, Nagaland, Manipur, Mizoram and Arunachal Pradesh in the N E region, West Bengal, Orissa, Andhra Pradesh, and Madhya Pradesh towards East and Maharashtra and Goa in Konkan region.

Non-traditional rubber growing areas in the North East are confronted with low temperature stress during November- February with temperature coming down to 4-5° C. The North Konkan region and the state of Orissa experience severe drought in association with high temperature and low humidity. Other constraints include low and depleted soils, high velocity winds and occasional hailstorms and high altitude effect in certain tracts of N East.

Rubber breeding in India for non-traditional regions was streamlined with the objective of utilization of available genetic variability in existing cultivated clones and polycross progenies to develop location specific clones. In NE region around 50 potential clones from traditional regions are being subjected to field evaluation in a phased manner. Data available from one set of clone trial revealed location specificity of certain popular clones. Among 25 clones in two clone trials at RRS

Agartala, high yielders include PB 235, RRIM 600, RRII 208, RRII 203, RRII 118 and SCATC 88-13 while vigorous clones identified include RRII 118, RRII 203, PB 235, RRIM 600, SCATC-88-13, and SCATC 93-114 (Priyadarshan *et al.*, 1998); (Priyadarshan *et al.*, 2000); (Priyadarshan, 2003).

Of the 20 popular clones evaluated in Assam, RRIM 600, RRII 105, PB 235, RRII 208 and RRII 118 were high yielders. The clones RRIC 102, RRII 118, RRIM 600 and RRII 203 were very vigorous in terms of growth (Mondal *et al.*, 1999, Varghese, 2003). Evaluation of 20 clones in high elevations of the state of Meghalaya, identified clones RRIM 600, PB 311, RRII 208 and RRII 118 as high yielders compared to RRII 105 in early mature phase while RRIM 600, PB 235, RRII 203 and RRII 118 recorded vigorous growth (Reju *et al.*, 2000, 2002).

Chinese clones tolerant to cold conditions performed well in Nagrakatta region in North of West Bengal, along with other popular clones. The clones SCATC 88/13, RRIM 600, PB 311, PB 235 and RRII 208 were early high yielders in this region while Chinese clones SCATC 93-114 and Haiken 1 followed by RRIM 612, RRIM 605 and RRIC 102 are among vigorous clones in this area. (Meti *et al.*, 1999; Varghese, 2003)

In dry regions of Dapchari in Maharastra, drought tolerant clones in terms of immature growth identified were of Sri Lankan and Indian origin *viz.*, RRII 208, RRIC 52, RRII 6, RRIC 102 and RRIC 100 (Chandrashekar *et al.*, 1998). This trend was continued in mature phase too with RRII 208 performing well both in terms of girth and yield. (RRII 2002). At Dhenkanal in Orissa, RRIM 600 and SCATC 88-13 were high yielders in early mature phase (Gupta and Edathil, 2001).

In all non-traditional rubber-growing areas, progeny derived from polyclonal seedlings, in general, performed quite well. Compared to clones, there exist high levels of genetic variability among seedlings. Seedling population was highly heterogeneous with respect to growth, yield and desirable secondary characters and offer considerable scope for selection. (Sasikumar *et al.*, 2001; Varghese, 2003) and there by adding to biodiversity of species.

Classification of Rubber Planting Materials

Planting materials used for cultivation in rubber could be broadly classified into clones and polyclonal seeds. A clone being a group of plants derived from a single tree using vegetative propagation, gives an identical genetic make up and there by exhibiting uniformity among clones. Among seeds, polyclonal seeds generated in special seed gardens comprising a mixture of selected clones are recommended planting material for specific situations. Clones generated through bud grafting, are the most popular planting material today while polyclonal seedlings are also recommended in non-traditional and marginal rubber growing areas in India. (Varghese, 2003).

Based on their origin, clones are classified into primary and hybrid clones. Primary clones such as GT 1, PR 107, PB 28/59, and RRII 5 are developed from mother trees of seedling origin and their parentage is unknown. On other hand since hybrid clones are developed through hybridization between two clones, parentage is known. Some of hybrid clones are RRII 105 (Tjir 1 × Gl 1), RRII 203 (PB 86 × Mil 3/2), RRIM 600 (Tjir1 × PB 86), RRIM 728 (GT 1 × RRIM 623), and PB 311 (RRIM 600 × PB 235).

Clones are usually named after the estates, institute or station from which they originated and such names are abbreviated to serve as identifying codes prefixed to the clone number. Abbreviations and full names of some of important clones in rubber are given in Table 18.3.

Table 18.3: Nomenclature of Various Clones of *Hevea brasiliensis*

Abbreviation	*Description*
AVROS	Alegemene Verneitging Rubber Planters Oostkust Sumatra, Indonesia.
GI	Glenshiel, Malaysia
GT	Gondang Tapen, Indonesia
Hil	Hilcroft, Srilanka
IAN	Institute Agronomico du Norte, Brazil
IRCA	Institute des Recherches sur le Caoutchouc, Ivory Coast
KRS	Kohong Rubber Estate, Thailand
LCB	S'Lands Caoutchouc Bedrijven Indonesia
Mil	Milakande, Srilanka
PB	Prang Basar, Malaysia
Pil	Pilmoor, Malaysia
PR	Profestation voor Rubber, Indonesia
RRIC/RRISL	Rubber Research Institute of Ceylon/Srilanka
RRII	Rubber Research Institute of India
RRIM	Rubber Research Institute of Malaysia
SCATC	South China Academy of Tropical Crops
Tjir	Tjirandji, Indonesia
Wagga	Wagga, Srilanka

Planting Materials Recommended in Rubber

In India, a multiclone planting is recommended by Rubber Board of India with cultivars grouped into three different categories, *viz.*,

Category I (for large scale planting, covering only up to 50 per cent of total rubber area in a holding) Clones RRII 105, PB 260, RRII 414 and RRII 430 for traditional areas and RRIM 600 and GT 1 for non-traditional areas.

Category II (Three or more of these cultivars may be used to plant upto 50 per cent of total area of any holding.) Clones RRIM 600, GT 1, RRII 5, RRII 203, PB 28/59, PB 217, RRII 417 and RRII 422.

Category III (Clones having limited data for restricted planting).

Clones RRII 50, RRII 51, RRII 52, RRII 118, RRII 176, RRII 208, RRII 300, RRII 429, PR 107, PR 255, PR 261, PB 86, PB 5/51, PB 235, PB 255, PB 280, PB 311, PB 312, PB 314, PB 330, RRIM 605, RRIM 701, RRIM 703, RRIM 712, RRIC 100, RRIC 102, RRIC 130, KRS 163, IRCA 111, IRCA 130, SCATC 88-13, SCATC 93-114, Haiken -1, BPM-24 and Polyclonal seeds.

Selections from any of these are approved for small scale planting not to exceed 15 per cent of total area, in aggregate (Rubber Board, 2006).

Characters of a Few Recommended Clones

RRII 105

This clone evolved by Rubber Research Institute of India is currently enjoying maximum popularity in the country. Parents are Tjir 1 and GI 1. Trunk is tall. Presence of more than one leader is observed in many trees. Branching habit is good with strong unions. Canopy is dense, mostly restricted to top. Foliage dark green leaflets long and glossy. Wintering and refoliation are early and partial, vigour before and after tapping average. Virgin bark and renewed bark thickness above average.

Yield is very good. Average yield obtained over the first fifteen years of tapping in on farm evaluation trials is 2210 kg/ha/yr. Latex is white with high dry rubber content. Clone has a fair degree of tolerance to abnormal leaf fall disease. Highly susceptible to pink disease, while incidence of powdery mildew is medium. Occurrence of tapping panel dryness high and therefore adherence to tapping under half spiral, once in three days system is essential. Free from serious wind damage if branch development is kept balanced. Small growers should take special care to avoid excessive or unbalanced manuring, which can accentuate tapping panel dryness and wind damage. Corrective pruning of unbalanced branches can reduce incidence of wind damage. The clone is susceptible to drought in terms of growth; hence not advisable for planting in drought prone areas. However if it survives and establishes well in such areas, yield is not much affected.

PB 260

This hybrid developed by Prang Basar Estates Ltd., Malaysia from parents PB 5/51 and PB 49 is also a high yielder. Trees have tall and straight trunk with light, balanced branching with a strong union. The clone has a dense canopy with pale green foliage. Clone is highly vigorous before tapping and an average vigorous clone, after commencement of tapping. Thickness of virgin and renewed bark is below average. On farm trials recorded a mean yield of 1631 kg/ha/yr over the first five years of tapping. Average yield from large-scale trial over 10 years is 47.23 g/tree/tap. The clone has moderate tolerance to pink, powdery mildew, abnormal leaf fall and a high incidence of taping panel dryness. Wind damage is below average.

RRII 414

This clone, recently included in Category I, is another high yielding clone developed by RRII. Parents are RRII 105 and RRIC 100. Trunk is tall, straight and cylindrical with open, broad canopy of dark green leaves, restricted to top. Girth at opening is high and girth increment on tapping is average. Shows 38 per cent yield improvement over RRII 105 during first eleven years of tapping in small-scale trial. Incidence of *Oidium* is moderate to severe. Incidence of all other diseases is comparatively lesser than that of RRII 105.

RRII 430

Parents of this clone are RRII 105 and RRIC 100. The clone has above average girth at opening. Canopy is open, broad and heavy with broad glossy leaves/. Virgin bark is of average thickness while renewed bark thickness is high. Increase of 20 per cent in yield was recorded over RRII 105 in small-scale trial, during first eleven years of tapping. Incidence of *Oidium* is moderate to severe while that of all other diseases is less than in RRII 105.

RRIM 600

A high yielding clone evolved by Rubber Research Institute of Malaysia and extensively grown in all rubber growing countries. Parents are Tjir 1 and PB 86. Trees have tall, straight trunk, moderate to fairly heavy branching and branch unions rather weak. Young plants show spindly growth and late branching with occasional learning. Narrow, broom shaped crown, foliage sparse with small yellowish green leaves; normal wintering and refoliation. Girth at opening is low while girth increment after opening high. Virgin bark thickness low while thickness of renewed bark high; usually results in bulging above tapping cut. The clone shows rising yield trend with above average initial yield and very high subsequent yield. Does not exhibit prominent yield depression during summer. Average annual yield in estates over first 20 years is 1349 kg per ha. Latex is unsuitable for concentration. Highly susceptible to diseases caused by *Phytophthora,* especially abnormal leaf fall disease and mild to severe incidence of pink disease. Requires effective control measures when planted in areas where these diseases are prevalent. Incidence of powdery mildew is mild.

GT 1

A primary clone developed in Indonesia and extensively planted in all rubber growing countries. Trunk is upright but slightly kinked and variable branching habit with long acute angled main branches, light secondary branches, narrow globular crown and dense dark green glossy foliage. Wintering and refoliation are late, which is often partial. Girth at opening is medium to high and girth increment of tapping medium and medium virgin and renewed bark thickness. This clone shows rising yield trend. Summer yield is fairly high. Average annual yield obtained in estates for 19 years is 1420 kg./ha. Latex is white. Incidence of tapping panel dryness and pink disease are mild and a fair level of wind fastness is observed. Incidence of abnormal leaf fall is mild to medium and that of powdery mildew medium to severe.

RRII 5

A primary clone developed by Rubber Research Institute of India selected from Malankara Estate, Thodupuzha. Clone is vigorous having straight and terete trunk, dense canopy and low branching with several branches arising at acute angles. Bark thickness high. High yield with rising yield trend. Mean yield from large scale trial over 16 years is 76.42 g/tree/tap, and that from estate trial over five years is 1352 kg/ha/yr. Susceptible to abnormal leaf fall and powdery mildew diseases while pink disease and wind damage low. Tapping panel dryness high. Clone yields relatively high quantity of timber yield.

RRII 203

A moderate yielding clone evolved by Rubber Research Institute of India. Parents are PB 86 and Mil 3/2. Trunk straight and tall, rather robust, canopy well distributed and balanced. Clone has an above average vigour at opening with average virgin and renewed bark thickness and average tolerance to disease. Clone has recorded medium initial and subsequent yield. Average yield over ten years of tapping from commercial planting is 1818 kg/ha/yr. Coagulum from latex and scrap show a discoloration. This however, does not affect quality of rubber. Timber yield is high.

PB 28/59

A Malaysian primary clone with fluted and crooked trunk, sometimes showing tendency for leaning with moderate to heavy branches and low branching. Girth at opening is average while girth increment on tapping is low. Virgin bark thickness is low but thickness on renewal high. Average annual commercial yield over 19 years is 1477 kg/ha. Summer yield is average. Susceptibility to wind damage is average. Incidence of tapping panel dryness is above average. Clone is highly prone to abnormal leaf fall, pink and powdery mildew diseases.

PB 217

Parents of this Malaysian clone are PB 5/51 and PB 6/9. Trunk is tall and straight with light branches and dense foliage. Wintering and refoliation range from normal to late. Girth at opening is average while girth increment on tapping is high. Virgin bark thickness is low but renewed bark is of average thickness. Initial yield is medium with arising trend in subsequent years. Average yield over first 15 years in estate sector is 1508 kg/ha/yr. Latex colour is light yellow. Wind damage is very low and tapping panel dryness mild. Although incidence of *Phytophthora* is severe in Malaysia, extent of abnormal leaf fall in India is low. Pink and powdery mildew disease affect the clone severely.

RRII 417

Parentage of this clone is RRII 105 x RRIC 100. Trunk is tall and straight with broad, open and heavy canopy with semi-glossy leaves. Thickness of virgin bark is average and that of renewed bark is high. This clone has above average tolerance to wind but susceptible to pink disease. Yield improvement over RRII 105 is 32 per cent during first ten years in a small-scale trial.

RRII 422

Parents of this clone are RRII 105 and RRIC 100. Girth at opening is above average. Stem is crooked and high branching with. Open, narrow canopy with dark green glossy leaves. This clone recorded 22 per cent yield increase over RRII 105 in small-scale trial during first ten years.

PB 235

This Malaysian clone has PB 5/51 and PB S/78 as parents. Very vigorously growing clone with very tall and straight stem. Branches light and spreading with dense foliage, which undergoes normal wintering and refoliation. Girth increment on tapping is average. Thickness of virgin bark is average and that of renewed bark low. Average annual yield obtained in estates over first 15 years of tapping is 1501 kg/ha/yr. Summer yield is average. Latex colour is pale yellow. Clone experiences high wind damage. Incidence of panel dryness is high. Incidence of abnormal leaf fall and pink diseases are medium but powdery mildew is severe. Stem bleeding and attack of bark feeding caterpillar are noted in certain localities.

RRIM 703

Parents of this clone are RRIM 600 and RRIM 500. It has an upright but slightly kinked trunk with a few heavy branches. Canopy is open and narrow. Wintering and refoliation occur early in the season. Girth at opening is high to average and girth increment on tapping low. Virgin bark thickness is high and renewed bark thickness average to high. Average estate level yield in India over the first 10 years is 1310 kg/ha/yr. Summer yield is low. Latex is light yellow. Wind damage tapping panel dryness high. Abnormal leaf fall is severe in India though reported to be only mild in Malaysia. Clone is subjected to severe pink disease. Occurrence of powdery mildew is mild.

PR 255

Parents of this clone developed in Indonesia are Tjir 1 and PR 107. Vigour at opening is above average with average girth increment on tapping. Trunk is tall and straight with dense and balanced canopy. Clone yields high with high summer yield. Mean yield in large-scale trial over ten years is 58.04 g//tree/tap. Incidence of abnormal leaf fall and pink disease is moderate. Incidence of powdery mildew is severe, while tapping panel dryness is moderate. Wind damage low.

PR 261

This is also an Indonesian clone with Tjir 1 and PR 107 as parents. An average vigourous clone with balanced crown, dense foliage and average bark thickness. Clone is high yielding with a mean yield of 51.34 g/tree/tap in large-scale trial over 10 years. Incidence of abnormal leaf fall and pink disease is moderate. Incidence of powdery mildew is moderate to severe while tapping panel dryness is moderate. Wind damage low.

PB 280

A Malaysian primary clone developed from Prang Besar isolation garden (PBIG) seedling. A clone with average vigour, bark thickness is high. This is a high yielding clone as per Malaysian reports. Yielding trend in India is also promising. Incidence of abnormal leaf fall and pink disease is medium while powdery mildew is severe. Wind damage is high and tapping panel dryness low.

PB 311

Parents are RRIM 600 and PB 235. The clone has a vigorous growth habit, trunk sometimes leaning with a heavy canopy and dense foliage. Thickness of virgin bark is low and that of renewed bark average. A high yielding clone with an average commercial yield of 1450/kg/ha/year over five years. Mean yield over six years in large-scale trials is 50.58 g/tree/tap. Summer yield is high. Incidence of powdery mildew and pink disease is mild to medium. Wind damage is very high and occurrence of panel dryness low.

PB 86

A Malaysian primary clone of slow growth, high crown and pale green leaves. Not easily prone to wind damage. Suitable for planting in exposed areas. Not tolerant to poor soils. Commercial yield in India over 25 years is 1165 kg/ha/yr. Latex white. Prolific seeder. Highly susceptible to abnormal leaf fall and shoot rot diseases. This clone performs well in Kanyakumari District where incidence of these diseases is very mild.

PR 107

This is a primary clone developed in Indonesia. Sturdy, wind resistant and of average vigour with good girth increment on tapping. The clone though a slow starter for yield, shows rising trend. In India, average yield over 15 years under commercial planting is 1043 kg/ha/yr. Yield gets slightly depressed during wintering. Withstands higher intensities of tapping. Clone is susceptible to *Phytophthora* and tolerant to powdery mildew.

RRIM 605

Parents of this clone are Tjir 1 and PB 49. Though its initial yield performance in Northern districts of Kerala is not encouraging, good performance reported in Punalur and Pathanamthitta areas. Growth is average. Average commercial yield over 20 years in India is 1135 kg/ha/yr.

RRIM 701

A moderate yielding clone with steady yield trend. Parents are 44/553 and RRIM 501. High vigour in early years. Average girth increment on tapping. Susceptible to pink and powdery mildew diseases and wind damage. Average commercial yield over 15 years in India is 1263 kg/ha/yr.

PB 5/51

A clone evolved in Malaysia by crossing PB 56 and PB 24. Stem straight and upright, branches light, horizontal and well distributed. Crown conical, light sparse foliage, small yellowish green leaves. Virgin bark thickness is average and renewed bark thickness low. Vigour is average before tapping and low after opening. Initial as well as subsequent yield is average. Commercial yield in India is 1389 kg/ha/yr, during first 20 years. Summer yield is above average. Clone is highly resistant to wind damage. Incidence of abnormal leaf fall disease is moderate; pink disease low; powdery mildew high and tapping panel dryness average.

RRII 118

A remarkably vigourous clone evolved by Rubber Research Institute of India. Parents are Mil 3/2 and Hil 28. Trunk is tall and stout with prominent heavy branches, long secondary branches and slightly drooping in young stage. Several branches arise almost at same level. Canopy is dense and crown balanced. Vigour at opening is high and average virgin and renewed bark thickness. Clone is an average initial yielder with rising trend. In commercial plantings, an average yield of 1164 kg/ha/yr is obtained during first ten years of tapping. Clone shows medium tolerance to diseases and average tapping panel dryness.

RRII 208

A clone yielding above average in the initial years of tapping. Parents are Mil 3/2 and AVROS 255. This clone shows high susceptibility to shoot rot and average tolerance to other diseases. Average yield from commercial planting over ten years of tapping is 1743 kg/ha/yr.

RRII 429

Parents are RRII 105 and RRIC 100. This clone is tall with straight and cylindrical trunk and a dense, heavy canopy with dark green leaves. Girth at opening is high and girth increment on tapping average. Clone is highly susceptible to pink and above average tapping panel dryness was observed in small scale trial. Yield improvement over RRII 105 during first ten years in small-scale trial is 42 per cent.

RRII 52

A primary clone selected from progeny of genetic variant having compact canopy. This clone is of normal morphotype. Mean yield over nine years of tapping in small-scale trial registered 21 per cent yield improvement over RRII 105. Yield trend during initial years of tapping in large-scale trial is also encouraging. Vigour is average. Trunk is straight and cylindrical; balanced branching with acute angled secondary branches. Clone has a medium sized open canopy with glossy leaves. Colour of latex is white.

Polyclonal seedlings: Seeds collected from clonal stands are known as clonal seeds. Monoclonal seeds are produced from a single clone while polyclonal seeds from different clones. Good quality polyclonal seeds are produced in specially designed gardens ensuring maximum cross pollination among component clones and also preventing pollen contamination from undesirable sources by providing an isolation belt of 100 metres width around the garden. Polyclonal seeds give rise to seedlings of good vigour and growth. In general, general yield levels are lower than that of selected modern clones with an average annual yield of 1200 to 1300 kg/ha/yr. Owing to inherent genetic variability, they are relatively less susceptible to wind damage and can tolerate undesirable climatic conditions better than clones (Rubber Board, 2006).

Genetics and Breeding

In *Hevea,* most of economically important characters are polygenically controlled. A clear understanding of magnitude of genetic variability estimates of genetic parameters, like heritability and genetic advance for desired characters is essential for an effective selection and utilization of genotypes for breeding programs. There are several reports from India on extent and nature of genetic variability and divergence existing in 'Wickham gene pool' (Premakumari, 1992; Mydin *et al.*, 1992, Licy *et al.*, 1992). Reports on additive gene action on the expression of yield and yield components of rubber suggest better scope for selection based on specific characters. Recombination of superior genes in a successful breeding program demands a clear understanding of extent of genetic divergence between parent clones as well as component clones of polyclonal seed gardens. Studies indicate that in spite of narrow genetic base, there exists sufficient genetic diversity in original Wickham material (Markose, 1984; Varghese *et al.*, 1997). Genetic diversity studies in Wickham clones, based on yield and various yield components, identified eight divergent clusters in a set of 40 clones (Mydin *et al.*, 1992) and nine divergent clusters in another group of 35 clones (Abraham *et al.*, 1997). In another similar study in wild accessions, 100 wild accessions were grouped into 5 divergent clusters (Abraham *et al.*, 1995) and another 80 wild accessions were grouped in to nine different clusters (Abraham, 2001). Selection of genetically unrelated parent clones offers greater chances of heterosis and superiority of resultant progeny. Use of molecular markers simplified such studies considerably.

Need for Broadening Genetic Base

Genetic improvement over past a few decades resulted in narrowing down of original 'Wickham gene pool' due to number of factors, like directional selection for yield, cyclical generation wise assortative breeding pattern and a wider adaptation of clonal propagation by budding. In all crop improvement programs, till recently, primary concern was improvement of productivity. Such unidirectional selection for yield, over years, ignoring genetic variability with regard to secondary characters (Wycherley, 1969), reduced genetic variability in population. Breeding in *Hevea* involves generation wise assortative mating (GAM), where the best clones in one cycle are used as parents for the next and so on (Simmonds, 1989). Hence, parentage of popular clones bred in various rubber growing countries can be traced back to just a handful of parent clones, like Tjir 1, Pil A 44, Pil B84, PB 24, PB 49, PB 56 and PB 86 (Tan, 1987; Varghese, 1992).

With development of high yielding clones, extensive areas are being planted with a limited number of modern clones propagated vegetatively by budding, which also had its share in decreasing genetic base. But this large-scale adoption of a few clones in rubber plantations in South East Asian countries put these plantations under potential threat of highly damaging disease South American Leaf Blight (SALB) caused by *Microcyclus ulei*, prevalent in American hemisphere (Chee and Holliday, 1986; Edathil, 1986). None of Wickham clones had resistance to SALB. There are also indications of erosion of genes controlling resistance to *Oidium* and *Gloesporium* in original Wickham material (Wycherley, 1977) and several minor diseases assuming epidemic proportions. Apart from problem of disease susceptibility, variability for resistance/tolerance to various abiotic stresses like drought, cold, high elevations and high velocity winds assumes significance in present day context of extension of rubber cultivation to marginal and non-traditional areas. To select clones adaptable to such specific locations, base material should contain ample genetic variability and a broad genetic base (Varghese and Abraham, 1999).

The 1981 Fresh Germplasm Collection

Hevea breeders across world realized need for broadening narrowing genetic base through introduction of wild germplasm into breeding pool. Recognizing the fact that international action was imperative to enrich available genetic variability of *Hevea*, International Rubber Research and Development Board (IRRDB) and Brazilian Government organized major collection expedition to

Amazon rain forests in 1981. The team collected 64,736 seeds and bud wood from 194 presumably high yielding mother trees not affected by *Microcyclus* and *Phytophthora,* from states of Acre, Mato Grosso and Rondonia (Ong *et al.,* 1983; Mohd Noor and Ibrahim, 1986). Malaysian and Ivory Coast germplasm centers established these and varying proportions of this wild germplasm were established in different rubber research institutes. India received a share of this germplasm during 1984–1990.

Management of Germplasm

Conservation

Both *in situ* or on the site conservation of germplasm in their original habitat in Amazon rain forests and *ex situ* or off the site conservation, where materials are introduced from centre of origin and maintained in special nurseries or fields are possible in case of *Hevea*. Since former poses several sociological, economical and logistic problems, *ex situ* conservation is generally adopted in secondary centers. In these centers, conservation of genetic resources is done with out losing genetic integrity and identity to make gene pool available as and when required. *Hevea* genetic resources are being conserved in field gene banks, both in conservation/source bush nurseries and in gardens or arboreta–the former intended for maintenance and multiplication, while latter is used mainly for breeding purposes. The 1981 IRRDB germplasm is being conserved in source bush nurseries/gene pool gardens in the Rubber Research Institutes of the different IRRDB member countries. In India, currently, a total of 4548 accessions including 90 ortet clones received from Malaysian center, were established in traditional and non-traditional areas (Varghese *et al.,* 2002). In conservation nurseries, each accession is represented by 5-10 plants, in 1 x 1m spacing, along with selected popular Wickham clones as controls. Concept of building up a core collection in large wild *Hevea* germplasm collections has also been initiated in India, using principal component scores for a set of quantitative data and Shannon-Weaver diversity index for a set of qualitative data (Abraham *et al.,* 2005).

Characterization and Evaluation

Characterization of genetic resources in rubber was given due importance to document wide variations in morphological characters of different clones using different identification tools. (Polhamus, 1962; Mercykutty *et al.,* 1991). The 1981 germplasm collection conserved in different Rubber Research Institutes is subjected to characterization and evaluation aimed at selection of potential accessions for utilization in genetic studies and crop improvement programs.

Hevea clones, from very young stage of growth have distinct and characteristic appearance and form and hence morphological characterization of clones could be done with a set of descriptors both in early growth phase and also in subsequent growth stages, for appropriate characters, as described by Rao *et al.* (2005a). Descriptors consist of passport data, plant type data at juvenile phase followed by data on relevant qualitative and quantitative characters of interest, in later stages of growth (Varghese and Abraham, 1999). Data on morphological characterization of a set of wild germplasm at the juvenile stage revealed interesting wide variability for most of characters (Abraham *et al.,* 1994). Interesting morphological variants for growth form, flower color and fruit shape were observed among wild genotypes (Madhavan *et al.,* 1996).

Major characterization effort is done at juvenile phase of growth for a set of easily identifiable and highly heritable characters like tree growth habit, leaf whorl characters like leaf whorl shape, size, separation etc. and leaf characters like leaf color, size, shape, texture, leaf lamina surface and leaf margin. Predominant classes of descriptors observed at juvenile growth phase include well-separated, hemispherical leaf storey, medium sized and green colored leaves with dull lustre, elliptic shape, acuminate leaf tip and smooth surfaced leaf blades. Frequency data on morphological characterization when subjected to Shannon- Weaver Diversity Index (SDI) (Jain *et al.*, 1975) gave high SDI values for most of characters indicating high level of diversity in wild collection (Varghese *et al.*, 2002). Descriptors for *Hevea* published for first time by Rubber Research Institute of India (Rao *et al.*, 2005b) also cover characters of clones in immature and mature phases on a host of characters like yield, tolerance to abiotic and biotic stresses, bark characters, bark anatomy, canopy characters and timber potential.

Field evaluation of wild germplasm is done in statistically laid out trials employing simple lattice/augmented design in two stages *viz.* a preliminary evaluation, where accessions are evaluated for agro-morphological and anatomical characters, and a further evaluation, where selections from preliminary evaluations are subjected to a more detailed study. In India, several studies (Varghese *et al.*, 1988, 1989; Abraham *et al.*, 1992, 1995; Abraham, 2001; Mercy *et al.*, 1993, 1995; Mercy, 2001; Madhavan *et al.*, 1993, 1996; Reghu *et al.*, 1996; Rao *et al.*, 1999; Rao and Reghu, 2000, Sanjith 2003, Suma, 2004 and Rao, *et al.*, 2005b) on characterization, variability estimates, character associations and genetic divergence in wild germplasm of *H. brasiliensis*, were reported in different sets of wild germplasm accessions. These studies revealed wide variability and significant genetic differences among accessions for most of agro morphologic traits, bark structural characters and juvenile yield indicating scope for selection of accessions with desirable characters. There were several individual accessions with values higher than that of control clone for several growth characters like girth, height, number of leaves, number of leaf whorls, and bark anatomical characters like bark thickness, number of latex vessel rows, density and diameter of latex vessels. Preliminary tapping trials, over three years, of germplasm materials in Sri Lanka revealed that GPB 1 (42-552) is a promising accession for yield (RRISL, 1995). Rondonian accessions in general ranked the highest for yield potential from among accessions from 16 provenances, both in African center and in Malaysian center (Clement-Demange *et al.*, 1997).

Wild germplasm of *Hevea* exhibits very wide variability, both within and between the provenances, with respect to several characters studied (George, 2000). Thirty new alleles for 14 loci were observed in wild germplasm using electrophoretic methods (Chevallier, 1988). Significant genetic distance based on leaf morphology studies and isozyme markers (Clement-Demange *et al.*, 1997), between 1981 germplasm and 1876 Wickham material had also been reported. These studies revealed distinct differences between Mato Grosso and Acre, with Rondonia falling in between. In general, Acre and Rondonian accessions were more vigorous in growth than those from Mato Grosso provenance. Similar results were reported from Malaysia that Acre accessions had in general long straight boles suitable for timber production (Masahuling *et al.*, 2002). Huat, *et al.* (1995) also reported a general superiority of accession from Acre and Rondonia in terms of girth in 1981 *Hevea* germplasm collection in Malaysia. Medium to high plant height combined with high branching is desirable for selection of potential timber/timber latex clones. In general, wild accessions recorded poor yield compared to domesticated control clones. However, a few individual accessions with comparable or even higher yield than popular clone RRII 105 were identified in early growth phase. Similar results were also reported from Malaysia, (IRRDB, 1996, Masahuling *et al.*, 2002), Ivory Coast (Clement-Demange *et al.*, 1997) and Vietnam (Lam *et al.*, 1997).

Provenance wise comparison of accessions for various morphological and anatomical characters and test tap yield showed accessions from Mato Grosso to be superior to those from other two provenances Acre and Rondonia (Abraham *et al.*, 1992, 2002; Reghu *et al.*, 1996 in India and Lam *et al.*, 1997 in Vietnam). Similarly Chapuset *et al.* (1995) reported better yield and timber production potential for accessions from Matogrosso than those from Rondonia and Acre in Ivory Coast. On other hand, in

Malaysia, accessions from Rondonia are reported better yielders in comparison with those from Mato Grosso and Acre (Masahuling *et al.*, 2002).

Moderate estimates of phenotypic and genotypic coefficients of variation with high estimates of broad sense heritability were recorded for girth, height, number of leaf flushes, total number of leaves, leaf area index, bark thickness, soft and hard bark thickness, number of latex vessel rows, density and diameter of latex vessels. Estimates of variation were high for test tap yield in a set of 80 wild accessions studied (Abraham, 2001). This indicates availability of high genetic variability and highly heritable nature of these characters. Genetic divergence studies in wild germplasm assume great importance in context of their utilization in hybridization programs.

Screening for Tolerance to Diseases

In case of emergence of a virulent strain of a particular pathogen in favorable environmental conditions, disease will spread and cause serious damage, if sufficient genetic variability for tolerance to diseases is not available in the population. Wild germplasm being a rich source of genes conferring resistance to diseases, screening of accessions for major diseases is a priority area of germplasm research. In India, field/laboratory screening of wild genotypes, for resistance to *Phytophthora, Odium and Corynespora* leaf diseases revealed wide variability among wild accessions indicating scope for selection of sources of resistance. In a study incorporating 20 wild and Wickham clones, eight genotypes *viz.* RRIC 52, AC/S/12 14/186, PR 261, RO/CM/10 44/7, RRIM 703, AC/S/12 40/59, PB 86 and IAN 45-873 showed a high degree of tolerance to *Oidium*. A field trial in French Guyana revealed the presence of continuous variability for resistance to *Microcyclus Ulei* (Clement-Demange *et al.*, 1997).

Wild Germplasm as a Source of Abiotic Stress Resistance

Screening for Drought Resistance

Extension of rubber cultivation to nontraditional lands confronted with drought along with other constraints necessitated development of clones capable of withstanding drought stress. Severe and prolonged soil moisture deficit coupled with very high temperatures and solar radiation and low atmospheric relative humidity for about five to six months every year, which inhibit growth and productivity of *Hevea* (Sethuraj *et al.*, 1989; Rao *et al.*, 1998). In India, Laboratory and field experiments on genotypic evaluation for assessment of genetic variability and screening of wild *Hevea* germplasm for drought tolerance were taken up. Early performance from field experiments for screening wild accessions in drought prone location at Dapchari in Maharashtra state is quite encouraging indicating possibility of identification of a good number of potential drought tolerant wild accessions. Mato Grosso accessions were in general superior compared to accessions from other two provenances. Wide variation in drought tolerance potential was reported in a set of wild *Hevea* accessions for biochemical parameters like total chlorophyll content, chlorophyll stability, epicuticular wax content and injury to cell membrane and ten potential accessions were identified based on these parameters (Mercy *et al.*, 2005).

Screening for Cold Resistance

Hevea, like other plants native to warm tropics, suffers from chilling injury when exposed to low temperatures during winter (Long *et al.*, 1994). Persistence of temperature as low as 5°C and below to causes cold injury (Jiang, 1984). In N. East India, chilling conditions set in by December and is at its peak by January, but do not prolong for more than a month or so as weather warms up by February. Survival of *Hevea* in cold conditions depends largely on intrinsic cold tolerance characteristics of clones. Young plants are more vulnerable to cold injury than mature trees. Early results from field experiments on screening wild germplasm for cold tolerance at cold prone location at Nagrakata, in sub Himalayan region of State of West Bengal led to identification of potential cold tolerant wild accessions.

Wild Germplasm as a Source of Timber-latex Traits

Contribution of rubber wood products to rubber industry is in the increase with rising demand of timber for wood based industries. Anticipating a scenario of wide spread cultivation of rubber for quality timber production in addition to latex yield, it is highly essential to develop high latex yielding clones combining high timber production potential with desirable quality and durability. Identification of desirable genotypes for qualitative and quantitative timber traits from wild germplasm was recognized as a priority area since rubber wood was accepted as an alternate source of commercial hardwood. Screening of wild genotypes for timber-latex traits was envisaged and initiated in India. Among the three provenances, accessions from Acre were more vigorous with long, straight trunks and less branching, which is a desirable feature for timber clones. On the other hand, growth habit of accessions from Mato Grosso resembled that of Wickham clones, with short and stout trunks and profuse branching. Twenty fast growing accessions were identified as potential clones for timber production, total wood volume ranging from 1.44-2.52 cu.m in Malaysia. Fifty vigorous clones were identified for timber value in Ivory Coast (Clement-Demange *et al.*, 1998). Studies on role of lignin, a phenolic polymer of cell wall and whose deposition is associated with mechanical strength and imperviousness to biodegradation of wood, is in progress in India. Reghu *et al.* (2005) reported 21.04 to 27.43 per cent in wild *Hevea* germplasm and 20.01-23.00 per cent in Wickham clones of *Hevea* with a significantly higher percentage of lignin in nine of the 18 wild accessions over the Wickham clones. Localization of Cinnamyl alcohol dehydrogenase (CAD) activity and quantification of lignin in juvenile growth phase can be used for early selection of *Hevea* germplasm for wood quality.

Further Broadening the Genetic Base by Incorporation of Wild Genes

Incorporation of superior wild accessions as one of the parents in Wickham x Amazonian hybridization programs was initiated, in various rubber research institutes all over the world including India. In India, among 12 cross combinations involving seven wild genotypes and two popular Wickham clones, hybrids of cross RRII 105 and RO/JP/3/6 had superior test tap yield at two years (RRII, 1994). Forty three hybrid clones resultant of above programme were planted in two small scale trials in 1995. Out of these, 21 clones recorded superior girth than control clone RRII 105 with standard heterosis ranging from 2.62–74.29 per cent (Saraswathy amma, 2003). Another set of nine superior genotypes was used in 14 cross combinations with two popular Wickham clones *viz.* RRII 105 and RRIM 600 and resultant 400 seedling progenies are under evaluation. Forty selected seedlings resultant of 1997 hybridization programme were cloned for further evaluation (RRII, 2003). Another 76 seedling progenies resultant of 2002 hybridization programme involving six selected wild accessions and three high yielding Wickham clones are under evaluation (Chandrasekhar *et al.*, 2004). Potential of wild accessions of rubber, to serve as parents in crop improvement programs was well understood now. Utilizing the powerful molecular tools, such useful genes should be identified and utilized in production of transgenic rubber plants for specific stress situations.

Future Prospects

In the present global scenario, NR stands to gain from increasing demand for it in the fast expanding automobile industry besides a host of other industries using different forms of rubber.

Meeting global demand in production and productivity and in quality standards of the product is the greatest challenge to Indian rubber industry. RRII and Rubber Board are committed to meet forthcoming challenges in the NR sector.

Conservation and utilization of plant genetic resources is a subject of high priority to international community in the context of ever diminishing per capita available land and water resources, expanding biotic and abiotic stresses compounded by incessant genetic erosion of plant genetic resources. Rampant deforestation in Amazon rain forests the center of origin of *Hevea* species is so alarming that the need to collect and conserve available germplasm collection of both Wickham and wild germplasm of *Hevea brasiliensis* is top priority to ensure sustainable growth of NR industry for future.

Inherent problems associated with conservation of *Hevea* germplasm in field nurseries should be minimized with development of alternate methods such as cryopreservation. Characterization of *Hevea* germplasm, morphologically by using standard set of descriptors and at DNA level by using DNA based molecular marker has to be given due importance to enable proper cataloging of accession characteristics and assessment of genetic diversity. A comprehensive and user-friendly computer database is essential for proper documentation and easy retrieval of the data as and when necessary.

In the context of rubber cultivation having been extended to non traditional and marginal areas, development of location specific clones capable of withstanding different environmental constraints prevailing in these areas such as drought, cold, high attitudes etc. assumes much significance. The large collection of wild germplasm needs to be exploited by breeder to identify useful genes capable of withstanding various stress situations.

Identification of genotypes having potential for higher timber out put and superior timber quality and durability traits along with high yield is another priority since value of rubber timber for various industrial applications is on increase. To cater to these needs, the base material should have broad genetic base containing sufficient genetic diversity.

Application of recent advances in molecular marker technologies helps for an efficient evaluation and utilization of gene pool. These molecular techniques serve as valuable adjuncts to conventional breeding procedures for better exploitation of available genetic variability in Wickham gene pool and wild germplasm collection. There is no doubt that available genetic resources of *Hevea* is the best raw material to realize current objectives of development of location specific clones capable of withstanding various stress situations also to meet future demands of industry.

Acknowledgement

Authors are grateful to Dr. N.M. Mathew, Director, Rubber Research Institute of India, for granting permission to contribute this chapter and the encouragement and support given during preparation of this manuscript.

References

Abraham, S. T., Reghu, C.P., Madhavan, J., George, P.J., Potty, S.N., Panikkar, A.O.N. and Saraswathy, P., 1992. Evaluation of *Hevea* germplasm 1: Variability in early growth phase. *Indian Journal of Natural Rubber Research*, 5(1&2): 195–198.

Abraham, S.T., 2001. Genetic parameters and divergence in certain wild genotypes of Hevea brasiliensis (Willd.ex Adr.de Juss.) Muell. Arg. *Ph.D. Thesis*, submitted to Mahatma Gandhi University, Kottayam, India.

Abraham, S.T., C.P. Reghu., George, P.J., Potty, S.N. and Saraswathi, P., 1995. Evaluation of *Hevea* germplasm: III. Genetic divergence in certain genotypes of *Hevea*. Paper presented in the 82nd Indian Science Congress, Calcutta, 1995.

Abraham, S.T., Nair, R.B. and Varghese, Y.A., 2005. Core collection in wild *Hevea* germplasm in India: the concept, method and application. *Preprints of papers. International Natural Rubber Conference*, India, 2005, 6–8 November, 2005, pp 68–78

Abraham, S.T., Panikkar, A.O.N., George, P.J., Reghu, C.P., Nair, R.B., 2002. Genetic evaluation of wild *Hevea* germplasm: early performance. (In) *Plantation Crops Research and Development in the new millennium* (Ed. P. Rethinam, H.H. Khan, P.K. Mandal, V.M. Reddy and K. Suresh. Coconut Development Board, Kerala, India. pp 274–279.

Abraham, S.T., Reghu, C.P., George, P.J. and Nair, R.B., 1997. Studies on genetic divergence in *Hevea brasiliensis*. Paper presented in the '*Symposium on Tropical Crop Research and Development. India–International*' held at Trichur from 9–12, September, 1997.

Abraham, S.T., Reghu, C.P., Panikkar A.O.N., George, P.J. and Potty, S.N., 1994. Juvenile characterisation of wild *Hevea* germplasm. *Indian Journal of Plant Genetics Resources*, 7(2): 157–164.

Baulkwill, W.J., 1989. The history of natural rubber production. (In) *Rubber* (Ed. C.C.Webster and W.J.Baulkwill). Longman Scientific and Technical, New York. pp1–56.

Chandrasekhar, T.T., Alice, J., Gireesh, T., Rao, G.P. and Saraswathyamma, C.K., 2004. Observations and Pollination, Fecundity/Siring ability and seed germination in *Hevea brasiliensis*. *J. Rubb. Res.*, 7(4), 265–280.

Chandrashekkar, T.R., Nazeer, M.A., Marattukalam, J.G., Prakash, C.P., Annamalainathan, K. and Thomas, J., 1998. An analysis of growth and drought tolerance in rubber during the immature phase in a dry subhumid climate. *Experimental Agriculture*, 34 : 1–14.

Chapuset, T., Legnate, H., Doumbia, A., Clement–Demange, A., Nicolas, D. and Kali, J., 1995. Agronomical characterisation of 1981 germplasm in Cote d' Ivoire: Growth, production, architecture and leaf disease sensibility. *IRRDB Symposium on Physiological and Molecular Aspects of Breeding of Hevea brasiliensis*, 6–7, November 1995, IRRDB, England, pp. 112–126.

Chee, K.H. and Holliday, P., 1986. South American Leaf Blight of *Hevea* rubber. *Malaysian Rubber Research and Development Borad Monograph*, No. 13, 50 p.

Chevallier, M.H., 1988. Genetic variability of *Hevea brasiliensis* germplasm using isozyme markers. *Journal of Natural Rubber Research*, 3: 42–53.

Clement–Demange, A., Legnate, H., Gnagne, Y.M, and Nicolas, D., 1990. Strategy for breeding *Hevea brasiliensis* germplasm in IRCA–Cote d' Ivorie. *Proceedings of the IRRDB Breeding Symposium*, 5–6 October 1990, Kunming, China.

Clement–Demange,A., H.Legnate., T.Chapuset., F.Pinard. and M.Seguin., 1997. Characterization and use of the IRRDB germplasm in Ivory Coast and French Guyana: Status in 1997, p71–88. In: Proceedings, *IRRDB Symposium on Natural Rubber. Vol–1.General, Soils and Fertilizations and Breeding and Selection session*. Ho Chi Minto City 14th and 15th October, 1997. pp. 71–88

Dean,W., 1987. *Brazil and the struggle for rubber: A study in environment history*. Cambridge University Press, Cambridge, 234p.

Edathil, T.T., 1986. South American Leaf Blight–a potential thread to the natural rubber industry in Asia and Africa. *Tropical Pest Management*. 32(4): 296–303.

George, P.J. (2000). Germplasm Resources In: *Natural Rubber–Agromanagement and Crop Processing* (Eds. P.J.George and C.K.Jacob), Rubber Research Institute of India, Kottayam. p–40.

Gupta and Edathil, 2001. Growth performance during the immature phase of a few rubber (*Hevea brasiliensis*) clones in Orissa. *Indian Journal of Natural Rubber Research*, 14(2) : 159–164.

Huat. O.S., Othman, R., Benong. Masahuling., 1995. Status report on the 1981 Hevea Germplasm collection–1995, IRRDB work shop, 1995, Malaysia.

IRRDB, 1996. *Quarterly Report, International Rubber Research and Development Board.*

Jain, S.K., Quarlset, C.O., Bhat G.M. and Wu, K.K. (1975). Geographical pattern of phenotypic diversity in a world collection of duran wheat. *Crop Sci.* 15 : 700–704.

Jiang, A., 1984. A geological study of rubber trees cultivationat high altitude in China. *Proceedings of the International Rubber Conference*, 1984, Colombo, Sri Lanka, pp. 117–130RRIM., 1996. *Annual report* 1995. Rubber Research Institute of Malaysia, Kuala Lumpur, p. 38.

Jones,K.P. and Allen, P.W., 1992. Historical development of the world rubber industry. In: *Natural Rubber: Biology, Cultivation and Technology*. (Ed. M.R.Sethuraj and N.M.Mathew). Elsevier, Amsterdam. pp 1–25.

Lam. L.V., Ha, T.T.T., Ha, V.T.T. and Hong, T., 1997. Studies of *Hevea* genetic resources in Vietnam: Results of evaluation and utilization. In: *Proceedings of IRRDB Symposium on Natural Rubber. Vol. 1–General, Soils and Fertilization and Breeding and Selection Sessions*. Ho Chi Minh City. 14th and 15th October, 1997. pp 89–100.

Lane, E.V.1953. The life and work of Sir Henry Wickham. *The Indian Rubber Journal*, 126: 25–27; 65–68; 95–98; 139–142; 177–180.

Licy, J., Panikkar, A.O.N., Premakumari, D., Varghese. Y.A. and Nazeer, M.A., 1992. Genetic parameters and heterosis in *Hevea brasiliensis*. 1. Hybrid clones of RRII 105 x RRIC 100. *Indian Journal of Natural Rubber Research*. 5(1&2): 51–56.

Licy, J., Saraswathyamma, C.K., Premakumari, D., Meenakumari, T., Rajeswary Meenattoor, J., and Nazeer, M.A., 2003. Genetic parameters and heterosis in rubber *Hevea brasiliensis* Muell Arg: Hybrid vigour for yield and yield components among the RRII 400 series clones in small scale trial. *Indian Journal of Natural Rubber Research*, 2003, 16 (1 and 2).

Long, S.P., Humphries, S. and Falkowski, P.G., 1994. Photo inhibition of photosynthesis in nature. *Annual Review of Plant Physiology and Plant Molecular Biology*, 45 : 633–662.

Madhavan, J., Reghu, C.P., Abraham, S.T., George, P.J. and Potty, S.N., 1996. Evaluation of *Hevea* germplasm :VII. Association analysis in wild *Hevea* germplasm. *Journal of Plantation Crops*. 24(Suppl.) : 453–457.

Madhavan, J.M., Reghu, C.P., Abraham, S.T., George, P.J. and Potty, S.N., 1993. Resources of *Hevea*. Paper presented in the '*ISPGR Dialogue in Plant Genetic Resources: Developing national policy*' held at NBPGR, New Delhi, 1993.

Marattukalam, J.G., Saraswathy Amma, C.K. and George, P.J., 1980. Crop improvement through ortet selection in India. *International Rubber Conference*, Kottayam, India.

Markose, V. C., 1984. *Biometric analysis of yield and certain yield attributes in the para rubber tree: Hevea brasiliensis* Muell. Arg. Ph.D. Thesis, Kerala Agricultural University.

Masahuling Berong, Rani L. Othman, Zarawi Ahd Ghani and Mohd Nazaruddin Mohd Aris., 2002. Status on the evaluation and utilization of the 1981 IRRDB Hevea germplasm, Malaysian Rubber Board.

Mercy, M.A., 2001. *Genotypic evaluation and screening for drought tolerence in wild Hevea germplasm. Ph.D. Thesis.* Dept. of Plant Breeding and Genetics, College of Horticulture, Kerala Agri. University, Trichur, Kerala.

Mercy, M. A., Abraham, S.T., George, P.J. and Potty, S.N., 1995. Evaluation of *Hevea* germplasm: Observations on certain prominent traits in a conservatory. *Indian Journal of Plant Genetic Resources*, 8(1): 35–39.

Mercy, M.A., Jose, G. and Varghese, Y.A., 2005. Screening of wild germplasm of rubber for drought tolerance based on chlorophyll and wax content, chlorophyll degradation and membrane stability. *Preprints of papers. International Natural Rubber Conference*, India–2005, 6–8 November, Kochi, Rubber Research Institute of India, Kottayam, India. pp 135–144.

Mercy, M.A., S.T.Abraham., P.J.George., S.N.Potty., M.R.Sethuraj and P.Saraswathi., 1993. Preliminary observations of the1981 IRRDB *Hevea* germplasm. II. Variability, dry matter and morphological characters. Jl. Plantation Crops 21(Supplement): 268–274.

Mercykutty, V.C., Varghese, Y.A., Licy, J. and Panikkar, A.O.N., 1991. Juvenile characters on seed morphology of certain modern *Hevea* clones. *Indian Journal of Natural Rubber Research*, 4(1) : 16–25.

Meti, S., Choudhuri, D and Varghese, Y. Annamma 1999. Growth performance of *Hevea* clones in Northern region of West Bengal. *Indian Journal of Natural Rubber Research*, 12(1&2) : 39–48.

Mohd Noor, A.G. and Naimah Ibrahim. (1986). Characterisation and evaluation of *Hevea* germplasm. *IRRDB Symposium*, 11–13 Nov., 1986, Malaysia.

Mondal *et al.*, 1999. Performance of *Periconia* leaf blight diseases on *Hevea* in North East India. *Indian Journal of Natural Ruber Research*, 11(1): 1–7.

Mydin, K. K., Nair, V. G., Sethuraj, M. R., Saraswathy, P. and Panikkar, A. O. N., 1992. Genetic divergence in *Hevea brasiliensis, Indian Journal of Natural Rubber Research*, 5(1&2): 120–124.

Ong, S.H., Ghani, M.N.A., Tan, A, M., and Tan, H., 1983. New *Hevea* germplasm: Its Introduction and potential. *Proceedings of the Rubber Research Institute of Malaysia, Planters Conference*, 1983, Kuala Lumpur, Malaysia, 3: 3–17.

Petch, T., 1914. Notes on the history of the plantation rubber industry in the east. *Annals of the Royal Botanic Gardens, Peradeniya*, 5: 440–487.

Polhamus, L.G., 1962. Rubber:Botany, production and utilization. Interscience Publishers, Inc., New York, 449 p.

Premakumari, D., 1992. *Variability, correlations and path co-efficient analysis for yield in relation of anatomical characters in Hevea brasiliensis* (Willd.ex Adr.de Juss.)Muell. Arg. Ph.D. Thesis, University of Kerala, Trivandrum, 1992, 260 p.

Priyadarshan, P.M., Sudhasowmyalatha, M.K., Sasikumar, S., Varghese, Y.Annamma and Dey, S.K., 1998. Relative performance of six *Hevea brasiliensis* clones during two yielding regimes in Tripura. *Indian J. Nat. Rubb. Res.* 11, 67–72.

Priyadarshan, P.M., Sudhasowmyalatha, M.K., Sasikumar, S., Varghese, Y.Annamma and Dey, S.K., 2000. Evaluation of *Hevea brasiliensis* clones for yielding trends in Tripura. *Indian J. Nat. Rubb. Res.* 13 (1&2): 56–63.

Priyadarshan, P.M., 2003. Breeding *Hevea brasiliensis* for environmental constrain. *Adv. Agronomy* 79, 351–400.

Rao, G.P. and Reghu, C.P., 2000. Variability and character association in wild *Hevea* germplasm. *Int. Conference on Managing Natural Resources for Sustainable Agricultural Production in the 21st Century*. Extended Summaries. Vol.4, pp.10–11.

Rao, P.S., Saraswathyamma, C.K. and Sethuraj, M.R., 1998. Studies on the relationship between yield and meteorological parameters of para rubber tree (*Hevea brasiliensis*). *Agricultural and Forest Meteorology*, 90: 235–245.

Rao, G.P., Reghu, C.P. and George, P.J., 1999. Evaluation of *Hevea* germplasm. VIII. Variability in certain juvenile characters of wild *Hevea* germplasm. *Journal of Cytology and Genetics*, 34 (2): 183–186.

Rao, G.P., Abraham, S.T., Reghu, C.P. and Varghese, Y.A., 2005a. Descriptors for Rubber. Rubber Research Institute of India, Kottayam. p.70

Rao, G.P., Madhavan, J., Suma, K., Nair, R.B. and Varghese, Y.A., 2005b. Genetic diversity in wild rubber (*Hevea brasiliensis*) germplasm. *Preprints of papers. International Natural Rubber Conference*, India, 2005, 6–8 November pp 128–134.

Reghu, C.P., Abraham, S.T., Madhavan, J., George, P.J., Potty, S.N. and Leelamma, K.P., 1996. Evaluation of *Hevea* germplasm: Variation in bark structure of wild Brazilian germplasm. *Indian Journal of Natural Rubber Research*, 9(1): 28–31.

Reghu, C.P., George, B.P., Iram, U. and Varghese, Y.A., 2005. Early detection of wood quality in Hevea germplasm using cinnamyl alcohol dehydrogenase (CAD) activity and lignification. *Preprints of papers. International Natural Rubber Conference*, India, 2005, 6–8 November, Rubber Research Institute of India, Kottayam, India. pp 54–61.

Reju, M.J., Taplial, A.P., Dekha, H, K. and Varghese Y. Annamma, 2002. Growth performance of *Hevea* clones in subtropical Meghalaya, *Journal of Plantation Crops* 29(1): 22–27.

Reju, M.J., Kumar, K.A., Thapliyal, A.P. and Varghese, Y. Annamma, 2000. Yield and yield components of certain *Hevea* clones at high elevation. *Recent Advances in Plantation Crop Research (Ed.N. Muraleedharan and R. Raj Kumar, Allied Publishers Ltd.*138–143.

RRII., 1994. *Annual Report 1993–1994*. Rubber Research Institute of India, Kottayam.

RRII., 2002. *Annual Report 2001–02*. Rubber Research Institute of India, Kottayam.

RRII 2003 *Annual Report 2002–03*. Rubber Research Institute of India, Kottayam. RRII.

RRISL 1995 *Annual Report 1995*. Rubber Research Institute of Sri Lanka, Sri Lanka

Rubber Board, 2005. *Indian Rubber Statistics*, Rubber Board, Kottayam, India.

Rubber Board, 2006. Approved cultivars, Rubber Growers Companion 2006, Rubber Board, Kottayam, India. pp. 68–72.

Sanjith, P.D. (2003). Characterization and preliminary evaluation of certain wild accessions of natural rubber, M.Sc. Dissertation. Dept. of Botany, Calicut University, Kerala. 76p.

Saraswathyamma, C.K., 2003. Advances in crop improvement in *Hevea* in the traditional rubber–growing tract in India. In: *Global competitiveness of Indian Rubber industry–Rubber Planters' Conference, India, 2003 Full text of papers*. (Eds. C. Kuruvilla Jacob and Kurien K. Thomas). Rubber Research Institute of India, Kottayam, Kerala, India. pp 101–116.

Saraswathyamma, C.K., 2004. Promising Indian Rubber Clones Indian/International Rubber Journal, July–Aug., 20004–56–57.

Sasikumar, S., Priyadarshan PM, Dey S.K and Varghese Y.Annamma., 2001. Evaluation of polyclonal seeding population of *Hevea brasiliensis* (Wild. Ex Adr, De Juss Muel Arg) in Tripura. *Ind. J.Nat. Rub.Res*. 14(2): 125–130.

Sethuraj, M.R., Potty, S.N., Vijayakumar, K.R., Krishanakumar, A.K., Rao, P.S., Thapliyal, A.P., Mohankrishna, T., Rao, C.G., Choudhuri, D., George, M.J., Soman, T.A. and Meenattoor, J.R., 1989. Growth performance of *Hevea* in the non–traditional regions of India. *Proceedings of the Rubber Research Institute of Malaysia, Rubber Growers' Conference*, 1989, Malacca, Malaysia, pp. 212–227.

Simmonds, N.W., 1989. Rubber Breeding. In : *Rubber* (Eds. C.C. Webster and W.J. Baulkwill) Longman Scientific and Technical, New York. pp 85–124.

Suma, 2004. Genetic Variability in a Population of Wild Germplasm of Natural Rubber. (*Hevea brasiliensis* (Wild. ex Adr. de Juss.) Muell. Arg.) M.Sc. Dissertation. Dept. of Botany, Calicut University, Kerala. 86 pp.

Tan, H., 1987. Strategies in rubber tree breeding. In: *Improving Vegetatively Propagated Crops* (Eds. A.J. Abbot and R.K. Atkin). Academic Press, London, pp. 27–62.

Tripura, *Indian J. Nat Rubb. Res*. 13, 56–63. Two yielding regimes in Tripura. *Indian J. Nat. Rubb. Res*. 11, 67–72.

Varghese, Y, Annamma., 1992. Germplasm resources and genetic improvement. In:*Natural Rubber: Bilogy, Cultivation and Technology* (Ed M.R. Sethuraj and N.M.Mathew). Elsevier, Amsterdam. pp 88–115.

Varghese, Y. Annamma, 2003. Crop Improvement in the non-traditional rubber growing areas in India. In: *Global Competitiveness of Natural Rubber Plantation Industry: Rubber Planters' Conference, India 2002:* RRII, Kottayam, India pp.117–131.

Varghese, Y. Annamma, Abraham, S.T., Mercy, M.A.; Madhavan, J.; Reghu, C.P.; Rao, G.P.; Sankariammal, L.; Idicula, S.P. and Joseph, A., 2002. Management of the 1981 IRRDB germplasm collection in India. IRRDB Joint Workshop on Plant Breeding, Agronomy and Socio–economics, 28 Aug–7 Sept., 2002, Malaysia.

Varghese, Y.A. and Abraham, S.T. (1999). Germplasm conservation, utilisation and improvement in rubber. *In: Improvement of plantation crops*. (Eds. M.J. Ratnambal, P.M. Kumaran, K. Muralidharan, V. Niral and V. Arunachalam). Central Plantation Crops Research Institute, Kasaragod, pp. 124–133.

Varghese, Y. Annamma., Knaak, C., Sethuraj, M.R. and Ecke, W., 1997 Evaluation of random amplified polymorphic DNA (RAPD) markers in *Hevea brasiliensis, Plant Breeding*, 116: 47–52.

Varghese, Y.A., Marattukalam. J.G., Premakumari, D. and Panikkar, A.O.N., 1988. Nursery evaluation of 100 Brazilian genotypes of *Hevea* in India. *Colloque Hevea* 88, IRRDB, Paris. p353–364.

Varghese, Y.A., Marattukulam J.G., George, P.J. and Panikkar, A.O.N., 1989. Nursery evaluation of some exotic genotypes of *Hevea brasiliensis* Muell Arg. *J.Plant. Crops*. 16 (supplement) 335–342.

Wycherley, P.R.1969. Breeding of *Hevea*. *Journal of Rubber Research Institute of Malaya*. 21 : 38–55.

Wycherley, P.R.1977 Motivation of *Hevea* germplasm collection and conservation. *Workshop on inter–collaboration in Hevea breeding and the collection and establishment of materials from the neo–tropics*, Kuala Lumpur. 5p.

Wycherly, P.R., 1968. Introduction of *Hevea* to the Orient. *The Planter*, 44: 127–137.

Wycherly, P.R., 1992. The genus *Hevea*–Botanical aspects. (In) *Natural Rubber: Biology, Cultivation and Technology*. (Ed. M.R.Sethuraj and N.M.Mathew). Elsevier, Amsterdam. pp50–66.